the finite-field version of the generalized hypergeometric function and prove identities in an analogous way to the Gauss function.

Chapter 6 contains various extensions of the Gauss and confluent hypergeometric functions. To this end, we first give the extensions of the gamma and beta function by introducing one or more parameters and then using the beta definition of Gauss and the confluent hypergeometric function, four extensions of Gauss and confluent hypergeometric functions are studied. We provide various results, namely integral representations, differential or difference equations and transformation and summation formulas to which these extensions of the Gauss and confluent hypergeometric function obey.

Chapter 7 contains the other three forms of the Gauss hypergeometric function that have been developed in recent years. In particular, we have given the (p, q)-analogue, the incomplete form and the discrete version of the Gauss hypergeometric function. Properties such as integral representations, differential or difference equations, transformation and summation formulas that these forms obey are also covered in this chapter.

The next chapter contains appendices with the basic results that are necessary for understanding the book. A bibliography concludes the book. Each chapter in this book begins with a brief outline summarizing the material presented in the chapter and is then divided into a number of sections.

This book is intended primarily as a reference work for various seemingly disparate groups of researchers and other users of hypergeometric function. In particular, teachers, researchers and graduate students in the field of mathematics and mathematical physics will find this book especially useful. They will be able to apply the various methods for determining the areas of convergence, integral representation, transformation formulas, differential formulas and recurrence relations of the special functions.

Many people have made important contributions to this book, both directly and indirectly. These contributions are duly acknowledged throughout the text and in the bibliography.

Finally, we would like to thank our family for their cooperation and support in this project.

Ravi Dwivedi
Swami Atmanand Govt English Medium
Model College Jagdalpur, Bastar (CG)

Contents

List of symbols

Sets, relations and functions

$\forall$	stands for "for all"
$\exists$	stands for "there exists"
$\mid$	stands for "such that"
$\in$	stands for "belongs to"
$\notin$	stands for "does not belong to"
$\mathbb{U}$	stands for "universal set"
$Q \subset P$	means "Q is a subset of P"
$P \cup Q$	means "union of P and Q"
$P \cap Q$	means "intersection of P and Q"
$P \setminus Q$	means "the elements of P are not in Q"
$P = Q$	"set equality"
$P \neq Q$	"set inequality"
P'	means "compliment of the set P"
$n(P)$ or $\mathrm{card}(P)$	stands for "the number of elements in the set P"
$\mathbb{N}$	denotes the "set of natural numbers"
$\mathbb{N}_0$	denotes the "set of natural numbers including 0"
$\mathbb{Z}$ or $\mathbb{I}$	denotes the "set of integers including 0"
$\mathbb{Z}^+$	denotes the "set of positive integers"
$\mathbb{Q}$	denotes the "set of rational numbers"
$\mathbb{R}$	denotes the "set of real numbers"
$\mathbb{C}$	denotes the "set of complex numbers"
$\mathbb{C}^* = \mathbb{C} \setminus \{0\}$	denotes the "set of complex numbers without zero"
$\mathbb{H}$	denotes the "set of quaternion"
$\iota = \sqrt{-1}$	stands for "pure imaginary number"
$\bar{z}$	stands for "complex conjugate of $z \in \mathbb{C}$"
$\lvert z \rvert$	stands for "modulus value of $z \in \mathbb{C}$"
$\mathbb{R}(z)$	stands for "real part of $z \in \mathbb{C}$"
$\mathfrak{I}(z)$	stands for "imaginary part of $z \in \mathbb{C}$"
$P \times Q$	"cartesian product"
$P\mathcal{R}Q$	relation on $P \times Q$
$\sum$	summation
$\prod$	product
$u \implies v$	means "u implies v"
$u \iff v$	means "u if and only if v"
$\max\{u_1, \dots, u_k\}$	used for "maximum of $u_1, \dots, u_k$"
$\min\{u_1, \dots, u_k\}$	used for "minimum of $u_1, \dots, u_k$"

Groups, rings and fields

$u \setminus v$	means "u divides v"
$u \not\setminus v$	means "u does not divide v"
$u \equiv v \pmod{k}$	means "k divides $(u - v)$ or $u = v + n \cdot k$"
$u \not\equiv v \pmod{k}$	means "k does not divide $(u - v)$"

https://doi.org/10.1515/9783111324586-202

δ	is the "Kronecker symbol"
$(u \setminus p)$	denotes the "Legendre symbol"
$\|G\|$	denotes the "cardinality of a group G"
S_n	is the "symmetric group"
$GL(n, \mathbb{C})$	is the "general linear group in n-dimension on the field $\mathbb{C}$"
$SL(n, \mathbb{C})$	is the "simple linear group in n-dimension on the field $\mathbb{C}$"
$(R, +, \cdot)$ or simply R	is used for a ring
$GF(q)$	denotes the "field with q-elements"
$GF(q)^*$	denotes the "field with nonzero $(q-1)$ elements"
$\mathbb{F}_p = \mathbb{Z}_p = GF(p)$	is the "field of integer modulo p (prime)"
$\mathbb{F}_p^*$	stands for "nonzero elements of $\mathbb{F}_p$"
$\phi, \psi, \zeta, \chi, A, \dots$	denotes the "multiplicative character"
ϵ	denotes the "trivial character"

Matrices

I	denotes the "identity matrix"
$\det(P)$ or $\|P\|$	denotes the "determinant of a square matrix of P"
$\mathrm{tr}(P)$	denotes the "trace of a matrix P"
P' or P^T	denotes the "transpose of a matrix P"
P^* or P^c	denotes the "transpose conjugate of a matrix P"
$\\|P\\|$	denotes the "2-norm of a matrix P"
$\\|P\\|_k$	denotes the "k-norm of a matrix P"
$\\|P\\|_\infty$	denotes the "∞-norm of a matrix P"
$\\|P\\|_F$	denotes the "Frobenius norm of a matrix P"
$\mu(P)$	denotes the "logarithmic norm of a matrix P"
$\alpha(P)$	denotes the "maximum value of the real parts of all eigenvalues"
$\beta(P)$	denotes the "minimum value of the real parts of all eigenvalues"
$J(U \to V)$	stands for "Jacobian of a linear transformation from U to V"
$C_k(S)$	denotes the "zonal polynomial"

Special functions

$k!$	denotes the "factorial of $k \in \mathbb{N}_0$"
e	denotes the "basis of Napierian logarithms"
$\Gamma(u)$	denotes the "classical gamma function"
$\Gamma_q(u)$	denotes the "q-gamma function"
$\Gamma_p(u)$	denotes the "multivariate gamma function"
$\Gamma(P)$	denotes the "gamma matrix function"
$\Gamma_q(P)$	denotes the "q-gamma matrix function"
$G(\phi)$	denotes the "Gauss sum" or "character sum analogue of gamma function"
$\Gamma_p(u)$	denotes the "generalized gamma function"
$\Gamma_p^{a_1,a_2}(u)$	denotes the "new generalized gamma function"
$\Gamma_{p,q}(u)$	denotes the "(p,q)-gamma function"
$\gamma(u,x)$ and $\Gamma(u,x)$	denotes the "incomplete gamma functions"
$\beta(u,v)$	denotes the "classical beta function"

$\beta_q(u,v)$	denotes the "q-beta function"
$\beta^P(u,v)$	denotes the "multivariate beta function"
$\beta(P,Q)$	denotes the "beta matrix function"
$\beta_q(P,Q)$	denotes the "q-beta matrix function"
$J(\phi,\psi)$	denotes the "Jacobi sum" or "character sum analogue of beta function"
$\beta(u,v,p)$	denotes the "extended beta function"
$\beta(u,v,p,m)$	denotes the "new generalized beta function"
$\beta_p^{a_1,a_2}(u,v)$	denotes the "new extended beta function"
$\beta(u,v,p,q)$	stands for "extension of extended beta function"
$\beta_{p,q}(u,v)$	denotes the "(p,q)-beta function"
${}_2F_1(u,v;w;z)$	denotes the "classical Gauss hypergeometric function"
${}_2\phi_1(u,v;w;q;z)$	denotes the "basic or q-Gauss hypergeometric function"
${}_2F_1(u,v;w;P)$	denotes the "Gauss hypergeometric function of matrix arguments"
${}_2F_1(P,Q;R;z)$	denotes the "Gauss hypergeometric matrix function"
${}_2\phi_1(P,Q;R;q;z)$	denotes the "basic or q-Gauss hypergeometric matrix function"
${}_2F_1(\phi,\psi;\chi;t)$	denotes the "character sum analogue of Gauss hypergeometric function"
$F_p(u,v;w;z)$	denotes the "extended Gauss hypergeometric function"
$F_p(u,v;w;z;m)$	denotes the "generalized extended Gauss hypergeometric function"
$F_p^{a_1,a_2}(u,v;w;z)$	denotes the "generalized Gauss hypergeometric function"
$F_{p,q}(u,v;w;z)$	denotes the "extension of extended Gauss hypergeometric function"
${}_2\psi_1(u,v;w;p,q;z)$	denotes the "(p,q)-Gauss hypergeometric function"
${}_2\gamma_1\left[\begin{smallmatrix}(u;x),v; \\ w;\end{smallmatrix} z\right]$	denotes the "incomplete Gauss hypergeometric function"
or ${}_2\Gamma_1\left[\begin{smallmatrix}[u;x],v; \\ w;\end{smallmatrix} z\right]$	also denotes the "incomplete Gauss hypergeometric function"
${}_2\mathcal{F}_1(u_1,u_2;v_1;u,n,z)$	denotes the "discrete Gauss hypergeometric function"

1 Classical Gauss hypergeometric function

This chapter is based on the study of the classical version of Gauss hypergeometric function. After the gamma and beta function, the Gauss hypergeometric function is another important series in the area of special function. Carl Friedrich Gauss, [63], studied this function systematically in 1813. Later, the mathematicians Euler, Gauss, Kummer, Riemann, Bailey, Slater, Rainville, Exton, Srivastava and Andrews enrich the properties of this function, see [3, 5–8, 53, 58, 107, 110, 128, 139, 140, 147, 148, 152]. We revisit the fundamental properties, namely regions of convergence, differential equations, integral representation and differential and contiguous formulas obeyed by the Gauss hypergeometric function.

1.1 Infinite series

When infinitely many quantities are added one after the other, they form an infinite series. Alternatively, we say infinite ordered sequence $(u_1, u_2, \dots)$ of numbers, functions or anything that can be added together forms an infinite series. In general, we represent an infinite series by

$$u_1 + u_2 + \cdots = \sum_{k=1}^{\infty} u_k. \tag{1.1}$$

For example,

$$\begin{aligned}
&1 + 2 + \cdots + k + \cdots \text{ up to } \infty;\\
&1 + \frac{1}{2} + \cdots + \frac{1}{k} + \cdots \text{ up to } \infty;\\
&1 + \frac{1\cdot 3}{2\cdot 4} + \frac{1\cdot 3\cdot 5}{2\cdot 4\cdot 6} + \cdots \text{ up to } \infty.
\end{aligned}$$

Sometimes an infinite series is summable or convergent if we can assign a value to it. In other words, a series is convergent if the limit

$$\lim_{k\to\infty}\left(S_k = \sum_{n=1}^{k} u_n\right)$$

exists, where S_k is called the partial sum of the series. Otherwise, the series is known as divergent.

Example 1. The series

$$1 + \frac{1}{2} + \frac{1}{2^2} + \frac{1}{2^3} + \frac{1}{2^4} + \cdots$$

is convergent.

https://doi.org/10.1515/9783111324586-001

Solution. The partial sum of the series is

$$S_k = 1 + \frac{1}{2} + \frac{1}{2^2} + \cdots + \frac{1}{2^{k-1}}, \tag{1.2}$$

which is clearly a geometric progression and the sum of a finite term can be calculated as

$$S_k = \frac{1(1-\frac{1}{2^k})}{1-\frac{1}{2}} = 2\left(1 - \frac{1}{2^k}\right).$$

Taking the limit as $k \to \infty$,

$$S = \lim_{k\to\infty} S_k = 2\left(1 - \frac{1}{2^k}\right) = 2,$$

gives a finite number. Hence, the series converges to the number 2.

Example 2. Show that the following infinite series:

$$3 + 6 + 9 + 12 + 15 + \cdots$$

is divergent.

Solution. The partial sum is

$$\begin{aligned} S_k &= 3 + 6 + 9 + \cdots + 3k \\ &= 3(1 + 2 + 3 + \cdots + k) \\ &= 3\,\frac{k(k+1)}{2}. \end{aligned}$$

Taking limit as $k \to \infty$ yields $\lim_{k\to\infty} S_k = \infty$. Hence, the series is divergent.

We now have a brief idea about the convergent behavior of infinite series instead of using the parent method, and we have a list of methods to identify the convergent behavior of the series. Some of them will be discussed in detail but many of them will be presented without proof.

Theorem 3. *The necessary condition for convergent of a positive infinite series* $u_1 + u_2 + \cdots + u_k + \cdots$ *is* $\lim_{k\to\infty} u_k = 0$.

Proof. Consider that the series $u_1 + u_2 + \cdots + u_k + \cdots$ is convergent, so we have

$$\lim_{k\to\infty} (S_k = u_1 + u_2 + \cdots + u_{k-1} + u_k) = l \quad \text{(a finite quantity)}. \tag{1.3}$$

Also,

$$\lim_{k\to\infty} (S_{k-1} = u_1 + u_2 + \cdots + u_{k-1}) = l. \tag{1.4}$$

Subtracting (1.4) from (1.3), we get

$$\lim_{k\to\infty}(S_k - S_{k-1}) = \lim_{k\to\infty} u_k = 0. \tag{1.5}$$

The condition is necessary and can be checked with the following example. The series

$$1 + \frac{1}{\sqrt{2}} + \frac{1}{\sqrt{3}} + \frac{1}{\sqrt{4}} + \cdots$$

is divergent while $\lim_{k\to\infty}(u_k = \frac{1}{\sqrt{k}}) = 0$. It completes the proof. □

Other tests are as follows:

1. *The p-test.* The infinite series

$$\sum \frac{1}{k^p} = \frac{1}{1^p} + \frac{1}{2^p} + \frac{1}{3^p} + \cdots$$

   is convergent if $p > 1$ and divergent if $p \leq 1$.
2. *Comparison test.* If $\sum u_k$ and $\sum v_k$ are two infinite series of positive terms such that

$$u_k \leq v_k, \quad \forall k \geq m;$$

   or

$$\frac{u_k}{v_k} \leq l, \quad \forall k \geq m, l > 0;$$

   or

$$\lim_{k\to\infty} \frac{u_k}{v_k} = l \quad \text{(a finite nonzero quantity)},$$

   and
   - $\sum v_k$ is convergent, then $\sum u_k$ is convergent;
   - $\sum u_k$ is divergent, then $\sum v_k$ is divergent.
3. *Ratio test.* Let $\sum u_k$ be the series of positive terms such that

$$\lim_{k\to\infty}\left|\frac{u_{k+1}}{u_k}\right| = l.$$

   Then
   (a) $\sum u_k$ is convergent if $l < 1$ and
   (b) $\sum u_k$ is divergent if $l > 1$.
4. *Cauchy n^{th} root test.* A series of positive terms $\sum u_n$ is convergent or divergent according as $\lim_{n\to\infty}(u_n)^{\frac{1}{n}} = l < 1$ or $l > 1$, respectively.

Note: The ratio test and Cauchy n^{th} root test fails at $l = 1$, and in such case, we cannot conclude the convergence or divergence of the series. In comparable situations, we can use the higher ratio test, Logarithmic test, Demorgan and Bertrand's test to name a few.

Power series

A series of the form

$$\sum_{k=0}^{\infty} a_k\,(z-u)^k$$

is called a power series centered at u, where a_k is the coefficient of the kth term and u is constant. More simply, we consider

$$\sum_{k=0}^{\infty} a_k\,z^k = a_0 + a_1\,z + a_2\,z^2 + \cdots$$

as a power series centred at 0. Here is a list of some power series

$$\sum_{k=0}^{\infty} z^k = 1 + z + z^2 + z^3 + \cdots,$$

$$\sum_{k=0}^{\infty} \frac{z^k}{k!} = 1 + \frac{z}{1!} + \frac{z^2}{2!} + \frac{z^3}{3!} + \cdots,$$

$$\sum_{k=0}^{\infty} (k!)\,z^k = 1 + z + 2\,z^2 + 6\,z^3 + \cdots,$$

$$\sum_{k=0}^{\infty} \frac{4^k}{k!}\,(z-2)^k = 1 + 4\,(z-2) + 8\,(z-2)^2 + \cdots.$$

To determine the radius of convergence r, we use

$$\frac{1}{r} = \lim_{k\to\infty} \sup |a_k|^{1\backslash k}$$

or

$$\frac{1}{r} = \lim_{k\to\infty} \left|\frac{a_{k+1}}{a_k}\right|.$$

The region of convergence is given by $|z-u| < r$. The series of the form $\sum_{k=0}^{\infty} a_k\,(z-u)^k$ converges absolutely in the region $|z-u| < r$.

Example 4. Find the radius of convergence of the following series:

(i) $\sum_{k=0}^{\infty} \frac{4^k}{k}\,(z-2)^k$,

(ii) $\sum_{k=0}^{\infty} \frac{(k-1)!}{k^k}\,z^k$,

(iii) $\sum_{k=0}^{\infty} \frac{(-1)^k}{(2\,k)!}\,z^{2k}$,

(iv) $\sum_{k=0}^{\infty} \frac{(-1)^{k+1}}{k}(z-1)^k$.

Solution. In the first series, we observe that $a_k = \frac{4^k}{k}$, and hence we have

$$\frac{1}{r} = \lim_{k\to\infty}\left|\frac{a_{k+1}}{a_k}\right| = \lim_{k\to\infty}\frac{4^{k+1}}{k+1}\cdot\frac{k}{4^k}$$
$$= \lim_{k\to\infty}\frac{4k}{k+1} = 4.$$

Therefore, $r = \frac{1}{4}$ is the radius of convergence and region of convergence is given by

$$|z-2| < \frac{1}{4} \implies -\frac{1}{4} < z-2 < \frac{1}{4} \implies \frac{7}{4} < z < \frac{9}{4}.$$

Thus, the series converges for all $z \in (\frac{7}{4}, \frac{9}{4})$. Similarly, the regions of convergence for other series can be evaluated and it is left to the reader.

Alternating series

An infinite series with alternate signs is called an alternating series that is a series of the form

$$\sum u_k = u_1 - u_2 + u_3 - u_4 + \cdots$$

is called an alternating series. For example,

$$\sum u_k = 1 - \frac{1}{2} + \frac{1}{3} - \frac{1}{4} + \cdots,$$
$$\sum u_k = \frac{1}{\sqrt{2}-1} - \frac{1}{\sqrt{2}+1} + \frac{1}{\sqrt{3}-1} - \frac{1}{\sqrt{3}+1} + \cdots.$$

An alternating series is convergent if
(i) $|u_k|$ is monotonically decreasing and
(ii) $\lim_{k\to\infty} u_k = 0$.

This test is sufficient only not necessary for convergence and is known as Leibnitz test.

Absolutely or conditionally convergent

An infinite series of real numbers $\sum u_k$ is called absolutely convergent if $\sum |u_k|$ is convergent. In case $\sum u_k$ is convergent, but $\sum |u_k|$ is not. Then $\sum u_k$ is called a conditionally convergent series. For example, the series

$$\sum u_k = 1 - \frac{1}{2} + \frac{1}{3} - \frac{1}{4} + \cdots$$

is convergent by the Leibnitz test but

$$\sum |u_k| = 1 + \frac{1}{2} + \frac{1}{3} + \frac{1}{4} + \cdots$$

is divergent by the p-test.

We now produce few results with respect to the manipulation of infinite series without their proofs.

Theorem 5. *If $\sum_{k=1}^{\infty} S(k,l)$ and $\sum_{l=1}^{\infty} S(k,l)$ are both absolutely convergent for every $k, l \in \mathbb{N}$, then the iterated series $\sum_{k=1}^{\infty}\sum_{l=1}^{\infty} S(k,l)$ and $\sum_{l=1}^{\infty}\sum_{k=1}^{\infty} S(k,l)$ are equal and absolutely convergent.*

Theorem 6. *If $[\cdot]$ denotes the integer part of a number, then the following useful conversions hold true,* [128]*:*

$$\begin{aligned}
\sum_{l=0}^{\infty}\sum_{k=0}^{\infty} S(k,l) &= \sum_{l=0}^{\infty}\sum_{k=0}^{l} S(k,l-k),\\
\sum_{l=0}^{\infty}\sum_{k=0}^{[l\backslash 2]} S(k,l) &= \sum_{l=0}^{\infty}\sum_{k=0}^{\infty} S(k,l+2k),\\
\sum_{l=0}^{\infty}\sum_{k=0}^{l} S(k,l) &= \sum_{l=0}^{\infty}\sum_{k=0}^{[l\backslash 2]} S(k,l-2k),\\
\sum_{l=0}^{\infty}\sum_{m=0}^{\infty}\sum_{k=0}^{\infty} S(l,m,k) &= \sum_{l=0}^{\infty}\sum_{m=0}^{l}\sum_{k=0}^{l-m} S(l-m-k,m,k).
\end{aligned} \tag{1.6}$$

1.2 Infinite product

For a given sequence u_k, where k is a positive integer, consider the finite product

$$p_n = \prod_{k=1}^{n}(1+u_k) = (1+u_1)(1+u_2)\cdots(1+u_n). \tag{1.7}$$

If $\lim_{n\to\infty} p_n$ exists and is equal to $p \neq 0$, we say that the infinite product $\prod_{k=1}^{\infty}(1+u_k)$ converges to the value p. If at least one of the factors of the infinite product $\prod_{k=1}^{\infty}(1+u_k)$ is zero and if the infinite product with the deleted zero factor is convergent. We say that the original product converges to zero. If the infinite product is not convergent, it is said to be divergent.

The necessary condition for convergence of an infinite series is $\lim_{k\to\infty} u_k = 0$, where u_k is the general term of the series. A similar result for an infinite product is given by the following.

Theorem 7. *If infinite product $\prod_{n=1}^{\infty}(1+u_n)$ is convergent, then $\lim_{n\to\infty} u_n = 0$.*

Proof. Consider $u_k \neq -1$ for all $k \geq 1$; i. e., there is no singular term among the factor of the infinite product. By assumption, $\prod_{k=1}^{\infty}(1+u_k)$ exists. Hence, we can define p as

$$p = \prod_{k=1}^{\infty}(1+u_k) = \lim_{n\to\infty}\prod_{k=1}^{n}(1+u_k).$$

As each of the factors is nonsingular, we can write

$$\begin{aligned}
1 = p^{-1}p &= \left[\lim_{n\to\infty}\prod_{k=1}^{n}(1+u_k)\right]^{-1}\left[\lim_{n\to\infty}\prod_{k=1}^{n}(1+u_k)\right] \\
&= \lim_{n\to\infty}\left[\prod_{k=1}^{n}(1+u_k)\right]^{-1}\lim_{n\to\infty}\left[\prod_{k=1}^{n}(1+u_k)\right] \\
&= \lim_{n\to\infty}\left\{\left[\prod_{k=1}^{n}(1+u_k)\right]^{-1}\left[\prod_{k=1}^{n}(1+u_k)\right]\right\} \\
&= \lim_{n\to\infty}\left\{(1+u_{k-1})^{-1}(1+u_{k-2})^{-1}\cdots(1+u_1)^{-1}\right\} \\
&\quad\times\left\{(1+u_1)\cdots(1+u_{k-1})(1+u_k)\right\} \\
&= \lim_{n\to\infty}(1+u_n) = 1 + \lim_{n\to\infty}u_n.
\end{aligned} \tag{1.8}$$

Therefore, $\lim_{n\to\infty}u_n = 0$. It completes the proof. □

An infinite product with no zero factor has been associated with the series of principal values of logarithms in the following sense.

Theorem 8. *If $u_k \neq -1$ for all $k \geq 1$, then $\prod_{k=1}^{\infty}(1+u_k)$ and $\sum_{k=1}^{\infty}\ln(1+u_k)$ converge or diverge together.*

Proof. By assumption $u_k \neq -1$ for all $k \geq 1$, and hence every term of the form $1+u_k$ is nonzero, and hence its logarithm exists. Let s_k and p_k denote the partial sum and partial product of $\sum_{k=1}^{\infty}\ln(1+u_k)$ and $\prod_{k=1}^{\infty}(1+u_k)$, respectively. Then we have

$$\begin{aligned}
e^{s_k} &= e^{\sum_{n=1}^{k}\log(1+u_n)} = e^{\log(1+u_1)}\cdots e^{\log(1+u_k)} \\
&= (1+u_1)(1+u_2)\cdots(1+u_k) \\
&= \prod_{n=1}^{k}(1+u_n) = p_k.
\end{aligned} \tag{1.9}$$

This yields

$$\lim_{k\to\infty}e^{s_k} = e^{\lim_{k\to\infty}s_k} = \lim_{k\to\infty}p_k.$$

Therefore, p_k has a limit if and only if s_k has a limit. Thus, $\prod_{k=1}^{\infty}(1+u_k)$ and $\sum_{k=1}^{\infty}\log(1+u_k)$ have the same character. It completes the proof. □

Assume that the infinite product $\prod_{n=1}^{\infty}(1+u_n)$ has no zero factors. The absolute convergence of the product can be utilized by the associated series of logarithms.

Theorem 9. *The infinite product $\prod_{n=1}^{\infty}(1+u_n)$ with no zero factors is said to be absolutely convergent if and only if the series $\sum_{n=1}^{\infty}\ln(1+u_n)$ is absolutely convergent.*

Proof. The proof follows from the above theorem. □

1.3 Gamma function and Pochhammer symbol

In the beginning of 18th century, there was a race among the mathematicians to generalize the factorial function in such a way that it can be defined on entire complex plane. Euler in 1729 finishes this and produces a generalization in terms of an integral known as Euler's second integral. The story behind the extension of factorial function to gamma function is interesting. One can read the article of Davis entitled "The history of the development of gamma function" for more detail on the development, [42].

We know that, for a number k, the factorial function is defined by

$$k! = 1\cdot 2\cdots(k-1)\cdot k = (k-1)!\,k. \tag{1.10}$$

The domain of this function is limited to nonnegative integers.

When Euler approached the problem, he produced two solutions in his letter to Goldbach. The first was written on October 13, 1729, in which the solution was obtained by dealing with interpolation problem. The second solution written on January 08, 1730, deals with the integration problem. He tied these two solutions and obtained the integral definition of the gamma function as

$$k! = \int_0^1 (-\log t)^k\,dt. \tag{1.11}$$

Later, Legendre modified the integral and gave the modern definition of gamma function. Let u be a complex number such that the real part of u is positive, i. e., $\Re(u) > 0$. Then the gamma function is denoted by the Greek letter "Γ" and defined in term of integral as

$$\Gamma(u) = \int_0^\infty t^{u-1}e^{-t}dt, \quad \Re(u) > 0, \tag{1.12}$$

by putting $-\log t = \mathrm{u}$, $\frac{-1}{t}dt = du$ and from (1.11), we get

$$k! = \int_\infty^0 \mathrm{u}^k\,(-e^{-\mathrm{u}}\,du) = \int_0^\infty \mathrm{u}^k e^{-\mathrm{u}}du. \tag{1.13}$$

On comparing the above equation with (1.12), one can deduce that

$$k! = \Gamma(k+1), \tag{1.14}$$

whenever k is not a negative integer. From this relation, we can see that $\Gamma(1) = 1$. Before proceeding further, let us recall some other forms of the gamma function.

Theorem 10. *The Euler definition of the gamma function as an infinite product is given by*

$$\Gamma(u) = \frac{1}{u} \prod_{m=1}^{\infty} \left[\left(1 + \frac{1}{m}\right)^{u} \left(1 + \frac{u}{m}\right)^{-1} \right] \tag{1.15}$$

and the product defined by Weierstrass is

$$\frac{1}{\Gamma(u)} = u e^{\gamma u} \prod_{m=1}^{\infty} \left[\left(1 + \frac{u}{m}\right) e^{\frac{-u}{m}} \right], \tag{1.16}$$

where $u \in \mathbb{C}$ and

$$\gamma = \lim_{k\to\infty} \left(\sum_{m=1}^{k} \frac{1}{m} - \log k \right) = 0.5772156649\ldots \tag{1.17}$$

is the Euler–Mascheroni's constant, [53].

Proof. One can calculate for a fixed integer m,

$$\lim_{k\to\infty} \frac{k!\,(k+1)^m}{(k+m)!} = 1.$$

We can extend this equation by replacing the arbitrary integer m with an arbitrary complex number u. Thus, we get

$$\lim_{k\to\infty} \frac{k!\,(k+1)^u}{(k+u)!} = 1.$$

Multiplying both sides by $(u-1)!$ give

$$\frac{1}{u} \lim_{k\to\infty} \frac{k!\,(k+1)^u\,u!}{(k+u)!} = (u-1)!.$$

Thus, we have

$$\begin{aligned} \Gamma(u) &= (u-1)! \\ &= \frac{1}{u} \lim_{k\to\infty} \frac{k!\,(k+1)^u\,u!}{(k+u)!} \end{aligned}$$

$$
\begin{aligned}
&= \frac{1}{u} \lim_{k\to\infty} (1\cdot 2\cdots k)\, \frac{1}{(u+1)\,(u+2)\cdots(u+k)} \left(\frac{2}{1}\cdot\frac{3}{2}\cdots\frac{k+1}{k}\right)^{u} \\
&= \frac{1}{u} \lim_{k\to\infty} \frac{1}{(1+u)\,(1+\frac{u}{2})\cdots(1+\frac{u}{k})} \left[\left(1+\frac{1}{1}\right)^{u}\left(1+\frac{1}{2}\right)^{u}\cdots\left(1+\frac{1}{k}\right)^{u}\right] \\
&= \frac{1}{u} \lim_{k\to\infty} \prod_{m=1}^{k} \left[\left(1+\frac{1}{m}\right)^{u}\left(1+\frac{u}{m}\right)^{-1}\right] \\
&= \frac{1}{u} \prod_{m=1}^{\infty} \left[\left(1+\frac{1}{m}\right)^{u}\left(1+\frac{u}{m}\right)^{-1}\right].
\end{aligned} \tag{1.18}
$$

It completes the proof of Euler's definition as an infinite product. From Euler's definition of the gamma function, we write

$$
\begin{aligned}
\frac{1}{\Gamma(u)} &= u \prod_{m=1}^{\infty}\left(1+\frac{1}{m}\right)^{-u}\left(1+\frac{u}{m}\right) \\
&= u \lim_{k\to\infty}\left(1+\frac{1}{k}\right)^{u} \lim_{k\to\infty}\prod_{m=1}^{k}\left(1+\frac{1}{m}\right)^{-u}\left(1+\frac{u}{m}\right) \quad \because \lim_{k\to\infty}\left(1+\frac{1}{k}\right)^{u} = 1 \\
&= u \lim_{k\to\infty}\left(\left(1+\frac{1}{k}\right)^{u}\prod_{m=1}^{k}\left(1+\frac{1}{m}\right)^{-u}\left(1+\frac{u}{m}\right)\right) \\
&= u \lim_{k\to\infty}\left(\prod_{m=1}^{k-1}\left(1+\frac{1}{m}\right)^{-u}\prod_{m=1}^{k}\left(1+\frac{u}{m}\right)\right) \\
&= u \lim_{k\to\infty}\left(k^{-u}\prod_{m=1}^{k}\left(1+\frac{u}{m}\right)\right) \\
&= u \lim_{k\to\infty}\left(\exp(-u\log(k))\prod_{m=1}^{k}\left(1+\frac{u}{m}\right)\right) \\
&= u \lim_{k\to\infty}\left(\exp\left(u\left(1-1+\frac{1}{2}-\frac{1}{2}+\cdots\frac{1}{k}-\frac{1}{k}-\log(k)\right)\right)\prod_{m=1}^{k}\left(1+\frac{u}{m}\right)\right) \\
&= u \lim_{k\to\infty}\exp\left(\frac{-u}{1}+\frac{-u}{2}+\cdots+\frac{u}{k}\right)\exp\left(u\left(1+\frac{1}{2}+\cdots\frac{1}{k}-\log(k)\right)\right) \\
&\quad\times\prod_{m=1}^{k}\left(1+\frac{u}{m}\right) \\
&= u \lim_{k\to\infty}\left(\exp\left(\frac{-u}{1}+\frac{-u}{2}+\cdots+\frac{-u}{k}\right)\prod_{m=1}^{k}\left(1+\frac{u}{m}\right)\right) \\
&\quad\times\lim_{k\to\infty}\exp\left(u\left(1+\frac{1}{2}+\cdots\frac{1}{k}-\log(k)\right)\right) \\
&= u\,e^{\gamma u}\prod_{m=1}^{\infty}\left[\left(1+\frac{u}{m}\right)e^{\frac{-u}{m}}\right].
\end{aligned} \tag{1.19}
$$

It completes the proof. □

We now produce a few elementary results for the gamma function.

Theorem 11. *The following fundamental relation holds true:*

$$\Gamma(u+1) = u\,\Gamma(u). \tag{1.20}$$

Proof.

$$\begin{aligned}\Gamma(u+1) &= \int_0^\infty t^u e^{-t}\,dt \\ &= [-t^u e^{-t}]_0^\infty + \int_0^\infty u\,t^{u-1}e^{-t}\,dt \\ &= u\int_0^\infty t^{u-1}e^{-t}\,dt \\ &= u\,\Gamma(u). \end{aligned} \tag{1.21}$$

It completes the proof. □

Theorem 12. *The Euler–Walle's formula for $k \in \mathbb{N}$ is*

$$\sin u = u\prod_{k=1}^{\infty}\left(1 - \frac{u^2}{k^2\,\pi^2}\right). \tag{1.22}$$

Proof. We know a well-known identity:

$$\sin 2u = 2\sin u\,\cos u = 2\sin u\,\sin\left(\frac{\pi}{2} + u\right). \tag{1.23}$$

With the help of this identity, we have

$$\sin u = 2\sin\left(\frac{u}{2}\right)\,\sin\left(\frac{\pi + u}{2}\right).$$

Again using (1.23), we get

$$\sin u = 2^3\,\sin\left(\frac{u}{2^2}\right)\,\sin\left(\frac{\pi + u}{2^2}\right)\,\sin\left(\frac{2\pi + u}{2^2}\right)\,\sin\left(\frac{3\pi + u}{2^2}\right).$$

By repeating this process k-times yields

$$\sin u = 2^{2^k-1}\,\sin\left(\frac{u}{2^k}\right)\,\sin\left(\frac{\pi + u}{2^k}\right)\,\sin\left(\frac{2\pi + u}{2^k}\right)\cdots\sin\left(\frac{(2^k-1)\pi + u}{2^k}\right),$$

which can be rewritten by using the identity $\sin(\frac{(2^k-1)\pi+u}{2^k}) = \sin(\pi - \frac{\pi-u}{2^k}) = \sin(\frac{\pi-u}{2^k})$ and $\sin(x+y)\,\sin(x-y) = \sin^2 x - \sin^2 y$ as

$$\begin{aligned}\sin u &= 2^{2^k-1}\sin\left(\frac{u}{2^k}\right)\left\{\sin\left(\frac{\pi+u}{2^k}\right)\sin\left(\frac{\pi-u}{2^k}\right)\right\}\left\{\sin\left(\frac{2\pi+u}{2^k}\right)\sin\left(\frac{2\pi-u}{2^k}\right)\right\}\cdots\\ &\quad\times\left\{\sin\left(\frac{(\frac{1}{2}2^k-1)\pi+u}{2^k}\right)\sin\left(\frac{(\frac{1}{2}2^k-1)\pi-u}{2^k}\right)\right\}\left\{\sin\left(\frac{(\frac{1}{2}2^k)\pi+u}{2^k}\right)\right\}\\ &= 2^{2^k-1}\sin\left(\frac{u}{2^k}\right)\left\{\sin^2\frac{\pi}{2^k}-\sin^2\frac{u}{2^k}\right\}\left\{\sin^2\frac{2\pi}{2^k}-\sin^2\frac{u}{2^k}\right\}\cdots\\ &\quad\times\left\{\sin^2\frac{(\frac{1}{2}2^k-1)\pi}{2^k}-\sin^2\frac{u}{2^k}\right\}\left\{\cos\frac{u}{2^k}\right\}.\end{aligned}\tag{1.24}$$

On dividing both sides by $\sin(\frac{u}{2^k})$ and using the limit $\lim_{u\to 0}\frac{\sin u}{\sin(\frac{u}{2^k})}=2^k$, we obtain

$$2^k = 2^{2^k-1}\sin^2\left(\frac{\pi}{2^k}\right)\sin^2\left(\frac{2\pi}{2^k}\right)\cdots\sin^2\left(\frac{(\frac{1}{2}2^k-1)\pi}{2^k}\right).\tag{1.25}$$

Again dividing (1.24) by (1.25) gives

$$\sin u = 2^k\sin\left(\frac{u}{2^k}\right)\left\{1-\frac{\sin^2\frac{u}{2^k}}{\sin^2\frac{\pi}{2^k}}\right\}\left\{1-\frac{\sin^2\frac{u}{2^k}}{\sin^2\frac{2\pi}{2^k}}\right\}\cdots\left\{1-\frac{\sin^2\frac{u}{2^k}}{\sin^2\frac{(\frac{1}{2}2^k-1)\pi}{2^k}}\right\}\cos\left(\frac{u}{2^k}\right).\tag{1.26}$$

Taking the limit as $k\to\infty$ yields

$$\sin u = u\left(1-\frac{u^2}{\pi^2}\right)\left(1-\frac{u^2}{2^2\pi^2}\right)\cdots = u\prod_{k=1}^{\infty}\left(1-\frac{u^2}{k^2\pi^2}\right).\tag{1.27}$$

It completes the proof. □

We use this Euler–Walli's formula to get Euler's reflection formula.

Theorem 13. *The Euler reflection formula for* $0<u<1$ *is* [53]

$$\Gamma(1-u)\,\Gamma(u) = \frac{\pi}{\sin\pi u}.\tag{1.28}$$

Proof. The Euler's product (1.15) and recurrence relation (1.20) together yield

$$\begin{aligned}&\frac{1}{\Gamma(1-u)\,\Gamma(u)}\\ &\quad=\frac{1}{(-u)\,\Gamma(-u)\,\Gamma(u)}\\ &\quad=\frac{-1}{u}\prod_{m=1}^{\infty}\left[u\left(1+\frac{1}{m}\right)^{u}\left(1+\frac{u}{m}\right)\right]\left[-u\left(1+\frac{1}{m}\right)^{-u}\left(1-\frac{u}{m}\right)\right]\\ &\quad=u\prod_{m=1}^{\infty}\left(1-\frac{u^2}{m^2}\right).\end{aligned}\tag{1.29}$$

The Euler–Walli's formula gives

$$\sin(\pi u) = \pi u \prod_{m=1}^{\infty}\left(1 - \frac{u^2}{m^2}\right). \tag{1.30}$$

On comparing the equations (1.29) and (1.30), we get

$$\Gamma(1-u)\,\Gamma(u) = \frac{\pi}{\sin \pi u}.$$

It completes the proof. □

Putting $u = \frac{1}{2}$ in (1.28) avail a very fruitful result as

$$\Gamma\left(\frac{1}{2}\right)\Gamma\left(1-\frac{1}{2}\right) = \frac{\pi}{\sin\frac{\pi}{2}} = \pi. \tag{1.31}$$

This implies

$$\Gamma\left(\frac{1}{2}\right) = \sqrt{\pi}. \tag{1.32}$$

Similarly, by putting $u = \frac{3}{2}$ gives

$$\Gamma\left(\frac{3}{2}\right)\Gamma\left(1-\frac{3}{2}\right) = \frac{\pi}{\sin\frac{3\pi}{2}} \tag{1.33}$$

This yields

$$\Gamma\left(-\frac{1}{2}\right) = -2\sqrt{\pi}. \tag{1.34}$$

Let u be a complex number. Then the Pochhammer symbol, also called a shifted factorial, is defined by

$$(u)_k = \begin{cases} u\,(u+1)\cdots(u+k-1), & \text{if } k \geq 1, \\ 1, & \text{if } k = 0. \end{cases} \tag{1.35}$$

In terms of a gamma function,

$$(u)_k = \frac{\Gamma(u+k)}{\Gamma(u)}. \tag{1.36}$$

Important relations related to the Pochhammer symbol are

(i) $(1)_n = n!$;

(ii) $(u)_{m+n} = (u)_m\,(u+m)_n = (u)_n\,(u+n)_m$;

(iii) $(u)_{-k} = \frac{(-1)^k}{(1-u)_k}$;

(iv) $(u)_{n-k} = \frac{(-1)^k (u)_n}{(1-u-n)_k}$, $0 \le k \le n$;

(v) $(-n)_k = \frac{(-1)^k n!}{(n-k)!}$;

(vi) $(-u)_k = (-1)^k (1+u-k)_k$;

(vii) $\binom{u}{k} = \frac{(-1)^k (-u)_k}{k!}$ or $\frac{\Gamma(u+1)}{\Gamma(k+1)\Gamma(u-k+1)}$;

(viii) $(u+n)_{k-n}(u+k)_{n-k} = 1$;

(ix) $(\frac{u}{2}+1)_k \setminus (\frac{u}{2})_k = \frac{u+2k}{u}$.

Proof. The proofs of (i) and (ii) follow directly from the definition of a shifted factorial (1.35). To prove (iii), we use the relation $(u)_k = \frac{\Gamma(u+k)}{\Gamma(u)}$ and get

$$\begin{aligned}(u)_{-k} &= \frac{\Gamma(u-k)}{\Gamma(u)} \\ &= \frac{\Gamma(u-k)}{(u-1)(u-2)\cdots\Gamma(u-k)} \\ &= \frac{1}{(-1)^k (1-u)(1-u+1)\cdots(1-u+k-1)} \\ &= \frac{1}{(-1)^k (1-u)_k}.\end{aligned}$$

This gives (iii). Again using (1.35), we get

$$\begin{aligned}(u)_{n-k} &= u(u+1)\cdots(u+n-k-1) \\ &= \frac{u(u+1)\cdots(u+n-k-1)(u+n-k)(u+n-k+1)\cdots(u+n-k+k-1)}{(u+n-k)(u+n-k+1)\cdots(u+n-k+k-1)} \\ &= \frac{(u)_n}{(-1)^k (1-u-n)(1-u-n+1)\cdots(1-u-n+k-1)} \\ &= \frac{(u)_n}{(-1)^k (1-u-n)_k} = \frac{(-1)^k (u)_n}{(1-u-n)_k}.\end{aligned}$$

This completes the proof of (iv). The proof of (v) can be seen as

$$\begin{aligned}(-n)_k &= (-n)(-n+1)\cdots(-n+k-1) \\ &= \frac{(-1)^k n(n-1)\cdots(n-k+1)(n-k)!}{(n-k)!} \\ &= \frac{(-1)^k n!}{(n-k)!}.\end{aligned}$$

Part (vi) can be proved in the following way:

$$\begin{aligned}(-u)_k &= (-u)(-u+1)\cdots(-u+k-1) \\ &= (-1)^k u(u-1)(u-2)\cdots(u-k+2)(u-k+1)\end{aligned}$$

$$= (-1)^k (1+u-k)(1+u-k+1)\cdots(1+u-k+k-2)(1+u-k+k-1)$$
$$= (-1)^k (1+u-k)_k.$$

Follow these steps to prove (vii):

$$\begin{aligned}\binom{u}{k} &= \frac{u!}{k!\,(u-k)!} = \frac{\Gamma(u+1)}{\Gamma(k+1)\,\Gamma(u-k+1)}\\ &= \frac{u\,(u-1)\cdots(u-k+1)\Gamma(u-k+1)}{\Gamma(k+1)\,\Gamma(u-k+1)}\\ &= \frac{(1+u-k)(1+u-k+1)\cdots(1+u-k+k-1)}{\Gamma(k+1)}\\ &= \frac{(1+u-k)_k}{\Gamma(k+1)} = \frac{(-1)^k(-u)_k}{\Gamma(k+1)}, \quad \text{by using (vi).}\end{aligned}$$

Using (iv), we have

$$\begin{aligned}(u+n)_{k-n}\cdot(u+k)_{n-k} &= \frac{(-1)^n (u+n)_k}{(1-u-n-k)_n}\,\frac{(-1)^k (u+k)_n}{(1-u-k-n)_k}\\ &= \frac{(-1)^{n+k}(u+n)_k\,(u+k)_n}{(-1)^k (u+k)_n\,(-1)^n (u+n)_k} = 1.\end{aligned}$$

It completes the proof of (viii). Using the definition of the Pochhammer symbol, one can easily obtain

$$\left(\frac{u}{2}+1\right)_k \Big\backslash \left(\frac{u}{2}\right)_k = \frac{u+2k}{u}.$$

Thus, we have completed the proofs. □

Now, in the following theorem, we establish the limiting definition of the gamma function in terms of a shifted factorial.

Theorem 14. *The following limiting definition holds true:*

$$\Gamma(u) - \lim_{k\to\infty}\frac{(k-1)!\,k^u}{(u)_k} \tag{1.37}$$

Proof. The Euler product for the gamma function gives

$$\begin{aligned}\Gamma(u) &= \frac{1}{u}\prod_{m=1}^{\infty}\left[\left(1+\frac{1}{m}\right)^u\left(1+\frac{u}{m}\right)^{-1}\right]\\ &= \frac{1}{u}\lim_{k\to\infty}\prod_{m=1}^{k-1}\left[\left(1+\frac{1}{m}\right)^u\left(1+\frac{u}{m}\right)^{-1}\right]\\ &= \frac{1}{u}\lim_{k\to\infty}\left[\frac{(\frac{2}{1})^u(\frac{3}{2})^u\cdots(\frac{k}{k-1})^u}{(u+1)(u+2)\cdots(u+k-1)}1\cdot 2\cdots(k-1)\right]\end{aligned}$$

$$= \lim_{k\to\infty} \frac{k^u\,(k-1)!}{u\,(u+1)\,(u+2)\cdots(u+k-1)}$$
$$= \lim_{k\to\infty} \frac{(k-1)!\,k^u}{(u)_k}.$$

It completes the proof. □

We can equally write (1.37) as

$$\Gamma(u) = \lim_{k\to\infty} \frac{k!\,k^u}{(u)_k}. \tag{1.38}$$

Lemma 15. *The following relation holds true:*

$$(u)_{2k} = 2^{2k}\left(\frac{u}{2}\right)_k\left(\frac{u+1}{2}\right)_k. \tag{1.39}$$

Proof. The definition of the Pochhammer symbol allows us to write

$$\begin{aligned}
(u)_{2k}
&= u\,(u+1)\cdots(u+k-1)\,(u+k)\cdots(u+2k-1)\\
&= 2^{2k}\left(\frac{u}{2}\right)\left(\frac{u+1}{2}\right)\left(\frac{u}{2}+1\right)\left(\frac{u+1}{2}+1\right)\cdots\left(\frac{u}{2}+k-1\right)\left(\frac{u+1}{2}+k-1\right)\\
&= 2^{2k}\left\{\left(\frac{u}{2}\right)\left(\frac{u}{2}+1\right)\cdots\left(\frac{u}{2}+k-1\right)\right\}\left\{\left(\frac{u+1}{2}\right)\left(\frac{u+1}{2}+1\right)\cdots\left(\frac{u+1}{2}+k-1\right)\right\}\\
&= 2^{2k}\left(\frac{u}{2}\right)_k\left(\frac{u+1}{2}\right)_k.
\end{aligned}$$

It completes the proof. □

The relation (1.39) can be generalized as

$$(u)_{nk} = n^{nk}\left(\frac{u}{n}\right)_k\left(\frac{u+1}{n}\right)_k \quad \left(\frac{u+n-1}{n}\right)_k. \tag{1.40}$$

Next, with the help of (1.39), we establish Legendre's duplication formula for the gamma function.

Theorem 16. *The following formula is known as the Legendre's duplication formula:*

$$\Gamma(2\,a) = 2^{2a-1}\,\pi^{-\frac{1}{2}}\,\Gamma(a)\,\Gamma\!\left(a+\frac{1}{2}\right). \tag{1.41}$$

Proof. In equation (1.39), substituting $u = 2\,a$, and thus we obtain

$$(2a)_{2k} = 2^{2k}\,(a)_k \left(a + \frac{1}{2}\right)_k$$

or

$$\frac{\Gamma(2a+2k)}{\Gamma(2a)} = 2^{2k}\,\frac{\Gamma(a+k)}{\Gamma(a)}\,\frac{(a+\frac{1}{2}+k)}{\Gamma(a+\frac{1}{2})}$$

or

$$\frac{\Gamma(2a)}{\Gamma(a)\,\Gamma(a+\frac{1}{2})} = \frac{\Gamma(2a+2k)}{2^{2k}\,\Gamma(a+k)\,(a+\frac{1}{2}+k)}.$$

Taking the limit as $k \to \infty$ yields

$$\begin{aligned}
&\frac{\Gamma(2a)}{\Gamma(a)\,\Gamma(a+\frac{1}{2})}\\
&\quad = \lim_{k\to\infty} \frac{\Gamma(2a+2k)}{2^{2k}\,\Gamma(a+k)\,(a+\frac{1}{2}+k)}\\
&\quad = \lim_{k\to\infty} \frac{\Gamma(2a+2k)}{(2k-1)!\,(2k)^{2a}} \cdot \frac{(k-1)!\,k^{a}}{\Gamma(a+k)} \cdot \frac{(k-1)!\,k^{a+\frac{1}{2}}}{\Gamma(a+\frac{1}{2}+k)} \cdot \frac{(2k-1)!\,2^{2a}}{2^{2k}\,k^{1\backslash 2}\,[(k-1)!]^2}.
\end{aligned}$$

Since $\lim_{k\to\infty} \frac{(k-1)!\,k^a}{\Gamma(a+k)} = 1$, we have

$$\frac{\Gamma(2a)}{\Gamma(a)\,\Gamma(a+\frac{1}{2})} = \lim_{k\to\infty} \frac{(2k-1)!\,2^{2a}}{2^{2k}\,k^{1\backslash 2}\,[(k-1)!]^2}.$$

It follows

$$\frac{\Gamma(2a)}{2^{2a}\,\Gamma(a)\,\Gamma(a+\frac{1}{2})} = c \quad \text{(some constant independent from } a\text{)}.$$

Substituting $a - \frac{1}{2}$ gives the value of c as

$$c - \frac{\Gamma(1)}{2\,\Gamma(1)\,\Gamma(\frac{1}{2})} = \frac{1}{2\,\sqrt{\pi}}$$

Thus, we obtain

$$\Gamma(2\,a) = 2^{2a-1}\,\pi^{-\frac{1}{2}}\,\Gamma(a)\,\Gamma\!\left(a + \frac{1}{2}\right). \tag{1.42}$$

It completes the proof. □

One can generalize this duplicate formula and obtain the Gauss multiplication formula. The proof is beyond the scope, and hence omitted.

Theorem 17. *The following formula is known as the Gauss multiplication formula:*

$$\Gamma(m\,u) = (2\,\pi)^{-\frac{1}{2}(m-1)}\, m^{m\,u-\frac{1}{2}} \prod_{r=0}^{m-1} \Gamma\left(u + \frac{r}{m}\right), \quad m = 2, 3, \ldots. \tag{1.43}$$

Another well-known result is the Stirling formula for gamma function, given in the following lemma.

Lemma 18. *Let u be a complex number such that $\mathbb{R}(u) \neq 0, -1, -2, \ldots$. Then the stirling formula, for gamma function, is*

$$\Gamma(u) \sim e^{-u}\, u^{u-\frac{1}{2}}\, \sqrt{2\pi}. \tag{1.44}$$

This formula is quite useful to obtain the approximate value of $\Gamma(u)$, when u is large enough.

Observing the Gauss and Weierstrass definitions of the gamma function, one can see that the gamma function is an analytic function of u with singularities at $u = 0, -1, -2, \ldots$. For the residue of the gamma function, we write

$$\Gamma(u) = \int_0^1 e^{-t}\, t^{u-1} dt + \int_1^\infty e^{-t}\, t^{u-1} dt = I_1(u) + I_2(u). \tag{1.45}$$

On solving $I_1(u)$, one can get

$$I_1(u) = \int_0^1 \sum_{k=0}^{\infty} \frac{(-1)^k}{k!}\, t^{u+k-1}\, dt, \tag{1.46}$$

which on integration gives

$$I_1(u) = \sum_{k=0}^{\infty} (-1)^k\, [k!\,(u+k)]^{-1}. \tag{1.47}$$

Also,

$$\lim_{u \to -k} (u+k) I_1(u) = \sum_{k=0}^{\infty} \frac{(-1)^k}{k!}. \tag{1.48}$$

Hence, $\frac{(-1)^k}{k!}$ is the residue of $\Gamma(u)$ at simple pole $u = -k$ $(k = 0, 1, 2, \ldots)$.

1.4 Beta function

As the gamma function is the generalization of a factorial function, we can say that the beta function is the genralization of binomial coefficient. Euler used two types of integrals while generalizing the factorial function. Legendre called the integral:

$$\int_0^1 x^e (1-x)^n\, dx, \tag{1.49}$$

as the first integral and the integral,

$$\int_0^1 (-\log t)^n\, dt, \tag{1.50}$$

as the second integral. The first integral was used by Euler to start his derivation for gamma function and it comes to an end in terms of the second integral. The definition of the beta function is motivated with the second integral.

Let u, v be complex numbers such that $\mathbb{R}(u), \mathbb{R}(v) > 0$. Then the beta function is defined by

$$\beta(u,v) = \int_0^1 t^{u-1}(1-t)^{v-1}\, dt. \tag{1.51}$$

Later, various properties of the beta function have been studied to very large extent. We list here the most important one and required in the subsequent sections and chapters. There is another integral, which can also be used to define the beta function. To get that, we substitute $t = \frac{\mathfrak{u}}{1+\mathfrak{u}}$ and $dt = \frac{1}{(1+\mathfrak{u})^2}\, d\mathfrak{u}$ in (1.51), and thus arrive at

$$\beta(u,v) = \int_0^\infty \frac{\mathfrak{u}^{a-1}}{(1+\mathfrak{u})^{a+b}}\, d\mathfrak{u}, \quad \mathbb{R}(u), \mathbb{R}(v) > 0. \tag{1.52}$$

Also, the beta function can be defined in terms of the gamma function as

$$\beta(u,v) = \frac{\Gamma(u)\,\Gamma(v)}{\Gamma(u+v)}, \quad \mathbb{R}(u), \mathbb{R}(v) > 0. \tag{1.53}$$

Proof. To prove (1.53), consider the product

$$\Gamma(u)\,\Gamma(v) = \left(\int_0^\infty e^{-t}\, t^{u-1} dt\right)\left(\int_0^\infty e^{-\mathfrak{u}}\, \mathfrak{u}^{v-1} d\mathfrak{u}\right)$$

$$= \int_0^\infty \int_0^\infty e^{-(t+\mathfrak{u})}\, t^{u-1}\, \mathfrak{u}^{v-1}\, dt\, d\mathfrak{u}, \quad \mathbb{R}(u), \mathbb{R}(v) > 0. \tag{1.54}$$

Now substituting $t = \mathfrak{v}\,\mathfrak{w}$, $\mathfrak{u} = (1-\mathfrak{v})\,\mathfrak{w}$ and

$$dt\, d\mathfrak{u} = \frac{\partial(t,\mathfrak{u})}{\partial(\mathfrak{v},\mathfrak{w})}\, d\mathfrak{v}\, d\mathfrak{w} = \mathfrak{w}\, d\mathfrak{v}\, d\mathfrak{w}, \tag{1.55}$$

where $\frac{\partial(t,\mathfrak{u})}{\partial(\mathfrak{v},\mathfrak{w})}$ is the Jacobian of $(t, \mathfrak{u})$ with respect to $(\mathfrak{v}, \mathfrak{w})$ and defined by

$$\frac{\partial(t,u)}{\partial(\mathfrak{v},\mathfrak{w})} = \begin{vmatrix} \frac{\partial t}{\partial \mathfrak{v}} & \frac{\partial t}{\partial \mathfrak{w}} \\ \frac{\partial u}{\partial \mathfrak{v}} & \frac{\partial u}{\partial \mathfrak{w}} \end{vmatrix} \text{ or } \begin{vmatrix} \frac{\partial t}{\partial \mathfrak{v}} & \frac{\partial u}{\partial \mathfrak{v}} \\ \frac{\partial t}{\partial \mathfrak{w}} & \frac{\partial u}{\partial \mathfrak{w}} \end{vmatrix}. \tag{1.56}$$

We get

$$\begin{aligned} \Gamma(u)\,\Gamma(v) &= \int_0^\infty \int_0^1 e^{-\mathfrak{w}}\,(\mathfrak{v}\,\mathfrak{w})^{u-1}\left((1-\mathfrak{v})\mathfrak{w}\right)^{v-1}\,\mathfrak{w}\,d\mathfrak{v}\,d\mathfrak{w} \\ &= \int_0^\infty \int_0^1 e^{-\mathfrak{w}}\,\mathfrak{w}^{u+v-1}\,\mathfrak{v}^{u-1}\,(1-\mathfrak{v})^{v-1}\,d\mathfrak{v}\,d\mathfrak{w} \\ &= \left(\int_0^\infty e^{-\mathfrak{w}}\,\mathfrak{w}^{u+v-1}\,d\mathfrak{w}\right)\left(\int_0^1 \mathfrak{v}^{u-1}\,(1-\mathfrak{v})^{v-1}\,d\mathfrak{v}\right) \\ &= \Gamma(u+v)\,\beta(u,v). \end{aligned} \tag{1.57}$$

It completes the proof of (1.53). ☐

With this fundamental relation between the gamma and beta function, one can observe that the beta function is symmetric in its arguments u and v. That is,

$$\beta(u,v) = \frac{\Gamma(u)\,\Gamma(v)}{\Gamma(u+v)} = \beta(v,u). \tag{1.58}$$

If u and v are positive integers, say $u = m$, $v = n$, then

$$\begin{aligned} \beta(m,n) &= \frac{\Gamma(m)\,\Gamma(n)}{\Gamma(m+n)} = \frac{(m-1)!\,(n-1)!}{(m+n-1)!} \\ &= \frac{1}{m\binom{m+n-1}{m}} \text{ or } \frac{1}{n\binom{m+n-1}{n}}, \end{aligned} \tag{1.59}$$

where $\binom{n}{k} = \frac{n!}{k!\,(n-k)!}$ is the binomial coefficient.

Lemma 19. *Some other integrals in terms of the beta function are as*

$$\beta(u,v) = 2\int_0^{\frac{\pi}{2}} \sin^{2u-1}\theta\,\cos^{2v-1}\theta\,d\theta \tag{1.60}$$

$$= (1+\alpha)^u \int_0^1 \mathfrak{v}^{u-1}\,(1-\mathfrak{v})^{v-1}\,(1+\alpha\mathfrak{v})^{-(u+v)}\,d\mathfrak{v}, \quad \alpha > -1, \Re(u), \Re(v) > 0 \tag{1.61}$$

$$= \alpha^v \int_0^\infty \mathfrak{v}^{v-1}\,(1+\alpha\mathfrak{v})^{-(u+v)}\,d\mathfrak{v}, \quad \alpha > 0, \Re(u), \Re(v) > 0 \tag{1.62}$$

$$= (\alpha_2-\alpha_1)^{1-u-v} \int_{\alpha_1}^{\alpha_2} (\mathfrak{v}-\alpha_1)^{u-1}\,(\alpha_2-\mathfrak{v})^{v-1}\,d\mathfrak{v}, \quad \alpha_2 > \alpha_1, \Re(u), \Re(v) > 0. \tag{1.63}$$

Proof. The equation (1.60) can be proved by substitution $u = \sin^2\theta$, $1-u = \cos^2\theta$ and $du = 2\sin\theta\cos\theta\, d\theta$. By substituting $u = \frac{(1+a)v}{1+av}$, $1-u = \frac{1-v}{1+av}$ and $du = \frac{(1+a)v}{(1+av)^2}\,dv$ and lead to the integral (1.61). The integral (1.62) can be obtained by using the symmetry of the beta function and putting $u = av$. The substitution $u = \frac{v-a_1}{a_2-a_1}$, $1-u = \frac{a_2-v}{a_2-a_1}$ and $du = \frac{dv}{a_2-a_1}$ give the integral form (1.63). □

We now list some functional relations in terms of the beta function.

Lemma 20. *The following relations hold true:*

$$\beta(u,v+1) = \frac{v}{u}\,\beta(u+1,v) = \frac{v}{u+v}\,\beta(u,v); \tag{1.64}$$

$$\beta(u,v)\,\beta(u+v,w) = \beta(v,w)\,\beta(v+w,u) = \beta(w,u)\,\beta(u+w,v) = \frac{\Gamma(u)\,\Gamma(v)\,\Gamma(w)}{\Gamma(u+v+w)}; \tag{1.65}$$

$$\beta(u,v)\,\beta(u+v,w)\,\beta(u+v+w,t) = \frac{\Gamma(u)\,\Gamma(v)\,\Gamma(w)\,\Gamma(t)}{\Gamma(u+v+w+t)}. \tag{1.66}$$

Proof. The proofs of these relations are simple and straightforward. Using the relation between the gamma and beta function $\beta(u,v) = \frac{\Gamma(u)\,\Gamma(v)}{\Gamma(u+v)}$, we can write

$$\begin{aligned}\beta(u,v+1) &= \frac{\Gamma(u)\,\Gamma(v+1)}{\Gamma(u+v+1)}\\ &= \frac{v}{u+v}\,\frac{\Gamma(u)\,\Gamma(v)}{\Gamma(u+v)} = \frac{v}{u+v}\,\beta(u,v)\end{aligned}$$

and

$$\begin{aligned}\beta(u+1,v) &= \frac{\Gamma(u+1)\,\Gamma(v)}{\Gamma(u+v+1)}\\ &= \frac{u}{u+v}\,\frac{\Gamma(u)\,\Gamma(v)}{\Gamma(u+v)} = \frac{u}{u+v}\,\beta(u,v).\end{aligned}$$

Combining these two equations together yield

$$\beta(u,v+1) = \frac{v}{u}\,\beta(u+1,v) = \frac{v}{u+v}\,\beta(u,v).$$

Again using the relation $\beta(u,v) = \frac{\Gamma(u)\,\Gamma(v)}{\Gamma(u+v)}$, we have

$$\beta(u+1,v) + \beta(u,v+1) = \frac{\Gamma(u+1)\,\Gamma(v)}{\Gamma(u+v+1)} + \frac{\Gamma(u)\,\Gamma(v+1)}{\Gamma(u+v+1)} \tag{1.67}$$

$$= \frac{(u+v)\,\Gamma(u)\,\Gamma(v)}{(u+v)\,\Gamma(u+v)} = \beta(u,v). \tag{1.68}$$

Similarly, one can prove the other identities. □

The relations (1.65) and (1.66) can be generalized as

$$\prod_{k=1}^{m} \beta\left(\sum_{j=1}^{k} u_j, u_{k+1}\right) = \frac{\Gamma(u_1)\cdots\Gamma(u_{k+1})}{\Gamma(u+\cdots+u_{k+1})}. \tag{1.69}$$

Example 21. The beta function and related properties are useful to solve the following integrals:

(i) $\int_0^1 \frac{t}{\sqrt{1-t^2}}\,dt$;

(ii) $\int_0^3 \frac{1}{\sqrt{9-t^2}}\,dt$;

(iv) $\int_0^1 t^4\,(1-t)^2\,dt$;

(v) $\int_0^2 t\,(8-t^3)^{1\backslash 3}\,dt$;

(vi) $\int_0^{\frac{\pi}{2}} \sin^5 t\,\cos^{\frac{7}{2}} t\,dt$.

Solutions. The simple substitution in the above examples change them into the standard forms of the beta and gamma function, and thus we will get the value of integrals. On substituting $t^2 = \mathrm{u}$ and $t\,dt = \frac{d\mathrm{u}}{2}$ in the first integral, so

$$\begin{aligned}\int_0^1 \frac{t}{\sqrt{1-t^2}}\,dt &= \frac{1}{2}\int_0^1 \frac{d\mathrm{u}}{\sqrt{1-\mathrm{u}}}\\ &= \frac{1}{2}\int_0^1 \mathrm{u}^{1-1}\,(1-\mathrm{u})^{\frac{1}{2}-1}\,d\mathrm{u}\\ &= \frac{1}{2}\beta\left(1,\frac{1}{2}\right) = \frac{1}{2}\frac{\Gamma(1)\,\Gamma(\frac{1}{2})}{\Gamma(\frac{3}{2})} = 1.\end{aligned}$$

To prove (ii), we substitute $t^2 = 9\,\mathrm{u}$, and we get

$$\begin{aligned}\int_0^3 \frac{1}{\sqrt{9-t^2}}\,dt &= \frac{3}{2}\int_0^1 \frac{d\mathrm{u}}{3\sqrt{1-\mathrm{u}}\,\sqrt{\mathrm{u}}}\\ &= \frac{1}{2}\int_0^1 \mathrm{u}^{\frac{1}{2}-1}\,(1-\mathrm{u})^{\frac{1}{2}-1}\,d\mathrm{u}\\ &= \frac{1}{2}\beta\left(\frac{1}{2},\frac{1}{2}\right).\end{aligned}$$

From the definition of the beta function, it will give the value of the integral. Similarly, the other integrals can be solved.

1.5 The hypergeometric differential equation

An equation in terms of the dependent variable, independent variable and their derivatives is known as a differential equation. The highest order of derivatives present in the equation represents the order of the differential equation. The exponent or power of the highest order derivative is known as the degree of the differential equation. On the basis of the number of independent variables, differential equations are divided in two parts. An equation is an ordinary differential equation or partial differential equation; accordingly, the number of independent variables is one or greater than one. A general second-order linear ordinary differential equation is written as

$$P_1(z)\, W''(z) + P_2(z)\, W'(z) + P_3(z)\, W(z) = P_4(z). \tag{1.70}$$

The equation (1.70) is linear because the dependent variable and each of its derivatives has degree 1. Also, no product of a dependent variable and its derivative occurs. The equation (1.70) is called homogeneous if $P_4(z) = 0$. Thus, a general second-order homogeneous linear differential equation is

$$W''(z) + P(z)\, W'(z) + Q(z)\, W(z) = 0. \tag{1.71}$$

A point $z = a$ is called a singular point if any of the limits,

$$\lim_{z\to a} P(z) \quad \text{or} \quad \lim_{z\to a} Q(z), \tag{1.72}$$

does not exist. The singular point $z = a$ is called regular or irregular; accordingly, the limits

$$\lim_{z\to a}(z-a)\, P(z) \quad \text{and} \quad \lim_{z\to a}(z-a)^2\, Q(z), \tag{1.73}$$

exist or not.

A second-order homogeneous linear ordinary differential equation of the form

$$z\,(1-z)\,\mathfrak{w}''(z) + \left[w - (u+v+1)\,z\right]\mathfrak{w}'(z) - u\,v\,\mathfrak{w}(z) = 0, \tag{1.74}$$

where u, v, w are complex numbers such that $w \neq 0, -1, -2, \ldots$ and independent from the complex variable z. The equation (1.74) is called a hypergeometric equation, and alternatively it can be written as

$$\left[\theta\,(\theta + w - 1) - z\,(\theta + u)\,(\theta + v)\right]\mathfrak{w}(z) = 0, \quad \theta = z\frac{d}{dz}. \tag{1.75}$$

On comparing the equation (1.74) with (1.71), one can observe that it has three regular kind of singularities at $z = 0, 1, \infty$. This equation is important because Poole described in a theorem that any second-order homogeneous ordinary differential equation with three regular singularities can be written in the form (1.74). Let us consider

$$\mathfrak{w}(z) = z^c \sum_{k=0}^{\infty} a_k z^k, \quad a_0 \neq 0, \tag{1.76}$$

be a solution of (1.74), which on differentiation with respect to z gives

$$\begin{aligned} \mathfrak{w}'(z) &= z^{c-1} \sum_{k=0}^{\infty} (c+k)\, a_k z^k, \\ \mathfrak{w}''(z) &= z^{c-2} \sum_{k=0}^{\infty} (c+k)(c+k-1)\, a_k z^k. \end{aligned} \tag{1.77}$$

On substituting the values of $\mathfrak{w}(z)$ and its derivatives in (1.74), we get

$$z(1-z)z^c \sum_{k=0}^{\infty} (c+k)(c+k-1)\, a_k z^{k-2} + \{w-(1+u+v)z\} z^c \sum_{k=0}^{\infty} (c+k)\, a_k z^{k-1} - uvz^c \sum_{k=0}^{\infty} a_k z^k = 0;$$

or

$$z^{c-1} \sum_{k=0}^{\infty} (c+k)(c+k-1)\, a_k z^k + z^{c-1} w \sum_{k=0}^{\infty} (c+k)\, a_k z^k - z^c \sum_{k=0}^{\infty} (c+k)(c+k-1)\, a_k z^k - (1+u+v) z^c \sum_{k=0}^{\infty} (c+k)\, a_k z^{k-2} - uvz^c \sum_{k=0}^{\infty} a_k z^k = 0;$$

or

$$z^{c-1} c(c+w-1)\, a_0 + z^{c-1} \sum_{k=1}^{\infty} (c+k)(c+k+w-1)\, a_k z^k - z^c \sum_{k=0}^{\infty} (c+k)(c+k-1)\, a_k z^k - (1+u+v) z^c \sum_{k=0}^{\infty} (c+k)\, a_k z^{k-2} - uvz^c \sum_{k=0}^{\infty} a_k z^k = 0;$$

or

$$z^{c-1} c(c+w-1)\, a_0 + z^c \sum_{k=0}^{\infty} (c+k+1)(c+k+w)\, a_{k+1} z^k - z^c \sum_{k=0}^{\infty} (c+k)(c+k-1)\, a_k z^k - (1+u+v) z^c \sum_{k=0}^{\infty} (c+k)\, a_k z^{k-2} - uvz^c \sum_{k=0}^{\infty} a_k z^k = 0;$$

or

$$z^{c-1} c(c+w-1)\, a_0 + \sum_{k=0}^{\infty} \{(c+k+1)(c+k+w)\, a_{k+1} - (c+k+u)(c+k+v)\, a_k\} z^{c+k} = 0.$$

Hence, we have the indicial equation,

$$c(c+w-1) = 0, \tag{1.78}$$

and in general,

$$(c+k+1)\,(c+k+w)\,a_{k+1} - (c+k+u)\,(c+k+v)\,a_k = 0. \tag{1.79}$$

The roots of the indicial equations are $c = 0$ and $c = 1 - w$. For each root, we have an independent solution of the differential equation (1.74). In the begining, we now discuss the solution in detail, when $c = 0$ and w is neither a zero nor a negative integer. The solution will be of the form

$$\mathfrak{w}_1(z) = \sum_{k=0}^{\infty} \lambda_k\, z^k, \tag{1.80}$$

which on differentiating with respect to z gives

$$\mathfrak{w}_1'(z) = \sum_{k=1}^{\infty} k\,\lambda_k\, z^{k-1}, \quad \mathfrak{w}_1''(z) = \sum_{k=2}^{\infty} k\,(k-1)\,\lambda_k\, z^{k-2}. \tag{1.81}$$

On substituting these values in (1.74), we get

$$\begin{aligned} &z \sum_{k=2}^{\infty} k\,(k-1)\,\lambda_k\, z^{k-2} - z^2 \sum_{k=2}^{\infty} k\,(k-1)\,\lambda_k\, z^{k-2} + w \sum_{k=1}^{\infty} k\,\lambda_k\, z^{k-1} \\ &\quad - (u+v+1) \sum_{k=1}^{\infty} k\,\lambda_k\, z^k - u\,v \sum_{k=0}^{\infty} \lambda_k\, z^k = 0, \end{aligned} \tag{1.82}$$

which yields

$$\begin{aligned} &\sum_{k=2}^{\infty} k\,(k-1)\,\lambda_k\, z^{k-1} - \sum_{k=2}^{\infty} k\,(k-1)\,\lambda_k\, z^k + w \sum_{k=1}^{\infty} k\,\lambda_k\, z^{k-1} - (u+v+1) \\ &\quad \times \sum_{k=1}^{\infty} k\,\lambda_k\, z^k - u\,v \sum_{k=0}^{\infty} \lambda_k\, z^k = 0. \end{aligned} \tag{1.83}$$

Now, equating the powers of z, and putting each coefficient of z equal to zero, the right-hand side of the equation is zero. The coefficients of z^0 or constant terms are

$$w\,\lambda_1 - u\,v\,\lambda_0 = 0. \tag{1.84}$$

This implies

$$\lambda_1 = \frac{u\,v}{w}\,\lambda_0. \tag{1.85}$$

Now, equating the power of z, one can get

$$2\lambda_2 + 2w\lambda_2 - (u+v+1)\lambda_1 - uv\lambda_1 = 0$$

or

$$\lambda_2 = \frac{(u+1)(v+1)}{2(w+1)}\lambda_1. \tag{1.86}$$

With the help of (1.85), we have

$$\lambda_2 = \frac{u(u+1)v(v+1)}{2\cdot 1w(w+1)}\lambda_0. \tag{1.87}$$

Proceeding in a similar manner yields

$$\lambda_k = \frac{u(u+1)\cdots(u+k-1)v(v+1)\cdots(v+k-1)}{w(w+1)\cdots(w+k-1)k!}\lambda_0. \tag{1.88}$$

Using the definition of Pochhammer symbol and for $\lambda_0 = 1$, we have

$$\lambda_k = \frac{(u)_k(v)_k}{(w)_k k!}. \tag{1.89}$$

Thus,

$$\mathfrak{w}_1(z) = {}_2F_1(u,v;w;z) = \sum_{k=0}^{\infty}\frac{(u)_k(v)_k}{(w)_k k!}z^k, \tag{1.90}$$

and is an analytic solution of the differential equation (1.74) at $z = 0$ with the initial condition $\mathfrak{w}_1(0) = {}_2F_1(u,v;w;0) = 1$.

Next, we discuss the solution, when $c = 1-w$ and w is not a positive integer greater than or equal to 2. Consider the second solution

$$\mathfrak{w}_2(z) = z^{1-w}\sum_{k=0}^{\infty}\lambda_k z^k. \tag{1.91}$$

The derivatives are

$$\begin{aligned}\mathfrak{w}_2'(z) &= \sum_{k=1}^{\infty}(1+k-w)\lambda_k z^{-w+k};\\ \mathfrak{w}_2''(z) &= \sum_{k=2}^{\infty}(1+k-w)(-w+k)\lambda_k z^{-w+k-1}.\end{aligned} \tag{1.92}$$

Now, putting all these values in (1.74), we get

$$\begin{aligned}z(1-z)&\sum_{k=2}^{\infty}(-w+k+1)(-w+k)\lambda_k z^{-w+k-1} + w\sum_{k=1}^{\infty}(-w+k+1)\lambda_k z^{-w+k}\\ &-(u+v+1)\sum_{k=1}^{\infty}(-w+k+1)\lambda_k z^{-w+k} - uv\sum_{k=0}^{\infty}\lambda_k z^{-w+k+1} = 0,\end{aligned}$$

which on simplification and some rearrangements give

$$z\,(1-z)\,\mathfrak{w}_2''(z) + [2 - w - \{(u - w + 1) + (v - w + 1) + 1\}z]\mathfrak{w}_2'(z)$$
$$- (u - w + 1)\,(v - w + 1)\mathfrak{w}_2(z) = 0. \tag{1.93}$$

The above differential equation is of the type (1.74). It will have a solution in the form

$${}_2F_1(\tilde{u}, \tilde{v}; \tilde{w}; z),$$

where:

1. $\tilde{u} = u - w + 1$;
2. $\tilde{v} = v - w + 1$;
3. $\tilde{w} = 2 - w$.

Thus, the second solution is given by

$$\mathfrak{w}_2(z) = z^{1-w}\,{}_2F_1(u - w + 1, v - w + 1; 2 - w; z). \tag{1.94}$$

We can summarize the whole findings as follow.

Theorem 22. *Let u, v, w be complex parameters such that $\mathbb{R}(w) \notin \mathbb{Z}^+ - \{1\}$. Then the complete solution of* (1.74) *will be given by*

$$\mathfrak{w}(z) = c_1\,\mathfrak{w}_1(z) + c_2\,\mathfrak{w}_2(z)$$
$$= c_1\,{}_2F_1(u, v; w; z) + c_2\,z^{1-w}\,{}_2F_1(u - w + 1, v - w + 1; 2 - w; z). \tag{1.95}$$

1.5.1 Kummer's 24 solutions

One can clearly observe that when $w = 1$, these two solutions $\mathfrak{w}_1$ and $\mathfrak{w}_2$ are equal. So, the second solution will be obtained by

$$\mathfrak{w}_2 = {}_2F_1(u, v; 1; z)\,\log z + \sum_{k=1}^{\infty}\left[\frac{\partial}{\partial c}\left\{\frac{(u+c)_k\,(v+c)_k}{(1+c)_k\,(1+c)_k}\right\}\right]_{c=0} z^k. \tag{1.96}$$

In case w is zero or a negative integer, then the condition for first solution is violated. So, the first solution will be replaced by

$$\mathfrak{w}_1 = z^{1-w}\,{}_2F_1(u + 1 - w, v + 1 - w; 2 - w; z)\,\log z$$
$$+ \sum_{k=0}^{\infty}\left[\frac{\partial}{\partial c}\left\{\frac{(u+c)_k\,(v+c)_k}{(w+c)_k\,(1+c)_k}\right\}\right]_{c=1-w} z^k. \tag{1.97}$$

Further, if w is a positive integer greater than or equal to 2, the condition of the second solution is violated, and in such case we obtain this as

$$\mathfrak{w}_2 = {}_2F_1(u, v; w; z) \log z + \sum_{k=0}^{\infty} \left[\frac{\partial}{\partial c} \left\{ \frac{(u+c)_k (v+c)_k}{(w+c)_k (1+c)_k} \right\} \right]_{c=0} z^k. \tag{1.98}$$

Next, we obtain another pair of solutions in the distinct regions of the complex plane. To do so, we begin with the substitution $(1-z)^n y(z)$ for $\mathfrak{w}(z)$ in (1.74). On simplification, it becomes

$$\begin{aligned} & z(1-z) \frac{d^2y}{dz^2} + [w - (u+v+1+2n) z] \frac{dy}{dz} \\ & + \left[\frac{n(n-1)z - n(w-(u+v+1)z)}{1-z} - uv \right] y = 0. \end{aligned} \tag{1.99}$$

The expression $n(n-1)z - n(w-(u+v+1)z)$ is completely divisible by $(1-z)$ if $n = 0$ or $n = w - u - v$. When $n = 0$, we get $\mathfrak{w}_1(z)$ and $\mathfrak{w}_2(z)$ as solutions of (1.74) but when $n = w - u - v$, the equation (1.74) is replaced by

$$\begin{aligned} & z(1-z) \frac{d^2y}{dz^2} + [w - (\tilde{u} + \tilde{v} + 1) z] \frac{dy}{dz} \\ & - \tilde{u}\tilde{v} y = 0, \end{aligned} \tag{1.100}$$

with $\tilde{u} = w - u$, $\tilde{v} = w - v$. The equation (1.100) is of the type (1.74) and a pair of solutions is given by

$$\begin{aligned} y_1 &= {}_2F_1(w-u, w-v; w; z), \quad \text{and} \\ y_2 &= z^{1-w} {}_2F_1(1-u, 1-v; 2-w; z). \end{aligned} \tag{1.101}$$

Hence, (1.74) has another pair of solutions as

$$\begin{aligned} \mathfrak{w}_3(z) &= (1-z)^{w-u-v} {}_2F_1(w-u, w-v; w; z), \quad \text{and} \\ \mathfrak{w}_4(z) &= z^{1-w} (1-z)^{w-u-v} {}_2F_1(1-u, 1-v; 2-w; z). \end{aligned} \tag{1.102}$$

Now substitute $z = 1 - x$ in (1.74), and after some manipulation, we get

$$x(1-x) \frac{d^2\mathfrak{w}}{dx^2} - [w - (u+v+1)(1-x)] \frac{d\mathfrak{w}}{dx} - uv\,\mathfrak{w} = 0. \tag{1.103}$$

On comparing with (1.74), the equation (1.103) has a set of four solutions in the region $|x| = |1-z| < 1$ as

$$\begin{aligned} \mathfrak{w}_5(z) &= {}_2F_1(u, v; 1+u+v-w; x) = {}_2F_1(u, v; 1+u+v-w; 1-z), \\ \mathfrak{w}_6(z) &= (1-z)^{w-u-v} {}_2F_1(w-u, w-v; 1-u-v+w; 1-z); \\ \mathfrak{w}_7(z) &= z^{1-w} {}_2F_1(1+u-w, 1+v-w; 1+u+v-w; 1-z); \\ \mathfrak{w}_8(z) &= z^{1-w} (1-z)^{w-u-v} {}_2F_1(1-u, 1-v; 1-u-v+w; 1-z), \end{aligned} \tag{1.104}$$

provided u, v, w are not a negative integer or zero. To obtain the solution in the region $|z| > 1$ and $|1 - z| > 1$, we put $z = \frac{1}{x}$ and $x^a y = \mathfrak{w}$. Thus, we get the equation

$$x(1-x)\frac{d^2y}{dx^2} + [1+u-v-(2u-w)x]\frac{dy}{dx} - u(1+u-w)y = 0. \tag{1.105}$$

In comparison with (1.74), the equation (1.103) has a pair of solutions that are given by

$$\begin{aligned} y_1 &= {}_2F_1(u, 1+u-w; 1+u-v; x), \quad \text{and} \\ y_2 &= x^{v-u}\, {}_2F_1(v, 1+v-w; 1+v-u; x). \end{aligned} \tag{1.106}$$

Corresponding to these two solutions, we have two sets of four solutions of (1.74) as

$$\begin{aligned} \mathfrak{w}_9(z) &= (-z)^{-u}\, {}_2F_1\left(u, 1+u-w; 1+u-v; \frac{1}{z}\right), \\ \mathfrak{w}_{10}(z) &= (-z)^{-v}\, {}_2F_1\left(v, 1+v-w; 1+v-u; \frac{1}{z}\right); \\ \mathfrak{w}_{11}(z) &= (-z)^{v-w}(1-z)^{w-u-v}\, {}_2F_1\left(1-v, w-v; 1+u-v; \frac{1}{z}\right); \\ \mathfrak{w}_{12}(z) &= (-z)^{u-w}(1-z)^{w-u-v}\, {}_2F_1\left(1-u, w-u; 1+v-u; \frac{1}{z}\right), \end{aligned} \tag{1.107}$$

and if we combined $1-z$ and $\frac{1}{z}$ in the above four solutions, we obtain another set of four solutions

$$\begin{aligned} \mathfrak{w}_{13}(z) &= (1-z)^{-u}\, {}_2F_1\left(u, w-v; 1+u-v; \frac{1}{1-z}\right), \\ \mathfrak{w}_{14}(z) &= (1-z)^{-v}\, {}_2F_1\left(v, w-u; 1+v-u; \frac{1}{1-z}\right); \\ \mathfrak{w}_{15}(z) &= (-z)^{1-w}(1-z)^{w-u-1}\, {}_2F_1\left(1+u-w, 1-v; 1+u-v; \frac{1}{1-z}\right); \\ \mathfrak{w}_{16}(z) &= (-z)^{1-w}(1-z)^{w-v-1}\, {}_2F_1\left(1+v-w, 1-u; 1+v-u; \frac{1}{1-z}\right). \end{aligned} \tag{1.108}$$

We now deduce solutions in the region $\mathbb{R}(z) > \frac{1}{2}$ and $\mathbb{R}(z) < \frac{1}{2}$. We substitute $z = \frac{x-1}{x}$, and thus obtain two sets of four solutions as

$$\begin{aligned} \mathfrak{w}_{17}(z) &= z^{-u}\, {}_2F_1\left(u, 1+u-w; 1+u+v-w; 1-\frac{1}{z}\right), \\ \mathfrak{w}_{18}(z) &= z^{-v}\, {}_2F_1\left(v, 1+v-w; 1+u+v-w; 1-\frac{1}{z}\right); \\ \mathfrak{w}_{19}(z) &= z^{v-w}(1-z)^{w-u-v}\, {}_2F_1\left(1-v, w-v; 1-u-v+w; 1-\frac{1}{z}\right); \end{aligned}$$

$$\mathfrak{w}_{20}(z) = z^{u-w}(1-z)^{w-u-v}\,{}_2F_1\left(1-u, w-u; 1-u-v+w; 1-\frac{1}{z}\right) \tag{1.109}$$

valid in the region $\mathbb{R}(z) > \frac{1}{2}$ and

$$\begin{aligned}
\mathfrak{w}_{21}(z) &= (1-z)^{-u}\,{}_2F_1\left(u, w-v; w; \frac{z}{z-1}\right),\\
\mathfrak{w}_{22}(z) &= (1-z)^{-v}\,{}_2F_1\left(v, w-u; w; \frac{z}{z-1}\right);\\
\mathfrak{w}_{23}(z) &= z^{1-w}(1-z)^{w-u-1}\,{}_2F_1\left(1+u-w, 1-v; 2-w; \frac{z}{z-1}\right);\\
\mathfrak{w}_{24}(z) &= z^{1-w}(1-z)^{w-v-1}\,{}_2F_1\left(1+v-w, 1-u; 2-w; \frac{z}{z-1}\right)
\end{aligned} \tag{1.110}$$

in the plane $\mathbb{R}(z) < \frac{1}{2}$.

Thus, for u, v, $w \in \mathbb{C}$ such that they are neither an integer nor zero, Kummer obtained 24 solutions of the hypergeometric equation (1.74) valid in any region of the complex plane.

The series ${}_2F_1(u, v; w; z) = \sum_{k=0}^{\infty} \frac{(u)_k (v)_k}{(w)_k\, k!} z^k$ as one of the solutions of the differential equation (1.74) and is called the hypergeometric series or Gauss hypergeometric series. The particular cases of the series, i. e., $u = w$, $v = 1$ or $u = 1$, $v = w$ reduce the series into

$${}_2F_1(w, 1; w; z) = 1 + z + \cdots, \tag{1.111}$$

a geometrical series and this is the reason of calling it hypergeometric.

For convergence, we use the D'almbert ratio test. Let us denote the Gauss hypergeometric series by

$${}_2F_1(u, v; w; z) = \sum_{k=0}^{\infty} \frac{(u)_k (v)_k}{(w)_k\, k!} z^k = \sum_{k=0}^{\infty} A_k z^k. \tag{1.112}$$

Then

$$\begin{aligned}
\left|\frac{A_{k+1}}{A_k}\right| &= \left|\frac{(u)_{k+1}(v)_{k+1}}{(w)_{k+1}(k+1)!}\,\frac{(w)_k\, k!}{(u)_k (v)_k}\right|\\
&= \left|\frac{(u+k)(v+k)}{(k+1)(w+k)}\right|.
\end{aligned} \tag{1.113}$$

Taking limit as $k \to \infty$,

$$\lim_{k\to\infty}\left|\frac{A_{k+1}}{A_k}\right| = \lim_{k\to\infty}\left|\frac{(u+k)(v+k)}{(k+1)(w+k)}\right| = 1. \tag{1.114}$$

So, the series is convergent in the unit disc $|z| < 1$, i. e., $-1 < z < 1$ and for $|z| > 1$, the series is divergent. Now, let us examine the nature of the series at $|z| = 1$. Let $\Re(w - u - v) = 2\delta$ for some positive number δ. Then

$$\begin{aligned} k^{1+\delta}\frac{(u)_k(v)_k}{(w)_k\,k!} &= \frac{k^{1+\delta}}{k!}\frac{(k-1)!\,k^u\,k^{-u}\,(u)_k}{(k-1)!}\frac{(k-1)!\,k^v\,k^{-v}\,(v)_k}{(k-1)!} \\ &\quad\times\frac{(k-1)!}{(k-1)!\,k^w\,k^{-w}\,(w)_k} \\ &= \frac{k^{1+\delta}}{k}\frac{k^{-u}\,(u)_k}{(k-1)!}\frac{k^{-v}\,(v)_k}{(k-1)!}\frac{(k-1)!}{k^{-w}\,(w)_k}k^{u+v-w} \\ &= \frac{k^{-u}\,(u)_k}{(k-1)!}\frac{k^{-v}\,(v)_k}{(k-1)!}\frac{(k-1)!}{k^{-w}\,(w)_k}k^{\delta+u+v-w}. \end{aligned} \tag{1.115}$$

By $\Gamma(a) = \lim_{k\to\infty}\frac{(k-1)!\,k^a}{(a)_k}$ and for $|z| = 1$, it follows that

$$\begin{aligned} \lim_{k\to\infty} k^{1+\delta}\left|\frac{(u)_k(v)_k}{(w)_k\,k!}z^k\right| &\leq \lim_{k\to\infty}\left|\frac{1}{k^{w-u-v-\delta}}\right|\left|\frac{k^{-u}\,(u)_k}{(k-1)!}\frac{k^{-v}\,(v)_k}{(k-1)!}\frac{(k-1)!}{k^{-w}\,(w)_k}\right| \\ &\leq \lim_{k\to\infty}\left|\frac{\Gamma(w)}{\Gamma(u)\,\Gamma(v)}\right|\left|\frac{1}{k^{w-u-v-\delta}}\right| \\ &\leq \left|\frac{\Gamma(w)}{\Gamma(u)\,\Gamma(v)}\right|\cdot 0 = 0. \end{aligned} \tag{1.116}$$

Since $\lim_{k\to\infty} k^{1+\delta}\,|\frac{(u)_k(v)_k}{(w)_k\,k!}z^k| = 0$, for $|z| = 1$, by the comparison theorem of numerical series of positive numbers, one concludes the absolute convergence of ${}_2F_1(u, v; w; z)$ at $z = 1$. We summarize the convergent region in the following theorem.

Theorem 23. *Let u, v, w be complex numbers such that $\Re(w) \neq 0, -1, \ldots$. Then the Gauss hypergeometric series converges absolutely when $|z| < 1$ and when $|z| = 1$ it converges absolutely if $\Re(w - u - v) > 0$; conditionally convergent if $-1 < \Re(w - u - v) \leq 0$; divergent if $\Re(w - u - v) \leq -1$.*

1.6 Differential formulae and contiguous relations

If we differentiate the Gauss hypergeometric series with respect to z, we get

$$\begin{aligned} \frac{d}{dz}\,{}_2F_1(u, v; w; z) &= \sum_{k=0}^{\infty}\frac{(u)_k\,(v)_k}{(w)_k}\frac{k\,z^{k-1}}{k!} \\ &= \sum_{k=1}^{\infty}\frac{(u)_k\,(v)_k}{(w)_k}\frac{z^{k-1}}{(k-1)!} \\ &= \sum_{k=0}^{\infty}\frac{(u)_{k+1}\,(v)_{k+1}}{(w)_{k+1}}\frac{z^k}{k!} \end{aligned}$$

$$= \sum_{k=0}^{\infty} \frac{u\,(u+1)_k\, v\,(v+1)_k}{w\,(w+1)_k} \frac{z^k}{k!}$$
$$= \frac{u\,v}{w}\, {}_2F_1(u+1, v+1; w+1; z). \tag{1.117}$$

Repeating this process, we find in general

$$\frac{d^n}{dz^n}\, {}_2F_1(u, v; w; z) = \frac{(u)_n\,(v)_n}{(w)_n}\, {}_2F_1(u+n, v+n; w+n; z). \tag{1.118}$$

Other differential formulas are:

1. $\frac{d^n}{dz^n}\,[z^{u+n-1}\,{}_2F_1(u, v; w; z)] = (u)_n\, z^{u-1}\, {}_2F_1(u+n, v; w; z)$,
2. $\frac{d^n}{dz^n}\,[(1-z)^{u+v-w}\,{}_2F_1(u, v; w; z)] = \frac{(w-u)_n\,(w-v)_n}{(w)_n}\,(1-z)^{u+v-w-n}\, {}_2F_1(u, v; w+n; z)$,
3. $\frac{d^n}{dz^n}\,[z^{w-1}\,{}_2F_1(u, v; w; z)] = (w-n)_n\, z^{w-n-1}\,{}_2F_1(u, v; w-n; z)$,
4. $\frac{d^n}{dz^n}\,[z^{w-u+n-1}\,(1-z)^{u+v-w}\,{}_2F_1(u, v; w; z)] = (w-u)_n\, z^{w-u-1}\,(1-z)^{u+v-w-n}\, {}_2F_1(u, v; w; z)$,
5. $\frac{d^n}{dz^n}\,[(1-z)^{u+n-1}\,{}_2F_1(u, v; w; z)] = \frac{(-1)^n\,(u)_n (w-v)_n\,(1-z)^{u-1}}{(w)_n}\, {}_2F_1(u+n, v; w+n; z)$,
6. $\frac{d^n}{dz^n}\,[z^{w-1}\,(1-z)^{v-w+n}\,{}_2F_1(u, v; w; z)] = (w-n)_n\, z^{w-n-1}\,(1-z)^{v-w}\, {}_2F_1(u-n, v; w-n; z)$,
7. $\frac{d^n}{dz^n}\,[z^{w-1}\,(1-z)^{u+v-w}\,{}_2F_1(u, v; w; z)] = (w-n)_n\, z^{w-n-1}\,(1-z)^{u+v+w-n}\, {}_2F_1(u-n, v-n; w-n; z)$,
8. $\frac{d^n}{dz^n}\,[z^{w+n-1}\,(1-z)^{n+u+v-w}\,{}_2F_1(u+n, v+n; w+n; z)] = (w)_n\, z^{w-1}\,(1-z)^{u+v-w}\, {}_2F_1(u, v; w; z)$,
9. $\frac{d^n}{dz^n}\,[z^{w+n-1}\,(1-z)^{v-w}] = (w)_n\, z^{w-1}\,(1-z)^{v-w-n}\, {}_2F_1(-n, v; w; z)$.

Proof. The proofs of these differential formulae are very simple and straightforward. Let us begin with the first one by differentiating the product $[z^{u+n-1}\,{}_2F_1(u, v; w; z)]$. For $n = 1$,

$$\frac{d}{dz}\,[z^{u}\,{}_2F_1(u, v; w; z)] = \frac{d}{dz}\left[z^u \sum_{m=0}^{\infty} \frac{(u)_m\,(v)_m}{(w)_m} \frac{z^m}{m!}\right]$$
$$= \sum_{m=0}^{\infty} \frac{(u)_m\,(v)_m}{(w)_m\, m!}\,(u+m)\, z^{u+m-1}. \tag{1.119}$$

Similarly, differentiating the product $[z^{u+n-1}\,{}_2F_1(u, v; w; z)]$ two times, for $n = 2$, yields

$$\frac{d^2}{dz^2}\,[z^{u+1}\,{}_2F_1(u, v; w; z)] = \sum_{m=0}^{\infty} \frac{(u)_m\,(v)_m}{(w)_m\, m!}\,(u+m+1)(u+m)\, z^{u+m-1}. \tag{1.120}$$

Repeating this process n-times, one can get

$$\frac{d^n}{dz^n}\,[z^{u+n-1}\,{}_2F_1(u, v; w; z)]$$
$$= \sum_{m=0}^{\infty} \frac{(u)_m\,(v)_m}{(w)_m\, m!}\,(u+m+n-1)(u+m+n-2)\cdots(u+m)\, z^{u+m-1}$$

$$
\begin{aligned}
&= \sum_{m=0}^{\infty} \frac{(u)_m\,(v)_m}{(w)_m\,m!}(u+m)_n\,z^{u+m-1} \\
&= \sum_{m=0}^{\infty} \frac{(u)_{m+n}\,(v)_m}{(w)_m\,m!}\,z^{u+m-1} \\
&= (u)_n\,z^{u-1}\sum_{m=0}^{\infty} \frac{(u+n)_m\,(v)_m}{(w)_m\,m!}\,z^m \\
&= (u)_n\,z^{u-1}\,{}_2F_1(u+n,v;w;z).
\end{aligned} \tag{1.121}
$$

It completes the proof of (1). To prove (2), we start differentiating the product

$$
\begin{aligned}
&\frac{d}{dz}[(1-z)^{u+v-w}{}_2F_1(u,v;w;z)] \\
&= (1-z)^{u+v-w}\sum_{m=0}^{\infty}\frac{(u+m)\,(v+m)}{(w+m)}\,\frac{(u)_m\,(v)_m}{(w)_m}\,\frac{z^m}{m!} \\
&\quad - (u+v-w)(1-z)^{u+v-w-1}\sum_{m=0}^{\infty}\frac{(u)_m\,(v)_m}{(w)_m}\,\frac{z^m}{m!} \\
&= (1-z)^{u+v-w-1}\sum_{m=0}^{\infty}\frac{(u)_m\,(v)_m}{(w)_m}\,\frac{z^m}{m!}\left(\frac{(u+m)\,(v+m)}{(w+m)}(1-z)-(u+v-w)\right) \\
&= (1-z)^{u+v-w-1}\sum_{m=0}^{\infty}\frac{(u)_m\,(v)_m}{(w)_m}\,\frac{z^m}{m!}\left[\left(m+(u+v-w)+\frac{(w-u)(w-v)}{(w+m)}\right)(1-z)\right] \\
&\quad - (1-z)^{u+v-w-1}\sum_{m=0}^{\infty}\frac{(u)_m\,(v)_m}{(w)_m}\,\frac{z^m}{m!}[(u+v-w)] \\
&= (1-z)^{u+v-w-1}\sum_{m=0}^{\infty}\frac{(u)_m\,(v)_m}{(w)_m}\,\frac{z^m}{m!}\left(m+\frac{(w-u)(w-v)}{(w+m)}-z\frac{(u+m)\,(v+m)}{(w+m)}\right) \\
&= (1-z)^{u+v-w-1}\sum_{m=0}^{\infty}\frac{(u)_m\,(v)_m}{(w)_m}\,\frac{(w-u)(w-v)}{(w+m)}\,\frac{z^m}{m!} \\
&= (1-z)^{u+v-w-1}\frac{(w-u)(w-v)}{w}\sum_{m=0}^{\infty}\frac{(u)_m\,(v)_m}{(w+1)_m}\,\frac{z^m}{m!} \\
&= (1-z)^{u+v-w-1}\frac{(w-u)(w-v)}{w}\,{}_2F_1(u,v;w+1;z).
\end{aligned} \tag{1.122}
$$

Now, differentiating (1.122) with respect to z, yields

$$
\begin{aligned}
&\frac{d^2}{dz^2}[(1-z)^{u+v-w}{}_2F_1(u,v;w;z)] \\
&= \frac{(w-u)(w-v)}{w}\,\frac{d}{dz}[(1-z)^{u+v-w-1}{}_2F_1(u,v;w+1;z)] \\
&= (1-z)^{u+v-w-2}\,\frac{(w-u)\,(w-u+1)\,(w-v)\,(w-v+1)}{w\,(w+1)}\,{}_2F_1(u,v;w+2;z).
\end{aligned} \tag{1.123}
$$

Repeating this process n-times, we get

$$\begin{aligned}
&\frac{d^n}{dz^n}\left[(1-z)^{u+v-w}{}_2F_1(u,v;w;z)\right]\\
&\quad=(1-z)^{u+v-w-n}\frac{(w-u)\cdots(w-u+n-1)\,(w-v)\cdots(w-v+n-1)}{w\cdots(w+n-1)}{}_2F_1(u,v;w+n;z)\\
&\quad=(1-z)^{u+v-w-n}\frac{(w-u)_n\,(w-v)_n}{(w)_n}{}_2F_1(u,v;w+n;z).
\end{aligned}\tag{1.124}$$

It completes the proof of (2). Now, we give the proof of (3):

$$\begin{aligned}
\frac{d}{dz}\left[z^{w-1}{}_2F_1(u,v;w;z)\right]&=\sum_{m=0}^{\infty}\frac{(u)_m\,(v)_m}{(w)_m\,m!}\frac{d}{dz}z^{w+m-1}\\
&=\sum_{m=0}^{\infty}\frac{(u)_m\,(v)_m}{(w)_m\,m!}(w+m-1)z^{w+m-2}\\
&=z^{w-2}\,(w-1)\sum_{m=0}^{\infty}\frac{(u)_m\,(v)_m}{(w-1)_m\,m!}z^m\\
&=z^{w-2}\,(w-1)\,{}_2F_1(u,v;w-1;z).
\end{aligned}\tag{1.125}$$

Repeating this process n-times

$$\begin{aligned}
\frac{d^n}{dz^n}\left[z^{w-1}{}_2F_1(u,v;w;z)\right]&=z^{w-n-1}\,(w-1)(w-2)\cdots(w-n)\sum_{m=0}^{\infty}\frac{(u)_m\,(v)_m}{(w-n)_m\,m!}z^m\\
&=z^{w-n-1}\,(w-n)_n\,{}_2F_1(u,v;w-n;z).
\end{aligned}\tag{1.126}$$

The proof of (4) is as follows:

$$\begin{aligned}
&\frac{d}{dz}\sum_{m=0}^{\infty}(1-z)^{u+v-w}\frac{(u)_m\,(v)_m}{(w)_m\,m!}z^{w-u+m}\\
&\quad=(1-z)^{u+v-w}\sum_{m=0}^{\infty}\frac{(u)_m\,(v)_m}{(w)_m\,m!}(w-u+m)z^{w-u+m-1}-(u+v-w)(1-z)^{u+v-w-1}\\
&\qquad\times\sum_{m=0}^{\infty}\frac{(u)_m\,(v)_m}{(w)_m\,m!}z^{w-u+m}\\
&\quad=z^{w-u-1}(1-z)^{u+v-w-1}\sum_{m=0}^{\infty}\frac{(u)_m\,(v)_m}{(w)_m}\frac{z^m}{m!}\left[(1-z)(w-u+m)-z(u+v-w)\right]\\
&\quad=z^{w-u-1}(1-z)^{u+v-w-1}\sum_{m=0}^{\infty}\frac{(u-1)_m\,(v)_m}{(w)_m}\frac{z^m}{m!}\left[\frac{u+m-1}{u-1}((w-u)+m-z(v+m))\right]\\
&\quad=z^{w-u-1}(1-z)^{u+v-w-1}\sum_{m=0}^{\infty}\frac{(u-1)_m\,(v)_m}{(w)_m}\frac{z^m}{m!}\left[\left(1+\frac{m}{u-1}\right)((w-u)+m-z(v+m))\right]\\
&\quad=z^{w-u-1}(1-z)^{u+v-w-1}\sum_{m=0}^{\infty}\frac{(u-1)_m\,(v)_m}{(w)_m}\frac{z^m}{m!}
\end{aligned}$$

$$\times\left[(w-u)+m-z(v+m)+\frac{m(w-u)}{(u-1)}+\frac{m^2}{u-1}-\frac{mz(v+m)}{u-1}\right]$$
$$= z^{w-u-1}(1-z)^{u+v-w-1}(w-u)\sum_{m=0}^{\infty}\frac{(u-1)_m\,(v)_m}{(w)_m}\frac{z^m}{m!}$$
$$+\frac{1}{u-1}\sum_{m=0}^{\infty}\frac{(u-1)_m\,(v)_m}{(w)_m}\frac{z^m}{m!}[m(w+m-1)-z(v+m)(u+m)+z(v+m)]$$
$$= z^{w-u-1}(1-z)^{u+v-w-1}(w-u)\sum_{m=0}^{\infty}\frac{(u-1)_m\,(v)_m}{(w)_m}\frac{z^m}{m!}$$
$$= z^{w-u-1}(1-z)^{u+v-w-1}(w-u)\,{}_2F_1(u-1,v;w;z). \tag{1.127}$$

Repeating this n-times yield the required formula (4). We now give the proof of (5)

$$\frac{d}{dz}[(1-z)^u\,{}_2F_1(u,v;w;z)]$$
$$=\frac{d}{dz}\left[(1-z)^u\sum_{m=0}^{\infty}\frac{(u)_m\,(v)_m}{(w)_m}\frac{z^m}{m!}\right]$$
$$=(1-z)^u\sum_{m=0}^{\infty}\frac{(u)_m\,(v)_m}{(w)_m}\frac{m\,z^{m-1}}{m!}-u(1-z)^{u-1}\sum_{m=0}^{\infty}\frac{(u)_m\,(v)_m}{(w)_m}\frac{z^m}{m!}$$
$$=(1-z)^{u-1}\sum_{m=0}^{\infty}\frac{(u)_m\,(v)_m}{(w)_m}\frac{z^{m-1}}{m!}[m-z(u+m)]$$
$$=(1-z)^{u-1}\sum_{m=1}^{\infty}\frac{(u)_m\,(v)_m}{(w)_m}\frac{z^{m-1}}{(m-1)!}-(1-z)^{u-1}\sum_{m=0}^{\infty}\frac{(u)_{m+1}\,(v)_m}{(w)_m}\frac{z^m}{m!}$$
$$=(1-z)^{u-1}\sum_{m=0}^{\infty}\frac{(u)_{m+1}\,(v)_{m+1}}{(w)_{m+1}}\frac{z^m}{m!}-(1-z)^{u-1}\sum_{m=0}^{\infty}\frac{(u)_{m+1}\,(v)_m}{(w)_m}\frac{z^m}{m!}$$
$$=-(1-z)^{u-1}\sum_{m=0}^{\infty}\frac{(u)_{m+1}\,(v)_m}{(w)_{m+1}}\frac{z^m}{m!}(w-v)$$
$$=-\frac{u(w-v)}{w}(1-z)^{u-1}{}_2F_1(u+1,v;w+1;z). \tag{1.128}$$

Repeating the process n-times give the required formula (5). The proofs of other formulas is left to the reader as an exercise. □

Now, we obtain the derivatives of the Gauss hypergeometric function with respect to parameters. We perform the derivative either with respect to numerator parameters or denominator. Since the numerator parameters u, v are symmetrical with respect to their places, we obtain the derivative with respect to u and w only. The nth derivative of ${}_2F_1(u,v;w;z)$ with respect to u and w are denoted by $G_u^{(n)}$ and $H_w^{(n)}$, respectively. Before we move further, let us see the derivative of a shifted factorial:

$$\begin{aligned}\frac{d}{du}(u)_n &= \frac{d}{du}\left(\frac{\Gamma(u+n)}{\Gamma(u)}\right) = \frac{\Gamma'(u+n)\,\Gamma(u) - \Gamma(u+n)\,\Gamma'(u)}{(\Gamma(u))^2}\\ &= \frac{\Gamma'(u+n)}{\Gamma(u)} - \frac{\Gamma(u+n)}{\Gamma(u)}\,\frac{\Gamma'(u)}{\Gamma(u)}\\ &= \frac{\Gamma'(u+n)}{\Gamma(u+n)}\,\frac{\Gamma(u+n)}{\Gamma(u)} - \frac{\Gamma(u+n)}{\Gamma(u)}\,\frac{\Gamma'(u)}{\Gamma(u)}\\ &= (u)_n\left[\psi(u+n) - \psi(u)\right]. \end{aligned} \tag{1.129}$$

In the above calculation, Γ' is the derivative of the gamma function and ψ is the digamma function defined as the logarithmic derivative of gamma function, i. e.,

$$\psi(u) = \frac{d}{du}\log\Gamma(u) = \frac{\Gamma'(u)}{\Gamma(u)}. \tag{1.130}$$

Using the expression,

$$\psi(u+n) - \psi(u) = \sum_{k=0}^{n-1}\frac{1}{u+k}, \tag{1.131}$$

in (1.129), we get

$$\frac{d}{du}(u)_n = \sum_{k=0}^{n-1}(u)_n\,\frac{1}{u+k}. \tag{1.132}$$

With the help of this functional identity, one can write the derivative of Gauss function with respect to the parameter u as

$$\begin{aligned}G_u^{(1)} &= \frac{d}{du}{}_2F_1(u,v;w;z) = \sum_{n=1}^{\infty}\frac{(u)_n\left[\psi(u+n)-\psi(u)\right](v)_n\,z^n}{(w)_n\qquad n!}\\ &= \sum_{n=0}^{\infty}\frac{(u)_{n+1}\,(v)_{n+1}\,z^{n+1}}{(w)_{n+1}\,(n+1)!}\left[\psi(u+n+1)-\psi(u)\right]\\ &= \sum_{n=0}^{\infty}\sum_{k=0}^{n}\frac{(u)_{n+1}\,(v)_{n+1}\,z^{n+1}}{(w)_{n+1}\,(n+1)!}\,\frac{1}{u+k}. \end{aligned} \tag{1.133}$$

Since the Gauss function is symmetrical in u and v, so one can obtain

$$G_v^{(1)} = \frac{d}{dv}\,{}_2F_1(u,v;w;z) = \sum_{n=0}^{\infty}\sum_{k=0}^{n}\frac{(u)_{n+1}\,(v)_{n+1}\,z^{n+1}}{(w)_{n+1}\,(n+1)!}\,\frac{1}{v+k}. \tag{1.134}$$

Now, using the derivative of inverse Pochhammer symbol,

$$\frac{d}{dw}\left(\frac{1}{(w)_n}\right)=\frac{d}{dw}\left(\frac{\Gamma(w)}{\Gamma(w+n)}\right)=\frac{1}{(w)_n}\left[\psi(w)-\psi(w+n)\right]$$
$$=-\frac{1}{(w)_n}\sum_{k=0}^{n-1}\frac{1}{w+k},\tag{1.135}$$

one will get the following expression:

$$H_w^{(1)}=\frac{d}{dw}\,{}_2F_1(u,v;w;z)=-\sum_{n=0}^{\infty}\sum_{k=0}^{n}\frac{(u)_{n+1}\,(v)_{n+1}\,z^{n+1}}{(w)_{n+1}\,(n+1)!}\,\frac{1}{w+k}.\tag{1.136}$$

If the parameters of two functions differ by one unit, they are said to be contiguous. Making the difference in the parameters of the Gaussian hypergeometric function by one unit yields six functions ${}_2F_1(u\pm1,v;w;z)$, ${}_2F_1(u,v\pm1;w;z)$, ${}_2F_1(u,v;w\pm1;z)$. These six functions are called connected functions of the hypergeometric function ${}_2F_1(u,v;w;z)$. A relation between a function and its connecting functions is called a connecting function relation.

If we use the following abbreviations F, $F(u\pm)$, $F(v\pm)$ and $F(w\pm)$ for ${}_2F_1(u,v;w;z)$, ${}_2F_1(u\pm1,v;w;z)$, ${}_2F_1(u,v\pm1;w;z)$ and ${}_2F_1(u,v;w\pm1;z)$, respectively, then the Gauss hypergeometric function and its contiguous functions are connected by the following relations:

$$(u-v)\,F=u\,F(u+)-v\,F(v+),\tag{1.137}$$
$$(u-w+1)\,F=u\,F(u+)-(w-1)\,F(w-),\tag{1.138}$$
$$\left[u+(v-w)z\right]F=u(1-z)\,F(u+)-z\,\frac{(w-u)\,(w-v)}{w}F(w+),\tag{1.139}$$
$$(1-z)F=F(u-)-\frac{w-v}{w}\,zF(w+).\tag{1.140}$$

Since u and v are interchangeable in the Gauss hypergeometric series, so (1.140) can be rewrite as

$$(1-z)F=F(v-)-\frac{w-u}{w}\,zF(w+).\tag{1.141}$$

The equations (1.137), (1.138), (1.139), (1.140) and (1.141) together are five contiguous relations of the Gauss hypergeometric series.

Proof. Lets start with

$$F(u+)=\sum_{k=0}^{\infty}\frac{(u+1)_k\,(v)_k}{(w)_k}\,\frac{z^k}{k!}$$
$$=\sum_{k=0}^{\infty}\frac{u+k}{u}\,\frac{(u)_k\,(v)_k}{(w)_k}\,\frac{z^k}{k!}.\tag{1.142}$$

Similarly,

$$F(v+) = \sum_{k=0}^{\infty} \frac{v+k}{v} \frac{(u)_k (v)_k}{(w)_k} \frac{z^k}{k!}; \tag{1.143}$$

$$F(w+) = \sum_{k=0}^{\infty} \frac{w}{w+k} \frac{(u)_k (v)_k}{(w)_k} \frac{z^k}{k!}; \tag{1.144}$$

$$F(u-) = \sum_{k=0}^{\infty} \frac{u-1}{u-1+k} \frac{(u)_k (v)_k}{(w)_k} \frac{z^k}{k!}; \tag{1.145}$$

$$F(v-) = \sum_{k=0}^{\infty} \frac{v-1}{v-1+k} \frac{(u)_k (v)_k}{(w)_k} \frac{z^k}{k!}; \tag{1.146}$$

$$F(w-) = \sum_{k=0}^{\infty} \frac{w-1+k}{w-1} \frac{(u)_k (v)_k}{(w)_k} \frac{z^k}{k!}. \tag{1.147}$$

Let $\theta = z\frac{d}{dz}$ be the differential operator. Then

$$(\theta + u)F = \sum_{k=0}^{\infty} (u+k) \frac{(u)_k (v)_k}{(w)_k} \frac{z^k}{k!} = u\,F(u+). \tag{1.148}$$

Also,

$$(\theta + v)F = \sum_{k=0}^{\infty} (v+k) \frac{(u)_k (v)_k}{(w)_k} \frac{z^k}{k!} = v\,F(v+). \tag{1.149}$$

$$(\theta + w - 1)F = \sum_{k=0}^{\infty} (w+k-1) \frac{(u)_k (v)_k}{(w)_k} \frac{z^k}{k!} = (w-1)\,F(w-). \tag{1.150}$$

From (1.148) and (1.149), we get

$$(u-v)F = u\,F(u+) - v\,F(v+).$$

From (1.148) and (1.150), we have

$$(u-w+1)F = u\,F(u+) - (w-1)\,F(w-).$$

Now, the action of differential operator $\theta = z\frac{d}{dz}$ gives

$$\begin{aligned}
\theta F &= \sum_{k=0}^{\infty} \frac{(u)_k (v)_k}{(w)_k} \frac{k z^k}{k!} \\
&= \sum_{k=1}^{\infty} \frac{(u)_k (v)_k}{(w)_k} \frac{z^k}{(k-1)!} \\
&= z \sum_{k=0}^{\infty} \frac{(u)_{k+1} (v)_{k+1}}{(w)_{k+1}} \frac{z^k}{k!} \\
&= z \sum_{k=0}^{\infty} \frac{(u+k)(v+k)}{(w+k)} \frac{(u)_k (v)_k}{(w)_k} \frac{z^k}{k!}.
\end{aligned}$$

The expression $\frac{(u+k)(v+k)}{(w+k)}$ can be written as

$$\frac{(u+k)(v+k)}{(w+k)} = k + (u+v-w) + \frac{(w-u)(w-v)}{w+k}.$$

Hence, we get

$$\begin{aligned}\theta F &= z\sum_{k=0}^{\infty} k\,\frac{(u)_k (v)_k}{(w)_k}\frac{z^k}{k!} + z(u+v-w)\sum_{k=0}^{\infty}\frac{(u)_k (v)_k}{(w)_k}\frac{z^k}{k!}\\ &\quad + z\,\frac{(w-u)(w-v)}{w}\sum_{k=0}^{\infty}\frac{w}{w+k}\,\frac{(u)_k (v)_k}{(w)_k}\frac{z^k}{k!}\\ &= z\,\theta F + z(u+v-w)F + z\,\frac{(w-u)(w-v)}{w}F(w+).\end{aligned}$$

This implies

$$(1-z)\,\theta F = z(u+v-w)F + z\,\frac{(w-u)(w-v)}{w}F(w+). \tag{1.151}$$

Multiplying (1.148) by $(1-z)$ gives

$$(1-z)\,\theta F = -(1-z)\,uF + u(1-z)F(u+). \tag{1.152}$$

Comparing (1.151) with (1.152) yields

$$\left[u + (v-w)z\right]F = u(1-z)F(u+) - z\,\frac{(w-u)(w-v)}{w}F(w+),$$

which is a third contiguous relation. Next, observing the action of differential operator θ on $F(u-)$ as

$$\begin{aligned}\theta F(u-) &= \sum_{k=0}^{\infty}\frac{(u-1)_k (v)_k}{(w)_k}\frac{k\,z^k}{k!}\\ &= \sum_{k=1}^{\infty}\frac{(u-1)_k (v)_k}{(w)_k}\frac{z^k}{(k-1)!}\\ &= z\sum_{k=0}^{\infty}\frac{(u-1)_{k+1} (v)_{k+1}}{(w)_{k+1}}\frac{z^k}{k!}\\ &= z(u-1)\sum_{k=0}^{\infty}\frac{v+k}{w+k}\,\frac{(u)_k (v)_k}{(w)_k}\frac{z^k}{k!}.\end{aligned}$$

The expression $\frac{v+k}{w+k}$ can be written as

$$\frac{v+k}{w+k} = 1 - \frac{w-v}{w+k},$$

so we have

$$\theta F(u-) = z\,(u-1)\sum_{k=0}^{\infty}\frac{(u)_k\,(v)_k}{(w)_k}\frac{z^k}{k!} - z\,(u-1)\,\frac{w-v}{w}\sum_{k=0}^{\infty}\frac{w}{w+k}\frac{(u)_k\,(v)_k}{(w)_k}\frac{z^k}{k!}$$
$$= z\,(u-1)\,F - z\,(u-1)\,\frac{w-v}{w}\,F(w+). \tag{1.153}$$

In equation (1.148), if we replace u by $u-1$, we obtain

$$\theta F(u-) = (u-1)\,F - (u-1)\,F(u-). \tag{1.154}$$

From (1.153) and (1.154), we get

$$(1-z)F = F(u-) - \frac{w-v}{w}\,zF(w+).$$

This is fourth contiguous relation. The fifth can be obtained by interchanging u and v in the above equation. □

The appropriate eliminations between these five contiguous relation lead to ten more contiguous relation, which are listed fairly in [128]. The other 10 contiguous relations, which can be obtained by some elimination from these five contiguous relations, are as follows:

1. $[2u-w+(v-u)\,z]F = u(1-z)\,F(u+) - (w-u)F(u-)$,
2. $(u+v-w)\,F = u(1-z)\,F(u+) - (w-v)\,F(v-)$,
3. $(w-u-v)\,F = (w-u)\,F(u-) - v(1-z)F(v+)$,
4. $(v-u)(1-z)\,F = (w-u)\,F(u-) - (w-v)F(v-)$,
5. $[1-u+(w-v-1)\,z]\,F = (w-u)\,F(u-) - (w-1)(1-z)F(w-)$,
6. $[2v-w+(u-v)\,z]\,F = v(1-z)F(v+) - (w-v)\,F(v-)$,
7. $[v+(u-w)\,z]\,F = v(1-z)F(v+) - z\,\frac{(w-u)\,(w-v)}{w}F(w+)$,
8. $(v-w+1)F = v\,F(v+) - (w-1)\,F(w-)$,
9. $[1-v+(w-u-1)z]F = (w-v)\,F(v-) - (w-1)(1-z)F(w-)$,
10. $[w-1+(u+v+1-2w)z]\,F = (w-1)(1-z)\,F(w-) - z\,\frac{(w-u)\,(w-v)}{w}F(w+)$.

1.7 Integral representations

One of the most important methods for describing the Gaussian hypergeometric function is the integral representation. Since the Pochhammer symbol is written in the form of the gamma function, we have thus

$$(v)_n\,(w)_n^{-1} = \frac{\Gamma(v+n)\,\Gamma(w)}{\Gamma(v)\,\Gamma(w+n)} = \frac{\Gamma(v+n)\,\Gamma\,(w)\,\Gamma(w-v)}{\Gamma(v)\,\Gamma(w-v)\,\Gamma(w+n)}$$
$$= \frac{\Gamma(w)}{\Gamma(v)\Gamma(w-v)}\,\beta(v+n, w-v). \tag{1.155}$$

Now, using the integral definition of beta function (1.51), one gets

$$(v)_n\,(w)_n^{-1} = \frac{\Gamma(w)}{\Gamma(v)\Gamma(w-v)} \int_0^1 t^{v+n-1}\,(1-t)^{w-v-1}\,dt. \tag{1.156}$$

Now, the equations (1.90) and (1.156) together yield

$$\begin{aligned}
{}_2F_1(u,v;w;z) &= \sum_{n=0}^{\infty} \frac{z^n\,(u)_n}{n!}\,\frac{\Gamma(w)}{\Gamma(v)\Gamma(w-v)} \int_0^1 t^{v+n-1}(1-t)^{w-v-1}\,dt \\
&= \frac{\Gamma(w)}{\Gamma(v)\Gamma(w-v)} \int_0^1 t^{v-1}(1-t)^{w-v-1} \sum_{n=0}^{\infty} \frac{(u)_n}{n!}(t\,z)^n \\
&= \frac{\Gamma(w)}{\Gamma(v)\Gamma(w-v)} \int_0^1 t^{v-1}(1-t)^{w-v-1}\,(1-zt)^{-u}\,dt.
\end{aligned} \tag{1.157}$$

We can summarize the result in the following theorem.

Theorem 24. *Let u, v, w be complex numbers such that $\Re(w-v) > \Re(v) > 0$. Then the Gauss hypergeometric function, for $|z| < 1$, can be put in the integral form as*

$${}_2F_1(u,v;w;z) = \frac{\Gamma(w)}{\Gamma(v)\Gamma(w-v)} \int_0^1 t^{v-1}(1-t)^{w-v-1}\,(1-zt)^{-u}\,dt. \tag{1.158}$$

Let us extend this result for $|z| = 1$, and from (1.158), we have

$$\begin{aligned}
{}_2F_1(u,v;w;1) &= \frac{\Gamma(w)}{\Gamma(v)\Gamma(w-v)} \int_0^1 t^{v-1}(1-t)^{w-v-1}\,(1-t)^{-u}\,dt \\
&= \frac{\Gamma(w)}{\Gamma(v)\Gamma(w-v)} \int_0^1 t^{v-1}(1-t)^{w-u-v-1}\,dt.
\end{aligned} \tag{1.159}$$

Now, using equation (1.156), the equation (1.159) yields

$${}_2F_1(u,v;w;1) = \frac{\Gamma(w)}{\Gamma(v)\Gamma(w-v)}\,\frac{\Gamma(v)\,\Gamma(w-u-v)}{\Gamma(w-u)} = \frac{\Gamma(w)\,\Gamma(w-u-v)}{\Gamma(w-u)\,\Gamma(w-v)}. \tag{1.160}$$

Another important integral form of the Gauss hypergeometric function is the contour integral or Berne's type integral. Before proceeding for Berne's integral of Gauss hypergeometric function, we need to understand the Mellin integral transform with few examples. Let $h(x)$ be a piecewise continuous function on a real line. Then the Mellin integral transform of $h(x)$ is given by

$$G(s) = \int_0^\infty x^{s-1}\, h(x)\, dx. \tag{1.161}$$

The above integral is convergent for $a < \Re(s) < b$. Conversely, if $G(s)$ is analytic in $a < \Re(s) < b$, then $h(x)$ can be obtained by the inversion formula,

$$h(x) = \frac{1}{2\pi\iota} \int_{c-\iota\infty}^{c+\iota\infty} x^{-s}\, G(s)\, ds. \tag{1.162}$$

Now, we can use this Mellin transformation to get the Berne's or the contour integral of special functions. We already know that the gamma function is defined by

$$\Gamma(s) = \int_0^\infty x^{s-1}\, e^{-x}\, dx, \quad \Re(s) > 0. \tag{1.163}$$

Instantly, one can observe that gamma function is the Mellin transform of e^{-x}. Hence, the inversion formula allows to get back the expression of e^{-x} as

$$e^{-x} = \frac{1}{2\pi\iota} \int_{c-\iota\infty}^{c+\iota\infty} x^{-s}\, \Gamma(s)\, ds, \quad c > 0. \tag{1.164}$$

One can also prove this directly using the Cauchy residue theorem. Let us consider the Contour C as a rectangle with vertices $c \pm \iota R$ and $-(N + \frac{1}{2}) \pm \iota R$, where $c > 0$ and N is a positive integer. It can be seen that the poles of $\Gamma(s)$ inside this contour are $0, -1, \dots, -N$. The residue of $\Gamma(s)$ at $s = 0$ is

$$\lim_{s\to 0} s\,\Gamma(s) = \lim_{s\to 0} \Gamma(s+1) = \Gamma(1) = 1. \tag{1.165}$$

At $s = -1$,

$$\lim_{s\to -1} (s+1)\Gamma(s) = \lim_{s\to -1} \frac{(s+1)}{s}\Gamma(s+1) = \lim_{s\to -1} \frac{\Gamma(s+2)}{s} = \frac{\Gamma(1)}{-1} = -1. \tag{1.166}$$

Similarly, at $s = -2$,

$$\lim_{s\to -2} (s+2)\Gamma(s) = \lim_{s\to -2} \frac{\Gamma(s+3)}{s\,(s+1)} = \frac{\Gamma(1)}{(-2)\,(-1)} = \frac{1}{2!}. \tag{1.167}$$

In general, we write the residues of $\Gamma(s)$ at the poles $0, -1, \dots, -N$ are $\frac{(-1)^k}{k!}$ for $k = 0, 1, \dots, N$, respectively. Hence, by the Cauchy residue theorem

$$\int_C x^{-s}\,\Gamma(s)\,ds = \sum_{k=0}^{N} \frac{(-1)^k}{k!}\,x^k. \tag{1.168}$$

The Stirling formula for the gamma function implies that the integral on C minus the line joining $c - \iota R$ and $c + \iota R$ tends to zero when N and R tends to ∞. Hence, we arrive at

$$\frac{1}{2\pi\iota}\int_{c-\iota\infty}^{c+\iota\infty} x^{-s}\,\Gamma(s)\,ds = \lim_{N\to\infty}\sum_{k=0}^{N} \frac{(-1)^k}{k!}\,x^k = \sum_{k=0}^{\infty} \frac{(-1)^k}{k!}\,x^k = e^{-x}. \tag{1.169}$$

Similarly, from the integral representation of beta function,

$$\beta(s,u) = \int_0^1 x^{s-1}\,(1-x)^{u-1}\,du, \quad \Re(s), \Re(u) > 0, \tag{1.170}$$

it can be observed that the beta function is the Mellin transform of the function

$$\phi(x) = \begin{cases} (1-x)^{u-1}, & \text{if } 0 < x < 1, \\ 0, & \text{if } x \geq 1. \end{cases} \tag{1.171}$$

Using the inversion formula, we have another way to express $(1-x)^{u-1}$ as

$$\begin{aligned} (1-x)^{u-1} &= \frac{1}{2\pi\iota}\int_{c-\iota\infty}^{c+\iota\infty} x^{-s}\,\beta(s,u)\,ds \\ &= \frac{\Gamma(u)}{2\pi\iota}\int_{c-\iota\infty}^{c+\iota\infty} x^{-s}\,\frac{\Gamma(s)}{\Gamma(s+u)}\,ds. \end{aligned} \tag{1.172}$$

Now, we use this to get the complex integral of the Gauss hypergeometric function ${}_2F_1(u,v;w;z)$. The Mellin transform of the Gauss function is given by

$$\begin{aligned} &\int_0^\infty z^{s-1}\,{}_2F_1(u,v;w;-z)\,dz \\ &= \int_0^\infty z^{s-1}\int_0^1 \frac{\Gamma(w)}{\Gamma(v)\,\Gamma(w-v)}\,\mathrm{u}^{v-1}\,(1-\mathrm{u})^{w-v-1}(1+\mathrm{u}\,z)^{-u}\,d\mathrm{u}\,dz \\ &= \frac{\Gamma(w)}{\Gamma(v)\,\Gamma(w-v)}\int_0^1 \mathrm{u}^{v-1}\,(1-\mathrm{u})^{w-v-1}\int_0^\infty \frac{z^{s-1}}{(1+z\,\mathrm{u})^{u}}\,dz\,d\mathrm{u}. \end{aligned} \tag{1.173}$$

On substituting $z\mathrm{u} = t$ and $\mathrm{u}\,dz = dt$, we get

$$\int_0^\infty z^{s-1} {}_2F_1(u, v; w; -z)\, dz$$

$$= \frac{\Gamma(w)}{\Gamma(v)\,\Gamma(w-v)} \int_0^1 u^{v-1} (1-u)^{w-v-1} u^{-s} \int_0^\infty \frac{t^{s-1}}{(1+t)^u}\, dt\, du$$

$$= \frac{\Gamma(w)}{\Gamma(v)\,\Gamma(w-v)} \int_0^1 u^{v-s-1} (1-u)^{w-v-1} \beta(s, u-s)\, du$$

$$= \frac{\Gamma(w)}{\Gamma(v)\,\Gamma(w-v)} \beta(v-s, w-v)\, \beta(s, u-s) = \frac{\Gamma(w)\,\Gamma(s)\,\Gamma(u-s)\,\Gamma(v-s)}{\Gamma(u)\,\Gamma(v)\,\Gamma(w-s)}. \tag{1.174}$$

With the assumption that $\Re(u), \Re(v), \Re(s) > 0$, one can apply the inversion formula and get

$$\frac{\Gamma(u)\,\Gamma(v)}{\Gamma(w)} {}_2F_1(u, v; w; -z)$$
$$= \frac{1}{2\pi\iota} \int_{k-\iota\infty}^{k+\iota\infty} \frac{\Gamma(s)\,\Gamma(u-s)\,\Gamma(v-s)}{\Gamma(w-s)} z^{-s}\, ds, \quad \min\{\Re(u), \Re(v)\} > k > 0, w \neq 0, -1, \ldots, \tag{1.175}$$

which can also be written as

$$\frac{\Gamma(u)\,\Gamma(v)}{\Gamma(w)} {}_2F_1(u, v; w; z) = \frac{1}{2\pi\iota} \int_{k-\iota\infty}^{k+\iota\infty} \frac{\Gamma(s)\,\Gamma(u-s)\,\Gamma(v-s)}{\Gamma(w-s)} (-z)^s\, ds. \tag{1.176}$$

This is the Berne's integral form of Gauss hypergeometric function. Another way to obtain the contour integral representation of Gauss hypergeometric function is given in the following theorem.

Theorem 25. *Let $u, v, w, z \in \mathbb{C}$ such that $w \neq 0, -1, \ldots$ and $\Re(u), \Re(v) > 0$. Then*

$$\frac{\Gamma(u)\,\Gamma(v)}{\Gamma(w)} {}_2F_1(u, v; w; z)$$
$$= \frac{1}{2\pi\iota} \int_{-\iota\infty}^{+\iota\infty} \frac{\Gamma(-s)\,\Gamma(u+s)\,\Gamma(v+s)}{\Gamma(w+s)} (-z)^s\, ds, \quad |\arg(-z)| < \pi. \tag{1.177}$$

Proof. Consider the integral

$$I_C = \int_C \frac{\Gamma(-s)\,\Gamma(u+s)\,\Gamma(v+s)}{\Gamma(w+s)} (-z)^s\, ds, \tag{1.178}$$

which has the only singularities as poles at $s = -u - n$, $s = -v - n$, $s = n$, for $n = 0, 1, \ldots$. We shall take a contour such that it excludes all the poles of $\Gamma(u+s)$ and $\Gamma(v+s)$ but

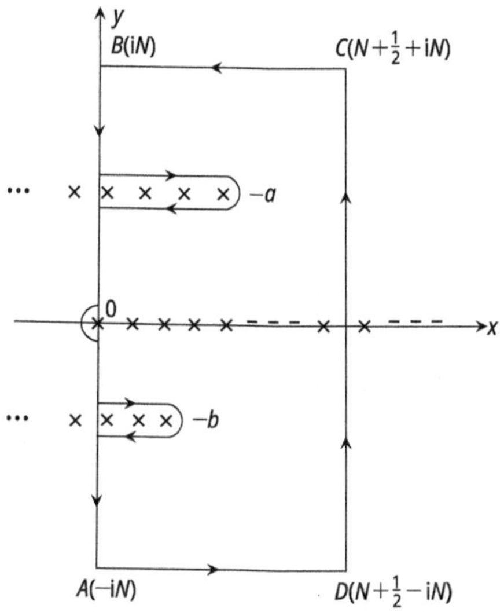

Figure 1.1

includes the poles of $\Gamma(-s)$. One such contour is given in Figure 1.1, where N is chosen such that

$$N \in \mathbb{Z}^{+} > |\Im(u)| + |\Im(v)|.$$

Now, the Cauchy residue theorem says

$$I_C = 2\pi\iota \times \text{sum of residues within the contour } C.$$

The only poles contained in the contour are $\Gamma(-s)$ and the residue of $\Gamma(-s)$ is

$$\frac{(-1)^n}{n!}, \quad \text{at pole } s = n. \tag{1.179}$$

We have

$$I_C = -2\pi\iota \sum_{n=0}^{N} \frac{\Gamma(u+n)\,\Gamma(v+n)}{\Gamma(w+n)} \frac{z^n}{n!}. \tag{1.180}$$

In another way, our contour consists of four-line integrals as

$$\begin{aligned} I_C = \int_C = -\int_{-\iota N}^{\iota N} + \int_{-\iota N}^{N+\frac{1}{2}-\iota N} + \int_{N+\frac{1}{2}-\iota N}^{N+\frac{1}{2}+\iota N} + \int_{N+\frac{1}{2}+\iota N}^{\iota N} \\ = -I_1 + I_2 + I_3 + I_4. \end{aligned} \tag{1.181}$$

The only thing to show that $I_2, I_3, I_4 \to 0$ as $N \to \infty$. For lasrge $|s|$ and $|\arg s| < \pi$, we have

$$\begin{aligned}\left|\frac{\Gamma(u+s)\,\Gamma(v+s)}{\Gamma(w+s)\,\Gamma(1+s)}\right| &= |s|^{\Re(u+v-w-1)}\exp\{-\Im(u+v-w-1)\arg s\}\{1+O(1)\}\\ &< A\,N^{\Re(u+v-w-1)};\quad A \text{ is a constant independent from } N.\end{aligned} \tag{1.182}$$

For I_2, we have $s = x - \iota N$. So,

$$\begin{aligned}|\Gamma(-s)\,\Gamma(1+s)| &= \frac{2\pi}{\exp(\pi(N+\iota x)) - \exp(-\pi(N+\iota x))}\\ &< 4\pi\,\exp(-\pi N).\end{aligned} \tag{1.183}$$

Also,

$$\begin{aligned}|(-z)^s| &= |z|^x\exp\{N\arg(-z)\}\\ &< |z|^x\exp\{(\pi-\epsilon)N\},\end{aligned} \tag{1.184}$$

where ϵ is small positive number. Thus,

$$\begin{aligned}|I_2| &< \int_0^{N+\frac{1}{2}} A\,N^{\Re(u+v-w-1)} 4\pi e^{-\pi N}\,|z|^x\,\exp\{(\pi-\epsilon)N\}\,dx\\ &< 4\pi e^{-\epsilon N} A\,N^{\Re(u+v-w-1)}\left(N+\frac{1}{2}\right).\end{aligned} \tag{1.185}$$

As $N \to \infty$, $e^{-\epsilon N} \to 0$ faster than $N^{\Re(u+v-w-1)} \to \infty$. Hence, $I_2 \to 0$ as $N \to \infty$.

For I_3, we have $s = N + \frac{1}{2} + \iota y$, $-N \le y \le N$. Then

$$|\Gamma(-s)\,\Gamma(1+s)| = \frac{\pi}{|\cosh(\pi y)|} < 2\pi\,\exp(-\pi\,|y|) \tag{1.186}$$

and

$$|(-z)^s| = |z|^{N+\frac{1}{2}}\exp\{-y\arg(-z)\}. \tag{1.187}$$

Hence,

$$\begin{aligned}|I_3| &< \int_{-N}^{N} A\,N^{\Re(u+v-w-1)}\,2\pi\,\exp(-\pi\,|y|)\,|z|^{N+\frac{1}{2}}\exp\{-y\arg(-z)\}dy\\ &< 2\pi A\,N^{\Re(u+v-w-1)}|z|^{N+\frac{1}{2}}\int_{-N}^{N} e^{-\epsilon|y|}\,dy\\ &< 4\pi A\,N^{\Re(u+v-w-1)}|z|^{N+\frac{1}{2}}\,N\,e^{-\epsilon N}.\end{aligned} \tag{1.188}$$

Now, $|z|^{N+\frac{1}{2}} \to 0$ is faster than $N^{\Re(u+v-w-1)} \to \infty$ as $N \to \infty$. Therefore, $I_3 \to 0$ as $N \to \infty$. A procedure similar to I_2 can be used to show that $I_4 \to 0$. Hence, $I_C = -I_1$, when $N \to \infty$ and, therefore,

$$-2\pi\iota \sum_{n=0}^{\infty} \frac{\Gamma(u+n)\,\Gamma(v+n)}{\Gamma(w+n)} \frac{z^n}{n!} = -\int_{-\iota\infty}^{+\iota\infty} \frac{\Gamma(-s)\,\Gamma(u+s)\,\Gamma(v+s)}{\Gamma(w+s)} (-z)^s\, ds \tag{1.189}$$

or

$$\frac{1}{2\pi\iota} \frac{\Gamma(w)}{\Gamma(u)\,\Gamma(v)} \int_{-\iota\infty}^{+\iota\infty} \frac{\Gamma(-s)\,\Gamma(u+s)\,\Gamma(v+s)}{\Gamma(w+s)} (-z)^s\, ds = {}_2F_1(u,v;w;z), \tag{1.190}$$

and completes the proof. □

Besides these, the Gauss function has a few more integral representations, which can be obtained by some simple substitutions. First, we list these integral representations and then we see how they can be derived.

Theorem 26. *The following integrals hold true for Gauss function ${}_2F_1(u,v;w;z)$:*

$$\begin{aligned} {}_2F_1(u,v;w;1-z) &= \frac{\Gamma(w)}{\Gamma(v)\,\Gamma(w-v)} \int_0^{\infty} \mathrm{u}^{v-1} (1+\mathrm{u})^{u-w} (1+\mathrm{u}z)^{-u}\, d\mathrm{u}, \\ &\quad \Re(w) > \Re(v) > 0, |\arg(z)| < \pi, \end{aligned} \tag{1.191}$$

$$\begin{aligned} {}_2F_1\left(u,v;w;\frac{1}{z}\right) &= \frac{\Gamma(w)}{\Gamma(v)\,\Gamma(w-v)} \int_1^{\infty} \mathrm{u}^{u-w} (\mathrm{u}-1)^{w-v-1} \left(\mathrm{u}-\frac{1}{z}\right)^{-u} d\mathrm{u}, \\ &\quad 1+\Re(u) > \Re(w) > \Re(v) > 0, |\arg(z-1)| < \pi, \end{aligned} \tag{1.192}$$

$$\begin{aligned} {}_2F_1(u,v;w;z) &= \frac{\Gamma(w)}{\Gamma(v)\,\Gamma(w-v)} \int_0^{\infty} e^{-v\mathrm{u}} (1-e^{-\mathrm{u}})^{w-v-1} (1-ze^{-\mathrm{u}})^{-u}\, d\mathrm{u}, \\ &\quad \Re(w) > \Re(v) > 0, \end{aligned} \tag{1.193}$$

$$\begin{aligned} {}_2F_1(u,v;w;z) &= \frac{2\,\Gamma(w)}{\Gamma(v)\,\Gamma(w-v)} \int_0^{\pi/2} \frac{(\sin\mathrm{u})^{2v-1} (\cos\mathrm{u})^{2w-2v-1}}{(1-z\sin^2\mathrm{u})^{u}}\, d\mathrm{u}, \\ &\quad \Re(w) > \Re(v) > 0, \end{aligned} \tag{1.194}$$

$$\begin{aligned} {}_2F_1(u,v;w;z) &= \frac{2^{1-w}\,\Gamma(w)}{\Gamma(v)\,\Gamma(w-v)} \int_0^{\pi} \frac{(\sin\mathrm{u})^{2v-1} (1+\cos\mathrm{u})^{w-2v}}{(1-\frac{1}{2}z+\frac{1}{2}z\cos\mathrm{u})^{u}}\, d\mathrm{u}, \\ &\quad \Re(w) > \Re(v) > 0, \end{aligned} \tag{1.195}$$

$${}_2F_1(u,v;w;z) = \frac{2\,\Gamma(w)}{\Gamma(v)\,\Gamma(w-v)} \int_0^{\infty} \frac{(\cosh\mathrm{u})^{2u-2w+1} (\sinh\mathrm{u})^{2w-2v-1}}{((\cosh\mathrm{u})^2-z)^{u}}\, d\mathrm{u},$$

$$\Re(w) > \Re(v) > 0, \tag{1.196}$$

$${}_2F_1(u, v; w; z) = \frac{2^{v-u}\,\Gamma(w)}{\Gamma(v)\,\Gamma(w-v)} \int_0^\infty \frac{(\cosh \mathrm{u} - 1)^{2w-u-v-1}\,(\sinh \mathrm{u})^{2u-2w+1}}{(\frac{1}{2} - z + \frac{1}{2}\cosh \mathrm{u})^u}\,d\mathrm{u},$$

$$\Re(w) > \Re(v) > 0, \tag{1.197}$$

$${}_2F_1(u, v; w; z) = \frac{2\,\Gamma(w)}{\Gamma(v)\,\Gamma(w-v)} \int_0^\infty \frac{(\sinh \mathrm{u})^{2v-1}\,(\cosh \mathrm{u})^{2u-2w+1}}{((\cosh \mathrm{u})^2 - z(\sinh \mathrm{u})^2)^u}\,d\mathrm{u},$$

$$\Re(w) > \Re(v) > 0. \tag{1.198}$$

Proof. From the integral of Gauss function, we have

$${}_2F_1(u, v; w; 1-z) = \frac{\Gamma(w)}{\Gamma(v)\Gamma(w-v)} \int_0^1 t^{v-1}(1-t)^{w-v-1}\,(1-(1-z)t)^{-u}\,dt. \tag{1.199}$$

Make substitution of $-\mathrm{u} = \frac{t}{t-1}$, i. e., $t = \frac{\mathrm{u}}{\mathrm{u}+1}$ and $dt = \frac{1}{(1+\mathrm{u})^2}\,d\mathrm{u}$. The limit of t is from 0 to 1, and hence of u is from 0 to ∞. Thus, we arrive at

$$\begin{aligned}
{}_2F_1(u, v; w; 1-z) &= \frac{\Gamma(w)}{\Gamma(v)\Gamma(w-v)} \int_0^\infty \left(\frac{\mathrm{u}}{1+\mathrm{u}}\right)^{v-1} \left(1 - \frac{\mathrm{u}}{1+\mathrm{u}}\right)^{w-v-1} \\
&\quad \times \left(1 - (1-z)\frac{\mathrm{u}}{1+\mathrm{u}}\right)^{-u} \left(\frac{1}{(1+\mathrm{u})^2}\right) d\mathrm{u} \\
&= \frac{\Gamma(w)}{\Gamma(v)\Gamma(w-v)} \int_0^\infty \mathrm{u}^{v-1}\,(1+\mathrm{u})^{u-w}\,(1+\mathrm{u}z)^{-u}\,d\mathrm{u}.
\end{aligned} \tag{1.200}$$

The proof is complete. The proof of the integrals from (1.192) to (1.198) is similar to (1.191), and we have to make the suitable substitution. To derive the integral from (1.192) to (1.198), we substitute $1/\mathrm{u} = t$; $e^{-\mathrm{u}} = t$; $\sin^2 \mathrm{u} = t$; $\frac{1}{2}\cos \mathrm{u} - \frac{1}{2} = t$; $\frac{1}{\cosh^2 \mathrm{u}} = t$; $\frac{1}{\frac{1}{2}\cosh \mathrm{u} + \frac{1}{2}} = t$ and $\tanh^2 \mathrm{u} = t$, respectively. □

1.8 Transformation and summation formulae

The transformation formulae associated with the Gauss hypergeometric function are given by

$$\begin{aligned}
&{}_2F_1(u, v; w; z) \\
&= (1-z)^{-u}\,{}_2F_1\left(u, w-v; w; \frac{z}{z-1}\right), \quad w \neq \mathbb{Z}_0^-, |\arg(1-z)| \leq \pi - \epsilon;
\end{aligned} \tag{1.201}$$

$$
\begin{aligned}
&{}_2F_1(u,v;w;z)\\
&\quad = (1-z)^{-v}{}_2F_1\left(v, w-u; w; \frac{z}{z-1}\right), \quad w \neq \mathbb{Z}_0^-, |\arg(1-z)| \leq \pi - \epsilon; \qquad (1.202)\\
&{}_2F_1(u,v;w;z)\\
&\quad = (1-z)^{w-u-v}{}_2F_1(w-u, w-v; w; z), \quad w \neq \mathbb{Z}_0^-, |\arg(1-z)| \leq \pi - \epsilon. \qquad (1.203)
\end{aligned}
$$

Proof. The right side of (1.201) can be written as

$$
(1-z)^{-u}{}_2F_1\left(u, w-v; w; \frac{z}{z-1}\right) = \sum_{k=0}^{\infty} \frac{(u)_k\,(w-v)_k}{(w)_k} \frac{(-z)^k}{k!} (1-z)^{-u-k}. \tag{1.204}
$$

Now, using the binomial identity $(1-z)^{-u-k} = \sum_{l=0}^{\infty} \frac{(u+k)_l}{l!} z^l$, we have

$$
\begin{aligned}
(1-z)^{-u}{}_2F_1\left(u, w-v; w; \frac{z}{z-1}\right) &= \sum_{k=0}^{\infty}\sum_{l=0}^{\infty} \frac{(u)_k\,(w-v)_k\,(u+k)_l}{(w)_k} \frac{(-1)^k z^{k+l}}{k!\,l!}\\
&= \sum_{k=0}^{\infty}\sum_{l=0}^{\infty} \frac{(u)_{k+l}\,(w-v)_k}{(w)_k} \frac{(-1)^k z^{k+l}}{k!\,l!}\\
&= \sum_{k=0}^{l}\sum_{l=0}^{\infty} \frac{(u)_l\,(w-v)_k}{(w)_k} \frac{(-1)^k z^l}{k!\,(l-k)!}\\
&= \sum_{k=0}^{l}\sum_{l=0}^{\infty} \frac{(u)_l\,(w-v)_k\,(-l)_k}{(w)_k\,k!} \frac{z^l}{l!}\\
&= \sum_{l=0}^{\infty} \frac{(u)_l\,z^l}{l!}\,{}_2F_1(-l, w-v; w; 1).
\end{aligned}
$$

Using the Gauss summation formula, we get

$$
\begin{aligned}
(1-z)^{-u}{}_2F_1\left(u, w-v; w; \frac{z}{z-1}\right) &= \sum_{l=0}^{\infty} \frac{(u)_l\,z^l}{l!} \frac{\Gamma(w)\,\Gamma(v+l)}{\Gamma(w+l)\,\Gamma(v)}\\
&= \sum_{l=0}^{\infty} \frac{(u)_l\,(v)_l\,z^l}{(w)_l\,l!} = {}_2F_1(u,v;w;z). \qquad (1.205)
\end{aligned}
$$

It completes the proof of (1.201). By the symmetry of u and v, interchanging of u and v in (1.201) yields (1.202) as

$$
{}_2F_1(u,v;w;z) = (1-z)^{-v}{}_2F_1\left(v, w-u; w; \frac{z}{z-1}\right).
$$

To prove (1.203), we again start with the right side and make use of (1.201). Thus, we get

$$
\begin{aligned}
(1-z)^{w-u-v}{}_2F_1(w-u,w-v;w;z) &= (1-z)^{w-u-v}(1-z)^{-(w-u)}{}_2F_1\left(w-u,v;w;\frac{z}{z-1}\right)\\
&= (1-z)^{-v}{}_2F_1\left(w-u,v;w;\frac{z}{z-1}\right)\\
&= (1-z)^{-v}{}_2F_1\left(v,w-u;w;\frac{z}{z-1}\right).
\end{aligned}
$$

Again using (1.201), we get

$$
(1-z)^{w-u-v}{}_2F_1(w-u,w-v;w;z) = {}_2F_1(u,v;w;z),
$$

and completes the proof. □

Theorem 27. *For $\Re(v) > 0$ and $k \in \mathbb{N}$, the following identity holds true:*

$$
{}_2F_1\left(\begin{matrix}-\frac{k}{2},\frac{1}{2}-\frac{k}{2}\\ v+\frac{1}{2}\end{matrix};1\right) = \frac{(v)_k\,2^k}{(2v)_k}. \tag{1.206}
$$

Proof. The Gauss summation formula (1.160) gives

$$
\begin{aligned}
{}_2F_1\left(\begin{matrix}-\frac{k}{2},\frac{1}{2}-\frac{k}{2}\\ v+\frac{1}{2}\end{matrix};1\right) &= \frac{\Gamma(v+\frac{1}{2})\,\Gamma(v+k)}{\Gamma(v+\frac{1}{2}+\frac{k}{2})\,\Gamma(v+\frac{k}{2})}\\
&= \frac{\Gamma(v+\frac{1}{2})\,(v)_k\,\Gamma(v)}{\Gamma(v+\frac{1}{2}+\frac{k}{2})\,\Gamma(v+\frac{k}{2})}.
\end{aligned} \tag{1.207}
$$

The Legendre duplication formula allows to write

$$
\Gamma\left(v+\frac{1}{2}\right)\Gamma(v) = 2^{1-2v}\sqrt{\pi}\,\Gamma(2v)
$$

and

$$
\Gamma\left(v+\frac{1}{2}+\frac{k}{2}\right)\Gamma\left(v+\frac{k}{2}\right) = 2^{1-2v-k}\sqrt{\pi}\,\Gamma(2v+k).
$$

On substituting in (1.207), we obtain

$$
{}_2F_1\left(\begin{matrix}-\frac{k}{2},\frac{1}{2}-\frac{k}{2}\\ v+\frac{1}{2}\end{matrix};1\right) = \frac{(v)_k\,2^k}{(2v)_k}. \tag{1.208}
$$

This completes the proof. □

Theorem 28. *The Chu–Vandermonde's identity, for $k \in \mathbb{N}$, is*

$$
{}_2F_1\left(\begin{matrix}-k,v\\ w\end{matrix};1\right) = \frac{(w-v)_k}{(w)_k}. \tag{1.209}
$$

Proof. The Gauss summation formula allows

$$ {}_2F_1\left(\begin{matrix}-k, v\\ w\end{matrix}; 1\right) = \frac{\Gamma(w)\,\Gamma(w-v+k)}{\Gamma(w+k)\,\Gamma(w-v)}. \tag{1.210}$$

Using $(u)_k = \frac{\Gamma(u+k)}{\Gamma(u)}$, we get

$$ {}_2F_1\left(\begin{matrix}-k, v\\ w\end{matrix}; 1\right) = \frac{(w-v)_k}{(w)_k}, \tag{1.211}$$

and completes the proof. □

1.9 Special cases

Most of the elementary functions, which occurs in mathematical physics, can be expressed in terms of the Gauss hypergeometric function. We list here a few of them as an example:

$$ {}_2F_1(u, v; v; z) = \sum_{k=0}^{\infty} (u)_k \frac{z^k}{k!} = {}_1F_0(u, -; z) = (1-z)^{-u}. \tag{1.212}$$

Similarly,

$$ z\,{}_2F_1(1, 1; 2; -z) = z - \frac{z^2}{2} + \frac{z^3}{3} - \cdots = \log(1+z), \tag{1.213}$$

$$ 2z\,{}_2F_1\left(\frac{1}{2}, 1; \frac{3}{2}; z^2\right) = \log\left(\frac{1+z}{1-z}\right). \tag{1.214}$$

Also,

$$\begin{aligned} {}_2F_1\left(1, v; 1; \frac{z}{v}\right) &= \sum_{k=0}^{\infty} \frac{(1)_k\,(v)_k}{(1)_k\,k!}\left(\frac{z}{v}\right)^k \\ &= 1 + \sum_{k=1}^{\infty} \frac{v\,(v+1)\cdots(v+k-1)}{v^k}\,\frac{z^k}{k!} \\ &= 1 + \sum_{k=1}^{\infty} 1\cdot\left(1+\frac{1}{v}\right)\left(1+\frac{2}{v}\right)\cdots\left(1+\frac{k-1}{v}\right)\frac{z^k}{k!}. \end{aligned} \tag{1.215}$$

Taking the limit as $v \to \infty$ yields

$$ \lim_{v\to\infty} {}_2F_1\left(1, v; 1; \frac{z}{v}\right) = 1 + \sum_{k=1}^{\infty} \frac{z^k}{k!} = e^z. \tag{1.216}$$

Various trigonometric and hyperbolic functions can also be deduced in the form of the Gauss hypergeometric function ${}_2F_1(u, v; w; z)$:

$$ {}_2F_1\left(\frac{1}{2}+\frac{1}{2}u, \frac{1}{2}-\frac{1}{2}u; \frac{3}{2}; (\sin z)^2\right) = \sin uz; \tag{1.217}$$

$$ {}_2F_1\left(\frac{1}{2}u, -\frac{1}{2}u; \frac{1}{2}; (\sin z)^2\right) = \cos uz; \tag{1.218}$$

$$ {}_2F_1\left(\frac{1}{2}+\frac{1}{2}u, \frac{1}{2}-\frac{1}{2}u; \frac{3}{2}; (\sinh z)^2\right) = \sinh uz; \tag{1.219}$$

$$ {}_2F_1\left(\frac{1}{2}u, -\frac{1}{2}u; \frac{1}{2}; (\sinh z)^2\right) = \cosh uz; \tag{1.220}$$

$$ z\,{}_2F_1\left(\frac{1}{2}, \frac{1}{2}; \frac{3}{2}; z^2\right) = \arcsin z; \tag{1.221}$$

$$ z\,{}_2F_1\left(\frac{1}{2}, 1; \frac{3}{2}; -z^2\right) = \arctan z; \tag{1.222}$$

$$ z\,{}_2F_1\left(\frac{1}{2}, \frac{1}{2}; \frac{3}{2}; -z^2\right) = \sinh^{-1} z. \tag{1.223}$$

Other elementary forms are

$$ {}_2F_1\left(u, u+\frac{1}{2}; \frac{1}{2}; z\right) = \frac{1}{2}(1+z^{\frac{1}{2}})^{-2u} + \frac{1}{2}(1-z^{\frac{1}{2}})^{-2u}; \tag{1.224}$$

$$ {}_2F_1\left(u-\frac{1}{2}, u; 2u; z\right) = \left\{\frac{1}{2}+\frac{1}{2}(1-z)^{\frac{1}{2}}\right\}^{1-2u}; \tag{1.225}$$

$$ {}_2F_1(2u, u+1; 2u; z) = \frac{(1+z)}{(1-z)^{2u+1}}. \tag{1.226}$$

The Legendre polynomial denoted by $P_k(z)$ is defined as the coefficient of x^k in the expansion of $(1-xz+x^2)^{-\frac{1}{2}}$. Legendre polynomial can be written in terms of Gauss hypergeometric function as

$$ \begin{aligned} {}_2F_1\left(-k, k+1; 1; \frac{1-z}{2}\right) &= \sum_{n=0}^{k} \frac{(-k)_n\,(k+1)_n}{(1)_n\,n!}\left(\frac{1-z}{2}\right)^n \\ &= \sum_{n=0}^{k} \frac{(-1)^n\,k!\,(k+1)_n}{(k-n)!\,n!\,n!}\left(\frac{1-z}{2}\right)^n = P_k(z). \end{aligned} \tag{1.227}$$

Alternatively, this can be obtained by

$$ \begin{aligned} P_k(z) &= \frac{1}{2^k\,k!}\frac{d^k}{dz^k}[(1-z)^k\{(1-z)-2\}^k] \\ &= \frac{(-1)^k}{2^k\,k!}\frac{d^k}{dz^k}\left\{\sum_{m=0}^{k} 2^{k-m}(1-z)^{k+m}\frac{(-k)_m}{m!}(-1)^m\right\} \end{aligned}$$

$$
\begin{aligned}
&= \frac{(-1)^k}{2^k\,k!}\sum_{m=0}^{k}\frac{(-k)_m}{m!}\,(-1)^m\,2^{k-m}\,(k+m-1)_m\,(1-z)^m\,(-1)^k \\
&= \sum_{m=0}^{k}\frac{(-k)_m\,(-1)^m}{m!\,k!\,2^m}\,(k+m-1)_m\,(1-z)^m \\
&= {}_2F_1\left(-k,k+1;1;\frac{1-z}{2}\right). \qquad (1.228)
\end{aligned}
$$

The Chebyshev polynomials of first and second kind can also be presented in terms of the Gauss hypergeometric function:

$$
{}_2F_1\left(-k,k;1;\frac{1-z}{2}\right) = T_n(z); \qquad (1.229)
$$

$$
(k+1)\,{}_2F_1\left(-k,k+2;\frac{3}{2};1-\frac{z}{2}\right) = U_n(z). \qquad (1.230)
$$

The Gagenbauer polynomial,

$$
\frac{(2\lambda)_k}{k!}\,{}_2F_1\left(-k,k+2\lambda;\lambda+\frac{1}{2};\frac{1-z}{2}\right) = C_k^{\lambda}(z). \qquad (1.231)
$$

The general Jacobi polynomial,

$$
\frac{(u+1)_k}{k!}\,{}_2F_1\left(-k,u+v+k+1;u+1;\frac{1-z}{2}\right) = P_k^{(u,v)}(z), \qquad (1.232)
$$

and so on. Thus, we have seen that the function having one variable and one or more parameters can be presented in terms of the Gauss hypergeometric function. These functions also include Bessel functions, the error functions, the Hermite polynomial, the airy function, the coloumb wave functions, the weber functions, Whittaker function and almost every function used in mathematical physics. Thus, the important Gauss function ${}_2F_1(u,v;w;z)$ can be seen with the fact that every result about the Gauss function will have interesting special cases and produce the same for corresponding elementary functions.

1.10 Generalizations of Gauss function

The generalizations of this function carried a great list of special functions. It can be seen in two ways, namely:

1. By increasing the number of parameters: When the number of parameters are increased, it takes the Gauss function to a generalized hypergeometric function. The generalized hypergeometric function is usually denoted by ${}_pF_q(u_1,\ldots,u_p;v_1,\ldots,v_q;z)$ or ${}_pF_q\left[{}^{u_1,\ldots,u_p}_{v_1,\ldots,v_q};z\right]$ and defined as [53, 128],

$$ {}_pF_q(u_1,\dots,u_p;v_1,\dots,v_q;z) = \sum_{k=0}^{\infty} \frac{(u_1)_k\dots(u_p)_k}{(v_1)_k\dots(v_q)_n} \frac{z^k}{k!}, \tag{1.233} $$

where the parameters $u_1,\dots,u_p$, $v_1,\dots,v_q$ and the variable z are in $\mathbb{C}$ such that $v_j \neq 0,-1,-2,\dots; j = 1,\dots,q$. The series ${}_pF_q$ converges for $|z| < \infty$ if $p \leq q$, for $|z| < 1$ if $p = q+1$ and diverges for all z, $z \neq 0$, if $p > q+1$.
Furthermore, if we set

$$ \omega = \sum_{j=1}^{q} v_j - \sum_{j=1}^{p} u_j, $$

then the ${}_pF_q$ series, with $p = q+1$, is
(a) absolutely convergent for $|z| = 1$ if $\mathbb{R}(\omega) > 0$;
(b) conditionally convergent for $|z| = 1$, $z \neq 1$, if $-1 < \mathbb{R}(\omega) \leq 0$;
(c) divergent for all $z = 1$, if $\mathbb{R}(\omega) \leq -1$.

2. By increasing the number of variables: The other way of generalization is by increasing the number of variables and we get a list of multivariable special functions. The set of four, two-variable Appell functions are given by, [6, 53, 147, 148]

$$ F_1(u,v,v';w;x,y) = \sum_{k_1,k_2=0}^{\infty} \frac{(u)_{k_1+k_2}(v)_{k_1}(v')_{k_2}}{(w)_{k_1+k_2}} \frac{x^{k_1}}{k_1!}\frac{y^{k_2}}{k_2!}, $$
$$ \max\{|x|,|y|\} \leq 1; \tag{1.234} $$
$$ F_2(u,v,v';w,w';x,y) = \sum_{k_1,k_2=0}^{\infty} \frac{(u)_{k_1+k_2}(v)_{k_1}(v')_{k_2}}{(w)_{k_1}(w')_{k_2}} \frac{x^{k_1}}{k_1!}\frac{y^{k_2}}{k_2!}, $$
$$ |x| + |y| \leq 1; \tag{1.235} $$
$$ F_3(u,u',v,v';w;x,y) = \sum_{k_1,k_2=0}^{\infty} \frac{(u)_{k_1}(u')_{k_2}(v)_{m_1}(b')_{m_2}}{(c)_{m_1+m_2}} \frac{x^{m_1}}{m_1!}\frac{y^{m_2}}{m_2!}, $$
$$ \max\{|x|,|y|\} \leq 1; \tag{1.236} $$
$$ F_4(u,v;w,w';x,y) = \sum_{m_1,m_2=0}^{\infty} \frac{(u)_{m_1+m_2}(v)_{m_1+m_2}}{(w)_{m_1}(w')_{m_2}} \frac{x^{m_1}}{m_1!}\frac{y^{m_2}}{m_2!}, $$
$$ \sqrt{|x|} + \sqrt{|y|} \leq 1. \tag{1.237} $$

The most generalized two-variable function, i. e., Kampé de Fériet hypergeometric function is given by

$$ F^{m_1:n_1;n_1'}_{m_2:n_2;n_2'}\left(\begin{matrix}(a):\ (b);\ (b')\\ (c):\ (d);\ (d')\end{matrix};x,y\right) = \sum_{m,n=0}^{\infty} \frac{\prod_{j=1}^{m_1}(a_j)_{m+n}\prod_{j=1}^{n_1}(b_j)_m\prod_{j=1}^{n_1'}(b_j')_n}{\prod_{j=1}^{m_2}(c_j)_{m+n}\prod_{j=1}^{n_2}(d_j)_m\prod_{j=1}^{n_2'}(d_j')_n} \frac{x^m\,y^n}{m!\,n!}, \tag{1.238} $$

where (a) abbreviates the sequence of m_1 parameters $a_1, \dots, a_{m_1}$, with similar interpretations for (b), (b'), etc., and where for convergence

(a) $m_1 + n_1 < m_2 + n_2 + 1$, $m_1 + n_1' < m_2 + n_2' + 1$, $|x| < \infty$, $|y| < \infty$, or

(b) $m_1 + n_1 = m_2 + n_2 + 1$, $m_1 + n_1' = m_2 + n_2' + 1$,

and

$$\begin{cases} |x|^{\frac{1}{m_1-m_2}} + |y|^{\frac{1}{m_1-m_2}} < 1, & \text{if } m_1 > m_2, \\ \max\{|x|, |y|\} < 1, & \text{if } m_1 \le m_2. \end{cases} \tag{1.239}$$

The Lauricella functions of three variables, [106, 147]:

$$F_A^{(3)}(a, b_1, b_2, b_3; c_1, c_2, c_3; x_1, x_2, x_3) = \sum_{m_1,m_2,m_3=0}^{\infty} \frac{(a)_{m_1+m_2+m_3} \prod_{i=1}^{3} (b_i)_{m_i}}{\prod_{i=1}^{3} (c_i)_{m_i}} \prod_{i=1}^{3} \frac{x_i^{m_i}}{m_i!}, \tag{1.240}$$

$$F_B^{(3)}(a_1, a_2, a_3, b_1, b_2, b_3; c; x_1, x_2, x_3) = \sum_{m_1,m_2,m_3=0}^{\infty} \frac{\prod_{i=1}^{3} (a_i)_{m_i} \prod_{i=1}^{3} (b_i)_{m_i}}{(c)_{m_1+m_2+m_3}} \prod_{i=1}^{3} \frac{x_i^{m_i}}{m_i!}, \tag{1.241}$$

$$F_C^{(3)}(a, b; c_1, c_2, c_3; x_1, x_2, x_3) = \sum_{m_1,m_2,m_3=0}^{\infty} \frac{(a)_{m_1+m_2+m_3} (b)_{m_1+m_2+m_3}}{\prod_{i=1}^{3} (c_i)_{m_i}} \prod_{i=1}^{3} \frac{x_i^{m_i}}{m_i!}, \tag{1.242}$$

$$F_D^{(3)}(a, b_1, b_2, b_3; c; x_1 x_2, x_3) = \sum_{m_1,m_2,m_3=0}^{\infty} \frac{(a)_{m_1+m_2+m_3} \prod_{i=1}^{3} (b_i)_{m_i}}{(c)_{m_1+m_2+m_3}} \prod_{i=1}^{3} \frac{x_i^{m_i}}{m_i!}, \tag{1.243}$$

$$F_E^{(3)}(a_1, a_1, a_1, b_1, b_2, b_2; c_1, c_2, c_3; x_1, x_2, x_3) = \sum_{m_1,m_2,m_3=0}^{\infty} \frac{(a_1)_{m_1+m_2+m_3} (b_1)_{m_1} (b_2)_{m_2+m_3}}{\prod_{i=1}^{3} (c_i)_{m_i}} \prod_{i=1}^{3} \frac{x_i^{m_i}}{m_i!}, \tag{1.244}$$

$$F_F^{(3)}(a_1, a_1, a_1, b_1, b_2, b_1; c_1, c_2, c_2; x_1, x_2, x_3) = \sum_{m_1,m_2,m_3=0}^{\infty} \frac{(a_1)_{m_1+m_2+m_3} (b_1)_{m_1+m_3} (b_2)_{m_2}}{(c_1)_{m_1} (c_2)_{m_2+m_3}} \prod_{i=1}^{3} \frac{x_i^{m_i}}{m_i!}, \tag{1.245}$$

$$F_G^{(3)}(a_1, a_1, a_1, b_1, b_2, b_3; c_1, c_2, c_2; x_1, x_2, x_3) = \sum_{m_1,m_2,m_3=0}^{\infty} \frac{(a_1)_{m_1+m_2+m_3} \prod_{i=1}^{3} (b_i)_{m_i}}{(c_1)_{m_1} (c_2)_{m_2+m_3}} \prod_{i=1}^{3} \frac{x_i^{m_i}}{m_i!}, \tag{1.246}$$

$$F_K^{(3)}(a_1, a_2, a_2, b_1, b_2, b_1; c_1, c_2, c_3; x_1, x_2, x_3) = \sum_{m_1,m_2,m_3=0}^{\infty} \frac{(a_1)_{m_1} (a_2)_{m_2+m_3} (b_1)_{m_1+m_3} (b_2)_{m_2}}{\prod_{i=1}^{3} (c_i)_{m_i}} \prod_{i=1}^{3} \frac{x_i^{m_i}}{m_i!}, \tag{1.247}$$

$$F_M^{(3)}(a_1, a_2, a_2, b_1, b_2, b_1; c_1, c_2, c_2; x_1, x_2, x_3) = \sum_{m_1,m_2,m_3=0}^{\infty} \frac{(a_1)_{m_1}(a_2)_{m_2+m_3}(b_1)_{m_1+m_3}(b_2)_{m_2}}{(c_1)_{m_1}(c_2)_{m_2+m_3}} \prod_{i=1}^{3} \frac{x_i^{m_i}}{m_i!}, \tag{1.248}$$

$$F_N^{(3)}(a_1, a_2, a_3, b_1, b_2, b_1; c_1, c_2, c_2; x_1, x_2, x_3) = \sum_{m_1,m_2,m_3=0}^{\infty} \frac{\prod_{i=1}^{3}(a_i)_{m_i}(b_1)_{m_1+m_3}(b_2)_{m_2}}{(c_1)_{m_1}(c_2)_{m_2+m_3}} \prod_{i=1}^{3} \frac{x_i^{m_i}}{m_i!}, \tag{1.249}$$

$$F_P^{(3)}(a_1, a_2, a_1, b_1, b_1, b_2; c_1, c_2, c_2; x_1, x_2, x_3) = \sum_{m_1,m_2,m_3=0}^{\infty} \frac{(a_1)_{m_1+m_3}(a_2)_{m_2}(b_1)_{m_1+m_2}(b_2)_{m_3}}{(c_1)_{m_1}(c_2)_{m_2+m_3}} \prod_{i=1}^{3} \frac{x_i^{m_i}}{m_i!}, \tag{1.250}$$

$$F_R^{(3)}(a_1, a_2, a_1, b_1, b_2, b_1; c_1, c_2, c_2; x_1, x_2, x_3) = \sum_{m_1,m_2,m_3=0}^{\infty} \frac{(a_1)_{m_1+m_3}(a_2)_{m_2}(b_1)_{m_1+m_3}(b_2)_{m_2}}{(c_1)_{m_1}(c_2)_{m_2+m_3}} \prod_{i=1}^{3} \frac{x_i^{m_i}}{m_i!}, \tag{1.251}$$

$$F_S^{(3)}(a_1, a_2, a_2, b_1, b_2, b_3; c_1, c_1, c_1; x_1, x_2, x_3) = \sum_{m_1,m_2,m_3=0}^{\infty} \frac{(a_1)_{m_1}(a_2)_{m_2+m_3}\prod_{i=1}^{3}(b_i)_{m_i}}{(c_1)_{m_1+m_2+m_3}} \prod_{i=1}^{3} \frac{x_i^{m_i}}{m_i!}, \tag{1.252}$$

$$F_T^{(3)}(a_1, a_2, a_2, b_1, b_2, b_1; c_1, c_1, c_1; x_1, x_2, x_3) = \sum_{m_1,m_2,m_3=0}^{\infty} \frac{(a_1)_{m_1}(a_2)_{m_2+m_3}(b_1)_{m_1+m_3}(b_2)_{m_2}}{(c_1)_{m_1+m_2+m_3}} \prod_{i=1}^{3} \frac{x_i^{m_i}}{m_i!}. \tag{1.253}$$

Besides, Srivastava's triple hypergeometric functions H_A, H_B and H_C are defined by, [147]

$$H_A(a, b_1, b_2; c_1, c_2; x_1, x_2, x_3) = \sum_{m_1,m_2,m_3=0}^{\infty} \frac{(a)_{m_1+m_3}(b_1)_{m_1+m_2}(b_2)_{m_2+m_3}}{(c_1)_{m_1}(c_2)_{m_2+m_3}} \prod_{i=1}^{3} \frac{x_i^{m_i}}{m_i!}, \tag{1.254}$$

$$H_B(a, b_1, b_2; c_1, c_2, c_3; x_1, x_2, x_3) = \sum_{m_1,m_2,m_3=0}^{\infty} \frac{(a)_{m_1+m_3}(b_1)_{m_1+m_2}(b_2)_{m_2+m_3}}{(c_1)_{m_1}(c_2)_{m_2}(c_3)_{m_3}} \prod_{i=1}^{3} \frac{x_i^{m_i}}{m_i!}, \tag{1.255}$$

$$H_C(a, b_1, b_2; c, ; x_1, x_2, x_3) = \sum_{m_1,m_2,m_3=0}^{\infty} \frac{(a)_{m_1+m_3}(b_1)_{m_1+m_2}(b_2)_{m_2+m_3}}{(c)_{m_1+m_2+m_3}} \prod_{i=1}^{3} \frac{x_i^{m_i}}{m_i!}. \tag{1.256}$$

For convergence of the above listed functions, we refer to [147, 148].

The generalized Lauricella functions:

$$F_A^{(k)}(a, b_1, \ldots, b_k; c_1, \ldots, c_k; x_1, \ldots, x_k) = \sum_{m_1,\ldots,m_k=0}^{\infty} \frac{(a)_{m_1+\cdots+m_k} \prod_{i=1}^k (b_i)_{m_i}}{\prod_{i=1}^k (c_i)_{m_i}} \prod_{i=1}^k \frac{x_i^{m_i}}{m_i!}, \tag{1.257}$$

$$F_B^{(k)}(a_1, \ldots, a_k, b_1, \ldots, b_k; c; x_1, \ldots, x_k) = \sum_{m_1,\ldots,m_k=0}^{\infty} \frac{\prod_{i=1}^k (a_i)_{m_i} \prod_{i=1}^k (b_i)_{m_i}}{(c)_{m_1+\cdots+m_k}} \prod_{i=1}^k \frac{x_i^{m_i}}{m_i!}, \tag{1.258}$$

$$F_C^{(k)}(a, b; c_1, \ldots, c_k; x_1, \ldots, x_k) = \sum_{m_1,\ldots,m_k=0}^{\infty} \frac{(a)_{m_1+\cdots+m_k} (b)_{m_1+\cdots+m_k}}{\prod_{i=1}^k (c_i)_{m_i}} \prod_{i=1}^k \frac{x_i^{m_i}}{m_i!}, \tag{1.259}$$

$$F_D^{(k)}(a, b_1, \ldots, b_k; c; x_1, \ldots, x_k) = \sum_{m_1,\ldots,m_k=0}^{\infty} \frac{(a)_{m_1+\cdots+m_k} \prod_{i=1}^k (b_i)_{m_i}}{(c)_{m_1+\cdots+m_k}} \prod_{i=1}^k \frac{x_i^{m_i}}{m_i!}. \tag{1.260}$$

1.11 Degeneration or confluent form of Gauss hypergeometric function

We called ${}_1F_1$ as the confluent hypergeometric function so far because ${}_1F_1$ can also be obtained as the confluent case of the Gauss hypergeometric function. Let u, v, w be complex numbers such that $\mathbb{R}(w) \neq 0, -1, -2, \ldots$. Then the Gauss hypergeometric function is defined by

$${}_2F_1(u, v;\ w;\ z) = \sum_{m=0}^{\infty} \frac{(u)_m\,(v)_m}{(w)_m\, m!} z^m. \tag{1.261}$$

Let us use $u = \frac{1}{c}$ and z as εz. Then we get

$${}_2F_1\left(\frac{1}{\varepsilon}, v\,; w;\ \varepsilon z\right) = \sum_{m=0}^{\infty} \frac{(\frac{1}{\varepsilon})_m\,(v)_m}{(w)_m\, m!} (\varepsilon z)^m, \quad |\varepsilon z| < 1. \tag{1.262}$$

Now, $(\frac{1}{\varepsilon})_m\, \varepsilon^m = (1 + \varepsilon) \cdots (1 + (m-1)\varepsilon)$. On letting ε tend to zero, we get

$$\lim_{\varepsilon \to 0} \left(\frac{1}{\varepsilon}\right)_m \varepsilon^m = 1. \tag{1.263}$$

Therefore,

$$\lim_{\varepsilon\to 0} {}_2F_1\left(\frac{1}{\varepsilon}, v; w; \varepsilon z\right) = \sum_{m=0}^{\infty} \frac{(v)_m}{(w)_m\, m!} z^m. \tag{1.264}$$

The series on the right-hand side is known as the confluent hypergeometric series, denoted by ${}_1F_1(v; w; z)$. Hence,

$$\lim_{\varepsilon\to 0} {}_2F_1\left(\frac{1}{\varepsilon}, v; w; \varepsilon z\right) = {}_1F_1(v; w; z).$$

The confluent hypergeometric function ${}_1F_1(v; w; z)$ constitutes a first degeneration of the Gauss hypergeometric function ${}_2F_1(a, b; c; z)$. The properties of ${}_1F_1(v; w; z)$ can be deduced by passing to the limit in the formulas relating to the function ${}_2F_1$. Consider, e. g., the integral representation of ${}_2F_1(\frac{1}{\varepsilon}, v; w; \varepsilon z)$ is given by

$$\begin{aligned} &{}_2F_1\left(\frac{1}{\varepsilon}, v; w; \varepsilon z\right) \\ &= \Gamma\left(\begin{matrix} w \\ v, w-v \end{matrix}\right) \int_0^1 (1-u\varepsilon z)^{-\frac{1}{\varepsilon}} u^{v-1}(1-u)^{w-v-1}\, du \\ &= \Gamma\left(\begin{matrix} w \\ v, w-v \end{matrix}\right) \int_0^1 \sum_{m=0}^{\infty} \frac{(\frac{1}{\varepsilon})_m}{m!} (\varepsilon u z)^m u^{v-1}(1-u)^{w-v-1}\, du. \end{aligned} \tag{1.265}$$

Letting $\varepsilon \to 0$ leads to

$$\lim_{\varepsilon\to 0} {}_2F_1\left(\frac{1}{\varepsilon}, v; w; \varepsilon z\right) = {}_1F_1(v; w; z) = \Gamma\left(\begin{matrix} w \\ v, w-v \end{matrix}\right) \int_0^1 u^{v-1}(1-u)^{w-v-1} e^{zu}\, du. \tag{1.266}$$

Similarly, the differential equation satisfied by the Gauss hypergeometric function is given by

$$[\theta(\theta + w - 1) - z(\theta + u)(\theta + v)]\,{}_2F_1(u, v; w; z) = 0, \tag{1.267}$$

where $\theta = z\frac{z}{dz}$. If u is replaced by $\frac{1}{\varepsilon}$ and z by εz, then we get

$$\left[\theta(\theta + w - 1) - \varepsilon z\left(\theta + \frac{1}{\varepsilon}\right)(\theta + v)\right] {}_2F_1\left(\frac{1}{\varepsilon}, v; w; \varepsilon z\right) = 0, \tag{1.268}$$

applying limit as $\varepsilon \to 0$, one can get

$$[\theta(\theta + w - 1) - z(\theta + v)]\,{}_1F_1(v; w; z) = 0, \tag{1.269}$$

which is the differential equation obeyed by the confluent hypergeometric function.

The region of convergence can also be concluded in the similar manner. The region of convergence of Gauss hypergeometric is given by $|z| < 1$. So, the convergence of

${}_2F_1(\frac{1}{\varepsilon}, v; w; \varepsilon z)$ is given by $|\varepsilon z| < 1$, which gives $|z| < \frac{1}{\varepsilon}$. Taking the limit as $\varepsilon \to 0$ gives the convergence of ${}_1F_1$, i. e., $|z| < \infty$.

Exercises

1. Test whether the following infinite series are convergent or divergent:
   (i) $2 + \frac{3x^2}{8} + \frac{4x^3}{27} + \cdots + \frac{(k+1)x^k}{k^3}$, $x > 0$;
   (ii) $x + \frac{3}{5}x^2 + \frac{8}{10}x^3 + \frac{15}{17}x^4 + \cdots + \frac{k^2-1}{k^2+1}x^k + \cdots$, $x > 0$;
   (iii) $\frac{1}{2\sqrt{1}} + \frac{x^2}{3\sqrt{2}} + \frac{x^4}{4\sqrt{3}} + \frac{x^6}{5\sqrt{4}} + \cdots$, $x > 0$;
   (iv) $1 + \frac{1}{2}x + \frac{1\cdot 3}{2\cdot 4}x^2 + \frac{1\cdot 3\cdot 5}{2\cdot 4\cdot 6}x^3$, $x > 0$.
2. Prove that the infinite series $\sum_{k=0}^{\infty}(u)_k \frac{z^k}{k!}$ converges absolutely for $|z| < 1$.
3. Prove that the infinite product $\prod_{k=1}^{\infty}[1 + \frac{(-1)^{k+1}}{k}]$ converges to unity.
4. Show that $\prod_{k=1}^{\infty} \frac{\sin(\frac{z}{k})}{\frac{z}{k}}$ is absolutely convergent for all finite z with the usual convention at $z = 0$.
5. With the usual definition of shifted factorial, prove that

$$(u)_{rk} = r^{kr}\left(\frac{u}{r}\right)_k \left(\frac{u+1}{r}\right)_k \cdots \left(\frac{u+r-1}{r}\right)_k.$$

6. Evaluate the following integrals:
   (i) $\int_0^{\infty} e^{-x^2}\,dx$;
   (ii) $\int_0^{\infty} 3^{-4t^2}\,dt$.
7. Prove that

$$\Gamma(u) = s^u \int_0^{\infty} e^{-su} u^{u-1}\,du.$$

8. Verify the following:
   (i) $\beta(a,b) = \frac{(\alpha_2-\alpha_3)^a(\alpha_1-\alpha_3)^b}{(\alpha_2-\alpha_1)^{a+b-1}} \int_{\alpha_1}^{\alpha_2} \frac{(v-\alpha_1)^{a-1}(\alpha_2-v)^{b-1}}{(v-\alpha_3)^{a+b}}\,dv$;
   (ii) $\beta(\frac{u}{w}, v) = w\int_0^{\infty} e^{-uu}(1-e^{-wu})^{v-1}\,du$, $\Re(u), \Re(v), \Re(w) > 0$;
   (iii) $\beta(\frac{u}{w}, v) = w\int_0^1 t^{u-1}(1-t^w)^{v-1}\,dt$, $\Re(u), \Re(v), \Re(w) > 0$.
9. Let u, v, w be complex numbers such that $\Re(u), \Re(v), \Re(w) > 0$. Then prove that

$$\iint u^{u-1} v^{v-1}(1-u-v)^{w-1}\,du\,dv = \frac{\Gamma(u)\Gamma(v)\Gamma(w)}{\Gamma(u+v+w)}.$$

10. Prove the following formulas:

$$(1-z)^{-u}\,{}_2F_1\left(\begin{matrix}u, v\\ w\end{matrix}; \frac{-z}{1-z}\right) = {}_2F_1\left(\begin{matrix}u, w-v\\ w\end{matrix}; z\right);$$

$$ {}_2F_1\left(\begin{matrix} u, v \\ \frac{1}{2}(u+v+1) \end{matrix}; \frac{1}{2}\right) = \frac{\Gamma(\frac{1}{2})\,\Gamma(\frac{1}{2}(u+v+1))}{\Gamma(\frac{1}{2}(u+1))\,\Gamma(\frac{1}{2}(v+1))}; $$

$$ {}_2F_1\left(\begin{matrix} u, 1-u \\ w \end{matrix}; \frac{1}{2}\right) = \frac{\Gamma(\frac{1}{2})\,\Gamma(\frac{1}{2}(w+1))}{\Gamma(\frac{1}{2}(w+u))\,\Gamma(\frac{1}{2}(w-u+1))}. $$

11. Verify Kummer's theorem:

$$ {}_2F_1\left(\begin{matrix} u, v \\ 1+u-v \end{matrix}; -1\right) = \frac{\Gamma(1+u-v)\,\Gamma(\frac{1}{2}(2+u))}{\Gamma(1+u)\,\Gamma(\frac{1}{2}(2+u-2v))}. $$

12. If the polygamma function is defined by

$$ \psi^{(k)}(u) = \frac{d^{k+1}}{du^{k+1}} \log\Gamma(u) = \frac{d^k}{du^k}\,\psi(u), $$

then prove that

$$ \psi^{(k)}(u) - \psi^{(k)}(u+1) = \frac{(-1)^{k+1}\,k!}{u^{k+1}}; $$

$$ \psi^{(k)}(m\,u) = m^{-k-1} \sum_{r=0}^{m-1} \psi^{(k)}\left(u + \frac{r}{m}\right). $$

13. If ${}_1F_1(u;w;z) = \sum_{k=0}^{\infty} \frac{(u)_k}{(w)_k} \frac{z^k}{k!}$ is the confluent hypergeometric function, then verify the following contiguous relations:

$$ (u-w+1)\,{}_1F_1(u;w;z) = u\,{}_1F_1(u+1;w;z) - (w-1)\,{}_1F_1(u;w-1;z); $$

$$ w(u+z)\,{}_1F_1(u;w;z) = uw\,{}_1F_1(u+1;w;z) - (u-w)\,z\,{}_1F_1(u;w+1;z); $$

$$ w\,{}_1F_1(u;w;z) = w\,{}_1F_1(u-1;w;z) + z\,{}_1F_1(u;w+1;z). $$

14. Show that the elementary integral representation of ${}_1F_1(u;w;z)$ can be transformed into the following integrals:

$$ {}_1F_1(u;w;z) = \frac{\Gamma(w)}{\Gamma(u)\,\Gamma(w-u)} z^{1-w} \int_0^z e^{\mathfrak{v}}\, \mathfrak{v}^{u-1} (z-\mathfrak{v})^{w-u-1}\, d\mathfrak{v}; $$

$$ {}_1F_1(u;w;z) = \frac{\Gamma(w)}{\Gamma(u)\,\Gamma(w-u)} e^{z} \int_0^1 e^{-\mathfrak{v}\,z}\, \mathfrak{v}^{w-u-1} (1-\mathfrak{v})^{u-1}\, d\mathfrak{v}; $$

$$ {}_1F_1(u;w;z) = \frac{\Gamma(w)}{\Gamma(u)\,\Gamma(w-u)} 2^{1-w} e^{z/2} \int_{-1}^1 e^{-1/2\,z\,\mathfrak{v}}\, (1+\mathfrak{v})^{w-2} \left(\frac{1-\mathfrak{v}}{1+\mathfrak{v}}\right)^{u-1} d\mathfrak{v}; $$

$$ {}_1F_1(u;w;z) = \frac{\Gamma(w)}{\Gamma(u)\,\Gamma(w-u)} 2^{1-w} e^{z/2} \int_{-1}^1 e^{1/2\,z\mathfrak{v}}\, (1-\mathfrak{v})^{w-2} \left(\frac{1+\mathfrak{v}}{1-\mathfrak{v}}\right)^{u-1} d\mathfrak{v}. $$

15. Let u, w be complex numbers such that $\Re(w) \neq 0, -1, -2, \ldots$. Then obtain the Kummer's first and second formulas:

$$ {}_1F_1(u; w; z) = e^z \, {}_1F_1(w-u; w; z) $$

and

$$ e^{-z} {}_1F_1(u;\ 2u;\ 2z) = {}_0F_1\left(-;\ u+\frac{1}{2};\ \frac{1}{4}z^2\right), \quad \Re(2u) \neq -1, -3, \ldots. \tag{1.270} $$

16. Verify the following differential formula that satisfies by confluent hypergeometric function:

$$ \frac{d}{dz}\,{}_1F_1(u;\ w;\ z) = \frac{u}{w}\,{}_1F_1(u+1;\ w+1; z); $$

$$ \frac{d^n}{dz^n}\,{}_1F_1(u;\ w;\ z) = \frac{(u)_n}{(w)_n}\,{}_1F_1(u+n;\ w+n; z); $$

$$ \frac{d^n}{dz^n}\left[z^{u+n-1}\,{}_1F_1(u;\ w;\ z)\right] = (u)_n z^{u-1}\,{}_1F_1(u+n;\ w; z); $$

$$ \frac{d^n}{dz^n}\left[z^{w-1}\,{}_1F_1(u;\ w;\ z)\right] = (-1)^n\,(1-w)_n\,z^{w-1-n}\,{}_1F_1(u;\ w-n;\ z); $$

$$ \frac{d^n}{dz^n}\left[e^{-z}\,{}_1F_1(u;\ w;\ z)\right] = \frac{(-1)^n\,(w-u)_n}{(w)_n}\,{}_1F_1(w-u+n;\ w+n;\ -z); $$

$$ \frac{d^n}{dz^n}\left[e^{-z}\,z^{w-u+n-1}\,{}_1F_1(u;\ w;\ z)\right] = (w-u)_n\,z^{w-u-1}\,{}_1F_1(w-u+n;\ w;\ -z); $$

$$ \frac{d^n}{dz^n}\left[e^{-z}\,z^{w-1}\,{}_1F_1(u;\ w;\ z)\right] = (-1)^n\,(1-w)_n\,z^{w-n-1}\,{}_1F_1(w-u;\ w-n;\ -z). $$

2 Basic Gauss hypergeometric function

As per the discussion in the previous chapter, the Gauss hypergeometric function ${}_2F_1(u, v, w; z)$ was first studied by Carl Friedrich Gauss in 1813. The function become popular as many applications were found, and hence mathematicians started studying this function from different aspects. One of the important directions is the q-analogue of this function, i. e., the basic hypergeometric function or q-hypergeometric function. After three decades from the birth of the Gauss function, Heine, [76], introduced the q-analogue of Gauss hypergeometric series in 1846. In this chapter, we give the q-analogues of various elementary functions including the Gauss hypergeometric function and study their fundamental properties. Before we dive into the main content of the chapter, we give some basic definitions and preliminaries required in the sequel.

2.1 The q-numbers

The q-analogue of a complex number c is given by, [62]

$$[c]_q = \frac{1-q^c}{1-q}, \quad q \neq 1, \tag{2.1}$$

and q-shifted factorial is defined by, [62]

$$(c;q)_n = \begin{cases} 1, & \text{if } n = 0, \\ (1-c)(1-qc)\cdots(1-q^{n-1}c), & \text{if } n \geq 1 \end{cases} \tag{2.2}$$

or

$$(c;q)_n = \begin{cases} 1, & \text{if } n = 0, \\ \prod_{r=0}^{n-1}(1-q^r c), & \text{if } n \geq 1. \end{cases} \tag{2.3}$$

For a positive integer n, the q-analogue of factorial function is also given by

$$[n]_q! = [n]_q\,[n-1]_q \cdots [1]_q. \tag{2.4}$$

The present definition (2.3) of the q-shifted factorial can be extended to infinity as

$$(c;q)_\infty = \lim_{n\to\infty} \prod_{r=0}^{n-1}(1-q^r c) = \prod_{r=0}^{\infty}(1-q^r c). \tag{2.5}$$

The relations from (2.1) to (2.5) tend to their ordinary one when $q \to 1^-$. The following abbreviated notaion will be used for a product of q-shifted factorials

$$(c_1, c_2, \ldots, c_n; q)_k = (c_1;q)_k\,(c_2;q)_k \cdots (c_n;q)_k.$$

https://doi.org/10.1515/9783111324586-002

Now, we list some identities in terms of the q-shifted factorial. For any $m < n$, one can conclude

$$(q^{-m};q)_n = (1-q^{-m})(1-q^{-m+1})\cdots(1-q^{-m+m})(1-q^{-m+m+1})\cdots(1-q^{-m+n-1}) = 0. \quad (2.6)$$

For a negative subscript, the q-shifted factorial is defined by

$$(c;q)_{-n} = \frac{1}{(1-cq^{-1})(1-cq^{-2})\cdots(1-cq^{-n})} = \frac{1}{(cq^{-n};q)_n} = \frac{(\frac{-q}{c})^n q^{\binom{n}{2}}}{(\frac{q}{c};q)_n}. \quad (2.7)$$

The q-shifted factorial satisfies a large number of identities. A few of them are very important to understand the q-series. We are listing them here:

1. $(c;q)_{m+k} = (c;q)_m\,(cq^m;q)_k = (c;q)_k\,(cq^k;q)_m$,
2. $(c;q)_m = \frac{(c;q)_\infty}{(cq^m;q)_\infty}$,
3. $(c;q)_{2m} = (c;q^2)_m\,(cq;q^2)_m$,
4. $(c^2;q^2)_m = (c;q)_m\,(-c;q)_m$.

Proof. From the definition of the q-shifted factorial,

$$\begin{aligned}(c;q)_{m+k} &= (1-c)\,(1-cq)\cdots(1-cq^{m-1})(1-cq^m)(1-cq^{m+1})\cdots(1-cq^{m+k-1})\\ &= [(1-c)\,(1-cq)\cdots(1-cq^{m-1})][(1-cq^m)(1-cq^{m+1})\cdots(1-cq^{m+k-1})]\\ &= (c;q)_m\,(cq^m;q)_k. \end{aligned} \quad (2.8)$$

Similarly, $(c;q)_{m+k} = (c;q)_k\,(cq^k;q)_m$. It completes the proof of first identity. To prove the second identity, let us consider the right-hand side,

$$\begin{aligned}\frac{(c;q)_\infty}{(cq^m;q)_\infty} &= \frac{(1-c)\,(1-cq)\cdots(1-cq^{m-1})(1-cq^m)(1-cq^{m+1})\cdots}{(1-cq^m)(1-cq^{m+1})\cdots}\\ &= (1-c)\,(1-cq)\cdots(1-cq^{m-1}) = (c;q)_m. \end{aligned} \quad (2.9)$$

For the third identity, using the q-shifted factorial,

$$\begin{aligned}(c;q)_{2m} &= (1-c)(1-cq)(1-cq^2)(1-cq^3)\cdots(1-cq^{2m-2})(1-cq^{2m-1})\\ &= (1-c)(1-cq^2)\cdots(1-cq^{2(m-1)})(1-cq)(1-cq^3)\cdots(1-cq^{2m-1})\\ &= (c;q^2)_m\,(cq;q^2)_m. \end{aligned} \quad (2.10)$$

Similarly, we can prove the fourth one:

$$\begin{aligned}(c^2;q^2)_m &= (1-c^2)(1-c^2q^2)\cdots(1-c^2q^{2(m-1)})\\ &= (1-c)(1+c)(1-cq)(1+cq)\cdots(1-cq^{m-1})(1+cq^{m-1})\end{aligned}$$

$$
\begin{aligned}
&= (1-c)(1-cq)\cdots(1-cq^{m-1})(1-(-c))(1-(-c)q)\cdots(1-(-c)q^{m-1}) \\
&= (c;q)_m\,(-c;q)_m. \qquad (2.11)
\end{aligned}
$$

It completes the proof. □

We have several other identities, which can be deduced from the q-shifted factorial:

$$(c^{-1}q^{1-m};q)_m = (c;q)_m(-c^{-1})^m q^{-\binom{m}{2}}, \quad \binom{m}{2} = \frac{m(m-1)}{2}; \tag{2.12}$$

$$(c;q)_{m-k} = \frac{(c;q)_m}{(c^{-1}q^{1-m};q)_k}(-qc^{-1})^k q^{\binom{k}{2}-mk}; \tag{2.13}$$

$$(cq^k;q)_{m-k} = \frac{(c;q)_m}{(c;q)_k}; \tag{2.14}$$

$$(cq^{2k};q)_{m-k} = \frac{(c;q)_m(cq^m;q)_k}{(c;q)_{2k}}; \tag{2.15}$$

$$(cq^{-m};q)_k = \frac{(c;q)_k(qc^{-1};q)_m}{(c^{-1}q^{1-k};q)_m}q^{-mk}. \tag{2.16}$$

Proof of (2.12). The definition of q-shifted factorial yields

$$
\begin{aligned}
(c^{-1}q^{1-m};q)_m &= (1-c^{-1}q^{1-m})(1-c^{-1}q^{1-m}q)\cdots(1-c^{-1}q^{1-m}q^{m-1}) \\
&= (1-c^{-1}q^{1-m})(1-c^{-1}q^{2-m})\cdots(1-c^{-1}) \\
&= (-c^{-1}q^{1-m})(1-cq^{m-1})(-c^{-1}q^{2-m})(1-cq^{m-2})\cdots(-c^{-1})(1-c) \\
&= (-1)^m\,(c^{-1})^m q^{-(1+2+\cdots+(m-1))}(1-c)(1-cq)\cdots(1-cq^{m-1}) \\
&= (-c^{-1})^m\,q^{-\binom{m}{2}}(c;q)_m. \qquad (2.17)
\end{aligned}
$$

Proof of (2.13). Again, using the definition of the q-shifted factorial, we have

$$
\begin{aligned}
(c;q)_{m-k} &= (1-c)(1-cq)\cdots(1-cq^{m-k-1}) \\
&= \frac{(1-c)(1-cq)\cdots(1-cq^{m-k-1})(1-cq^{m-k})\cdots(1-cq^{m-k+k-1})}{(1-cq^{m-k})\cdots(1-cq^{m-k+k-1})} \\
&= \frac{(c;q)_m}{(-cq^{m-k})(1-c^{-1}q^{-m+k})(-cq^{m-k+1})(1-c^{-1}q^{k-m-1})\cdots(-cq^{m-1})(1-c^{-1}q^{1-m})} \\
&= \frac{(c;q)_m}{(-c)^k q^{mk} q^{-(1+2+\cdots+(k-1)+k)}(c^{-1}q^{1-m};q)_k} \\
&= \frac{(c;q)_m}{(-cq^{-1})^k q^{mk} q^{-(1+2+\cdots+(k-1))}(c^{-1}q^{1-m};q)_k} \\
&= \frac{(c;q)_m}{(-cq^{-1})^k q^{mk} q^{-\binom{k}{2}}(c^{-1}q^{1-m};q)_k} \\
&= \frac{(c;q)_m(-qc^{-1})^k q^{-mk+\binom{k}{2}}}{(c^{-1}q^{1-m};q)_k}. \qquad (2.18)
\end{aligned}
$$

The identities (2.14) and (2.15) can be proved similarly. Now, we will see the proof of (2.16). Starting with the right-hand side of (2.16),

$$\begin{aligned}
&\frac{(c;q)_k(qc^{-1};q)_n}{(c^{-1}q^{1-k};q)_n}q^{-nk}\\
&=\frac{(1-c)(1-cq)\cdots(1-cq^{k-1})(1-qc^{-1})(1-qc^{-1}q)\cdots(1-qc^{-1}q^{n-1})}{(1-c^{-1}q^{1-k})(1-c^{-1}q^{1-k}q)\cdots(1-c^{-1}q^{1-k}q^{n-1})}q^{-nk}\\
&=\frac{(1-c)(1-cq)\cdots(1-cq^{k-1})(1-qc^{-1})(1-qc^{-1}q)\cdots(1-qc^{-1}q^{n-1})}{(-c^{-1}q^{1-k})(1-cq^{k-1})(-c^{-1}q^{2-k})(1-cq^{k-2})\cdots(-c^{-1}q^{n-k})(1-cq^{k-n})}q^{-nk}\\
&=\frac{(1-c)(1-cq)\cdots(1-cq^{k-n-1})(-qc^{-1})(1-cq^{-1})(-q^2c^{-1})(1-cq^{-2})\cdots(-q^nc^{-1})(1-cq^{-n})}{(-c^{-1}q^{1-k})(-c^{-1}q^{2-k})\cdots(-c^{-1}q^{n-k})}q^{-nk}\\
&=(1-cq^{-n})(1-cq^{-n+1})\cdots(1-cq^{-1})(1-c)(1-cq)\cdots(1-cq^{-n+k-1})\\
&=(cq^{-n};q)_k.
\end{aligned}\tag{2.19}$$

Now, we write identities, which can be deduced from the above two sets of identities:

$$\frac{(cq^{-m};q)_m}{(bq^{-m};q)_m}=\frac{(\frac{q}{c};q)_m}{(\frac{q}{b};q)_m}\left(\frac{c}{b}\right)^m,\tag{2.20}$$

$$\frac{(c;q)_{m-k}}{(b;q)_{m-k}}=\frac{(c;q)_m}{(b;q)_m}\frac{(q^{1-m}b^{-1};q)_k}{(q^{1-m}c^{-1};q)_k}\left(\frac{b}{c}\right)^k,\tag{2.21}$$

$$(cq^{-m};q)_{m-k}=\frac{(\frac{q}{c};q)_m}{(\frac{q}{c};q)_k}\left(\frac{-c}{q}\right)^{m-k}q^{\binom{k}{2}-\binom{m}{2}},\tag{2.22}$$

$$(c;q)_{3m}=(c,cq,cq^2;q^3)_m,\tag{2.23}$$

$$(c^3;q^3)_m=(c,c\omega,c\omega^2;q)_m,\quad \omega=e^{\frac{2\pi i}{3}}.\tag{2.24}$$

The identities (2.14), (2.15) can be generalized as

$$(cq^{jk};q)_{m-k}=\frac{(c;q)_m(cq^m;q)_{(j-1)k}}{(c;q)_{jk}}.\tag{2.25}$$

Further, the identities (3) and (2.23) can be extended as

$$(c;q)_{km}=(c,cq,\dots,cq^{k-1};q^k)_m.\tag{2.26}$$

At last, the identities (4) and (2.24) can be generalized as

$$(c^k;q^k)_m=(c,c\omega_k,\dots,c\omega_k^{k-1};q)_m,\quad \omega_k=e^{\frac{2\pi i}{k}}.\tag{2.27}$$

2.2 The q-binomial theorem

The classical form of binomial theorem says that

$$ {}_2F_1(u,v;v;z) = {}_1F_0(u,-;z) = \sum_{k=0}^{\infty} \frac{(u)_k}{k!} z^k = (1-z)^{-u}, \quad |z|<1. \tag{2.28}$$

We now derive the q-analogue of the binomial theorem.

Theorem 29. *The following formula is the q-analogue of* (2.28), [62, 99]*:*

$$ {}_2\phi_1(u,v;v;q;z) = {}_1\phi_0(u,-;q;z) = \sum_{k=0}^{\infty} \frac{(u;q)_k}{(q;q)_k} z^k = \frac{(u\,z;q)_\infty}{(z;q)_\infty}, \quad |z|<1, |q|<1, \tag{2.29}$$

where ${}_2\phi_1(u,v;v;q;z)$ *and* ${}_1\phi_0(u,-;q;z)$ *are the q-analogues of their classical ones, defined by*

$$ {}_2\phi_1(u,v;w;q;z) = \sum_{k=0}^{\infty} \frac{(u;q)_k(v;q)_k}{(q;q)_k(w;q)_k} z^k \tag{2.30}$$

and

$$ {}_1\phi_0(u;-;q;z) = \sum_{k=0}^{\infty} \frac{(u;q)_k\, z^k}{(q;q)_k}. \tag{2.31}$$

Proof. Let us say

$$ f_u(z) = \sum_{k=0}^{\infty} \frac{(u;q)_k}{(q;q)_k} z^k, \quad f_u(0)=1. \tag{2.32}$$

This gives

$$\begin{aligned} f_u(z) - f_{qu}(z) &= \sum_{k=1}^{\infty} \frac{(u;q)_k \quad (qu;q)_k}{(q;q)_k} z^k \\ &= \sum_{k=1}^{\infty} \frac{(uq;q)_{k-1}(1-u-(1-uq^k))}{(q;q)_k} z^k \\ &= -u \sum_{k=1}^{\infty} \frac{(uq;q)_{k-1}}{(q;q)_{k-1}} z^k \\ &= -u\,z \sum_{k=0}^{\infty} \frac{(uq;q)_k}{(q;q)_k} z^k = -u\,z\,f_{qu}(z). \end{aligned} \tag{2.33}$$

Again from (2.32),

$$
\begin{aligned}
f_u(z) - f_u(qz) &= \sum_{k=1}^{\infty} \frac{(u;q)_k}{(q;q)_k} (1-q^k) z^k \\
&= \sum_{k=1}^{\infty} \frac{(u;q)_k}{(q;q)_{k-1}} z^k \\
&= z \sum_{k=0}^{\infty} \frac{(u;q)_{k+1}}{(q;q)_k} z^k \\
&= z(1-u) \sum_{k=0}^{\infty} \frac{(uq;q)_k}{(q;q)_k} z^k = z(1-u) f_{qu}(z).
\end{aligned} \tag{2.34}
$$

From (2.33) and (2.34), we get

$$
f_u(z) = \frac{1-uz}{1-z} f_u(qz). \tag{2.35}
$$

Iterating this process $(k-1)$ times gives

$$
\begin{aligned}
f_u(z) &= \frac{(1-uz)(1-uqz)\cdots(1-uq^{k-1}z)}{(1-z)(1-qz)\cdots(1-q^{k-1}z)} f_u(q^k z) \\
&= \frac{(uz;q)_k}{(z;q)_k} f_u(q^k z).
\end{aligned} \tag{2.36}
$$

Taking limit as $k \to \infty$, we obtain

$$
f_u(z) = \frac{(uz;q)_\infty}{(z;q)_\infty} f_u(0) = \frac{(uz;q)_\infty}{(z;q)_\infty}, \quad \text{as } f_u(0) = 1. \tag{2.37}
$$

Thus, we derive

$$
\sum_{k=0}^{\infty} \frac{(u;q)_k}{(q;q)_k} z^k = \frac{(uz;q)_\infty}{(z;q)_\infty}, \quad |z| < 1, |q| < 1. \tag{2.38}
$$

□

It completes the proof.

In a special case if $u = q^{-k}$, $k = 0, 1, 2, \ldots$, then we get

$$
{}_1\phi_0(q^{-k}, -; q; z) = \frac{(q^{-k} z;q)_\infty}{(z;q)_\infty} = (-z)^k q^{-\frac{k(k+1)}{2}} \left(\frac{q}{z};q\right)_k. \tag{2.39}
$$

The q-analogue of two binomial expansions $(1+y)^m$ and $(1-y)^{-m}$ are given by

$$
(1+y)_q^m = \sum_{j=0}^{m} q^{j(j-1)/2} \binom{m}{j}_q y^j, \tag{2.40}
$$

$$
\frac{1}{(1-y)_q^m} = \sum_{j=0}^{m} \frac{[m]_q [m+1]_q \cdots [m+j-1]_q}{[j]_q!} y^j, \tag{2.41}
$$

where $\binom{m}{j}_q = \frac{[m]_q!}{[j]_q!\,[m-j]_q!}$, which has

$$\lim_{m\to\infty}\binom{m}{j}_q = \lim_{m\to\infty}\frac{(1-q^m)(1-q^{m-1})\cdots(1-q^{m-j+1})}{(1-q)(1-q^2)\cdots(1-q^j)} = \frac{1}{(1-q)(1-q^2)\cdots(1-q^j)}. \tag{2.42}$$

Taking the limit as $m\to\infty$ in (2.40) and (2.41) and using (2.42), we have

$$(1+y)_q^\infty = \sum_{j=0}^\infty q^{j(j-1)/2}\frac{y^j}{(1-q)(1-q^2)\cdots(1-q^j)} = \sum_{j=0}^\infty \frac{q^{\binom{j}{2}}(\frac{y}{1-q})^j}{[j]_q!}, \tag{2.43}$$

$$\frac{1}{(1-y)_q^\infty} = \sum_{j=0}^\infty \frac{y^j}{(1-q)(1-q^2)\cdots(1-q^j)} = \sum_{j=0}^\infty \frac{(\frac{y}{1-q})^j}{[j]_q!}. \tag{2.44}$$

A q-analogue of the classical exponential function e^y is

$$e_q^y = \sum_{j=0}^\infty \frac{(y)^j}{[j]_q!}. \tag{2.45}$$

Another q-analogue of the exponential function is

$$E_q^y = \sum_{j=0}^\infty \frac{q^{\binom{j}{2}}(y)^j}{[j]_q!}. \tag{2.46}$$

In view of the (2.43) and (2.44), equations (2.45) and (2.46) can be rewritten in the form

$$e_q^{\frac{y}{1-q}} = \sum_{j=0}^\infty \frac{(\frac{y}{1-q})^j}{[j]_q!} = \frac{1}{(1-y)_q^\infty} \tag{2.47}$$

and

$$E_q^{\frac{y}{1-q}} = \sum_{j=0}^\infty \frac{(q^{\binom{j}{2}}\frac{y}{1-q})^j}{[j]_q!} = (1+y)_q^\infty. \tag{2.48}$$

Also,

$$e_q^y = \frac{1}{(1-(1-q)y)_q^\infty} \tag{2.49}$$

and

$$E_q^y = (1+(1-q)y)_q^\infty. \tag{2.50}$$

From (2.49) and (2.50), one can deduce directly that

$$e_q^y E_q^{-y} = e_q^{-y} E_q^y = 1. \tag{2.51}$$

2.3 The q-calculus

The forward and backward q-difference operators, denoted by Δ_q and ∇_q, are defined by [99]

$$\Delta_q f(z) = f(qz) - f(z), \quad \nabla_q f(z) = f(q^{-1}z) - f(z). \tag{2.52}$$

The symmetric q-difference operator δ_q is defined by

$$\delta_q f(z) = f(zq^{1/2}) - f(zq^{-1/2}). \tag{2.53}$$

The q-derivative is defined by

$$D_q f(z) = \frac{\Delta_q f(z)}{\Delta(z)} = \frac{f(qz) - f(z)}{(q-1)z}. \tag{2.54}$$

Note that $\lim_{q\to 1} D_q f(x) = \frac{d}{dx} f(x)$, provided the function $f(x)$ is differentiable.

If T_q denotes the q-dilation operator and defined by $T_q f(x) = f(qx)$, then the q-derivative can also be written as

$$D_q f(z) = \frac{f(qz) - f(z)}{(q-1)z} = \frac{(T_q - 1)f(z)}{(q-1)z}. \tag{2.55}$$

The q-derivative operator D_q is a linear operator. In other words, for any constants a and b,

$$D_q(af(x) \pm bg(x)) = aD_q f(x) \pm bD_q g(x). \tag{2.56}$$

Example 30. The q-derivative of $f(x) = x^n$, where n is a positive integer, is given by

$$D_q x^n = \frac{(qx)^n - x^n}{(q-1)x} = \frac{1-q^n}{1-q} x^{n-1} = [n]_q x^{n-1}. \tag{2.57}$$

For the q-derivative of the product and quotient of two functions $f(x)$ and $g(x)$ can be determined as

$$\begin{aligned} D_q(f(x).g(x)) &= \frac{\Delta_q[f(x).g(x)]}{(q-1)x} \\ &= \frac{f(qx).g(qx) - f(x).g(x)}{(q-1)x} \end{aligned}$$

$$= \frac{f(qx).g(qx) - f(qx).g(x) + f(qx).g(x) - f(x).g(x)}{(q-1)x}$$
$$= \frac{f(qx)[g(qx) - g(x)] + g(x)[f(qx) - f(x)]}{(q-1)x}$$
$$= \frac{f(qx)\Delta_q g(x) + g(x)\Delta_q f(x)}{(q-1)x}$$
$$= f(qx)\, D_q\, g(x) + g(x)\, D_q\, f(x). \tag{2.58}$$

By symmetry, we can interchange f and g,

$$D_q(f(x).g(x)) = f(x)\, D_q\, g(x) + g(qx)\, D_q\, f(x). \tag{2.59}$$

For the q-derivative of the quotient of $f(x)$ and $g(x)$, write $f(x) = g(x)\frac{f(x)}{g(x)}$ and apply the q-derivative formula (2.58), we get

$$\begin{aligned} D_q f(x) &= D_q \left[g(x)\frac{f(x)}{g(x)} \right] \\ &= g(qx)\, D_q\left(\frac{f(x)}{g(x)}\right) + \frac{f(x)}{g(x)} D_q g(x). \end{aligned} \tag{2.60}$$

Rearranging the terms of (2.60), we have

$$D_q\left(\frac{f(x)}{g(x)}\right) = \frac{g(x)\, D_q f(x) - f(x)\, D_q\, g(x)}{g(x)\, g(qx)}. \tag{2.61}$$

If we use (2.59) instead of (2.58), we get

$$D_q\left(\frac{f(x)}{g(x)}\right) = \frac{g(qx)\, D_q f(x) - f(qx)\, D_q\, g(x)}{g(x)\, g(qx)}. \tag{2.62}$$

If $|f(x)x^{\alpha}|$ is bounded on interval $(0, A]$ for some $0 \leq \alpha < 1$ and $0 < q < 1$, then there exists a continuous function $F(x)$, such that $F(0) = 0$ is a q-antiderivative of $f(x)$. In other words, the function $F(x)$ is a q-antiderivative of $f(x)$ if $D_q F(x) = f(x)$. It is denoted by

$$\int f(x)\, d_q x.$$

Let us find the expression for $\int f(x) d_q x$. The expression

$$D_q F(x) = \frac{F(qx) - F(x)}{(q-1)x} = f(x) \tag{2.63}$$

implies

$$\frac{T_q F(x) - F(x)}{(q-1)x} = f(x) \tag{2.64}$$

and

$$F(x) = \frac{(1-q)xf(x)}{1-T_q} = (1-q)(1-T_q)^{-1}xf(x)$$
$$= (1-q)\sum_{j=0}^{\infty} T_q^j xf(x)$$
$$= (1-q)\,x\sum_{j=0}^{\infty} q^j f(q^j x). \tag{2.65}$$

Thus, we obtain

$$\int f(x)d_q x = (1-q)\,x\sum_{j=0}^{\infty} q^j f(q^j x). \tag{2.66}$$

The series in (2.66) is called the Jackson's integral, which can be generalized as

$$\int f(x)d_q g(x) = \sum_{j=0}^{\infty} f(q^j x)(g(q^j x) - g(q^{j+1}x)). \tag{2.67}$$

Now, we apply the Jackson integral (2.66) to define the definite q-integral. Suppose $0 < a < b$, then the definite q-integral is defined as

$$\int_0^b f(x)d_q(x) = (1-q)b\sum_{j=0}^{\infty} q^j f(q^j b) \tag{2.68}$$

and

$$\int_a^b f(x)d_q(x) = \int_0^b f(x)d_q(x) - \int_0^a f(x)d_q(x). \tag{2.69}$$

In general,

$$\int_0^b f(x)d_q g(x) = \sum_{j=0}^{\infty} f(q^j b)(g(q^j b) - g(q^{j+1}b)). \tag{2.70}$$

The improper q-integral of $f(x)$ on $[0,\infty)$ is defined as

$$\int_0^{\infty} f(x)d_q(x) = \begin{cases} \sum_{j=-\infty}^{\infty} \int_{q^{j+1}}^{q^j} f(x)d_q x, & \text{if } 0 < q < 1, \\ \sum_{j=-\infty}^{\infty} \int_{q^j}^{q^{j+1}} f(x)d_q x, & \text{if } q > 1. \end{cases} \tag{2.71}$$

The fundamental theorem of q-calculus says if $F(x)$ is an q-antiderivative of $f(x)$ and $F(x)$ is continuous at $x = 0$, then

$$\int_a^b f(x)d_qx = F(b) - F(a), \quad 0 \le a \le b < \infty. \tag{2.72}$$

The product rule of q-derivative of a function allows to write

$$\int_a^b D_q(u)\{f(u)\,g(u)\} = \int_a^b f(u)\,D_q(u)\{g(u)\}\,d_q(u) + \int_a^b D_q(u)\{f(u)\}\,g(qu)\,d_q(u); \tag{2.73}$$

Or

$$f(b)\,g(b) - f(a)\,g(a) = \int_a^b f(u)\,d_qg(u) + \int_a^b g(qu)\,d_qf(u). \tag{2.74}$$

Which gives the by part rule of q-integration as

$$\int_a^b f(u)\,d_qg(u) = f(b)\,g(b) - f(a)\,g(a) - \int_a^b g(qu)\,d_qf(u). \tag{2.75}$$

2.4 The q-gamma and q-beta function

Heine in 1847 defined the q-gamma function for a nonnegative integer n by

$$\Gamma_q(n+1) = 1(1+q)(1+q+q^2)\cdots(1+q+\cdots+q^{n-1}), \tag{2.76}$$

while approaching $q \to 1$,

$$\begin{aligned}\lim_{q\to 1^-} \Gamma_q(n+1) &= \lim_{h\to 0} \Gamma_{1-h}(n+1) = 1(1+1)(1+1+1)\cdots(1+1+\cdots+1)\\ &= 1\cdot 2\cdots n = n! = \Gamma(n+1).\end{aligned} \tag{2.77}$$

This definition of the q-gamma function is extended by Jackson [1904e] for $|q| < 1$ and $x \in \mathbb{C}$ as

$$\Gamma_q(x) = \frac{(q;q)_\infty}{(q^x;q)_\infty}(1-q)^{1-x}, \quad 0 < q < 1. \tag{2.78}$$

Lemma 31. *Let x be any complex number and $|q| < 1$. Then the q-gamma function defined in* (2.78) *approaches to an ordinary gamma function when $q \to 1^-$, i. e.,*

$$\lim_{q\to 1^-} \Gamma_q(x) = \Gamma(x). \tag{2.79}$$

Proof. From the definition of the q-gamma function,

$$\begin{aligned}\Gamma_q(x+1) &= \frac{(q;q)_\infty}{(q^{x+1};q)_\infty}(1-q)^{-x} \\ &= \frac{\prod_{n=1}^{\infty}(1-q^n)}{\prod_{n=0}^{\infty}(1-q^{x+n+1})}(1-q)^{-x} \\ &= \prod_{n=1}^{\infty}\frac{(1-q^n)}{(1-q^{x+n})}\frac{(1-q^{n+1})^x}{(1-q^n)^x}. \end{aligned} \tag{2.80}$$

Taking the limit as $q \to 1^-$, we get

$$\begin{aligned}\lim_{q\to 1^-}\Gamma_q(x+1) &= \prod_{n=1}^{\infty}\frac{n}{(x+n)}\left(\frac{n+1}{n}\right)^x \\ &= x\left[x^{-1}\prod_{n=1}^{\infty}\left(1+\frac{x}{n}\right)^{-1}\left(1+\frac{1}{n}\right)^x\right]. \end{aligned} \tag{2.81}$$

Using the Weirstrass definition of the gamma function, one can easily get $\lim_{q\to 1^-}\Gamma_q(x+1) = x\Gamma(x) = \Gamma(x+1)$. It completes the proof. □

Like the ordinary gamma function, the q-gamma function also satisfies the fundamental identity presented in the following lemma.

Lemma 32. $\Gamma_q(x+1) = [x]_q\,\Gamma_q(x)$.

Proof. From the definition of q-gamma function,

$$\begin{aligned}\Gamma_q(x+1) &= \frac{(q;q)_\infty}{(q^{x+1};q)_\infty}(1-q)^{-x} \\ &= \frac{(q;q)_\infty(1-q)^{-x+1}(1-q^x)}{(q^x;q)_\infty(1-q)} \\ &= \frac{(q;q)_\infty(1-q)^{-x+1}}{(q^x;q)_\infty}\frac{1-q^x}{1-q} = \Gamma_q(x)\,[x]_q. \end{aligned} \tag{2.82}$$

It completes the proof. □

The q-gamma function has no zeroes. It has only poles as singularities at $x = 0, -1, -2, \ldots$. The residue at $x = -n$ is

$$\begin{aligned}\lim_{x\to -n}(x+n)\Gamma_q(x) &= \frac{(1-q)^{n+1}}{(1-q^{-n})(1-q^{-n+1})\cdots(1-q^{-1})}\lim_{x\to -n}\frac{(x+n)}{1-q^{x+n}} \\ &= \frac{(1-q)^{n+1}}{(q^{-n};q)_n(\log q^{-1})}. \end{aligned} \tag{2.83}$$

Since the q-gamma has only poles at $x = 0, -1, -2, \ldots$, its reciprocal is an entire function with zeroes at $x = 0, -1, -2, \ldots$ defined by

$$\frac{1}{\Gamma_q(x)} = (1-q)^{x-1} \prod_{n-0}^{\infty} \frac{1-q^{n+x}}{1-q^{n+1}}. \tag{2.84}$$

Using the functional identity of the q-gamma function, we have

$$\Gamma_q(x+2) = [x+1]_q \Gamma_q(x+1) = [x]_q \, [x+1]_q \, \Gamma_q(x). \tag{2.85}$$

Repeating this process, we get

$$\Gamma_q(x+n) = [x]_q \, [x+1]_q \cdots \Gamma_q(x+n-1)\Gamma_q(x). \tag{2.86}$$

This implies

$$\Gamma_q(x+n)\, \Gamma_q^{-1}(x) = [x]_q \, [x+1]_q \cdots \Gamma_q(x+n-1). \tag{2.87}$$

Using the definition of a q-shifted factorial function, it can be written as

$$(q^x; q)_n = (1-q)^n \frac{\Gamma_q(x+n)}{\Gamma_q(x)}, \qquad n = 1, 2, \ldots. \tag{2.88}$$

One of the natural ways to define the q-beta function, for $x, y \in \mathbb{C}$ is

$$\beta_q(x,y) = \frac{\Gamma_q(x)\, \Gamma_q(y)}{\Gamma_q(x+y)}. \tag{2.89}$$

Using the definition of the q-gamma function, we can rewrite the q-beta function as

$$\begin{aligned} \beta_q(x,y) &= \frac{(q;q)_\infty (1-q)^{1-x} (q;q)_\infty (1-q)^{1-y} (q^{x+y};q)_\infty}{(q^x;q)_\infty (q^y;q)_\infty (q;q)_\infty (1-q)^{1-x-y}} \\ &= \frac{(1-q)(q;q)_\infty (q^{x+y};q)_\infty}{(q^x;q)_\infty (q^y;q)_\infty} \\ &= \frac{(1-q)(q;q)_\infty}{(q^y;q)_\infty} \sum_{n=0}^{\infty} \frac{(q^y;q)_n}{(q;q)_n} q^{nx} \\ &= (1-q) \sum_{n=0}^{\infty} \frac{(q^{n+1};q)_\infty}{(q^{y+n};q)_\infty} q^{nx}. \end{aligned} \tag{2.90}$$

The q-integral representation of the q-gamma function is given by

$$\Gamma_q(x) = \int_0^{\frac{1}{1-q}} t^{x-1} E_q(-qt)\, d_q t, \qquad \Re(x) > 0. \tag{2.91}$$

The q-beta function can be represented in the form of the q-integral as

$$\beta_q(x,y) = \int_0^1 (tq;q)_\infty \,(tq^y;q)_\infty^{-1}\, t^{x-1} d_q t, \tag{2.92}$$

where $q^{-n} \neq (q^x), (q^y)$, for all integers $n \geq 0$.

2.5 Basic Gauss hypergeometric function

Now, let us dive into the main content of the chapter. As we discussed earlier, the basic idea of getting the q-analogue of an ordinary function is that it must reach its ordinary one when we apply limit on q. Following this idea, Heine introduced the series on the base q as

$$1 + \frac{(1-q^u)(1-q^v)}{(1-q)(1-q^w)} + \frac{(1-q^u)(1-q^{u+1})(1-q^v)(1-q^{v+1})}{(1-q)(1-q^2)(1-q^w)(1-q^{w+1})} + \cdots, \tag{2.93}$$

with the assumption that $q \neq 1$, $w \neq 0, -1, \ldots$ and considered this as the q-analogue of Gauss series because it meets with the Gauss series when $q \to 1$. Usually, this series is denoted by

$$\phi(u,v;w;q;z) \quad \text{or} \quad {}_2\phi_1(u,v;w;q;z) \quad \text{or} \quad {}_2\phi_1\left[\begin{matrix} u,v \\ w \end{matrix};q;z\right], \tag{2.94}$$

and called the Heine series, basic hypergeometric series or q-series. So, we have

$${}_2\phi_1(u,v;w;q;z) = \sum_{n=0}^{\infty} \frac{(u;q)_n (v;q)_n}{(q;q)_n (w;q)_n} z^n. \tag{2.95}$$

The series (2.95) converges absolutely for $|z| < 1$ when $|q| < 1$. The proof is simple and straightforward, and one can apply the ratio test

$$\begin{aligned}
\lim_{n\to\infty} \left| \frac{U_{n+1}}{U_n} \right| &= \lim_{n\to\infty} \left| \frac{(u;q)_{n+1}(v;q)_{n+1}}{(q;q)_{n+1}(w;q)_{n+1}} z^{n+1} \frac{(q;q)_n (w;q)_n}{(u;q)_n (v;q)_n z^n} \right| \\
&= \lim_{n\to\infty} \left| \frac{(1-uq^n)(1-vq^n)}{(1-q^{n+1})(1-wq^n)} z \right| \\
&\leq \lim_{n\to\infty} \frac{(1+|u|\,|q|^n)(1+|v|\,|q|^n)}{(1-|q|^{n+1})(1-|w|\,|q|^n)} |z|.
\end{aligned} \tag{2.96}$$

Thus, $\lim_{n\to\infty} |\frac{U_{n+1}}{U_n}| \to 0$ as $n \to \infty$ for $|q| < 1$ and $|z| < 1$. When $|q| > 1$, let $q^{-1} = p$ such that $|p| < 1$. Then the series defined in (2.93) can be written as

$$\begin{aligned}
&{}_2\phi_1(u,v;w;p^{-1};z) \\
&= 1 + \frac{(1-p^{-u})(1-p^{-v})}{(1-p^{-w})(1-p^{-1})} z + \frac{(1-p^{-u})(1-p^{-u-1})(1-p^{-v})(1-p^{-v-1})}{(1-p^{-w})(1-p^{-w-1})(1-p^{-1})(1-p^{-2})} z^2 + \cdots
\end{aligned}$$

$$= 1 + \frac{(1-p^u)(1-p^v)}{(1-p^w)(1-p)} z\, p^{1+w-u-v} + \frac{(1-p^u)(1-p^{u+1})(1-p^v)(1-p^{v+1})}{(1-p^w)(1-p^{w+1})(1-p)(1-p^2)}$$
$$\times z^2\, p^{2(1+w-u-v)} + \cdots, \tag{2.97}$$

which is again a series of the form (2.93) with p in place of q and $z\,p^{(1+w-u-v)}$ in place of z. Hence, the series converges absolutely for $|z\,p^{(1+w-u-v)}| < 1$ and $|p| < 1$. This implies $|z| < 1$. Thus, the process of inversion for q^{-1} to p will reduce the Heine series into a similar series with $|p| < 1$. So, without loss of generality, we always consider $|q| < 1$ until stated.

It can be easy to verify that the basic Gauss hypergeometric function $\mathfrak{w}(z) = {}_2\phi_1(u, v; w; q; z)$, for $|z| < 1$, satisfies the following second-order q-difference equation:

$$z(w - uvqz)D_q^2\mathfrak{w} + \left[\frac{1-w}{1-q} + \frac{(1-u)(1-v) - (1-uvq)}{1-q} z\right] D_q\mathfrak{w} - \frac{(1-u)(1-v)}{(1-q)^2}\mathfrak{w} = 0. \tag{2.98}$$

If we replace u, v, w by q^u, q^v, q^w, respectively, then the difference equation (2.98) can be rewritten as

$$z\,q^w\, D_q^2\mathfrak{w}(z) - z^2\, q^{u+v+1}\, D_q^2\mathfrak{w}(z) - z\,[u]_q\, q^v\, D_q\mathfrak{w}(z) + [w]_q\, D_q\mathfrak{w}(z)$$
$$- z\,q^u\,[v+1]_q\, D_q\mathfrak{w}(z) - [u]_q\,[v]_q\,\mathfrak{w}(z) = 0. \tag{2.99}$$

Further, if $\delta_q(z) = z\,D_q(z)$, then alternate form of the equation (2.99) is

$$[\delta_q(q^{w-1}\,\delta_q + [w-1]_q) - z\,(q^u\delta_q + [u]_q)\,(q^v\,\delta_q + [v]_q)]\mathfrak{w}(z) = 0. \tag{2.100}$$

Letting $q \to 1^-$ the q-difference equations (2.99) and (2.100) satisfied by basic Gauss hypergeometric function tends to the second-order differential relation

$$z(1-z)\mathfrak{w}'' + [w - (u+v+1)z]\mathfrak{w}' - u\,v\mathfrak{w} = 0, \tag{2.101}$$

for the hypergeometric function $\mathfrak{w}(z) = {}_2F_1(u, v; w; z)$, where $|z| < 1$. In the next theorem, we find the possible solutions of the q-difference equation (2.99) in the region $|z| < 1$.

Theorem 33. *For $|z| < 1$ and $|q| < 1$, the q-difference equation,*

$$z\,q^w\, D_q^2\mathfrak{w}(z) - z^2\, q^{u+v+1}\, D_q^2\mathfrak{w}(z) - z\,[u]_q\, q^v\, D_q\mathfrak{w}(z) + [w]_q\, D_q\mathfrak{w}(z)$$
$$- z\,q^u\,[v+1]_q\, D_q\mathfrak{w}(z) - [u]_q\,[v]_q\,\mathfrak{w}(z) = 0, \tag{2.102}$$

has

$$\mathfrak{w}_1(z) = {}_2\phi_1(q^u, q^v; q^w; q; z) \tag{2.103}$$

and

$$\mathfrak{w}_2(z) = z^{1-w}{}_2\phi_1(q^{u+w-1}, q^{v+w-1}; q^{2-w}; q; z) \tag{2.104}$$

as solutions with $\mathfrak{w}_1(0) = {}_2\phi_1(q^u, q^v; q^w; q; 0) = 1$.

Proof. To find the solution, we use the well-nown Frobenius method. Consider

$$\mathfrak{w}_1(z) = \sum_{k=0}^{\infty} a_k z^k, \tag{2.105}$$

as one of the solutions. Taking the q-derivatives, we get

$$\begin{aligned} \mathfrak{w}_1'(z) &= \sum_{k=1}^{\infty} [k]_q a_k z^{k-1}; \\ \mathfrak{w}_1''(z) &= \sum_{k=2}^{\infty} [k]_q [k-1]_q a_k z^{k-2}. \end{aligned} \tag{2.106}$$

Substituting these values in (2.102), we have

$$\begin{aligned} q^w \sum_{k=2}^{\infty} [k]_q [k-1]_q a_k z^{k-1} - q^{u+v+1} \sum_{k=2}^{\infty} [k]_q [k-1]_q a_k z^k - [u]_q q^v \sum_{k=1}^{\infty} [k]_q a_k z^k \\ + [w]_q \sum_{k=1}^{\infty} [k]_q a_k z^{k-1} - q^u [v+1]_q \sum_{k=1}^{\infty} [k]_q a_k z^k - [u]_q [v]_q \sum_{k=0}^{\infty} a_k z^k = 0. \end{aligned} \tag{2.107}$$

In comparing the constant term that is the coefficient of z^0, we get

$$[w]_q [1]_q a_1 - [u]_q [v]_q a_0 = 0.$$

This implies

$$a_1 = \frac{[u]_q [v]_q}{[w]_q [1]_q} a_0.$$

Similarly, comparing the coefficients of z, we have

$$\begin{aligned} & q^w [1]_q [2]_q a_2 + [2]_q [w]_q a_2 - q^v [1]_q [u]_q a_1 - q^u [1]_q [v+1]_q a_1 - [u]_q [v]_q a_1 = 0 \\ & \implies [2]_q [w+1]_q a_2 - [u]_q [v+1]_q a_1 - q^u [v+1]_q a_1 = 0 \\ & \implies [2]_q [w+1]_q a_2 = [u+1]_q [v+1]_q a_1 \\ & \implies a_2 = \frac{[u+1]_q [v+1]_q}{[2]_q [w+1]_q} a_1 = \frac{[u+1]_q [v+1]_q [u]_q [v]_q}{[1]_q [2]_q [w]_q [w+1]_q} a_0. \end{aligned}$$

In general, comparing the coefficient of z^k yields

$$q^w [k]_q [k+1]_q a_{k+1} + [w]_q [k+1]_q a_{k+1}$$
$$= q^v [u]_q [k]_q a_k + q^u [v+1]_q [k]_q a_k q^{u+1} q^v [k]_q [k-1]_q a_k + [u]_q [v]_q a_k. \quad (2.108)$$

This can be rewritten as

$$[w+k]_q [k+1]_q a_{k+1} = [u+k]_q [v+k]_q a_k.$$

Using the above equations, we obtain

$$a_{k+1} = \frac{[u+k]_q [v+k]_q}{[w+k]_q [k+1]_q} a_k$$
$$= \frac{[u+k]_q \cdots [u]_q [v+k]_q \cdots [v]_q}{[w+k]_q \cdots [w]_q [k+1]_q \cdots [1]_q} a_0 = \frac{(q^u;q)_{k+1} (q^v;q)_{k+1}}{(q^w;q)_{k+1} (q;q)_{k+1}} a_0. \quad (2.109)$$

Hence, there exists a solution $\mathfrak{w}_1(z)$ of (2.102) defined in $|z| < 1$ satisfying $\mathfrak{w}_1(0) = 1$, given by

$$\mathfrak{w}_1(z) = {}_2\phi_1(q^u, q^v; q^w; q; z).$$

Suppose the second solution is of the form

$$\mathfrak{w}_2(z) = z^{1-w} \mathfrak{v}(z), \quad z^{1-w} = \exp((1-w) \ln z). \quad (2.110)$$

The q-derivatives are given by

$$D_q \mathfrak{w}_2(z) = [1-w]_q z^{-w} \mathfrak{v}(z) + (qz)^{1-w} d_q \mathfrak{v}(z),$$
$$D_q^2 \mathfrak{w}_2(z) = [1-w]_q [-w]_q z^{-w-1} \mathfrak{v}(z) + [1-w]_q qz^{-w} D_q \mathfrak{v}(z) + (q^2 z)^{1-w} D_q^2 \mathfrak{v}(z)$$
$$+ q^{1-w} [1-w]_q z^{-w} D_q \mathfrak{v}(z). \quad (2.111)$$

Substituting these values in (2.102) and simplification yield

$$z q^{2-w} D_q^2 \mathfrak{v}(z) - z^2 q^{u+1-w+1} q^{v+1-w} D_q^2 \mathfrak{v}(z) - z [u+1-w]_q q^{v+1-w} D_q \mathfrak{v}(z) + [2-w]_q D_q \mathfrak{v}(z)$$
$$- z q^{u+1-w} [v+1-w+1]_q D_q \mathfrak{v}(z) - [u+1-w]_q [v+1-w]_q \mathfrak{v}(z) = 0, \quad (2.112)$$

which is a q-difference equation of the type (2.102) with parameters

$$\tilde{u} = u+1-w, \quad \tilde{v} = v+1-w, \quad \tilde{w} = 2-w.$$

Hence,

$$\mathfrak{v}(z) = {}_2\phi_1(q^{u+1-w}, q^{v+1-w}; q^{2-w}; q; z), \quad |z| < 1$$

is a solution of (2.112). Thus, we have two fundamental solutions of the q-hypergeometric equation (2.102) as

$$\mathfrak{w}_1(z) = {}_2\phi_1(q^u, q^v; q^w; q; z);$$
$$\mathfrak{w}_2(z) = z^{1-w}{}_2\phi_1(q^{u+w-1}, q^{v+w-1}; q^{2-w}; q; z), \quad |z| < 1.$$

Any solution $\mathfrak{w}(z)$ of (2.102) can be written as a combination of $\mathfrak{w}_1(z)$ and $\mathfrak{w}_2(z)$,

$$\mathfrak{w}(z) = c_1\,{}_2\phi_1(q^u, q^v; q^w; q; z) + c_2\, z^{1-w}{}_2\phi_1(q^{u+w-1}, q^{v+w-1}; q^{2-w}; q; z), \tag{2.113}$$

for some arbitrary constants c_1 and c_2. □

2.6 The q-integral representation

For complex numbers u, v, w such that $w \neq 0, -1, -2, \ldots$. The basic Gauss hypergeometric function can be put in the following q-integral:

$${}_2\phi_1(q^u, q^v; q^w; q; z) = \frac{\Gamma_q(w)}{\Gamma_q(v)\Gamma_q(w-v)} \int_0^1 t^{v-1} \frac{(tzq^u, tq; q)_\infty}{(tz, tq^{w-v}; q)_\infty} d_q t. \tag{2.114}$$

Proof. To get the q-integral for basic Gauss hypergeometric series, consider

$$\begin{aligned}\frac{(q^v; q)_n}{(q^w; q)_n} &= \frac{(1-q)^n\,\Gamma_q(v+n)\,\Gamma_q(w)}{\Gamma_q(v)\,(1-q)^n\,\Gamma_q(w+n)} \\ &= \frac{\Gamma_q(v+n)\,\Gamma_q(w)\,\Gamma_q(w-v)}{\Gamma_q(v)\,\Gamma_q(w+n)\,\Gamma_q(w-v)} \\ &= \frac{\Gamma_q(w)}{\Gamma_q(v)\,\Gamma_q(w-v)}\,\beta_q(v+n, w-v).\end{aligned} \tag{2.115}$$

Now, using the q-integral representation of q-beta function

$$\frac{(q^v; q)_n}{(q^w; q)_n} = \frac{\Gamma_q(w)}{\Gamma_q(v)\,\Gamma_q(w-v)} \int_0^1 t^{v+n-1} \frac{(tq; q)_\infty}{(t\,q^{w-v}; q)_\infty} d_q t. \tag{2.116}$$

Substituting this into the Heine series, one can get

$$\begin{aligned}{}_2\phi_1(q^u, q^v; q^w; q; z) &= \frac{\Gamma_q(w)}{\Gamma_q(v)\,\Gamma_q(w-v)} \sum_{n=0}^{\infty} (q^u; q)_n \int_0^1 t^{v+n-1} \frac{(tq; q)_\infty}{(t\,q^{w-v}; q)_\infty} \frac{z^n}{(q; q)_n} d_q t \\ &= \frac{\Gamma_q(w)}{\Gamma_q(v)\,\Gamma_q(w-v)} \sum_{n=0}^{\infty} \int_0^1 t^{v-1} \frac{(tq; q)_\infty}{(t\,q^{w-v}; q)_\infty} \frac{(q^u; q)_n\,(zt)^n}{(q; q)_n} d_q t.\end{aligned}$$

Interchange of summation and integration leads to

$$ {}_2\phi_1(q^u, q^v; q^w; q; z) = \frac{\Gamma_q(w)}{\Gamma_q(v)\Gamma_q(w-v)} \int_0^1 t^{v-1} \frac{(tzq^u, tq; q)_\infty}{(tz, tq^{w-v}; q)_\infty} d_q t. \tag{2.117} $$

□

It completes the proof.

Letting $q \to 1^-$, the q-integral (2.114) gives the Euler's integral representation of the Gauss hypergeometric function ${}_2F_1(u, v; w; z)$.

2.7 The q-contiguous and q-recursion relations

The following abbreviations will be helpful in the finding of q-contiguous function relations satisfied by the q-Gauss hypergeometric function, [130]:

$$ \phi = {}_2\phi_1(q^u, q^v; q^w; q; z); \tag{2.118} $$

$$ \phi(u\pm) = {}_2\phi_1(q^{u\pm1}, q^v; q^w; q; z); \tag{2.119} $$

$$ \phi(v\pm) = {}_2\phi_1(q^u, q^{v\pm1}; q^w; q; z); \tag{2.120} $$

$$ \phi(w\pm) = {}_2\phi_1(q^u, q^v; q^{w\pm1}; q; z). \tag{2.121} $$

Hence,

$$ \begin{aligned} \phi(u+) &= {}_2\phi_1(q^{u+1}, q^v; q^w; q; z) \\ &= \sum_{k=0}^{\infty} \frac{(q^{u+1}; q)_k\,(q^v; q)_k}{(q; q)_k\,(q^w; q)_k} z^k \\ &= \sum_{k=0}^{\infty} \frac{1-q^{u+k}}{1-q^u} \frac{(q^u; q)_k\,(q^v; q)_k}{(q; q)_k\,(q^w; q)_k} z^k. \end{aligned} \tag{2.122} $$

Similarly,

$$ \phi(u-) = \sum_{k=0}^{\infty} \frac{1-q^u}{1-q^{u+k-1}} \frac{(q^u; q)_k\,(q^v; q)_k}{(q; q)_k\,(q^w; q)_k} z^k; \tag{2.123} $$

$$ \phi(v+) = \sum_{k=0}^{\infty} \frac{1-q^{v+k}}{1-q^v} \frac{(q^u; q)_k\,(q^v; q)_k}{(q; q)_k\,(q^w; q)_k} z^k; \tag{2.124} $$

$$ \phi(v-) = \sum_{k=0}^{\infty} \frac{1-q^v}{1-q^{v+k-1}} \frac{(q^u; q)_k\,(q^v; q)_k}{(q; q)_k\,(q^w; q)_k} z^k; \tag{2.125} $$

$$ \phi(w+) = \sum_{k=0}^{\infty} \frac{1-q^w}{1-q^{w+k}} \frac{(q^u; q)_k\,(q^v; q)_k}{(q; q)_k\,(q^w; q)_k} z^k; \tag{2.126} $$

$$ \phi(w-) = \sum_{k=0}^{\infty} \frac{1-q^{w+k-1}}{1-q^{w-1}} \frac{(q^u; q)_k\,(q^v; q)_k}{(q; q)_k\,(q^w; q)_k} z^k. \tag{2.127} $$

If T_z denotes the q-shifted operator defined by $T_z f(z) = f(qz)$, then

$$(1 - q^u T_z)\phi = \sum_{k=0}^{\infty}(1 - q^{u+k})\frac{(q^u;q)_k (q^v;q)_k}{(q;q)_k (q^w;q)_k} z^k = (1 - q^u)\phi(u+). \tag{2.128}$$

Also,

$$(1 - q^v T_z)\phi = (1 - q^v)\phi(v+); \tag{2.129}$$

$$(1 - q^{w-1} T_z)\phi = (1 - q^{w-1})\phi(w-). \tag{2.130}$$

Eliminating T_z from (2.128) and (2.129) gives

$$(q^v - q^u)\phi = q^v(1 - q^u)\phi(u+) - q^u(1 - q^v)\phi(v+),$$

which can also be written as

$$([u]_q - [v]_q)\phi = q^v [u]_q \phi(u+) - q^u [v]_q \phi(v+).$$

In the same way, removing T_z from (2.128) and (2.130) yields

$$(q^{w-1} - q^u)\phi = q^{w-1}(1 - q^u)\phi(u+) - q^u(1 - q^{w-1})\phi(w-)$$

or

$$([u]_q - [w-1]_q)\phi = q^{w-1}[u]_q \phi(u+) - q^u [w-1]_q \phi(w-).$$

In the above discussed manner, one can determine more contiguous relations obeyed by the q-Gauss hypergeometric function. Now, we determine recursion relations in terms of the q-Gauss function.

Theorem 34. *The following recursion relations are held by the basic Gauss hypergeometric function:*

$${}_2\phi_1\left[\begin{matrix} uq^r, v \\ w \end{matrix}; q; z\right] = {}_2\phi_1\left[\begin{matrix} u, v \\ w \end{matrix}; q; z\right] + uz\frac{(v;q)_1}{(w;q)_1}\sum_{m=1}^{r} q^{m-1}\,{}_2\phi_1\left[\begin{matrix} q^m u, qv \\ qw \end{matrix}; q; z\right]; \tag{2.131}$$

$${}_2\phi_1\left[\begin{matrix} uq^{-r}, v \\ qw \end{matrix}; q; z\right] = {}_2\phi_1\left[\begin{matrix} u, v \\ w \end{matrix}; q; z\right] - uz\frac{(v;q)_1}{(w;q)_1}\sum_{m=1}^{r} q^{-m}\,{}_2\phi_1\left[\begin{matrix} q^{-m+1}u, qv \\ qw \end{matrix}; q; z\right]. \tag{2.132}$$

Proof. From the definition of the q-shifted factorial, we have

$$\begin{aligned}(uq;q)_k &= (1 - uq)(1 - uq^2)\cdots(1 - uq^k) \\ &= \frac{1}{(1-u)}(1 - u)(1 - uq)\cdots(1 - uq^k)\end{aligned}$$

$$= (u;q)_k \left[1 + \frac{u(1-q^k)}{1-u}\right]. \tag{2.133}$$

Using (2.133) in (2.95), we get

$$\begin{aligned}
{}_2\phi_1\left[\begin{matrix} qu, v \\ w \end{matrix}; q; z\right] &= \sum_{k=0}^{\infty} \frac{(qu;q)_k (v;q)_k}{(q;q)_k (w;q)_k} z^k \\
&= \sum_{k=0}^{\infty} \frac{(v;q)_k z^k}{(q;q)_k (w;q)_k} (u;q)_k \left[1 + \frac{u(1-q^k)}{1-u}\right] \\
&= \sum_{k=0}^{\infty} \frac{(u;q)_k (v;q)_k z^k}{(q;q)_k (w;q)_k} + \frac{u}{1-u} \sum_{k=1}^{\infty} \frac{(u;q)_k (v;q)_k z^k}{(q;q)_{k-1} (w;q)_k} \\
&= {}_2\phi_1\left[\begin{matrix} u, v \\ w \end{matrix}; q; z\right] + \frac{u}{1-u} \sum_{k=0}^{\infty} \frac{(u;q)_{k+1} (v;q)_{k+1} z^{k+1}}{(q;q)_k (w;q)_{k+1}} \\
&= {}_2\phi_1\left[\begin{matrix} u, v \\ w \end{matrix}; q; z\right] + \frac{z\,u(1-v)}{1-w} \sum_{k=0}^{\infty} \frac{(qu;q)_k (qv;q)_k z^k}{(q;q)_k (qw;q)_k} \\
&= {}_2\phi_1\left[\begin{matrix} u, v \\ w \end{matrix}; q; z\right] + u\,z \frac{(v;q)_1}{(w;q)_1} {}_2\phi_1\left[\begin{matrix} q\,u, qv \\ qw \end{matrix}; q; z\right].
\end{aligned} \tag{2.134}$$

Replacing u by qu in (2.134) and again using (2.134), we get

$$ {}_2\phi_1\left[\begin{matrix} u\,q^2, v \\ w \end{matrix}; q; z\right] = {}_2\phi_1\left[\begin{matrix} u, v \\ w \end{matrix}; q; z\right] + u\,z \frac{(v;q)_1}{(w;q)_1} \sum_{m=1}^{2} q^{m-1} {}_2\phi_1\left[\begin{matrix} q^m\,u, qv \\ qw \end{matrix}; q; z\right]. \tag{2.135}$$

Iterating this process r-times yields

$$ {}_2\phi_1\left[\begin{matrix} u\,q^r, v \\ w \end{matrix}; q; z\right] = {}_2\phi_1\left[\begin{matrix} u, v \\ w \end{matrix}; q; z\right] + u\,z \frac{(v;q)_1}{(w;q)_1} \sum_{m=1}^{r} q^{m-1} {}_2\phi_1\left[\begin{matrix} q^m\,u, qv \\ qw \end{matrix}; q; z\right]. \tag{2.136}$$

To prove (2.132), replace u by $q^{-1}u$ in (2.134), we get

$$ {}_2\phi_1\left[\begin{matrix} u, v \\ w \end{matrix}; q; z\right] = {}_2\phi_1\left[\begin{matrix} q^{-1}u, v \\ w \end{matrix}; q; z\right] + q^{-1}u\,z \frac{(v;q)_1}{(w;q)_1} {}_2\phi_1\left[\begin{matrix} q\,u, qv \\ qw \end{matrix}; q; z\right] \tag{2.137}$$

or

$$ {}_2\phi_1\left[\begin{matrix} q^{-1}u, v \\ w \end{matrix}; q; z\right] = {}_2\phi_1\left[\begin{matrix} u, v \\ w \end{matrix}; q; z\right] - q^{-1}u\,z \frac{(v;q)_1}{(w;q)_1} {}_2\phi_1\left[\begin{matrix} q\,u, qv \\ qw \end{matrix}; q; z\right]. \tag{2.138}$$

Iterating this process gives

$$ {}_2\phi_1\left[\begin{matrix} u\,q^{-r}, v \\ w \end{matrix}; q; z\right] = {}_2\phi_1\left[\begin{matrix} u, v \\ w \end{matrix}; q; z\right] - u\,z \frac{(v;q)_1}{(w;q)_1} \sum_{m=1}^{r} q^{-m} {}_2\phi_1\left[\begin{matrix} q^{-m+1}u, qv \\ qw \end{matrix}; q; z\right].$$

It completes the proof. □

Using (2.131) and (2.132), we obtain another form of recursion formulas as given in the following theorem.

Theorem 35. *The following recursion formulas hold true:*

$$ {}_2\phi_1\left[\begin{matrix} u\,q^r, v \\ w \end{matrix}; q; z\right] = \sum_{m=0}^{r} \begin{bmatrix} r \\ m \end{bmatrix}_q \frac{(v;q)_m\, z^m\, u^m\, q^{m(m-1)}}{(w;q)_m}\, {}_2\phi_1\left[\begin{matrix} u\,q^m, v\,q^m \\ w\,q^m \end{matrix}; q; z\right]; \tag{2.139} $$

$$ {}_2\phi_1\left[\begin{matrix} u\,q^{-r}, v \\ w \end{matrix}; q; z\right] = \sum_{m=0}^{r} \begin{bmatrix} r \\ m \end{bmatrix}_{q^{-1}} \frac{(v;q)_m\, (-z)^m\, u^m\, q^{-\frac{m(m+1)}{2}}}{(w;q)_m}\, {}_2\phi_1\left[\begin{matrix} u, v\,q^m \\ w\,q^m \end{matrix}; q; z\right], \tag{2.140} $$

where $\left[\begin{smallmatrix} r \\ m \end{smallmatrix}\right]_q$ *is the binomial coefficient defined by*

$$ \begin{bmatrix} r \\ m \end{bmatrix}_q = \frac{(q;q)_r}{(q;q)_m\,(q;q)_{r-m}}, \qquad m = 0, 1, \ldots, r. $$

Proof. We prove (2.139) by applying the mathematical induction on r. For $r = 1$, the result (2.139) is obvious. Assume that (2.139) is true for $r = s$, i. e.,

$$ {}_2\phi_1\left[\begin{matrix} u\,q^s, v \\ w \end{matrix}; q; z\right] = \sum_{m=0}^{s} \begin{bmatrix} s \\ m \end{bmatrix}_q \frac{(v;q)_m\, z^m\, u^m\, q^{m(m-1)}}{(w;q)_m}\, {}_2\phi_1\left[\begin{matrix} u\,q^m, v\,q^m \\ w\,q^m \end{matrix}; q; z\right]. \tag{2.141} $$

Replacing u with uq in the above equation and using (2.134), we get

$$ \begin{aligned} {}_2\phi_1\left[\begin{matrix} u\,q^{s+1}, v \\ w \end{matrix}; q; z\right] = \sum_{m=0}^{s} \begin{bmatrix} s \\ m \end{bmatrix}_q \frac{(v;q)_m\, z^m\, u^m\, q^{m^2}}{(w;q)_m} \Bigg\{ & {}_2\phi_1\left[\begin{matrix} u\,q^m, v\,q^m \\ w\,q^m \end{matrix}; q; z\right] \\ & + u\,z\,q^m \frac{(v\,q^m;q)_1}{(cq^m;q)_1}\, {}_2\phi_1\left[\begin{matrix} u\,q^{m+1}, v\,q^{m+1} \\ w\,q^{m+1} \end{matrix}; q; z\right]\Bigg\}. \end{aligned} \tag{2.142} $$

After simplification,

$$ \begin{aligned} {}_2\phi_1\left[\begin{matrix} u\,q^{s+1}, v \\ w \end{matrix}; q; z\right] = \sum_{m=0}^{s} \left(\begin{bmatrix} s \\ m \end{bmatrix}_q q^m + \begin{bmatrix} s \\ m-1 \end{bmatrix}_q\right) & \frac{(v;q)_m\, z^m\, u^m\, q^{m(m-1)}}{(w;q)_m} \\ \times\, {}_2\phi_1 & \left[\begin{matrix} u\,q^m, v\,q^m \\ w\,q^m \end{matrix}; q; z\right]. \end{aligned} \tag{2.143} $$

Using the q-pascal identity,

$$ \begin{bmatrix} s \\ m \end{bmatrix}_q q^m + \begin{bmatrix} s \\ m-1 \end{bmatrix}_q = \begin{bmatrix} s+1 \\ m \end{bmatrix}_q, $$

we get

$$ {}_2\phi_1\left[\begin{matrix}u\,q^{s+1}, v\\ w\end{matrix};q;z\right] = \sum_{m=0}^{s+1}\begin{bmatrix}s+1\\ m\end{bmatrix}_q \frac{(v;q)_m\, z^m\, u^m\, q^{m(m-1)}}{(w;q)_m}\, {}_2\phi_1\left[\begin{matrix}u\,q^m, v\,q^m\\ w\,q^m\end{matrix};q;z\right]. $$

Hence, it is true for $r = s+1$ therefore it is true for every positive integer r. Similarly, the other identity (2.140) can be proved. □

Theorem 36. *The following recursion formulas hold true for the denominator parameter w of the q-hypergeometric function:*

$$ {}_2\phi_1\left[\begin{matrix}u, v\\ w\,q^{-r}\end{matrix};q;z\right] = {}_2\phi_1\left[\begin{matrix}u, v\\ w\end{matrix};q;z\right] + w\,z\,(1-u)\,(1-v) \times \sum_{m=1}^{r} \frac{q^{m-1}}{(q^m - w)\,(q^{m-1} - w)}\, {}_2\phi_1\left[\begin{matrix}u\,q, v\,q\\ w\,q^{2-m}\end{matrix};q;z\right]. \tag{2.144} $$

Equivalently,

$$ {}_2\phi_1\left[\begin{matrix}u, v\\ w\,q^{-r}\end{matrix};q;z\right] = \sum_{m=0}^{r}\begin{bmatrix}r\\ m\end{bmatrix}_{q^{-1}} \frac{(u;q)_m\,(v;q)_m\, z^m\, w^m\, q^{-m}}{(w;q)_m\,(w\,q^{-r};q)_m}\, {}_2\phi_1\left[\begin{matrix}u\,q^m, v\,q^m\\ w\,q^m\end{matrix};q;z\right]. \tag{2.145} $$

Proof. Using the relation,

$$ \frac{1}{(w\,q^{-1};q)_k} = \left[1 + \frac{w\,(1-q^m)}{q-w}\right]\frac{1}{(w;q)_m} \tag{2.146} $$

into the definition of the q-hypergeometric function, we get

$$ {}_2\phi_1\left[\begin{matrix}u, v\\ w\,q^{-1}\end{matrix};q;z\right] = {}_2\phi_1\left[\begin{matrix}u, v\\ w\end{matrix};q;z\right] + w\,z\frac{(1-u)\,(1-v)}{(1-w)\,(q-w)}\, {}_2\phi_1\left[\begin{matrix}u\,q, v\,q\\ w\,q\end{matrix};q;z\right]. \tag{2.147} $$

Replacing w by $w\,q^{-1}$ in (2.147) and again using (2.147), we have

$$ {}_2\phi_1\left[\begin{matrix}u, v\\ w\,q^{-2}\end{matrix};q;z\right] = {}_2\phi_1\left[\begin{matrix}u, v\\ w\end{matrix};q;z\right] + w\,z\,(1-u)\,(1-v)\sum_{m=1}^{2}\frac{q^{m-1}}{(q^m - q)\,(q^{m-1} - w)} \times {}_2\phi_1\left[\begin{matrix}u\,q, v\,q\\ w\,q^{2-m}\end{matrix};q;z\right]. \tag{2.148} $$

Continuing this process r-times yields

$$ {}_2\phi_1\left[\begin{matrix}u, v\\ w\,q^{-r}\end{matrix};q;z\right] = {}_2\phi_1\left[\begin{matrix}u, v\\ w\end{matrix};q;z\right] + w\,z\,(1-u)\,(1-v) \times \sum_{m=1}^{r} \frac{q^{m-1}}{(q^m - w)\,(q^{m-1} - w)}\, {}_2\phi_1\left[\begin{matrix}u\,q, v\,q\\ w\,q^{2-m}\end{matrix};q;z\right]. \tag{2.149} $$

It completes the proof of (2.144). To prove (2.145), we use mathematical induction. For $r = 1$, the formula (2.145) is obvious. Assume that the result is true for $r = s$, i. e.,

$$
\begin{aligned}
&{}_2\phi_1\left[\begin{matrix}u, v\\ w\,q^{-s}\end{matrix}; q; z\right]\\
&= \sum_{m=0}^{s}\begin{bmatrix}s\\ m\end{bmatrix}_{q^{-1}} \frac{(u;q)_m\,(v;q)_m\,z^m\,w^m\,q^{-m}}{(w;q)_m\,(w\,q^{-s};q)_m}\,{}_2\phi_1\left[\begin{matrix}u\,q^m, v\,q^m\\ w\,q^m\end{matrix}; q; z\right].
\end{aligned}
\tag{2.150}
$$

Replacing w with $w\,q^{-1}$ in (2.150) and using (2.145), we get

$$
\begin{aligned}
&{}_2\phi_1\left[\begin{matrix}u, v\\ w\,q^{-s-1}\end{matrix}; q; z\right]\\
&= \sum_{m=0}^{s}\begin{bmatrix}s\\ m\end{bmatrix}_{q^{-1}} \frac{(u;q)_m\,(v;q)_m\,z^m\,w^m\,q^{-2m}}{(wq^{-1};q)_m\,(w\,q^{-s-1};q)_m}\,{}_2\phi_1\left[\begin{matrix}u\,q^m, v\,q^m\\ w\,q^m\end{matrix}; q; z\right]\\
&\quad + \sum_{m=0}^{s}\begin{bmatrix}s\\ m\end{bmatrix}_{q^{-1}} \frac{(u;q)_{m+1}\,(v;q)_{m+1}\,z^{m+1}\,w^{m+1}\,q^{-m-1}}{(wq^{-1};q)_{m+2}\,(w\,q^{-s-1};q)_m}\,{}_2\phi_1\left[\begin{matrix}u\,q^{m+1}, v\,q^{m+1}\\ w\,q^{m+1}\end{matrix}; q; z\right].
\end{aligned}
\tag{2.151}
$$

This gives

$$
\begin{aligned}
&{}_2\phi_1\left[\begin{matrix}u, v\\ w\,q^{-s-1}\end{matrix}; q; z\right]\\
&= {}_2\phi_1\left[\begin{matrix}u, v\\ w\end{matrix}; q; z\right]\\
&\quad + \sum_{m=1}^{s}\frac{(q^{-1};q^{-1})_s}{(q^{-1};q^{-1})_m\,(q^{-1};q^{-1})_{s+1-m}}\left(\frac{q^{-m}(1-q^{-s+m-1})}{1-w\,q^{-s+m-2}} + \frac{1-q^{-m}}{1-w\,q^{m-1}}\right)\\
&\quad \times \frac{(u;q)_m\,(v;q)_m z^m\,w^m\,q^{\;m}}{(w\,q^{-1};q)_m\,(w\,q^{-s-1};q)_{m-1}}\,{}_2\phi_1\left[\begin{matrix}u\,q^m, v\,q^m\\ w\,q^m\end{matrix}; q; z\right]\\
&\quad + \frac{(u;q)_{s+1}\,(v;q)_{s+1} z^{s+1}\,w^{s+1}\,q^{-s-1}}{(w;q)_{s+1}\,(w\,q^{-s-1};q)_{s+1}}\,{}_2\phi_1\left[\begin{matrix}u\,q^{s+1}, v\,q^{s+1}\\ w\,q^{s+1}\end{matrix}; q; z\right]\\
&= \sum_{m=0}^{s+1}\begin{bmatrix}s+1\\ m\end{bmatrix}_{q^{-1}} \frac{(u;q)_m\,(v;q)_m z^m\,w^m\,q^{-m}}{(w\,q^{-1};q)_m\,(w\,q^{-s-1};q)_{m-1}}\,{}_2\phi_1\left[\begin{matrix}u\,q^m, v\,q^m\\ w\,q^m\end{matrix}; q; z\right].
\end{aligned}
\tag{2.152}
$$

This proves (2.145) for $r = s + 1$. Thus, it is true for each r. □

2.8 The q-derivative formulas

Theorem 37. *Let $D_q(z)$ be the q-derivative operator with respect to z. Then*

$$
[D_q(z)]^r\,{}_2\phi_1\left[\begin{matrix}u, v\\ w\end{matrix}; q; z\right] = \frac{1}{(1-q)^r}\,\frac{(u;q)_r\,(v;q)_r}{(w;q)_r}\,{}_2\phi_1\left[\begin{matrix}u\,q^r, v\,q^r\\ w\,q^r\end{matrix}; q; z\right].
\tag{2.153}
$$

Proof. Taking the q-derivative of the function ${}_2\phi_1[{}^{u\,q^r, v\,q^r}_{w\,q^r}; q; z]$, we get

$$\begin{aligned}
D_q(z)\, {}_2\phi_1\left[\begin{matrix} u, v \\ w \end{matrix}; q; z\right] &= \sum_{k=0}^{\infty} \frac{(u;q)_k\,(v;q)_k}{(w;q)_k\,(q;q)_k} D_{z,q}\left(z^k\right) \\
&= \sum_{k=0}^{\infty} \frac{(u;q)_k\,(v;q)_k}{(w;q)_k\,(q;q)_k} \frac{(1-q^k)}{1-q} z^{k-1} \\
&= \frac{1}{1-q} \sum_{k=1}^{\infty} \frac{(u;q)_k\,(v;q)_k}{(w;q)_k\,(q;q)_{k-1}} z^{k-1} \\
&= \frac{1}{1-q} \sum_{k=0}^{\infty} \frac{(u;q)_{k+1}\,(v;q)_{k+1}}{(w;q)_{k+1}\,(q;q)_k} z^{k} \\
&= \frac{1}{1-q} \frac{(1-u)\,(1-v)}{(1-w)} \sum_{k=0}^{\infty} \frac{(uq;q)_k\,(vq;q)_k}{(wq;q)_k\,(q;q)_k} z^k \\
&= \frac{1}{(1-q)} \frac{(u;q)_1\,(v;q)_1}{(w;q)_1} {}_2\phi_1\left[\begin{matrix} u\,q, v\,q \\ w\,q \end{matrix}; q; z\right].
\end{aligned}$$

Again differentiating with respect to z yields

$$\begin{aligned}
\left[D_q(z)\right]^2 {}_2\phi_1\left[\begin{matrix} u, v \\ w \end{matrix}; q; z\right] &= \frac{1}{(1-q)^2} \frac{(1-u)(1-uq)\,(1-v)(1-vq)}{(1-w)(1-wq)} \sum_{k=0}^{\infty} \frac{(uq^2;q)_k\,(vq^2;q)_k}{(wq^2;q)_k\,(q;q)_k} z^k \\
&= \frac{1}{(1-q)^2} \frac{(u;q)_2\,(v;q)_2}{(w;q)_2} {}_2\phi_1\left[\begin{matrix} u\,q^2, v\,q^2 \\ w\,q^2 \end{matrix}; q; z\right].
\end{aligned} \tag{2.154}$$

Repeating this process r-times yields

$$\left[D_q(z)\right]^r {}_2\phi_1\left[\begin{matrix} u, v \\ w \end{matrix}; q; z\right] = \frac{1}{(1-q)^r} \frac{(u;q)_r\,(v;q)_r}{(w;q)_r} {}_2\phi_1\left[\begin{matrix} u\,q^r, v\,q^r \\ w\,q^r \end{matrix}; q; z\right]. \tag{2.155}$$

It completes the proof. □

Now, we determine the q-derivative of the basic Gauss hypergeometric function with respect to the parameters in the following theorem.

Theorem 38. *Let $D_q(u)$ and $D_q(v)$ be the q-derivative operators with respect to parameters u and v respectively. Then, the following derivative formulas hold true* [64]

$$\left[D_q(u)\right]^r {}_2\phi_1\left[\begin{matrix} u, v \\ w \end{matrix}; q; z\right] = \frac{(-z)^r\, q^{\binom{r}{2}}}{(1-q)^r} \frac{(v;q)_r}{(w;q)_r} {}_2\phi_1\left[\begin{matrix} u\,q^r, v\,q^r \\ w\,q^r \end{matrix}; q; z\right]; \tag{2.156}$$

$$\left[D_q(v)\right]^r {}_2\phi_1\left[\begin{matrix} u, v \\ w \end{matrix}; q; z\right] = \frac{(-z)^r\, q^{\binom{r}{2}}}{(1-q)^r} \frac{(u;q)_r}{(w;q)_r} {}_2\phi_1\left[\begin{matrix} u\,q^r, v\,q^r \\ w\,q^r \end{matrix}; q; z\right]. \tag{2.157}$$

Proof. Acting the q-difference operator $D_q(u)$ on ${}_2\phi_1[{u,v \atop w};q;z]$, we get

$$
\begin{aligned}
[D_q(u)]\,{}_2\phi_1\left[{u,v \atop w};q;z\right] &= \sum_{k=0}^{\infty} \frac{([D_q(u)]\,(u;q)_k)\,(v;q)_k}{(w;q)_k\,(q;q)_k} z^k \\
&= \sum_{k=0}^{\infty} \frac{(u;q)_k-(uq;q)_k}{(1-q)\,u} \frac{(v;q)_k}{(w;q)_k\,(q;q)_k} z^k \\
&= \sum_{k=0}^{\infty} \frac{(uq;q)_{k-1}[1-u-(1-uq^k)]}{(1-q)\,u} \frac{(v;q)_k}{(w;q)_k\,(q;q)_k} z^k \\
&= \sum_{k=0}^{\infty} \frac{(uq;q)_{k-1}(uq^k-u)}{(1-q)\,u} \frac{(v;q)_k}{(w;q)_k\,(q;q)_k} z^k \\
&= \frac{-1}{1-q} \sum_{k=1}^{\infty} \frac{(uq;q)_{k-1}\,(v;q)_k}{(w;q)_k\,(q;q)_{k-1}} z^k \\
&= \frac{-1}{1-q} \sum_{k=0}^{\infty} \frac{(uq;q)_k\,(v;q)_{k+1}}{(w;q)_{k+1}\,(q;q)_k} z^{k+1} \\
&= \frac{-z}{1-q} \frac{1-v}{1-w}\,{}_2\phi_1\left[{uq,vq \atop wq};q;z\right] \\
&= \frac{-z}{1-q} \frac{(v;q)_1}{(w;q)_1}\,{}_2\phi_1\left[{uq,vq \atop wq};q;z\right].
\end{aligned}
$$

Again taking the derivative, we have

$$
\begin{aligned}
[D_q(u)]^2\,{}_2\phi_1\left[{u,v \atop w};q;z\right] &= \frac{(-z)^2 q}{(1-q)^2} \frac{(1-v)\,(1-vq)}{(1-w)\,(1-wq)}\,{}_2\phi_1\left[{uq^2,vq^2 \atop wq^2};q;z\right] \\
&= \frac{(-z)^2 q}{(1-q)^2} \frac{(v;q)_2}{(w;q)_2}\,{}_2\phi_1\left[{uq^2,vq^2 \atop wq^2};q;z\right].
\end{aligned}
$$

Continuing this process r-times yields

$$
[D_q(u)]^r\,{}_2\phi_1\left[{u,v \atop w};q;z\right] = \frac{(-z)^r q^{\binom{r}{2}}}{(1-q)^r} \frac{(v;q)_r}{(w;q)_r}\,{}_2\phi_1\left[{uq^r,vq^r \atop wq^r};q;z\right].
$$

Since the q-hypergeometric function ${}_2\phi_1[{u,v \atop w};q;z]$ is symmetric about u and v, we can interchange u and v in the above result, and thus get

$$
[D_q(v)]^r\,{}_2\phi_1\left[{u,v \atop w};q;z\right] = \frac{(-z)^r q^{\binom{r}{2}}}{(1-q)^r} \frac{(u;q)_r}{(w;q)_r}\,{}_2\phi_1\left[{uq^r,vq^r \atop wq^r};q;z\right].
$$

It completes the proof. □

Next, we take the derivative of ${}_2\phi_1\left[\begin{smallmatrix}u,v\\w\end{smallmatrix};q;z\right]$ with respect to w as

$$D_q(w)\,{}_2\phi_1\left[\begin{matrix}u,v\\w\end{matrix};q;z\right]=\sum_{k=0}^{\infty}\frac{(u;q)_k\,(v;q)_k}{(q;q)_k}\,z^k\,D_q(w)\,\frac{1}{(w;q)_k} \tag{2.158}$$

$$=\sum_{k=0}^{\infty}\frac{(u;q)_k\,(v;q)_k}{(q;q)_k}\,z^k\,\frac{1}{(1-q)\,w}\left[\frac{1}{(w;q)_k}-\frac{1}{(wq;q)_k}\right] \tag{2.159}$$

$$=\sum_{k=0}^{\infty}\frac{(u;q)_k\,(v;q)_k}{(q;q)_k}\,z^k\,\frac{1}{(1-q)\,w}\left[\frac{(1-wq^k)-(1-w)}{(1-w)\,(1-wq)\,(wq^2;q)_{k-1}}\right] \tag{2.160}$$

$$=\sum_{k=0}^{\infty}\frac{(u;q)_k\,(v;q)_k\,z^k}{(q;q)_k}\,\frac{(1-q^k)}{(1-q)\,(1-w)\,(1-wq)\,(wq^2;q)_{k-1}} \tag{2.161}$$

$$=\frac{1}{(1-q)\,(1-w)\,(1-wq)}\sum_{k=1}^{\infty}\frac{(u;q)_k\,(v;q)_k\,z^k}{(wq^2;q)_{k-1}\,(q;q)_{k-1}} \tag{2.162}$$

$$=\frac{1}{(1-q)\,(1-w)\,(1-wq)}\sum_{k=0}^{\infty}\frac{(u;q)_{k+1}\,(v;q)_{k+1}\,z^{k+1}}{(wq^2;q)_k\,(q;q)_k} \tag{2.163}$$

$$=\frac{z\,(1-u)\,(1-v)}{(1-q)\,(1-w)\,(1-wq)}\,{}_2\phi_1\left[\begin{matrix}u\,q,v\,q\\w\,q^2\end{matrix};q;z\right] \tag{2.164}$$

$$=\frac{z\,(u;q)_1\,(v;q)_1}{(1-q)\,(w;q)_2}\,{}_2\phi_1\left[\begin{matrix}u\,q,v\,q\\w\,q^2\end{matrix};q;z\right]. \tag{2.165}$$

Iterating the above procedure does not accumulate an elegant form. Hence, we omit them.

Theorem 39. *The following relations hold true:*

$$[D_q(u)]^r\,{}_2\phi_1\left[\begin{matrix}u,v\\w\end{matrix};q;z\right]=\frac{(-z)^r\,q^{\binom{r}{2}}}{(u;q)_r}\,[D_q(z)]^r\,{}_2\phi_1\left[\begin{matrix}u,v\\w\end{matrix};q;z\right]; \tag{2.166}$$

$$[D_q(v)]^r\,{}_2\phi_1\left[\begin{matrix}u,v\\w\end{matrix};q;z\right]=\frac{(-z)^r\,q^{\binom{r}{2}}}{(v;q)_r}\,[D_q(z)]^r\,{}_2\phi_1\left[\begin{matrix}u,v\\w\end{matrix};q;z\right]. \tag{2.167}$$

Proof. These two relations can be established easily using (2.153), (2.156) and (2.157). □

Further, we have the following q-difference formulas presented in the next theorem.

Theorem 40. *For the q-difference operator $D_q(z)$, the following q-difference formulas are satisfied by the basic Gauss hypergeometric function:*

$$[D_q(z)]^r\left\{z^{w-1}\,{}_2\phi_1\left[\begin{matrix}q^u,q^v\\q^w\end{matrix};q;z\right]\right\}=(-q^w)^r\,q^{-\binom{r+1}{2}}\,\frac{(q^{1-w};q)_r}{(1-q)^r}\,z^{w-(r+1)}\times{}_2\phi_1\left[\begin{matrix}q^u,q^v\\q^{w-r}\end{matrix};q;z\right]; \tag{2.168}$$

$$[z^2 D_q(z)]^r \left\{ z^{u-(r-1)} {}_2\phi_1 \left[\begin{matrix} q^u, q^v \\ q^w \end{matrix} ; q; z \right] \right\} = \frac{(q^u;q)_r}{(1-q)^r} z^{u+r} {}_2\phi_1 \left[\begin{matrix} q^{u+r}, q^v \\ q^w \end{matrix} ; q; z \right]; \tag{2.169}$$

$$[z^2 D_q(z)]^r \left\{ z^{v-(r-1)} {}_2\phi_1 \left[\begin{matrix} q^u, q^v \\ q^w \end{matrix} ; q; z \right] \right\} = \frac{(q^v;q)_r}{(1-q)^r} z^{v+r} {}_2\phi_1 \left[\begin{matrix} q^u, q^{v+r} \\ q^w \end{matrix} ; q; z \right]. \tag{2.170}$$

Proof. With the usual definition of q-difference operator $D_q(z)$ and ${}_2\phi_1[\begin{smallmatrix} q^u,q^v \\ q^w \end{smallmatrix};q;z]$, we have

$$\begin{aligned} D_q(z) \left\{ z^{w-1} {}_2\phi_1 \left[\begin{matrix} q^u, q^v \\ q^w \end{matrix} ; q; z \right] \right\} &= \sum_{k=0}^{\infty} \frac{(q^u;q)_k (q^v;q)_k}{(q^w;q)_k (q;q)_k} D_q(z)(z^{w+k-1}) \\ &= \sum_{k=0}^{\infty} \frac{(q^u;q)_k (q^v;q)_k}{(q^w;q)_k (q;q)_k} \frac{1-q^{w+k-1}}{1-q} z^{w+k-2} \\ &= \frac{1}{1-q} \sum_{k=0}^{\infty} \frac{(q^u;q)_k (q^v;q)_k}{(q^w;q)_{k-1} (q;q)_k} z^{w+k-2} \\ &= \frac{1-q^{w-1}}{1-q} \sum_{k=0}^{\infty} \frac{(q^u;q)_k (q^v;q)_k}{(q^{w-1};q)_k (q;q)_k} z^{w+k-2} \\ &= \frac{(-1)\, q^{w-1} (1-q^{1-w})}{1-q} z^{w-2} {}_2\phi_1 \left[\begin{matrix} q^u, q^v \\ q^{w-1} \end{matrix} ; q; z \right]. \end{aligned}$$

Repeating this once gives

$$\begin{aligned} &[D_q(z)]^2 \left\{ z^{w-1} {}_2\phi_1 \left[\begin{matrix} q^u, q^v \\ q^w \end{matrix} ; q; z \right] \right\} \\ &= \frac{(-1)^2 q^{w-1} q^{w-2} (1-q^{1-w})(1-q^{2-w})}{(1-q)^2} z^{w-3} {}_2\phi_1 \left[\begin{matrix} q^u, q^v \\ q^{w-2} \end{matrix} ; q; z \right] \\ &= \frac{(-1)^2 q^{2w} q^{-(1+2)} (q^{1-w};q)_2}{(1-q)^2} z^{w-3} {}_2\phi_1 \left[\begin{matrix} q^u, q^v \\ q^{w-2} \end{matrix} ; q; z \right]. \end{aligned}$$

We generalize this as

$$\begin{aligned} &[D_q(z)]^r \left\{ z^{w-1} {}_2\phi_1 \left[\begin{matrix} q^u, q^v \\ q^w \end{matrix} ; q; z \right] \right\} \\ &= \frac{(-1)^r q^{rw} q^{-(1+2+\cdots+r)} (q^{1-w};q)_r}{(1-q)^r} z^{w-(r+1)} {}_2\phi_1 \left[\begin{matrix} q^u, q^v \\ q^{w-r} \end{matrix} ; q; z \right] \\ &= \frac{(-1)^r q^{rw} q^{-(\frac{r(r+1)}{2})} (q^{1-w};q)_r}{(1-q)^r} z^{w-(r+1)} {}_2\phi_1 \left[\begin{matrix} q^u, q^v \\ q^{w-r} \end{matrix} ; q; z \right] \\ &= (-q^w)^r q^{-\binom{r+1}{2}} \frac{(q^{1-w};q)_r}{(1-q)^r} z^{w-(r+1)} {}_2\phi_1 \left[\begin{matrix} q^u, q^v \\ q^{w-r} \end{matrix} ; q; z \right]. \end{aligned}$$

To prove (2.169), we start in a similar way to (2.168), and thus get

$$[z^2 D_q(z)]\left\{z^u\,{}_2\phi_1\left[\begin{matrix}q^u, q^v\\ q^w\end{matrix}; q; z\right]\right\} = \frac{(q^u;q)_1}{(1-q)}\, z^{u+1}\,{}_2\phi_1\left[\begin{matrix}q^{u+1}, q^v\\ q^w\end{matrix}; q; z\right],$$

which on iterating r-times give

$$[z^2 D_q(z)]^r\left\{z^{u-(r-1)}\,{}_2\phi_1\left[\begin{matrix}q^u, q^v\\ q^w\end{matrix}; q; z\right]\right\} = \frac{(q^u;q)_r}{(1-q)^r}\, z^{u+r}\,{}_2\phi_1\left[\begin{matrix}q^{u+r}, q^v\\ q^w\end{matrix}; q; z\right].$$

Due to the symmetry of a q-Gauss function in u and v, the formula (2.170) can be obtained by interchanging u and v in (2.169). □

2.9 Transformation and summation formulae

Theorem 41.

$$ {}_2\phi_1\left[\begin{matrix}u, v\\ w\end{matrix}; q; z\right] = \frac{(uz;q)_\infty\,(v;q)_\infty}{(w;q)_\infty\,(z;q)_\infty}\,{}_2\phi_1\left[\begin{matrix}\frac{w}{v}, z\\ uz\end{matrix}; q; v\right], \quad |z|, |v| < 1. \tag{2.171}$$

Proof. From the q-binomial theorem, we observe that

$$\frac{(wq^k;q)_\infty}{(vq^k;q)_\infty} = \sum_{m=0}^{\infty}\frac{(\frac{w}{v};q)_m}{(q;q)_m}(vq^k)^m. \tag{2.172}$$

This yields

$$\begin{aligned}
{}_2\phi_1\left[\begin{matrix}u, v\\ w\end{matrix}; q; z\right] &= \sum_{k=0}^{\infty}\frac{(u;q)_k\,(v;q)_k}{(w;q)_k\,(q;q)_k} z^k\\
&= \frac{(v;q)_\infty}{(w;q)_\infty}\sum_{k=0}^{\infty}\frac{(u;q)_k\,(wq^k;q)_\infty}{(vq^k;q)_\infty\,(q;q)_k} z^k\\
&= \frac{(v;q)_\infty}{(w;q)_\infty}\sum_{k=0}^{\infty}\frac{(u;q)_k}{(q;q)_k} z^k \sum_{m=0}^{\infty}\frac{(\frac{w}{v};q)_m}{(q;q)_m}(vq^k)^m\\
&= \frac{(v;q)_\infty}{(w;q)_\infty}\sum_{m=0}^{\infty}\frac{(\frac{w}{v};q)_m}{(q;q)_m} v^m \sum_{k=0}^{\infty}\frac{(u;q)_k}{(q;q)_k}(z\,q^m)^k\\
&= \frac{(v;q)_\infty}{(w;q)_\infty}\sum_{m=0}^{\infty}\frac{(\frac{w}{v};q)_m}{(q;q)_m} v^m \frac{(uzq^m;q)_\infty}{(zq^m;q)_\infty}\\
&= \frac{(uz;q)_\infty\,(v;q)_\infty}{(w;q)_\infty\,(z;q)_\infty}\sum_{m=0}^{\infty}\frac{(\frac{w}{v};q)_m\,(z;q)_m}{(uz;q)_m\,(q;q)_m} v^m\\
&= \frac{(uz;q)_\infty\,(v;q)_\infty}{(w;q)_\infty\,(z;q)_\infty}\,{}_2\phi_1\left[\begin{matrix}\frac{w}{v}, z\\ uz\end{matrix}; q; v\right].
\end{aligned} \tag{2.173}$$

It completes the proof. □

In the next theorem, we see the q-analogue of the Euler's transformation formula

$$ {}_2F_1(u,v;w;z) = (1-z)^{w-u-v}\,{}_2F_1(w-u,w-v;w;z). \tag{2.174}$$

Theorem 42.

$$ {}_2\phi_1\left[\begin{matrix}u,v\\ w\end{matrix};q;z\right] = \frac{(\frac{uv}{w}z;q)_\infty}{(z;q)_\infty}\,{}_2\phi_1\left[\begin{matrix}v,\frac{uv}{w}z\\ vz\end{matrix};q;\frac{w}{v}\right]. \tag{2.175}$$

Proof. The subsequent use of equation (2.171) allows to write

$$\begin{aligned}
{}_2\phi_1\left[\begin{matrix}u,v\\ w\end{matrix};q;z\right] &= \frac{(uz;q)_\infty\,(v;q)_\infty}{(w;q)_\infty\,(z;q)_\infty}\,{}_2\phi_1\left[\begin{matrix}\frac{w}{v},z\\ uz\end{matrix};q;v\right]\\
&= \frac{(uz;q)_\infty\,(v;q)_\infty}{(w;q)_\infty\,(z;q)_\infty}\,{}_2\phi_1\left[\begin{matrix}z,\frac{w}{v}\\ uz\end{matrix};q;v\right]\\
&= \frac{(uz;q)_\infty\,(v;q)_\infty}{(w;q)_\infty\,(z;q)_\infty}\,\frac{(vz;q)_\infty\,(w\setminus v;q)_\infty}{(uz;q)_\infty\,(v;q)_\infty}\,{}_2\phi_1\left[\begin{matrix}\frac{uvz}{w},v\\ vz\end{matrix};q;\frac{w}{v}\right]\\
&= \frac{(vz;q)_\infty\,(w\setminus v;q)_\infty}{(w;q)_\infty\,(z;q)_\infty}\,{}_2\phi_1\left[\begin{matrix}v,\frac{uvz}{w}\\ vz\end{matrix};q;\frac{w}{v}\right]\\
&= \frac{(vz;q)_\infty\,(w\setminus v;q)_\infty}{(w;q)_\infty\,(z;q)_\infty}\,\frac{(\frac{uvz}{w};q)_\infty\,(w;q)_\infty}{(w\setminus v;q)_\infty\,(vz;q)_\infty}\,{}_2\phi_1\left[\begin{matrix}\frac{w}{u},\frac{w}{v}\\ w\end{matrix};q;\frac{uvz}{w}\right]\\
&= \frac{(uvz\setminus w;q)_\infty}{(z;q)_\infty}\,{}_2\phi_1\left[\begin{matrix}v,\frac{uv}{w}z\\ vz\end{matrix};q;\frac{w}{v}\right].
\end{aligned} \tag{2.176}$$

It completes the proof. □

Again using (2.171) with $z=\frac{w}{uv}$ gives

$$\begin{aligned}
{}_2\phi_1\left[\begin{matrix}u,v\\ w\end{matrix};q;\frac{w}{uv}\right] &= \frac{(\frac{w}{v};q)_\infty\,(v;q)_\infty}{(w;q)_\infty\,(\frac{w}{uv};q)_\infty}\,{}_2\phi_1\left[\begin{matrix}\frac{w}{v},\frac{w}{uv}\\ \frac{w}{v}\end{matrix};q;v\right]\\
&= \frac{(\frac{w}{v};q)_\infty\,(v;q)_\infty}{(w;q)_\infty\,(\frac{w}{uv};q)_\infty}\sum_{k=0}^{\infty}\left(\frac{w}{uv};q\right)_k\frac{v^k}{k!}.
\end{aligned} \tag{2.177}$$

Now, the implication of the q-binomial theorem yields

$$\begin{aligned}
{}_2\phi_1\left[\begin{matrix}u,v\\ w\end{matrix};q;\frac{w}{uv}\right] &= \frac{(\frac{w}{v};q)_\infty\,(v;q)_\infty}{(w;q)_\infty\,(\frac{w}{uv};q)_\infty}\,\frac{(\frac{w}{u};q)_\infty}{(v;q)_\infty}\\
&= \frac{(\frac{w}{u};q)_\infty\,(\frac{w}{v};q)_\infty}{(w;q)_\infty\,(\frac{w}{uv};q)_\infty}.
\end{aligned} \tag{2.178}$$

The above relation is known as the q-analogue of the Euler's summation theorem

$$ {}_2F_1(u,v;w;1)=\frac{\Gamma(w-u)\,\Gamma(w-v)}{\Gamma(w)\,\Gamma(w-u-v)}. \tag{2.179} $$

Setting $u=q^{-k}$ in (2.178), we have

$$ {}_2\phi_1\left[\begin{matrix}q^{-k},v\\ w\end{matrix};q;\frac{w\,q^k}{v}\right]=\frac{(w\,q^k;q)_\infty\,(\frac{w}{v};q)_\infty}{(w;q)_\infty\,(\frac{wq^k}{v};q)_\infty}=\frac{(\frac{w}{v};q)_k}{(w;q)_k}. \tag{2.180} $$

Theorem 43.

$$ {}_2\phi_1\left[\begin{matrix}u,v\\ w\end{matrix};q;z\right]=\frac{(uz;q)_\infty}{(z;q)_\infty}\sum_{n=0}^{\infty}\frac{(u;q)_n\,(w\setminus v;q)_n}{(uz;q)_n\,(w;q)_n\,(q;q)_n}(-vz)^n\,q^{\binom{n}{2}}. \tag{2.181} $$

Proof. From (2.180),

$$ \frac{(v;q)_k}{(w;q)_k}=\sum_{n=0}^{k}\frac{(q^k;q)_n\,(w\setminus v;q)_n}{(w;q)_n\,(q;q)_n}(vq^k)^n. \tag{2.182} $$

Hence,

$$ \begin{aligned} {}_2\phi_1\left[\begin{matrix}u,v\\ w\end{matrix};q;z\right]&=\sum_{k=0}^{\infty}\frac{(u;q)_k}{(q;q)_k}z^k\sum_{n=0}^{k}\frac{(q^k;q)_n\,(w\setminus v;q)_n}{(w;q)_n\,(q;q)_n}(vq^k)^n\\ &=\sum_{k=0}^{\infty}\sum_{n=0}^{k}\frac{(u;q)_k}{(q;q)_n}z^k\frac{(w\setminus v;q)_n}{(w;q)_n\,(q;q)_{n-k}}v^n\\ &=\sum_{k=0}^{\infty}\sum_{n=0}^{\infty}\frac{(u;q)_{k+n}}{(q;q)_n}z^k\frac{(w\setminus v;q)_n}{(w;q)_n\,(q;q)_k}(vz)^n\\ &=\sum_{n=0}^{\infty}\frac{(u;q)_n\,(w\setminus v;q)_n}{(w;q)_n\,(q;q)_n}(-vz)^n\,q^{\binom{n}{2}}\sum_{k=0}^{\infty}\frac{(uq^n;q)_k}{(q;q)_k}z^k\\ &=\frac{(uz;q)_\infty}{(z;q)_\infty}\sum_{n=0}^{\infty}\frac{(u;q)_n\,(w\setminus v;q)_n}{(uz;q)_n\,(w;q)_n\,(q;q)_n}(-vz)^n\,q^{\binom{n}{2}}. \end{aligned} \tag{2.183} $$

It completes the proof. □

Exercises

1. Verify the following:

$$ \begin{aligned} \frac{z}{1-q}\,{}_2\phi_1(q,q;q^2;q;z)&=\frac{z}{1-q}+\frac{z^2}{1-q^2}+\frac{z^3}{1-q^3}+\cdots;\\ \frac{z}{1-q^{\frac{1}{2}}}\,{}_2\phi_1(q,q^{\frac{1}{2}};q^{\frac{3}{2}};q;z)&=\frac{z}{1-q^{\frac{1}{2}}}+\frac{z^2}{1-q^{\frac{3}{2}}}+\frac{z^3}{1-q^{\frac{5}{2}}}+\cdots; \end{aligned} $$

$$ {}_2\phi_1(q,-1;-q;q;z) = 1 + \frac{2\,z}{1+q} + \frac{2\,z^2}{1+q^2} + \frac{2\,z^3}{1+q^3} + \cdots. $$

2. Verify the q-analogue of Vandermonde's formula, given by

$$ {}_2\phi_1(u,q^{-n};v;q;z) = \frac{(\frac{v}{u};q)_n\,u^n}{(v;q)_n}. $$

3. Verify

$$ {}_2\phi_1\left(u,q^{-n};w;q;\frac{w}{u}\,q^n\right) = \frac{(\frac{w}{u};q)_n}{(w;q)_n}. $$

4. If the q-binomial coefficient is defined by $\left[{m \atop j}\right]_q = \frac{(q;q)_m}{(q;q)_j\,(q;q)_{m-j}}$, then prove that

$$ \begin{bmatrix} m \\ j \end{bmatrix}_q = \begin{bmatrix} m \\ m-j \end{bmatrix}_q; $$

$$ \begin{bmatrix} u+j \\ j \end{bmatrix}_q = \frac{(q^{u+1};q)_j}{(q;q)_j}; $$

$$ \begin{bmatrix} u \\ j \end{bmatrix}_q = \frac{(q^{-u};q)_j}{(q;q)_j}\,(-q^u)^j\,q^{-\binom{j}{2}}; $$

$$ \begin{bmatrix} u+1 \\ j \end{bmatrix}_q = \begin{bmatrix} u \\ j \end{bmatrix}_q q^j + \begin{bmatrix} u \\ j-1 \end{bmatrix}_q = \begin{bmatrix} u \\ j \end{bmatrix}_q + \begin{bmatrix} u \\ j-1 \end{bmatrix}_q q^{u+1-j}. $$

5. Using the notation given from (2.118) to (2.121), show that

$$ \begin{aligned} & q^u\,(1-q^{v-w})\,\phi(v-) - q^v\,(1-q^{u-w})\,\phi(u-) = (q^u - q^v)\,(1-q^{u+v-w-1}\,z)\,\phi; \\ & q\,(1-q^{u-w})\,\phi(u-) + (1-q^u)\,(1-q^{u+v-w}z)\,\phi(u+) \\ & \quad = [1+q-q^u-q^{u-w+1}+q^{2u-w}\,z\,(1-q^{v-u})]\,\phi; \\ & (1-q^w)\,(q-q^w)\,(q^{u+v}z-q^w)\,\phi(w-) + (q^w-q^u)\,(q^w-q^v)\,z\,\phi(w+) \\ & \quad = (q^w-1)\,[q^w(q-q^w) + (q^{w+u}+q^{w+v}-q^{u+v}-q^{u+v+1})\,z]\,\phi. \end{aligned} $$

6. Prove that

$$ \sum_{k=0}^{\infty} \frac{(q^u;q)_k}{(q;q)_k}\,q^{\frac{k(k+1)}{2}} = (-q;q)_\infty\,(q^{u+1};q^2)_\infty. $$

7. Prove that if $\mathbb{R}(u) > 0$ and $0 < q < 1$, then

$$ \Gamma_q(u) = (q;q)_\infty\,(1-q)^{1-u} \sum_{k=0}^{\infty} \frac{q^{ku}}{(q;q)_k}; $$

$$\frac{1}{\Gamma_q(u)} = \frac{(1-q)^{u-1}}{(q;q)_\infty} \sum_{k=0}^{\infty} \frac{(-1)^k q^{ku}}{(q;q)_k} q^{\binom{k}{2}}.$$

8. Show that if $q > 1$, then the q-gamma function is defined by

$$\Gamma_q(u) = \frac{(q^{-1};q^{-1})_\infty}{(q^{-u};q^{-1})_\infty} (q-1)^{1-u} q^{\frac{u(u-1)}{2}}.$$

   Also, prove that $\lim_{q\to 1^+} \Gamma_q(u) = \Gamma(u)$.
9. If u, v are complex numbers such that $u\,v = q\,v\,u$, q commutes with u, v and the associative law holds. Then prove that

$$\begin{aligned} e_q(v)\,e_q(u) &= e_q(u+v); \\ e_q(u)\,e_q(v) &= e_q(u+v-v\,u); \\ E_q(u)\,E_q(v) &= E_q(u+v); \\ E_q(v)\,E_q(u) &= E_q(u+v+uv). \end{aligned}$$

3 Gauss hypergeometric function with matrix arguments

In the consecutive two chapters, we see the matrix analogues of the Gauss hypergeometric function. One can find the matrix version of a special function either by taking the variable or parameters as matrices. In this chapter, we elaborate the Gauss hypergeometric function with matrix arguments and its various properties. We are assuming that the readers of these two chapters are familiar with the basic matrix theory. For smooth reading of the chapter, we suggest to visit Appendix B and specific preliminaries given in this chapter. We start this chapter with the following definitions.

3.1 Positive definite matrix

A p-square complex matrix P is called a positive definite if $z^*Pz > 0$ for all nonzero vector $z \in \mathbb{C}^p$. Alternatively,

$$P \text{ is positive definite} \iff z^*Pz > 0 \quad \text{for all } z \in \mathbb{C}^p \setminus \{0\}.$$

A p-square complex matrix P is called a positive semidefinite if $z^*Pz \geq 0$ for all $z \in \mathbb{C}^p$. That is,

$$P \text{ is positive semidefinite} \iff z^*Pz \geq 0 \quad \text{for all } z \in \mathbb{C}^p.$$

There are some conditions that can be used to characterize a positive definite or positive semidefinite matrix:

1. A matrix P is positive definite or positive semidefinite if and only if it is Hermitian ($P = P^*$) and all the eigenvalues of P are nonnegative. It is positive definite if and only if all its eigenvalues are positive.
2. A matrix P is positive definite or positive semidefinite if and only if it is Hermitian and all its principal minors are nonnegative. It is positive definite if and only if all its principal minors are positive.
3. A matrix P is positive definite or positive semidefinite if and only if $P = Q^*Q$ for some matrix Q. Matrix P is positive definite if and only if Q is invertible.
4. The matrix P is positive definite or positive semidefinite if and only if $P = Q^2$ for some positive definite or positive semidefinite matrix Q.
5. A 2×2 real symmetric matrix is positive definite if and only if diagonal entries and determinant are positive.

Example 44. In the symmetric matrix $P = \left[\begin{smallmatrix} 1 & 3 \\ 3 & 4 \end{smallmatrix}\right]$, the diagonal entries are positive but the determinant $4 - 9 = -5$ is negative. Hence, P is not positive definite.

https://doi.org/10.1515/9783111324586-003

Example 45. The matrix $\begin{bmatrix} 1 & -2 \\ -2 & 5 \end{bmatrix}$ is symmetric with positive diagonal entries and determinant, so it is positive definite.

3.2 Matrix differentiation

The derivative of a matrix or scalar valued matrix functions can be computed in the following ways:

- The derivative of a matrix with respect to a scalar.
- The derivative of scalar valued matrix function with respect to a vector.
- The derivative of scalar valued matrix function with respect to a matrix.
- The derivative of a matrix with respect to an element of the matrix.

We now elaborate on each of the matrix differentiations given above in a detailed manner.

1. *Derivative of a matrix with respect to a scalar:* For a matrix $U = (u_{ij})$, where $u_{ij}'s$ are functions of the real scalar variable u, the derivative of U with respect to u is the matrix of derivative of the elements, obtained by

$$\frac{\partial U}{\partial u} = \left(\frac{\partial u_{ij}}{\partial u}\right).$$

Example 46. The derivative of the matrix

$$U = \begin{bmatrix} u^3 & 8u & \log u \\ e^u & u^2 & \sin u \\ a^u & -2u & u^4 \end{bmatrix}$$

with respect to the u is given by

$$\frac{\partial U}{\partial u} = \begin{bmatrix} \frac{d}{du}u^3 & \frac{d}{du}8u & \frac{d}{du}\log u \\ \frac{d}{du}e^u & \frac{d}{du}u^2 & \frac{d}{du}\sin u \\ \frac{d}{du}a^u & \frac{d}{du}(-2u) & \frac{d}{du}u^4 \end{bmatrix} = \begin{bmatrix} 3u^2 & 8 & \frac{1}{u} \\ e^u & 2u & \cos u \\ a^u \log a & -2 & 4u^3 \end{bmatrix}.$$

2. *The derivative of a scalar valued matrix function with respect to a vector:* Let U be a $p \times 1$ real vector. Then the partial derivative operator $\frac{\partial}{\partial u}$ is defined as the column vector of partial derivative operators with respect to the elements $u_i's$ of U. That is,

$$\frac{\partial}{\partial U'} = \left(\frac{\partial}{\partial U}\right)' = \left(\frac{\partial}{\partial u_1}, \frac{\partial}{\partial u_2}, \dots, \frac{\partial}{\partial u_p}\right).$$

The second-order derivative $\frac{\partial^2}{\partial U^2}, \frac{\partial^2}{\partial U \partial U'}, \frac{\partial^2}{\partial U' \partial U}$ can be obtained by applying the partial derivative operators $\frac{\partial}{\partial U}$ and $\frac{\partial}{\partial U'}$ twice as per the requirements. As such, the derivative $\frac{\partial^2}{\partial U \partial U'}$ is given by the following matrix:

$$\frac{\partial^2}{\partial U\,\partial U'} = \begin{bmatrix} \frac{\partial^2}{\partial u_1^2} & \cdots & \frac{\partial^2}{\partial u_1\,\partial u_p} \\ \vdots & \cdots & \vdots \\ \frac{\partial^2}{\partial u_p\,\partial u_1} & \cdots & \frac{\partial^2}{\partial u_p^2} \end{bmatrix}.$$

Example 47. If $\mathrm{u} = 3u_1 - 4u_2 + 2u_3$, then find $\frac{\partial \mathrm{u}}{\partial U}$, where $U = \begin{bmatrix} u_1 \\ u_2 \\ u_3 \end{bmatrix}$.

Solution. We can write $\mathrm{u} = U'u = u'U$, where $U' = (u_1, u_2, u_3)$ and $u' = (3, -4, 2)$. So, by the definition of the derivative of scalar valued matrix function with respect to a vector, we get

$$\left(\frac{\partial \mathrm{u}}{\partial U}\right)' = \left(\frac{\partial}{\partial u_1}, \frac{\partial}{\partial u_2}, \frac{\partial}{\partial u_3}\right)\mathrm{u} = (3, -4, 2) = u'.$$

Hence, the value of the derivative $(\frac{\partial \mathrm{u}}{\partial U})$ is $u = \begin{bmatrix} 3 \\ -4 \\ 2 \end{bmatrix}$.

Example 48. If $\mathrm{u} = 4u_1^2 + 2u_2^2 - u_3^2 + 6u_1 u_2 - 4u_1 u_3$, then find the value of $(\frac{\partial \mathrm{u}}{\partial U})$ and $(\frac{\partial^2 \mathrm{u}}{\partial U\,\partial U'})$, where $U = \begin{bmatrix} u_1 \\ u_2 \\ u_3 \end{bmatrix}$.

Solution. By definition, we have

$$\begin{aligned} \left(\frac{\partial \mathrm{u}}{\partial U}\right)' &= \left(\frac{\partial}{\partial u_1}, \frac{\partial}{\partial u_2}, \frac{\partial}{\partial u_3}\right)\mathrm{u} \\ &= \left(\frac{\partial \mathrm{u}}{\partial u_1}, \frac{\partial \mathrm{u}}{\partial u_2}, \frac{\partial \mathrm{u}}{\partial u_3}\right) \\ &= (8u_1 + 6u_2 - 4u_3, 4u_2 + 6u_1, -2u_3 - 4u_1). \end{aligned} \tag{3.1}$$

This gives

$$\left(\frac{\partial \mathrm{u}}{\partial U}\right) = \begin{bmatrix} 8u_1 + 6u_2 - 4u_3 \\ 4u_2 + 6u_1 \\ -2u_3 - 4u_1 \end{bmatrix}.$$

For $(\frac{\partial^2 \mathrm{u}}{\partial U\,\partial U'})$, we obtain first $(\frac{\partial \mathrm{u}}{\partial U'})$, which will be given by

$$\left(\frac{\partial \mathrm{u}}{\partial U'}\right) = (8u_1 + 6u_2 - 4u_3, 4u_2 + 6u_1, -2u_3 - 4u_1).$$

Again operating the partial derivative operator $\frac{\partial \mathrm{u}}{\partial U}$, we get

$$\begin{aligned} \frac{\partial^2 \mathrm{u}}{\partial U\,\partial U'} &= \left(\frac{\partial}{\partial U}(8u_1 + 6u_2 - 4u_3), \frac{\partial}{\partial U}(4u_2 + 6u_1), \frac{\partial}{\partial U}(-2u_3 - 4u_1)\right) \\ &= \begin{bmatrix} 8 & 6 & -4 \\ 6 & 4 & 0 \\ -4 & 0 & -2 \end{bmatrix} = 2\begin{bmatrix} 4 & 3 & -2 \\ 3 & 2 & 0 \\ -2 & 0 & -1 \end{bmatrix}. \end{aligned} \tag{3.2}$$

3. *The derivative of a scalar valued matrix function with respect to this matrix:* Let $f(U)$ be a scalar function of the real matrix $U = (u_{ij})$. Then the derivative of $f(U)$ with respect to U is given by

$$\frac{\partial f}{\partial U} = \left(\frac{\partial f}{\partial u_{ij}}\right).$$

Example 49. If $P = \begin{bmatrix} 1 & 2 & 0 \\ 1 & 3 & 2 \\ 0 & 1 & 1 \end{bmatrix}$, $U = \begin{bmatrix} u_{11} & u_{12} & u_{13} \\ u_{21} & u_{22} & u_{23} \\ u_{31} & u_{32} & u_{33} \end{bmatrix}$, $u = \begin{bmatrix} 1 \\ 0 \\ 0 \end{bmatrix}$ and $v = \begin{bmatrix} 1 \\ 1 \\ 1 \end{bmatrix}$, then find

(i) $\frac{\partial}{\partial U}\operatorname{tr}(U)$ (ii) $\frac{\partial}{\partial U}\operatorname{tr}(PU)$ (iii) $\frac{\partial}{\partial U}(u'Uv)$.

Solution. (i) The trace of U is given by

$$\operatorname{tr}(U) = u_{11} + u_{22} + u_{33},$$

and by definition, we get

$$\begin{aligned}\frac{\partial}{\partial U}\operatorname{tr}(U) &= \begin{bmatrix} \frac{\partial}{\partial u_{11}}\operatorname{tr}(U) & \frac{\partial}{\partial u_{12}}\operatorname{tr}(U) & \frac{\partial}{\partial u_{13}}\operatorname{tr}(U) \\ \frac{\partial}{\partial u_{21}}\operatorname{tr}(U) & \frac{\partial}{\partial u_{22}}\operatorname{tr}(U) & \frac{\partial}{\partial u_{23}}\operatorname{tr}(U) \\ \frac{\partial}{\partial u_{31}}\operatorname{tr}(U) & \frac{\partial}{\partial u_{32}}\operatorname{tr}(U) & \frac{\partial}{\partial u_{33}}\operatorname{tr}(U) \end{bmatrix} \\ &= \begin{bmatrix} 1 & 0 & 0 \\ 0 & 1 & 0 \\ 0 & 0 & 1 \end{bmatrix} = I_3. \end{aligned} \tag{3.3}$$

(ii) The matrix product PU is obtained as

$$PU = \begin{bmatrix} u_{11} + 2u_{21} & u_{12} + 2u_{22} & u_{13} + 2u_{23} \\ u_{11} + 3u_{21} + 2u_{31} & u_{12} + 3u_{22} + 2u_{32} & u_{13} + 3u_{23} + 2u_{33} \\ u_{21} + u_{31} & u_{22} + u_{32} & u_{23} + u_{33} \end{bmatrix}, \tag{3.4}$$

and trace is given as

$$\operatorname{tr}(PU) = u_{11} + 2u_{21} + u_{12} + 3u_{22} + 2u_{32} + u_{23} + u_{33}.$$

Then

$$\begin{aligned}\frac{\partial}{\partial U}\operatorname{tr}(PU) &= \begin{bmatrix} \frac{\partial}{\partial u_{11}}\operatorname{tr}(PU) & \frac{\partial}{\partial u_{12}}\operatorname{tr}(PU) & \frac{\partial}{\partial u_{13}}\operatorname{tr}(PU) \\ \frac{\partial}{\partial u_{21}}\operatorname{tr}(PU) & \frac{\partial}{\partial u_{22}}\operatorname{tr}(PU) & \frac{\partial}{\partial u_{23}}\operatorname{tr}(PU) \\ \frac{\partial}{\partial u_{31}}\operatorname{tr}(PU) & \frac{\partial}{\partial u_{32}}\operatorname{tr}(PU) & \frac{\partial}{\partial u_{33}}\operatorname{tr}(PU) \end{bmatrix} \\ &= \begin{bmatrix} 1 & 1 & 0 \\ 2 & 3 & 1 \\ 0 & 2 & 1 \end{bmatrix}. \end{aligned} \tag{3.5}$$

For (iii), the product $u'Uv$ is given by

$$u'Uv = u_{11} + u_{12} + u_{13},$$

so

$$\frac{\partial}{\partial U}(u'Uv) = \begin{bmatrix} 1 & 1 & 1 \\ 0 & 0 & 0 \\ 0 & 0 & 0 \end{bmatrix} = u \cdot v'.$$

Example 50. If $U = \begin{bmatrix} u_{11} & u_{12} \\ u_{21} & u_{22} \end{bmatrix}$, $V = \begin{bmatrix} u_{11} & u_{12} \\ u_{12} & u_{22} \end{bmatrix}$ and $T = \begin{bmatrix} u_{11} & \frac{1}{2}u_{12} \\ \frac{1}{2}u_{12} & u_{22} \end{bmatrix}$ are invertible matrices, then evaluate

(i) $\frac{\partial}{\partial U}(\det U)$ (ii) $\frac{\partial}{\partial V}(\det V)$ (iii) $\frac{\partial}{\partial T}(\det T)$ (iv) $\frac{\partial^*}{\partial T}(\det T)$,

where $\frac{\partial^*}{\partial T} = \begin{bmatrix} \frac{\partial}{\partial u_{11}} & \frac{\partial}{\partial u_{12}} \\ \frac{\partial}{\partial u_{12}} & \frac{\partial}{\partial u_{22}} \end{bmatrix}$.

Solution. (i) The determinant of U and definition of the derivative give

$$\frac{\partial}{\partial U}(\det U) = \begin{bmatrix} \frac{\partial}{\partial u_{11}} \det U & \frac{\partial}{\partial u_{12}} \det U \\ \frac{\partial}{\partial u_{21}} \det U & \frac{\partial}{\partial u_{22}} \det U \end{bmatrix} = \begin{bmatrix} u_{22} & -u_{21} \\ -u_{12} & u_{11} \end{bmatrix}.$$

We can also write this as

$$\frac{\partial}{\partial U}(\det U) = \det(U)\,(U^{-1})'.$$

Similarly, for (ii),

$$\frac{\partial}{\partial V}(\det V) = \begin{bmatrix} u_{22} & -2u_{12} \\ -2u_{12} & u_{11} \end{bmatrix} = \det(V)\,\{2V^{-1} - \operatorname{diag}(V')\}.$$

(iii)

$$\frac{\partial}{\partial T}(\det T) = \begin{bmatrix} \frac{\partial}{\partial u_{11}} \det T & \frac{\partial}{\partial(\frac{1}{2}u_{12})} \det T \\ \frac{\partial}{\partial(\frac{1}{2}u_{12})} \det T & \frac{\partial}{\partial u_{22}} \det T \end{bmatrix}$$
$$= \begin{bmatrix} u_{22} & -u_{12} \\ -u_{12} & u_{11} \end{bmatrix} = \det(T)\,\{2T^{-1} - \operatorname{diag}(T^{-1})\}. \quad (3.6)$$

(iv)

$$\frac{\partial^*}{\partial T}(\det T) = \begin{bmatrix} \frac{\partial}{\partial u_{11}} \det T & \frac{\partial}{\partial u_{12}} \det T \\ \frac{\partial}{\partial u_{12}} \det T & \frac{\partial}{\partial u_{22}} \det T \end{bmatrix}$$
$$= \begin{bmatrix} u_{22} & -\frac{1}{2}u_{12} \\ -\frac{1}{2}u_{12} & u_{11} \end{bmatrix} = \det(T) \cdot T^{-1}. \quad (3.7)$$

4. *The derivative of a matrix with respect to an element of the matrix:* As a consequence of the above concept, we can determine the derivative of a matrix with respect to the element of the same matrix. Let U be a real matrix of functionally independent elements u_{ij}. Then the derivative of U with respect to an element u_{rs} is denoted by $\frac{\partial U}{\partial u_{rs}}$ and given by a matrix Δ of same order with 1 as $(r,s)^{\text{th}}$ place and zero elsewhere. Then

$$\frac{\partial U}{\partial u_{rs}} = \Delta.$$

Now, we give some lemmas and theorems related to the derivative of a matrix. The proofs of these theorems are beyond the scope of the book, and hence are omitted.

Theorem 51. *Let U be a $p \times 1$ vector of functionally independent real variables $u_1, \ldots, u_p$ and $a' = (a_1, \ldots, a_p)$ a vector of constants. Then*

$$y_1 = a'U = U'a = a_1u_1 + \cdots + a_pu_p \implies \frac{\partial y_1}{\partial U} = a,$$
$$y_2 = U'U = u_1^2 + \cdots + u_p^2 \implies \frac{\partial y_2}{\partial U} = 2U,$$

and

$$y_3 = U'PU = \sum_{i=1}^{p}\sum_{j=1}^{p} a_{ij}\, u_i\, u_j \implies \frac{\partial y_3}{\partial U} = 2PU,$$

where $P = (a_{ij})$ is a symmetric matrix of constants.

Theorem 52. *Let U be a matrix of functionally independent real variables and $P = (a_{ij})$ be a matrix of constants. Then:*
1. $f(U) = \operatorname{tr}(U)$, $U(p \times p) \implies \frac{\partial f}{\partial U} = I$,
2. $f(U) = \operatorname{tr}(P'U)$, $U(p \times q)$, $P(p \times q) \implies \frac{\partial f}{\partial U} = P$,
3. $f(U) = a'Ub$, $U(p \times p)$, $a(p \times 1)$, $b(p \times 1) \longrightarrow \frac{\partial f}{\partial U} = ab'$, *where a and b are vectors of constants,*
4. $f(U) = \operatorname{tr}(UP)$, $U(p \times p)$, $P(p \times p)$, $U' \neq U \implies \frac{\partial f}{\partial U} = P'$,
5. $\frac{\partial f}{\partial U} = P + P' - \operatorname{diag}(P)$.

Theorem 53. *Let U be a $p \times p$ nonsingular matrix of functionally independent real variables. Then*

$$\begin{aligned}\frac{\partial |U|}{\partial U} &= |U|(U^{-1})', \quad \textit{for a general } U\\ &= |U|(2U^{-1} - \operatorname{diag}(U^{-1})), \quad \textit{for } U = U',\end{aligned} \tag{3.8}$$

where $\operatorname{diag}(U^{-1})$ *means the diagonal matrix formed with the diagonal elements of* U^{-1}.

Theorem 54. *For a $p \times p$ matrix U of functionally independent real scalar variables,*

$$\frac{\partial(\mathrm{tr}(U^2))}{\partial U} = 2\,U'.$$

Theorem 55. *For a $p \times p$ matrix U of functionally independent real variables and a $q \times q$ constant matrix P,*

$$\frac{\partial(\mathrm{tr}(UPU'))}{\partial U} = U(P + P').$$

Theorem 56. *For a $p \times p$ real nonsingular matrix U,*

$$\begin{aligned}\frac{\partial |U|^k}{\partial U} &= k\,|U|^k\,(U^{-1})', \quad \text{for a general } U,\\ &= k\,|U|^k\,(2\,U^{-1} - \mathrm{diag}(U^{-1})), \quad \text{for } U = U'; \end{aligned} \tag{3.9}$$

$$\frac{\partial(e^{\mathrm{tr}(U^2)})}{\partial U} = 2\,e^{\mathrm{tr}(U^2)}\,U'; \tag{3.10}$$

$$\begin{aligned}\frac{\partial(\ln |U|)}{\partial U} &= (U^{-1})', \quad \text{for a general } U,\\ &= 2\,U^{-1} - \mathrm{diag}(U^{-1}), \quad \text{for } U = U'. \end{aligned} \tag{3.11}$$

Theorem 57. *Let P and Q be matrices of constants, U a matrix of functionally independent real variables and u_{ij} be an element of U. The orders of matrices are set at each stage so that the product in each statement makes sense. Then*

$$\begin{aligned} V = U\,P &\implies \frac{\partial V}{\partial u_{ij}} = \Delta\,P,\\ V = P\,U &\implies \frac{\partial V}{\partial u_{ij}} = P\,\Delta,\\ V = U'\,P &\implies \frac{\partial V}{\partial u_{ij}} = \Delta'\,P,\\ V = P\,U' &\implies \frac{\partial V}{\partial u_{ij}} = P\,\Delta',\\ V = P\,U\,Q &\implies \frac{\partial V}{\partial u_{ij}} = P\,\Delta\,Q,\\ V = P\,U'\,Q &\implies \frac{\partial V}{\partial u_{ij}} = P\,\Delta'\,Q. \end{aligned} \tag{3.12}$$

Theorem 58. *For a matrix U of functionally independent real variables, we have*

$$V = U^2 \implies \frac{\partial V}{\partial u_{ij}} = \Delta\,U + U\,\Delta,$$

$$
\begin{aligned}
V = U'\,U &\implies \frac{\partial V}{\partial u_{ij}} = \Delta'\,U + U'\,\Delta,\\
V = U\,U' &\implies \frac{\partial V}{\partial u_{ij}} = \Delta\,U' + U\,\Delta',\\
V = U'\,U' &\implies \frac{\partial V}{\partial u_{ij}} = \Delta'\,U' + U'\,\Delta'.
\end{aligned} \tag{3.13}
$$

Theorem 59. *Let U be a matrix of functionally independent real variables and P, Q be matrices of constants so that the following product makes sense. Then*

$$
\begin{aligned}
V = U\,U'\,U &\implies \frac{\partial V}{\partial u_{ij}} = \Delta\,U'\,U + U\,\Delta'\,U + U\,U'\,\Delta,\\
V = P\,U\,U'\,Q &\implies \frac{\partial V}{\partial u_{ij}} = P\,\Delta\,U'\,Q + P\,U\,\Delta'\,Q.
\end{aligned} \tag{3.14}
$$

Theorem 60. *For a nonsingular matrix U of distinct elements,*

$$
V = U^{-1} \implies \frac{\partial V}{\partial u_{ij}} = -U^{-1}\,\Delta\,U^{-1}. \tag{3.15}
$$

Theorem 61. *Let P, Q, R, S be matrices of constants and U be a matrix of functionally independent real variables so that the products are defined and $Q\,U\,R$ is nonsingular. Then*

$$
V = P(Q\,U\,R)^{-1}S \implies \frac{\partial V}{\partial u_{ij}} = -P(Q\,U\,R)^{-1}Q\,\Delta\,R(Q\,U\,R)^{-1}\,S. \tag{3.16}
$$

3.3 Jacobian of transformations

For any two matrices U and V with same number of independent variables $u_1, \dots, u_p$ and $v_1, \dots, v_p$, respectively, consider the matrix transformation $V = F(U)$. Then the Jacobian of the transformation from $U \to V$ is defined as

$$
J(U \to V) = \operatorname{mod}\det \begin{pmatrix} \frac{\partial u_1}{\partial v_1} & \cdots & \frac{\partial u_1}{\partial v_p} \\ \vdots & \vdots & \vdots \\ \frac{\partial u_p}{\partial v_1} & \cdots & \frac{\partial u_p}{\partial v_p} \end{pmatrix}. \tag{3.17}
$$

Now, we list some well-known results for the Jacobian of transformations without their proofs:

1. $du_1 \cdots du_p = J(U \to V) dv_1 \cdots dv_p$.
2. If $J(U \to V) \neq 0$, then $J(V \to U) = \{J(U \to V)\}^{-1}$.
3. If $V = F(W)$ and $W = G(U)$, then $J(U \to V) = J(U \to W)J(W \to V)$.
4. If $V = F(U)$ and $W = G(Z)$, then $J(U, Z \to V, W) = J(U \to V)J(Z \to W)$.

5. $J(U \to V) = J((dU) \to (dV))$, where (dU) and (dV) are the matrix of differentials of elements of U and V eespectively.
6. If $v_i = f_i(u_1, \ldots, u_m, u_{m+1}, \ldots, u_{m+n})$, $i = 1, \ldots, m$, where $u_1, \ldots, u_{m+n}$ are subjects to constraints $f_i(u_1, \ldots, u_{m+n}) = 0$, $i = m+1, \ldots, m+n$, then

$$J(v_1, \ldots, v_m \to u_1, \ldots, u_m) = \frac{J(f_1, \ldots, f_{m+n} \to u_1, \ldots, u_{m+n})}{J(f_{m+1}, \ldots, f_{m+n} \to u_{m+1}, \ldots, u_{m+n})}.$$

There are several other transformations like Jacobian of linear transformations, inverse transformations, quadratic transformations and orthogonal transformations. We list a few of them:

1. For $v(p \times 1)$, $u(p \times 1)$ and $P(p \times p)$ if $v = P\,u$, then

$$J(v \to u) = \det(P).$$

2. For $V(p \times q)$, $U(p \times q)$ and $P(p \times p)$ if $V = P\,U$, then

$$J(V \to U) = \det(P)^q.$$

3. For $V(p \times q)$, $U(p \times q)$ and $Q(q \times q)$ if $V = U\,Q$, then

$$J(V \to U) = \det(Q)^p.$$

4. For $V(p \times q)$, $U(p \times q)$, $P(p \times p)$ and $Q(q \times q)$ if $V = P\,U\,Q$, then

$$J(V \to U) = \det(P)^q \det(Q)^p.$$

5. For $V(p \times p)$, $U(p \times p)$ symmetric and $P(p \times p)$ if $V = P\,U\,P'$, then

$$J(V \to U) = \det(P)^{p+1}.$$

6. For $V(p \times q)$, $U(p \times q)$ and a scalar a if $V = a\,U$, then

$$J(V \to U) = a^{pq}.$$

7. For lower triangular matrices $V(p \times p)$, $U(p \times p)$ and $P = (a_{ij})$ if $V = P\,U$, then

$$J(V \to U) = \prod_{i=1}^{p} a_{ii}^{i}.$$

8. For upper triangular matrices $V(p \times p)$, $U(p \times p)$ and $P = (a_{ij})$ if $V = P\,U$, then

$$J(V \to U) = \prod_{i=1}^{p} a_{ii}^{p-i+1}.$$

9. For a symmetric positive definite matrix $V(p \times p)$ and a lower triangular matrix $T(p \times p) = (t_{ij})$, if $V = T\,T'$, then

$$J(V \to T) = 2^p \prod_{i=1}^{p} t_{ii}^{p-i+1}. \tag{3.18}$$

10. For lower triangular matrices $V(p\times p)$, $U(p\times p)$, $P(p\times p) = (a_{ij})$ and $Q(p\times p) = (b_{ij})$ if $V = P\,U\,Q$, then

$$J(V \to U) = \prod_{i=1}^{p} a_{ii}^{i}\, b_{ii}^{p-i+1}.$$

11. For a symmetric matrices $V(p\times p)$, lower triangular matrix $U(p\times p)$ and $P(p\times p) = (a_{ij})$ if $V = P\,U' + U\,P'$, then

$$J(V \to U) = 2^p \prod_{i=1}^{p} a_{ii}^{p-i+1}. \tag{3.19}$$

12. For a symmetric matrices $V(p\times p)$, upper triangular matrix $U(p\times p)$ and $P(p\times p) = (a_{ij})$ if $V = P\,U' + U\,P'$, then

$$J(V \to U) = 2^p \prod_{i=1}^{p} a_{ii}^{i}.$$

13. For nonsingular matrices $V(p\times p)$ and $U(p\times p)$ if $V = U^{-1}$, then

$$J(V \to U) = \det(U)^{-2p}.$$

14. For nonsingular symmetric matrices $V(p\times p)$ and $U(p\times p)$ if $V = U^{-1}$, then

$$J(V \to U) = \det(U)^{-(p+1)}.$$

3.4 Matrix integral transforms

Most of the hypergeometric functions with a matrix argument are defined in terms of integrals, where integrands are some matrix functions. Throughout the chapter, we see the integral of the form

$$\int_R f(U)\, dU,$$

where $f(U)$ is the scalar function of matrix U and $dU = d(u_{ij})$. Such integrals are defined as the iterated integral of $f(U)$ with respect to the each element of U separately over a region R in the space defined by the simplex bounding the ranges of the elements of U.

Herz in 1955 studied these integrals in a detailed manner. One of the consequences of such integrals is the Laplace transforms given in the following definition.

Definition 62. Let $f(P)$ be a function of $P(p \times p) > 0$ and $W = U + \iota V$, $\iota = \sqrt{-1}$ be a $p \times p$ complex matrix. Then the Laplace transform $g(W)$ of $f(P)$ is defined as

$$g(W) = \int_{P>0} e^{\operatorname{tr}(-WP)} f(P)\, dP,$$

where the integral is assumed to be absolutely convergent in the right half-plane $\Re(W) = U > U_0 > 0$.

The Laplace transform $g(W)$ of $f(P)$ defined above is an analytic function of W in the right half-plane $\Re(W) = U > U_0 > 0$. In addition, if

$$\int_{-\infty<V=V'<\infty} \left|g(U + \iota V)\right| dV < \infty, \quad U > U_0 > 0$$

and

$$\lim_{U\to 0} \int_{-\infty<V=V'<\infty} \left|g(U + \iota V)\right| dV = 0,$$

then the unique inverse Laplace transform $f(P)$ of $g(W)$ is given by

$$f(P) = \frac{2^{\frac{1}{2}p(p-1)}}{(2\pi\iota)^{\frac{1}{2}p(p+1)}} \int_{\Re(W)=U>U_0>0} \operatorname{etr}(WP)\, g(W)\, dW.$$

Another important property of the Laplace transformation is the convolution result. If g_1 and g_2 are the respective Laplace transforms of f_1 and f_2, then $g_1 \cdot g_2$ is the Laplace transform of f_3 where

$$f_3(Q) = \int_{0<P<Q} f_1(Q - P) f_2(P)\, dP.$$

M-transformation

Consider the class of $p \times p$ real symmetric definite matrices. If S is the member of the class and α is a complex number, such that $\Re(\alpha) > \frac{p-1}{2}$. Also, $f(S)$ is a symmetric function then the M-transform of $f(S)$, denoted by $M_\alpha(f)$, is defined as

$$M_\alpha(f) = \int_{U=U'>0} |U|^{\alpha - \frac{p+1}{2}} f(U)\, dU,$$

provided the $M_\alpha(f)$ exists.

3.5 Zonal polynomial

Let $P(p \times p)$ be a symmetric matrix and V_n be the vector space of homogeneous polynomials $\phi(x)$ of degree n in $\frac{1}{2}p(p+1)$ distinct elements of P. The space V_n can be decomposed into a direct sum of irreducible invariant subspaces V_κ where $\kappa = (n_1, \ldots, n_p)$, $n_1 + \cdots + n_p = \kappa$, $n_1 \geq \cdots \geq n_p \geq 0$. Then the polynomial $(\operatorname{tr} P)^n \in V_n$ has the unique decomposition into polynomials $C_\kappa(P) \in V_\kappa$. Thus, the zonal polynomial $C_\kappa(P)$ is the component of $(\operatorname{tr} P)^k \in V_n$ in the subspace V_κ.

The zonal polynomial $C_\kappa(P)$ is defined for all n and p, but for a partition κ of n into more than p parts, it is identically zero. These polynomials are invariant under orthogonal transformation, i. e.,

$$C_\kappa(P) = C_\kappa(R\,P\,R'), \quad R \in O(p).$$

Hence, $C_\kappa(P)$ is a symmetric homogeneous polynomial in the characteristic roots of P. Also, if Q is a symmetric positive definite matrix, then

$$C_\kappa(Q\,P) = C_\kappa(Q^{\frac{1}{2}}\,P\,Q^{\frac{1}{2}}),$$

where $Q^{\frac{1}{2}}$ is the unique symmetric positive definite square root of R. Khatri in 1971 proves that

$$\left|C_\kappa(P)\right| \leq C_\kappa(P_0),$$

where $P_0 = \operatorname{diag}(|a_1|, \ldots, |a_p|)$, and $a_i, i = 1, \ldots, p$ are the characteristic roots of P. If $P = I_p$ and the partition κ of n has r nonzero parts, then Constantine in 1963 and James in 1964 have shown that

$$C_\kappa(I_p) = \frac{2^{2n} n!\,(\frac{1}{2}p)_\kappa \prod_{i<j}^{r}(2n_i - 2n_j - i + j)}{\prod_{i=1}^{r}(2n_i + r - i)!},$$

where $(\frac{1}{2}p)_\kappa = \prod_{i=1}^{r}(\frac{1}{2}(p - i + 1))_{n_i}$ with $(u)_n = u\,(u+1)\cdots(u+n-1)$, $(u)_0 = 1$.

3.6 Multivariate gamma and beta functions

The multivariate gamma function, denoted by $\Gamma_p(u)$, is defined as

$$\Gamma_p(u) = \int\limits_{P>0} \operatorname{etr}(-P)\,\det(P)^{u-\frac{1}{2}(p+1)}\,dP, \tag{3.20}$$

where $\Re(u) > \frac{1}{2}(p-1)$, and the integral is over the space of $p \times p$ symmetric positive definite matrices.

The multivariate gamma function $\Gamma_p(u)$ can be written in terms of ordinary gamma functions as given in the following theorem.

Theorem 63. *Let u be a cumplex number such that* $\Re(u) > \frac{1}{2}(p-1)$. *Then*

$$\Gamma_p(u) = \pi^{\frac{1}{4}p(p-1)} \prod_{i=1}^{p} \Gamma\left[u - \frac{1}{2}(i-1)\right]. \tag{3.21}$$

Proof. By the definition of multivariate gamma function

$$\Gamma_p(u) = \int_{P>0} \operatorname{etr}(-P) \det(P)^{u-\frac{1}{2}(p+1)}\, dP. \tag{3.22}$$

Substitute $P = TT'$, where T is a lower triangular matrix with $t_{ii} > 0$, $i = 1, \ldots, p$. Then $\operatorname{tr}(P) = \operatorname{tr}(TT') = \sum_{j\le i}^{p}$, $\det(P) = \det(TT') = \det(T)^2 = \prod_{i=1}^{p} t_{ii}^2$, and from (3.19), we have

$$J(P \to T) = 2^p \prod_{i=1}^{p} t_{ii}^{p-i+1}.$$

Hence, the equation (3.22) yields

$$\begin{aligned}
\Gamma_p(u) &= 2^p \underbrace{\int \cdots \int}_{-\infty<t_{ij}<\infty, t_{ii}>0} \prod_{i=1}^{p} (t_{ii}^2)^{u-\frac{1}{2}i} \exp\left(-\sum_{j\le i}^{p} t_{ij}^2\right) \prod_{j\le i}^{p} dt_{ij} \\
&= \left[\prod_{j<i}^{p} \int_{-\infty<t_{ij}<\infty} \exp(-t_{ij}^2)\, dt_{ij}\right]\left[\prod_{i=1}^{p} 2 \int_{t_{ii}>0} \exp(-t_{ii}^2)(t_{ii}^2)^{u-\frac{1}{2}i} dt_{ii}\right] \\
&= \pi^{\frac{1}{4}p(p-1)} \prod_{i=1}^{p} \Gamma\left[u - \frac{1}{2}(i-1)\right].
\end{aligned} \tag{3.23}$$

It completes the proof. □

For distinct values of u and p, we have the outcomes in Table 3.1.

These outcomes can be obtained easily by putting particular values of u and p in the (3.21). A particular Laplace transform, which is quite useful, is

$$\int_{\mathcal{P}>0} \operatorname{etr}(-\mathcal{P}Z) \det(\mathcal{P})^{u-\frac{1}{2}(p+1)} d\mathcal{P} = \Gamma_p(u) \det(Z)^{-u}. \tag{3.24}$$

Herz in 1955 proved that the above integral is absolutely convergent for $\Re(Z) > 0$ and $\Re(u) > \frac{1}{2}(p-1)$. Hence, for $\Re(Z) > 0$, substituting $P = Z^{\frac{1}{2}}\mathcal{P}Z^{\frac{1}{2}}$ with the Jacobian $J(\mathcal{P} \to P) = \det(Z)^{-\frac{1}{2}(p+1)}$, transforms the above integral in the form

Table 3.1: Particular cases of the multivariate gamma function.

Particular values	Deduced functions
For $p = 1$	$\Gamma_p(u) = \Gamma(u)$
For $p = 1, u = 1$	$\Gamma_p(u) = 1$
For $p = 2, u = 1$	$\Gamma_p(u) = \pi$
For $p = 1, u = 2$	$\Gamma_p(u) = 1$
For $p = 2, u = 2$	$\Gamma_p(u) = \frac{\pi}{2}$
For $p = 3, u = 2$	$\Gamma_p(u) = \frac{\pi^2}{2}$
For $p = 2, u = 3$	$\Gamma_p(u) = \frac{3}{2}\pi$

$$\int_{\mathcal{P}>0} \operatorname{etr}(-\mathcal{P} Z) \det(\mathcal{P})^{u-\frac{1}{2}(p+1)} d\mathcal{P} = \det(Z)^{-u} \int_{P>0} \operatorname{etr}(-P) \det(P)^{u-\frac{1}{2}(p+1)} dP$$
$$= \Gamma_p(u) \det(Z)^{-u}. \tag{3.25}$$

It completes the proof of (3.24) for real Z. For complex Z, it follows by analytical continuation since $\Re(Z) > 0, \det(Z) \neq 0$ and $\det(Z)^u$ are well-defined by continuation. The inverse formuka of (3.24) is given by Herz as

$$f(P) = \frac{2^{\frac{1}{2}p(p-1)}}{(2\pi\iota)^{\frac{1}{2}p(p+1)}} \int_{\Re(Z)=X>X_0>0} \operatorname{etr}(Z\mathcal{P}) \det(Z)^{-u-\frac{1}{2}(p+1)} dZ$$
$$= \frac{\det(\mathcal{P})^u}{\Gamma_p(u + \frac{1}{2}(p+1))}, \quad \mathcal{P} > 0.$$

Definition 64. The multivariate beta function, denoted by $\beta^p(u, v)$ is defined by

$$\beta^p(u, v) = \int_{0<P<I_p} \det(P)^{u-\frac{1}{2}(p+1)} \det(I_p - P)^{v-\frac{1}{2}(p+1)} dP, \tag{3.26}$$

where $\Re(u) > \frac{1}{2}(p - 1)$ and $\Re(v) > \frac{1}{2}(p - 1)$.

In the next theorem, we produce the relation between multivariate beta and gamma functions.

Theorem 65. *Let u and v be complex numbers such that $\Re(u) > \frac{1}{2}(p - 1)$ and $\Re(v) > \frac{1}{2}(p - 1)$. Then*

$$\beta^p(u, v) = \frac{\Gamma_p(u)\Gamma_p(v)}{\Gamma_p(u + v)} \tag{3.27}$$
$$= \beta^p(v, u). \tag{3.28}$$

Proof.

$$\begin{aligned}\Gamma_p(u)\,\Gamma_p(v) &= \int_{P>0} \operatorname{etr}(-P)\,\det(P)^{u-\frac{1}{2}(p+1)}\,dP \int_{Q>0} \operatorname{etr}(-Q)\,\det(Q)^{v-\frac{1}{2}(p+1)}\,dQ \\ &= \int_{P>0}\int_{Q>0} \operatorname{etr}(-(P+Q))\,\det(P)^{u-\frac{1}{2}(p+1)}\,\det(Q)^{v-\frac{1}{2}(p+1)}\,dP\,dQ.\end{aligned}$$

Now making the transformation $W = P + Q, Z = (P + Q)^{-\frac{1}{2}}\,P\,(P + Q)^{-\frac{1}{2}}$, where $(P + Q)^{\frac{1}{2}}$ is the symmetric root of $P + Q$ with Jacobian $J(P, Q \to Z, W) = \det(W)^{\frac{1}{2}(p+1)}$, we get

$$\begin{aligned}\Gamma_p(u)\,\Gamma_p(v) &= \Big(\int_{W>0} \operatorname{etr}(-W)\,\det(W)^{u+v-\frac{1}{2}(p+1)}\,dW\Big) \\ &\quad\times\Big(\int_{0<Z<I_p} \det(Z)^{u-\frac{1}{2}(p+1)}\,\det(I_p - Z)^{v-\frac{1}{2}(p+1)}\,dZ\Big) \\ &= \Gamma_p(u+v)\,\beta^p(u,v).\end{aligned}$$

It completes the proof. □

Next, we define the generalized hypergeometric coefficient, which frequently occurs in integrals involving zonal polynomials. Let $\kappa = (n_1, \ldots, n_p)$, $n_1 \geq \cdots \geq n_p \geq 0$, $n_1 + \cdots + n_p = \kappa$. Then

$$(u)_\kappa = \prod_{j=1}^{p}\Big(u - \frac{1}{2}(j-1)\Big)_{n_j}. \tag{3.29}$$

Using the notation,

$$\Gamma_p(u,\kappa) = \pi^{\frac{1}{4}p(p-1)}\prod_{j=1}^{p}\Gamma\Big(u + n_j - \frac{1}{2}(j-1)\Big), \quad \Re(u) \geq \frac{1}{2}(p-1) - n_p, \tag{3.30}$$

so that $\Gamma_p(u,0) = \Gamma_p(u)$, we can write (3.29) as

$$(u)_\kappa = \frac{\Gamma_p(u,\kappa)}{\Gamma_p(u)}. \tag{3.31}$$

Khatri in 1966 used the notation

$$\Gamma_p(u,-\kappa) = \pi^{\frac{1}{4}p(p-1)}\prod_{j=1}^{p}\Gamma\Big(u - n_j - \frac{1}{2}(j-1)\Big), \quad \Re(u) \geq \frac{1}{2}(p-1) + n_1. \tag{3.32}$$

Alternatively, we can write

$$\Gamma_p(u,-\kappa)=\frac{(-1)^n\,\Gamma_p(u)}{(-u+\frac{1}{2}(p+1))_\kappa}.$$

We now produce certain integrals showing a connection between zonal polynomials and the multivariate gamma function, without their proofs.

Lemma 66. *Let $Z(p\times p)$ be a complex symmetric matrix such that $\Re(Z)>0$ and $T(p\times p)$ be a complex symmetric matrix. Then*

$$\begin{aligned}&\int_{P>0}\operatorname{etr}(-Z\,P)\,\det(P)^{u-\frac{1}{2}(p+1)}\,C_\kappa(T\,P)\,dP\\&\quad=\Gamma_p(u,\kappa)\,\det(Z)^{-u}\,C_\kappa(T\,Z^{-1}),\quad \Re(u)>\frac{1}{2}(p-1)\end{aligned}\tag{3.33}$$

and

$$\begin{aligned}&\int_{P>0}\operatorname{etr}(-Z\,P)\,\det(P)^{u-\frac{1}{2}(p+1)}\,C_\kappa(T\,P^{-1})\,dP\\&\quad=\Gamma_p(u,-\kappa)\,\det(Z)^{-u}\,C_\kappa(T\,Z),\quad \Re(u)>\frac{1}{2}(p-1)+n_1.\end{aligned}\tag{3.34}$$

Lemma 67. *Let $T(p\times p)$ be a symmetric matrix. Then*

$$\begin{aligned}&\int_{0<P<I_p}\det(P)^{u-\frac{1}{2}(p+1)}\,\det(I_p-P)^{v-\frac{1}{2}(p+1)}\,C_\kappa(T\,P)\,dP\\&\quad=\frac{\Gamma_p(u,\kappa),\Gamma_p(v)}{\Gamma_p(u+v,\kappa)}\,C_\kappa(T),\quad \Re(u)>\frac{1}{2}(p-1),\Re(v)>\frac{1}{2}(p-1)\end{aligned}\tag{3.35}$$

and

$$\begin{aligned}&\int_{0<P<I_p}\det(P)^{u-\frac{1}{2}(p+1)}\,\det(I_p-P)^{v-\frac{1}{2}(p+1)}C_\kappa(T\,P^{-1})\,dP\\&\quad=\frac{\Gamma_p(u,-\kappa),\Gamma_p(v)}{\Gamma_p(u+v,-\kappa)}\,C_\kappa(T),\quad \Re(u)>\frac{1}{2}(p-1)+n_1,\Re(v)>\frac{1}{2}(p-1).\end{aligned}\tag{3.36}$$

Lemma 68. *Let $T(p\times p)$ be a complex symmetric matrix. Then*

$$\begin{aligned}&\int_{P>0}\det(P)^{u-\frac{1}{2}(p+1)}\,\det(I_p+P)^{-(u+v)}\,C_\kappa(T\,P)\,dP\\&\quad=\frac{\Gamma_p(u,\kappa),\Gamma_p(v,-\kappa)}{\Gamma_p(u+v)}\,C_\kappa(T),\quad \Re(u)>\frac{1}{2}(p-1),\Re(v)>\frac{1}{2}(p-1)+n_1\end{aligned}\tag{3.37}$$

and

$$\int_{P>0} \det(P)^{u-\frac{1}{2}(p+1)} \det(I_p + P)^{-(u+v)} C_\kappa(T P^{-1})\, dP$$
$$= \frac{\Gamma_p(u, -\kappa), \Gamma_p(v, \kappa)}{\Gamma_p(u + v)} C_\kappa(T), \quad \Re(u) > \frac{1}{2}(p-1) + n_1, \Re(v) > \frac{1}{2}(p-1). \tag{3.38}$$

It may be noted that (3.33) is the Laplace transform of $\det(P)^{u-\frac{1}{2}(p+1)}$. Thus, using inverse Laplace transform, we get

$$\frac{2^{\frac{1}{2}p(p-1)}}{(2\pi\iota)^{\frac{1}{2}p(p+1)}} \int_{\Re(Z)>0} \operatorname{etr}(Z P) \det(Z)^{-u} C_\kappa(T Z^{-1})\, dZ$$
$$= \frac{1}{\Gamma_p(u, \kappa)} \det(P)^{u-\frac{1}{2}(p+1)} C_\kappa(T P), \quad \Re(u) > \frac{1}{2}(p-1). \tag{3.39}$$

Similarly, from (3.34) we get

$$\frac{2^{\frac{1}{2}p(p-1)}}{(2\pi\iota)^{\frac{1}{2}p(p+1)}} \int_{\Re(Z)>0} \operatorname{etr}(Z P) \det(Z)^{-u} C_\kappa(T Z)\, dZ$$
$$= \frac{1}{\Gamma_p(u, -\kappa)} \det(P)^{u-\frac{1}{2}(p+1)} C_\kappa(T P^{-1}), \quad \Re(u) > \frac{1}{2}(p-1) + n_1. \tag{3.40}$$

3.7 Gauss hypergeometric function of the matrix arguments

There are mainly two ways of defining hypergeometric functions with matrix arguments via the Laplace transformation and zonal polynomial. Constantine (1963) gave the power series representation involving zonal polynomials as given below.

Definition 69. The Gauss hypergeometric function of the matrix argument is defined by

$${}_2F_1(u_1, u_2; v_1; P) = \sum_{k=0}^{\infty} \sum_{\kappa} \frac{(u_1)_\kappa (u_2)_\kappa}{(v_1)_\kappa} \frac{C_\kappa(P)}{k!}, \tag{3.41}$$

where u_1, u_2 and v_1 are complex numbers, $P(p \times p)$ is a complex symmetric matrix and $\sum_\kappa$ denotes the summation over all partitions κ.

The series (3.41) converges

- if v_1 is nonzero, an integer or half-integer less than or equal to $\frac{1}{2}(p-1)$;
- for all $P(p \times p)$ such that $\|P\| < 1$, where $\|P\|$ denotes the maximum absolute value of the characteristic roots of P.

Definition 70. The Gauss hypergeometric function for two symmetric matrices $P(p \times p)$ and $Q(p \times p)$ is defined by

$$ {}_2F_1^{(p)}(u_1, u_2; v_1; P, Q) = \sum_{k=0}^{\infty} \sum_{\kappa} \frac{(u_1)_\kappa (u_2)_\kappa}{(v_1)_\kappa} \frac{C_\kappa(P)\, C_\kappa(Q)}{C_\kappa(I_p)\, k!}. \tag{3.42} $$

The series (3.42) converges for $\|P\| < 1$ or $\|Q\| < 1$. It converges more rapidly if both of the matrices P and Q are such that $\|P\| < 1$ and $\|Q\| < 1$.

These two definitions of the Gauss hypergeometric function of the matrix argument can be generalized by increasing the number of numerator and denominator parameters as the following.

Definition 71. The generalized hypergeometric function of the matrix argument is defined by

$$ {}_mF_n(u_1, \ldots, u_m; v_1, \ldots, v_n; P) = \sum_{k=0}^{\infty} \sum_{\kappa} \frac{(u_1)_\kappa \cdots (u_m)_\kappa}{(v_1)_\kappa \cdots (v_n)_\kappa} \frac{C_\kappa(P)}{k!}, \tag{3.43} $$

where $u_1, \ldots, u_m$ and $v_1, \ldots, v_n$ are complex numbers.

The series (3.43) converges

- if v_i is nonzero, for $i = 1, \ldots, n$, an integer or half-integer less than or equal to $\frac{1}{2}(p-1)$;
- for all $P(p \times p)$ if $m < n + 1$,
- if $m = n+1$, the series converges for all $P(p \times p)$ such that $\|P\| < 1$, where $\|P\|$ denotes the maximum absolute value of the characteristic roots of P,
- unless the series terminates, it diverges for all $P \neq 0$ if $m > n + 1$.

Definition 72. The generalized hypergeometric function of two symmetric matrices $P(p \times p)$ and $Q(p \times p)$ is defined by

$$ {}_mF_n^{(p)}(u_1, \ldots, u_m; v_1, \ldots, v_n; P, Q) = \sum_{k=0}^{\infty} \sum_{\kappa} \frac{(u_1)_\kappa \cdots (u_m)_\kappa}{(v_1)_\kappa \cdots (v_n)_\kappa} \frac{C_\kappa(P)\, C_\kappa(Q)}{C_\kappa(I_p)\, k!}. \tag{3.44} $$

Conditions for convergence of (3.44) are similar to the conditions for convergence of (3.43)except that for $m = n+1$ the series converges for $\|P\| < 1$ or $\|Q\| < 1$. It converges more rapidly, if both the matrices P and Q are such that $\|P\| < 1$ and $\|Q\| < 1$.

Now, some elementary hypergeometric functions can be defined with the help of these generalized hypergeometric functions of the matrix argument. From (3.43), it follows that

$$
\begin{aligned}
{}_0F_0(P) &= \sum_{k=0}^{\infty} \sum_{\kappa} \frac{C_\kappa(P)}{k!} \\
&= \sum_{k=0}^{\infty} \frac{(\operatorname{tr}(P))^k}{k!} \\
&= \operatorname{etr}(P)
\end{aligned} \tag{3.45}
$$

and

$$
{}_1F_1(u_1; v_1; P) = \sum_{k=0}^{\infty} \sum_{\kappa} \frac{(u_1)_\kappa}{(v_1)_\kappa} \frac{C_\kappa(P)}{k!}. \tag{3.46}
$$

From (3.44), we have

$$
{}_1F_1^{(p)}(u_1; v_1; P, Q) = \sum_{k=0}^{\infty} \sum_{\kappa} \frac{(u_1)_\kappa}{(v_1)_\kappa} \frac{C_\kappa(P)\, C_\kappa(Q)}{C_\kappa(I_p)\, k!}. \tag{3.47}
$$

We now give the integral form of the Gauss hypergeometric function with matrix arguments in the theorem given below.

Theorem 73. *Let $Z(p \times p)$ be a complex symmetric matrix of which the real part is positive definite and $T(p \times p)$, $R(p \times p)$ be complex symmetric matrices. Then*

$$
\begin{aligned}
&\int_{P>0} \operatorname{etr}(-Z\,P)\, \det(P)^{u-\frac{1}{2}(p+1)}\, {}_1F_1(u_1; v_1; PT)\, dP \\
&\quad = \Gamma_p(u)\, \det(Z)^{-u}\, {}_2F_1(u_1, u_2; v_1; Z^{-1}\,T), \quad \Re(u) > \frac{1}{2}(p-1),
\end{aligned} \tag{3.48}
$$

and

$$
\begin{aligned}
&\int_{P>0} \operatorname{etr}(-Z\,P)\, \det(P)^{u-\frac{1}{2}(p+1)}\, {}_1F_1^{(p)}(u_1; v_1; PT, Q)\, dP \\
&\quad = \Gamma_p(u)\, \det(Z)^{-u}\, {}_2F_1^{(p)}(u_1, u_2; v_1; Z^{-1}\,T, Q), \quad \Re(u) > \frac{1}{2}(p-1),
\end{aligned} \tag{3.49}
$$

where ${}_1F_1(u_1; v_1; PT)$ and ${}_1F_1^{(p)}(u_1; v_1; PT, Q)$ are the confluent hypergeometric function of the matrix arguments given by (3.46) *and* (3.47), *respectively.*

The following corollary will be helpful in the derivation of the integral form of generalized hypergeometric function with matrix arguments.

Corollary 74. *For $\|Z\| < 1$,*

$$
{}_1F_0(u; Z) = \det(I_p - Z)^{-u}. \tag{3.50}
$$

Proof. From (3.48), letting $T = I_p$ and replacing Z by Z^{-1}, we get

$$ {}_1F_0(u;Z) = \frac{\det(Z)^{-u}}{\Gamma_p(u)} \int\limits_{P>0} \operatorname{etr}(-Z^{-1}P) \det(P)^{u-\frac{1}{2}(p+1)} {}_0F_0(P)\,dP. \tag{3.51} $$

Now substituting $Z^{-\frac{1}{2}} P Z^{-\frac{1}{2}} = S$ with the Jacobian $J(P \to S) = \det(Z)^{\frac{1}{2}(p+1)}$ and using ${}_0F_0(ZS) = \operatorname{etr}(Z\,S)$, we get

$$ {}_1F_0(u;Z) = \frac{1}{\Gamma_p(u)} \int\limits_{P>0} \operatorname{etr}\{-S(I_p - Z)\} \det(S)^{u-\frac{1}{2}(p+1)}\,dS = \det(I_p - Z)^{-u}. \tag{3.52} $$

It completes the proof. □

Theorem 75. *Let $R(p \times p)$ be a symmetric matrix. Then*

$$ \int\limits_{0<P<I_p} \det(P)^{u-\frac{1}{2}(p+1)} \det(I_p - P)^{v-\frac{1}{2}(p+1)} {}_mF_n(u_1,\dots,u_m; v_1,\dots,v_n; R\,P)\,dP $$
$$ = \frac{\Gamma_p(u)\,\Gamma_p(v)}{\Gamma_p(u+v)} {}_{m+1}F_{n+1}(u_1,\dots,u_m,u; v_1,\dots,v_n,u+v; R). \tag{3.53} $$

The particular cases of the above theorem give the integral representations for confluent and Gauss hypergeometric functions.

Corollary 76. *Let $R(p \times p)$ be a symmetric matrix and $\Re(u) > \frac{1}{2}(p-1)$, $\Re(v) > \frac{1}{2}(p-1)$, $\Re(v-u) > \frac{1}{2}(p-1)$. Then*

$$ {}_1F_1(u;v;R) $$
$$ = \frac{\Gamma_p(v)}{\Gamma_p(u)\,\Gamma_p(v-u)} \int\limits_{0<P<I_p} \det(P)^{u-\frac{1}{2}(p+1)} \det(I_p - P)^{v-u-\frac{1}{2}(p+1)} \operatorname{etr}(R\,P)\,dP. \tag{3.54} $$

Proof. Substituting $m = n = 0$ and replacing v with $v-u$ in the above theorem yields

$$ {}_1F_1(u;v;R) $$
$$ = \frac{\Gamma_p(v)}{\Gamma_p(u)\,\Gamma_p(v-u)} \int\limits_{0<P<I_p} \det(P)^{u-\frac{1}{2}(p+1)} \det(I_p - P)^{v-u-\frac{1}{2}(p+1)} {}_0F_0(R\,P)\,dP. $$

As we know ${}_0F_0(R\,P) = \operatorname{etr}(R\,P)$. Thus, we get

$$ {}_1F_1(u;v;R) $$
$$ = \frac{\Gamma_p(v)}{\Gamma_p(u)\,\Gamma_p(v-u)} \int\limits_{0<P<I_p} \det(P)^{u-\frac{1}{2}(p+1)} \det(I_p - P)^{v-u-\frac{1}{2}(p+1)} \operatorname{etr}(R\,P)\,dP, $$

which is the required result. □

Corollary 77. *Let $R(p\times p)$ be a symmetric matrix such that $\Re(R) < I_p$ and $\Re(w) > \frac{1}{2}(p-1)$, $\Re(u) > \frac{1}{2}(p-1)$, $\Re(w-u) > \frac{1}{2}(p-1)$. Then*

$$
\begin{aligned}
&{}_2F_1(u, v; w; R) \\
&= \frac{\Gamma_p(w)}{\Gamma_p(u)\,\Gamma_p(w-u)} \int\limits_{0<P<I_p} \det(P)^{u-\frac{1}{2}(p+1)} \det(I_p - P)^{w-u-\frac{1}{2}(p+1)} \det(I_p - R\,P)^{-v}\, dP. \quad (3.55)
\end{aligned}
$$

Proof. Substituting $m = 1$, $n = 0$, $u_1 = v$ and $v = w - u$ in Theorem 75 yields

$$
\begin{aligned}
&{}_2F_1(u, v; w; R) \\
&= \frac{\Gamma_p(w)}{\Gamma_p(u)\,\Gamma_p(w-u)} \int\limits_{0<P<I_p} \det(P)^{u-\frac{1}{2}(p+1)} \det(I_p - P)^{w-u-\frac{1}{2}(p+1)}\, {}_1F_0(v; R\,P)\, dP.
\end{aligned}
$$

We already proved that ${}_1F_0(u; Z) = \det(I_p - Z)^{-u}$. Thus, we get

$$
\begin{aligned}
&{}_2F_1(u, v; w; R) \\
&= \frac{\Gamma_p(w)}{\Gamma_p(u)\,\Gamma_p(w-u)} \int\limits_{0<P<I_p} \det(P)^{u-\frac{1}{2}(p+1)} \det(I_p - P)^{w-u-\frac{1}{2}(p+1)} \det(I_p - R\,P)^{-v}\, dP.
\end{aligned}
$$

It completes the proof. □

The presented chapter is a motivation of the work carried by Muirhead, James, Herz, Mathai to name a few. For more detail, we suggest the reader to visit the following literature and references there in, [32, 33, 69–71, 75, 77, 87–89, 111, 114, 117, 118, 125]

Exercises

1. Check which of the following matrices are positive definite?

$$
\begin{bmatrix} 1 & \frac{1}{4} \\ \frac{1}{4} & \frac{1}{4} \end{bmatrix}, \quad \begin{bmatrix} 3 & 1-i \\ 1+i & 1 \end{bmatrix}, \quad \begin{bmatrix} 1 & -1 & 4 \\ 2 & -1 & 1 \\ 3 & -1 & 1 \end{bmatrix}.
$$

2. Check whether the sum and product of two positive definite matrices is positive definite or not?
3. If $u = 6\,u_1^2 + 2\,u_2^2 + 8\,u_1\,u_2 - 6\,u_1\,u_3$, then find the value of $\frac{\partial u}{\partial U}$, $\frac{\partial u}{\partial U'}$ and $\frac{\partial^2 u}{\partial U\,\partial U'}$, where $U = \left[\begin{smallmatrix} u_1 \\ u_2 \\ u_3 \end{smallmatrix}\right]$.
4. If $P = \left[\begin{smallmatrix} 3 & 2 \\ 1 & 1 \end{smallmatrix}\right]$, $U = \left[\begin{smallmatrix} u_{11} & u_{12} \\ u_{21} & u_{22} \end{smallmatrix}\right]$, $u = \left[\begin{smallmatrix} 0 \\ 1 \end{smallmatrix}\right]$ and $v = \left[\begin{smallmatrix} 1 \\ -1 \end{smallmatrix}\right]$, then find

$$
\text{(i)}\ \frac{\partial}{\partial U}\operatorname{tr}(U) \qquad \text{(ii)}\ \frac{\partial}{\partial U}\operatorname{tr}(PU) \qquad \text{(iii)}\ \frac{\partial}{\partial U}\,(u'Uv).
$$

5. Prove that for a symmetric positive definite matrix $V(p\times p)$ and a lower triangular matrix $T(p\times p) = (t_{ij})$, if $V = T\,P\,T'$, then

$$J(V \to T) = 2^p \prod_{i=1}^{p} (t_{ii}^{p-i+1} \det(P^{[i]})),$$

where $P(p\times p) = (a_{jk})$ is nonsingular and $P^{[i]} = (a_{jk}), j, k = 1, \dots, i$.

6. Prove that for a symmetric positive definite matrix $V(p\times p)$ and an upper triangular matrix $T(p\times p) = (t_{ij})$, if $V = T\,T'$, then

$$J(V \to T) = 2^p \prod_{i=1}^{p} t_{ii}^{i}$$

and if $V = T\,P\,T'$, then

$$J(V \to T) = 2^p \prod_{i=1}^{p} (t_{ii}^{i} \det(P_{[i]})),$$

where $P(p\times p) = (a_{jk})$ is nonsingular and $P_{[i]} = (a_{jk}), j, k = p - i + 1, \dots, p$.

7. Let $T(p\times p)$ be a complex symmetric matrix. Then prove that

$$\int_{P>0} \operatorname{etr}(-P) \det(P)^{u-\frac{1}{2}(p+1)} (\operatorname{tr} P)^j C_\kappa(T\,P)\, dP$$
$$= \frac{\Gamma_p(u,\kappa), \Gamma(pu+j+k)}{\Gamma(pu+k)} C_\kappa(T), \quad \Re(u) > \frac{1}{2}(p-1)$$

and

$$\int_{P>0} \operatorname{etr}(-P) \det(P)^{u-\frac{1}{2}(p+1)} (\operatorname{tr} P)^j C_\kappa(T\,P^{-1})\, dP$$
$$= \frac{\Gamma_p(u,-\kappa), \Gamma(pu+j-k)}{\Gamma(pu-k)} C_\kappa(T), \quad \Re(u) > \frac{1}{2}(p-1) + n_1.$$

8. Show that

$$\beta^p(u,v,w) = \beta^p(u,v)\,\beta^p(u+v,w) = \frac{\Gamma_p(u)\,\Gamma_p(v)\,\Gamma_p(w)}{\Gamma_p(u+v+w)}.$$

9. Let Q be a q-square symmetric matrix and P be a p-square symmetric positive definite matrix with $q > p$. Also, let $U(p\times q)$ be a matrix with real entries. Then prove that

$$\int_{UU'=V} \exp(\operatorname{Tr} PUQU')\, dU = (\pi)^{\frac{1}{2}pq} \left\{\Gamma_p\left(\frac{1}{2}q\right)\right\}^{-1} |V|^{\frac{1}{2}(q-p-1)}\, {}_0F_0^q(PV, Q).$$

10. Verify that the hypergeometric functions ${}_1F_1$ and ${}_2F_1$ satisfy the following relations:

$$\begin{aligned}
{}_1F_1(u;w;P) &= \operatorname{etr}(P)\,{}_1F_1(w-u;w;-P);\\
{}_2F_1(u,v;w;P) &= \det(I_p-P)^{-v}\,{}_2F_1(w-u,v;w;-P(I_p-P)^{-1})\\
&= \det(I_p-P)^{w-u-v}\,{}_2F_1(w-u,w-v;w;P).
\end{aligned}$$

11. Using the zonal polynomial expansion, establish the following confluence relations:

$$\begin{aligned}
\lim_{u\to\infty} {}_1F_1\left(u;w;\frac{1}{u}P\right) &= {}_0F_1(w;P);\\
\lim_{u\to\infty} {}_2F_1\left(u,v;w;\frac{1}{u}P\right) &= {}_1F_1(v,w;P).
\end{aligned}$$

4 Gauss hypergeometric matrix function

This chapter deals with the hypergeometric function when the parameters are chosen from a collection of p-square complex matrices. We discuss the various forms of the hypergeometric matrix function and its properties in detail. In particular, we look at the Gauss hypergeometric matrix function, the matrix-valued Gauss hypergeometric function and the basic Gauss hypergeometric matrix function. We list various properties of these forms, namely regions of convergence, differential equations, integral representations and recurrence relations to name a few. Before we proceed, let us look at the matrix terminology and concepts needed in this chapter. For more detail, see [10, 34, 38, 39, 44, 61, 65, 78–81, 94–96, 132, 134, 153–156] and references therein.

4.1 Matrix functional calculus

Let P be a matrix in $\mathbb{C}^{p\times p}$. Then the spectrum of P, denoted by $\sigma(P)$, is the set of all eigenvalues of P. Also,

$$\begin{aligned} \alpha(P) &= \max\{\Re(z) \mid z \in \sigma(P)\}, \\ \beta(P) &= \min\{\Re(z) \mid z \in \sigma(P)\}, \end{aligned} \tag{4.1}$$

where $\Re(z)$ denotes the real part of z. A matrix P is called positive stable if the real part of each eigenvalue of P is positive, i. e., $\beta(P) > 0$.

Example 78. Find the spectrum, $\alpha(P)$ and $\beta(P)$ for a matrix $P = \left[\begin{smallmatrix} 2 & -1 \\ -1 & 2 \end{smallmatrix}\right]$.

Solution. The characteristic equation is given by

$$\det(P - \lambda I) = \begin{vmatrix} 2-\lambda & -1 \\ -1 & 2-\lambda \end{vmatrix} = 0, \tag{4.2}$$

which on solving gives the characteristic values of the matrix P as $\lambda_1 = 1, \lambda_2 = 3$. Thus, we have the following:

$$\text{spectrum of } P = \sigma(P) = \{1, 3\}, \quad \alpha(P) = 3, \quad \beta(P) = 1.$$

One can easily verify the following results.

Lemma 79. *Let P be a p-square complex matrix. Then:*
1. $\alpha(-P) = -\beta(P)$;
2. $\alpha(P + kI) = \alpha(P) + k$;
3. $\beta(P + kI) = \beta(P) + k$.

If $f(z)$ and $g(z)$ are holomorphic functions of the complex variable z, which are defined in an open set Ω of the complex plane, and P is a matrix in $\mathbb{C}^{p\times p}$ with $\sigma(P) \subset \Omega$, then from

https://doi.org/10.1515/9783111324586-004

the properties of the matrix functional calculus, it follows that the image of $f(z)$ and $g(z)$ with respect to the matrix P are well-defined matrices. Also, they commute with each other, i. e.,

$$f(P)g(P) = g(P)f(P). \tag{4.3}$$

Furthermore, if $Q \in \mathbb{C}^{p\times p}$ is a matrix for which $\sigma(Q) \subset \Omega$, and if $PQ = QP$, then

$$f(P)g(Q) = g(Q)f(P). \tag{4.4}$$

There are several ways of defining the matrix function motivated from their complex versions, namely through the Cauchy integral theorem, the Jordan based definition, Schur decomposition approach, an eigenvector approach and approximation method to name a few. All these methods have their individual important and can be applied as per the requirement. According to chapter need, we describe few of them in very short way.

1. *By direct substitution:* if P be a p-square complex matrix and $g(z)$ be a function defined on $\sigma(P)$, then the matrix function $g(P)$ can be defined by replacing z with P. For example, if

$$g(z) = \frac{1 + z + z^2}{2 - z}, \quad 2 \notin \sigma(P), \tag{4.5}$$

then

$$g(P) = (I + P + P^2)\,(2I - P)^{-1}. \tag{4.6}$$

2. *Through the Cauchy integral theorem:* The line integral approach for a matrix function can be seen as the matrix version of the Cauchy integral theorem. Let $g(z)$ be a complex valued function and analytic inside, and on the closed contour C. Also, let $\sigma(P) \subset C$, for $P \in \mathbb{C}^{p\times p}$. Then the matrix analogue of $g(z)$ is a well-defined matrix, denoted by $g(P)$, given by

$$g(P) = \frac{1}{2\pi\iota}\int_C g(z)\,(zI - P)^{-1}dz. \tag{4.7}$$

3. *Jordan based definition:* Another way to define the matrix version of a function is using the Jordan canonical characterization. Let P be matrix in $\mathbb{C}^{p\times p}$ and also let

$$Q^{-1}PQ = \operatorname{diag}(J_1, \ldots, J_n)$$

be the Jordan canonical form of P, where

$$J_i = \begin{bmatrix} \lambda_i & 1 & 0 & \cdots & 0 \\ 0 & \lambda_i & 1 & \cdots & 0 \\ \vdots & \vdots & \vdots & \vdots & \vdots \\ 0 & \cdots & \cdots & \cdots & \lambda_i \end{bmatrix},$$

is a Jordan block of order m_i. If $g(z)$ is an analytic function in an open set Ω such that $\sigma(P) \subset \Omega$, then

$$g(P) = Q\,\mathrm{diag}(g(J_1)\cdots g(J_n))\,Q^{-1},$$

where

$$g(J_i) = \begin{bmatrix} g(\lambda_i) & g'(\lambda_i) & \cdots & \frac{g^{(m_i-1)}(\lambda_i)}{(m_i-1)!} \\ 0 & g(\lambda_i) & \cdots & \frac{g^{(m_i-2)}(\lambda_i)}{(m_i-2)!} \\ \vdots & \vdots & \vdots & \vdots \\ 0 & \cdots & \cdots & g(\lambda_i) \end{bmatrix}.$$

The superscript on each function $g(\lambda_i)$ represents the derivative of $g(z)$ at $z = \lambda_i$. In particular, if each Jordan block in P is of order 1, i. e., P is a diagonal matrix, then $g(P)$ can be evaluated directly as

$$g(P) = Q^{-1}\,\mathrm{diag}(g(\lambda_1), \ldots, g(\lambda_n))\,Q.$$

4. *An eigenvector approach:* If $P \in \mathbb{C}^{p\times p}$ such that $P = X\ \mathrm{diag}(\lambda_1, \ldots, \lambda_p)X^{-1}$ and $g(P)$ is defined, then

$$g(P) = X\ \mathrm{diag}(g(\lambda_1), \ldots, g(\lambda_p))X^{-1}.$$

5. *Schur decomposition approach:* If $P = Q\,T\,Q^*$ is the Schur decomposition of P, then

$$g(P) = Q\,g(T)\,Q^*.$$

6. *Approximation method:* If $P \in \mathbb{C}^{p\times p}$ and $g(z)$ approximate $h(z)$ on $\sigma(P)$, then $g(P)$ approximate $h(P)$.

We now list elementary matrix functions and their expansions. These matrix functions will be useful in the subsequent sections of this chapter.

1. *Exponential matrix function:* Let P be a p-square real or complex matrix. Then the exponential function e^P is defined by

$$e^P = \sum_{k=0}^{\infty} \frac{1}{k!}P^k = I + \frac{P}{1!} + \frac{P^2}{2!} + \frac{P^3}{3!} + \cdots. \tag{4.8}$$

Equivalently,

$$e^P = \lim_{k\to\infty}\left(I + \frac{P}{k}\right)^k. \tag{4.9}$$

2. *Logarithm of a matrix:* The logarithm of a matrix is defined as the inverse of an exponential matrix function. If the exponential of a matrix P is Q, then Q will be the matrix logarithm of P. The power series expansion of the logarithm of a matrix P is given by

$$\log(I + P) = \sum_{k=1}^{\infty} \frac{(-1)^k}{k} P^k = P - \frac{P^2}{2} + \frac{P^3}{3} - \cdots, \quad \|P\| < 1. \tag{4.10}$$

Alternatively,

$$\begin{aligned}\log(P) &= \sum_{k=1}^{\infty} \frac{(-1)^k}{k} (P - I)^k \\ &= (P - I) - \frac{(P - I)^2}{2} + \frac{(P - I)^3}{3} - \cdots, \quad \|P - I\| < 1.\end{aligned} \tag{4.11}$$

3. *Trigonometric sine and cosine matrix functions:* For a matrix $P \in \mathbb{C}^{p\times p}$, the sine and cosine matrix functions are defined by

$$\sin P = P - \frac{P^3}{3!} + \frac{P^5}{5!} - \frac{P^7}{7!} + \cdots, \tag{4.12}$$

$$\cos P = I - \frac{P^2}{2!} + \frac{P^4}{4!} - \frac{P^6}{6!} + \cdots. \tag{4.13}$$

4.2 Norm of a matrix

By the norm of a matrix, we mean that a function g defined from the set of $(p\times n)$ ordered matrices to a scalar value, i. e.,

$$g : M_{p\times n} \to \mathbb{R},$$

and having these three properties:
1. $g(P) > 0$ and $g(P) = 0$ if and only if $P = 0$, for every $P \in M_{p\times n}$;
2. $g(P + Q) \le g(P) + g(Q), P, Q \in M_{p\times n}$;
3. $g(\alpha P) = |\alpha|\, g(P)$.

We usually denote a norm by $\|\cdot\|$ and subscripts to designate the norm. There are several norms that arise frequently in the literature. For example:

1. *Frobenius norm.* For a matrix $P(p \times n)$, the Frobenius norm or Euclidean norm are defined as the square root of the sum of the absolute square of its elements:

$$\|P\|_F = \sqrt{\sum_{i=1}^{n}\sum_{j=1}^{n} |c_{ij}|^2},$$

which is also defined as the square root of the trace of $P P^*$, where P^* denotes the conjugate transpose of P, i. e.,

$$\|P\|_F = \sqrt{\mathrm{Tr}(P \cdot P^*)}.$$

2. *The k-norm.* The k-norm, for a matrix $P(p \times n)$, is defined by

$$\|P\|_k = \sup_{x \neq 0} \frac{\|P x\|_k}{\|x\|_k},$$

where $\|x\|_k$ is vector k-norm given by

$$\|x\|_k = (x_1^k + x_2^k + \cdots + x_n^k)^{1/k}.$$

In view of the definition for $\|x\|_k$, the k-norm of the matrix can be rewritten as

$$\|P\|_k = \sup_{x \neq 0} \left\| P\left(\frac{x}{\|x\|_p}\right)\right\| = \max_{\|x\|_k = 1} \|P x\|_k.$$

3. In particular, the 2-norm is defined by

$$\|P\|_2 = \sup_{x \neq 0} \frac{\|P x\|_2}{\|x\|_2}.$$

Also, we can evaluate the 2-norm by the square root of the largest eigenvalue of $P^T P$, i. e.,

$$\|P\|_2 = \sqrt{\lambda_{\max}(P^T P)}.$$

4. There are few other norms known as 1-norm and ∞-norms defined by

$$\|P\|_1 = \max_{1 \leq j \leq n} \sum_{i=1}^{p} |c_{ij}|$$

and

$$\|P\|_\infty = \max_{1 \leq i \leq p} \sum_{j=1}^{n} |c_{ij}|,$$

respectively.

Furthermore, the logarithmic norm of a matrix $P \in \mathbb{C}^{p\times p}$ is defined as [81, 154]

$$\mu(P) = \lim_{l\to 0} \frac{\|I + lP\| - 1}{l} = \max\left\{z \;\middle|\; z \in \sigma\left(\frac{P+P^*}{2}\right)\right\}, \tag{4.14}$$

where I is the $p \times p$ identity matrix. Let the number $\tilde{\mu}$ be such that [137]

$$\tilde{\mu}(P) = -\mu(-P) = \min\left\{z \;\middle|\; z \in \sigma\left(\frac{P+P^*}{2}\right)\right\}. \tag{4.15}$$

Then $\alpha(P) \le \mu(P)$ and $\beta(P) \ge \tilde{\mu}(P)$.

By [78], it follows that $\|e^{uP}\| \le e^{u\mu(P)}$ for $u \ge 0$. Hence, $u^P = \exp(P \ln u)$, where ln is the principal branch of the logarithmic function and it satisfies

$$\|u^P\| \le \begin{cases} u^{\mu(P)}, & \text{if } u \ge 1, \\ u^{\tilde{\mu}(P)}, & \text{if } 0 \le u \le 1. \end{cases} \tag{4.16}$$

One can verify the following relations, for $P \in \mathbb{C}^{p\times n}$, among the distinct norms:
1. $\max_{i,j} |c_{ij}| \le \|P\|_2 \le \sqrt{pn}\, \max_{i,j} |c_{ij}|$;
2. $\|P\|_2 \le \|P\|_F \le \sqrt{n}\, \|P\|_2$;
3. $\frac{1}{\sqrt{n}} \|P\|_\infty \le \|P\|_2 \le \sqrt{p}\, \|P\|_\infty$;
4. $\frac{1}{\sqrt{p}} \|P\|_1 \le \|P\|_2 \le \sqrt{n}\, \|P\|_1$.

There are several other results on the norm of matrices. It is easy to verify the following properties of the 2-norm defined on matrices. Let P, Q be matrices in $\mathbb{C}^{p\times p}$ and $u \in \mathbb{C}$. Then one can have the following properties [9]:
1. $\|uP\| = |u|\|P\|$.
2. $\|PQ\| \le \|P\|\|Q\|$.
3. $\|P + Q\| \le \|P\| + \|Q\|$.
4. $\|P\| = 0$ if and only if $P = 0$.
5. $\|I_{p\times p}\| = 1$.

Example 80. For a matrix $P = \left[\begin{smallmatrix} 2 & 3 \\ 4 & 1 \end{smallmatrix}\right]$, evaluate $\|P\|_F$, $\|P\|_1$ and $\|P\|_\infty$.

Solution. The Frobenius norm can be obtained as

$$\begin{aligned} \|P\|_F &= \sqrt{\sum_{i=1}^{2}\sum_{i=1}^{2} |c_{ij}|^2} \\ &= \sqrt{2^2 + 3^2 + 4^2 + 1^2} \\ &= \sqrt{30}. \end{aligned}$$

The 1-norm is

$$\begin{aligned}\|P\|_1 &= \max_{1\le j\le 2}\left(\sum_{i=1}^{2}|c_{ij}|\right)\\ &= \max_{1\le j\le 2}(|c_{1j}|+|c_{2j}|)\\ &= \max(|c_{11}|+|c_{21},|c_{12}|+|c_{22})\\ &= \max(2+4,3+1)=6.\end{aligned}$$

Similarly, we get $\|P\|_\infty = 5$.

The other result known as the perturbation lemma is being presented below. The perturbation lemma will be useful in the findings of the chapter.

Lemma 81. *If P is a p-square real matrix and $\|P\|_k < 1$, then $I - P$ is nonsingular and*

$$(I-P)^{-1} = \sum_{m=0}^{\infty} P^m \tag{4.17}$$

with

$$\|(I-P)^{-1}\|_k \le \frac{1}{1-\|P\|_k}. \tag{4.18}$$

Proof. Consider that $I - P$ is singular, and it follows $(I - P)\,x = 0$, for some $x \neq 0$. But $\|x\|_k = \|x\,P\|_k \implies \|P\|_k \ge 1$, which is a contradiction to $\|P\|_k < 1$, and hence $I - P$ is nonsingular.

Since $\|P\|_k < 1$ and $\|P^n\|_k \le \|P\|_k^n$, one can observe that $\lim_{n\to\infty} P^n = 0$. Consider

$$\left(\sum_{m-0}^{n} P^m\right)(I-P) = I - P^{n+1}. \tag{4.19}$$

Taking the limit $n \to \infty$ on both sides gives

$$\lim_{n\to\infty}\left(\sum_{m=0}^{n} P^m\right)(I-P) = \lim_{n\to\infty}(I-P^{n+1}). \tag{4.20}$$

This gives

$$(I-P)^{-1} = \sum_{m=0}^{\infty} P^m. \tag{4.21}$$

Implicating the k-norm yields

$$\|(I-P)^{-1}\|_k = \left\|\sum_{m=0}^{\infty} P^m\right\| \le \sum_{m=0}^{\infty}\|P\|_k^m = (1-\|P\|_k)^{-1}. \tag{4.22}$$

It completes the proof. □

The immediate use of this lemma produces the following identities:

1. If P is nonsingular and $\|P^{-1}Q\|_k < 1$, then $P + Q$ is nonsingular and
$$\|(P+Q)^{-1} - P^{-1}\|_k \le \frac{\|Q\|_k \, \|P^{-1}\|_k^2}{1 - \|P^{-1}Q\|_k}. \tag{4.23}$$

2. If P and Q are invertible matrices in $\mathbb{C}^{p\times p}$, then for $\|P - Q\| < \|Q^{-1}\|^{-1}$, one can get
$$\|P^{-1} - Q^{-1}\| \le \frac{\|Q^{-1}\|^2 \, \|P - Q\|}{1 - \|Q^{-1}\| \, \|P - Q\|}. \tag{4.24}$$

Van Loan in [156] obtained the norm of the exponential matrix function using Schur decomposition. We present this in the following lemma.

Lemma 82. *Let P be a matrix in $\mathbb{C}^{p\times p}$. Then*
$$\|e^{tP}\| \le e^{t\,\alpha(P)} \sum_{k=0}^{p-1} \frac{(\|P\|\, p^{\frac{1}{2}}\, t)^k}{k!}, \qquad t \ge 0. \tag{4.25}$$

Proof. For a square matrix P, according to Schur decomposition, there exists an unitary matrix $Q \in \mathbb{C}^{p\times p}$ such that
$$Q^* P Q = D + N,$$

where D is the diagonal matrix with diagonal entries as the eigenvalues of P and $N \in \mathbb{C}^{p\times p}$ is strictly an upper triangular matrix. Now, using the identity,
$$e^{(A+B)t} = e^{At} + \int_0^t e^{A(t-s)}\, B\, e^{(A+B)s}\, ds, \tag{4.26}$$

one can write
$$e^{(D+N)t} = e^{Dt} + \int_0^t e^{D(t-t_1)}\, N\, e^{(D+N)t_1}\, dt_1. \tag{4.27}$$

Again, using the formula (4.26) to expand $e^{(D+N)t_1}$, we get
$$\begin{aligned} e^{(D+N)t} &= e^{Dt} + \int_0^t e^{D(t-t_1)}\, N \left[e^{Dt_1} + \int_0^{t_1} e^{D(t_1-t_2)}\, N\, e^{(D+N)t_2}\, dt_2 \right] dt_1 \\ &= e^{Dt} + \int_0^t e^{D(t-t_1)}\, N\, e^{Dt_1}\, dt_1 + \int_0^t \int_0^{t_1} e^{D(t-t_1)}\, N\, e^{D(t_1-t_2)}\, N\, e^{(D+N)t_2}\, dt_2\, dt_1. \end{aligned} \tag{4.28}$$

On repeating this process, we obtain

$$e^{(D+N)t} = e^{Dt} + \sum_{k=1}^{p-1} A_k + R_p(t), \tag{4.29}$$

where

$$A_k = \int_0^t \int_0^{t_1} \cdots \int_0^{t_{k-1}} e^{D(t-t_1)} N e^{D(t_1-t_2)} N \cdots e^{D(t_{k-1}-t_k)} N e^{D t_k} \, dt_k \cdots dt_1 \tag{4.30}$$

and

$$R_p(t) = \int_0^t \int_0^{t_1} \cdots \int_0^{t_{p-1}} e^{D(t-t_1)} N e^{D(t_1-t_2)} N \cdots e^{D(t_{p-1}-t_p)} N e^{(D+N) t_p} \, dt_p \cdots dt_1. \tag{4.31}$$

Since N is an upper triangular matrix, $e^{D(t-t_1)} N$ and $R_p(t)$ contain the product

$$e^{D(t-t_1)} N e^{D(t_1-t_2)} N \cdots e^{D(t_{p-1}-t_p)} N$$

of p-strictly upper triangular matrices. Hence, $R_p(t) = 0$, and thus we have

$$e^{(D+N)t} = e^{Dt} + \sum_{k=1}^{p-1} \int_0^t \int_0^{t_1} \cdots \int_0^{t_{k-1}} e^{D(t-t_1)} N e^{D(t_1-t_2)} N \cdots e^{D(t_{k-1}-t_k)} N e^{D t_k} \, dt_k \cdots dt_1. \tag{4.32}$$

By taking the norm, we get

$$\begin{aligned} \|e^{Pt}\| \le \|e^{Dt}\| &+ \int_0^t \|e^{D(t-t_1)}\| \, \|N\| \, \|e^{D t_1}\| \, dt_1 + \int_0^t \int_0^{t_1} \|e^{D(t-t_1)}\| \, \|N\| \, \|e^{D(t_1-t_2)}\| \, \|N\| \\ &\times \|e^{D t_2}\| \, dt_2 \, dt_1 + \int_0^t \int_0^{t_1} \cdots \int_0^{t_{p-2}} \|e^{D(t-t_1)}\| \, \|N\| \, \|e^{D(t_1-t_2)}\| \\ &\times \|N\| \cdots \|e^{D(t_{p-2}-t_{p-1})}\| \, \|N\| \, \|e^{D t_{p-1}}\| \, dt_{p-1} \cdots dt_1. \end{aligned} \tag{4.33}$$

Using $\|e^{Ds}\| = e^{\alpha(P)s}$, we have

$$\begin{aligned} \|e^{Pt}\| &\le e^{\alpha(P)t} \left[1 + \int_0^t \|N\| \, dt_1 \right] \\ &\quad + e^{\alpha(P)t} \left[\int_0^t \int_0^{t_1} \|N\|^2 \, dt_2 \, dt_1 + \cdots + \int_0^t \int_0^{t_1} \cdots \int_0^{t_{p-2}} \|N\|^{p-1} \, dt_{p-1} \cdots dt_1 \right] \\ &\le e^{\alpha(P)t} \left[1 + \|N\| t + \|N\|^2 \frac{t^2}{2!} + \cdots + \|N\|^{p-1} \frac{t^{p-1}}{(p-1)!} \right] \end{aligned}$$

$$\leq e^{\alpha(P)\,t} \sum_{k=0}^{p-1} \frac{\|N\|^k\, t^k}{k!}$$

$$\leq e^{\alpha(P)\,t} \sum_{k=0}^{p-1} \frac{\|N\, t\|^k}{k!}. \tag{4.34}$$

The identity $\|N\| \leq p^{1/2}\, \|P\|$ yields

$$\|e^{t\,P}\| \leq e^{t\,\alpha(P)} \sum_{k=0}^{p-1} \frac{(\|P\|\, p^{\frac{1}{2}}\, t)^k}{k!}, \qquad t \geq 0. \tag{4.35}$$

It completes the proof. □

An immediate application of the above lemma gives the following result:

$$\|t^P\| \leq \|e^{P \ln t}\| \leq t^{\alpha(P)} \sum_{k=0}^{p-1} \frac{(\|P\|\, \sqrt{p} \ln t)^k}{k!}, \qquad t \geq 1, \tag{4.36}$$

$$\|t^P\| \leq \|e^{-P \ln t^{-1}}\| \leq t^{\beta(P)} \sum_{k=0}^{p-1} \frac{(-\|P\|\, \sqrt{p} \ln t)^k}{k!}, \qquad 0 < t \leq 1. \tag{4.37}$$

If numerical range of a matrix $P \in \mathbb{C}^{p\times p}$ is defined by

$$W(P) = \left\{ \frac{x^*\, P\, x}{x^*\, x} \;\middle|\; x \in \mathbb{C}^p, x \neq 0 \right\}, \tag{4.38}$$

then the numerical abscissa of P is given by

$$w(P) = \sup\{\Re(z) \mid z \in W(P)\}. \tag{4.39}$$

4.3 Infinite matrix product

Let $\{P_k\}_{k=1}^{\infty}$ be a sequence of matrices in $\mathbb{C}^{p\times p}$. Then the finite matrix product

$$U_n = \prod_{k=1}^{n} P_k \tag{4.40}$$

is called the nth partial product of the sequence. If $\lim_{n\to\infty} U_n$ exists and its value is an invertible matrix denoted by $U \in \mathbb{C}^{p\times p}$, we say that the infinite matrix product $\prod_{k=1}^{\infty} P_k$ exists and the matrix U is its value. Thus, we write

$$\prod_{k=1}^{\infty} P_k = U. \tag{4.41}$$

The matrix P_k in (4.41) is called the general term of the infinite product.

In general, an infinite product $\prod_{k=1}^{\infty} P_k$ of $p \times p$ matrices converges (invertibly) if there is an integer N such that P_n is invertible for $n \geq N$ and $Q = \lim_{n\to\infty} \prod_{k=N}^{n} P_k$ exists invertibly. In this case, we define $U = \prod_{k=1}^{\infty} P_k = Q \prod_{k=1}^{N-1} P_k$. Thus, if there is a finite number of singular factors in (4.41) and the remainder of the product converges to an invertible matrix, then due to the continuity of the determinant function, the original product converges to the zero matrix. If only a finite number of factors are singular, we will exclude them from the matrix product and treat the rest of the infinite matrix product.

Now, we will establish the following necessary condition for the convergence of an infinite matrix product, which is analogous to the corresponding one for infinite series.

Theorem 83. *Let $\{P_k\}_{k=1}^{\infty}$ be a sequence of matrices in $\mathbb{C}^{p\times p}$. If the infinite matrix product $\prod_{k=1}^{\infty}(I + P_k)$ exists, then*

$$\lim_{k\to\infty} P_k = 0 \in \mathbb{C}^{p\times p}.$$

Proof. First, we deal with the case that $-1 \notin \sigma(P_k)$ for all $k \geq 1$; i. e., there is no singular term among the factor of the infinite product. By assumption, $\prod_{k=1}^{\infty}(I+P_k)$ exists. Hence, we can define U as

$$U = \prod_{k=1}^{\infty}(I + P_k) = \lim_{k\to\infty} \prod_{n=1}^{k}(I + P_n).$$

Thus, taking into account the continuity of the inverse function as well as the inverse of a finite product of invertible matrices, one gets

$$\begin{aligned}
I = U^{-1}U &= \left[\lim_{k\to\infty} \prod_{n=1}^{k}(I + P_n)\right]^{-1} \left[\lim_{k\to\infty} \prod_{n=1}^{k}(I + P_n)\right] \\
&= \lim_{k\to\infty} \left[\prod_{n=1}^{k}(I + P_n)\right]^{-1} \lim_{k\to\infty} \left[\prod_{n=1}^{k}(I + P_n)\right] \\
&= \lim_{k\to\infty} \left\{\left[\prod_{n=1}^{k}(I + P_n)\right]^{-1} \left[\prod_{n=1}^{k}(I + P_n)\right]\right\} \\
&= \lim_{k\to\infty} \{(I + P_{k-1})^{-1}(I + P_{k-2})^{-1} \cdots (I + P_1)^{-1}\} \\
&\quad \times \{(I + P_1) \cdots (I + P_{k-1})(I + P_k)\} \\
&= \lim_{k\to\infty} (I + P_k) = I + \lim_{k\to\infty} P_k.
\end{aligned} \tag{4.42}$$

So, $I = I + \lim_{k\to\infty} P_k$ therefore $\lim_{k\to\infty} P_k = 0 \in \mathbb{C}^{p\times p}$. If there are singular factors in the matrix infinite product, we discard them and repeat the previous argument. □

Next, we show that every infinite product associated with an infinite matrix series whose terms are logarithms have the same character.

Theorem 84. *Let $\{P_k\}_{k=1}^{\infty}$ be a sequence of matrices in $\mathbb{C}^{p\times p}$ such that $P_k P_l = P_l P_k$ for every $k, l \geq 1$. Then $\prod_{k=1}^{\infty}(I+P_k)$ and $\sum_{k=1}^{\infty}\log(I+P_k)$ have the same character that is both convergent or divergent together.*

Proof. Since we are assuming that $-1 \notin \sigma(P_k)$ for all $k \geq 1$, so every matrix term of the form $I + P_k$ is invertible, and hence its logarithm exists. If the infinite product has a finite number of singular factors, we remove them and proceed with the remaining product. Let S_k and U_k denote the partial sum and partial product of $\sum_{k=1}^{\infty}\log(I+P_k)$ and $\prod_{k=1}^{\infty}(I+P_k)$, respectively, as

$$S_k = \sum_{n=1}^{k}\log(I+P_n), \quad U_k = \prod_{n=1}^{k}(I+P_n).$$

Since $P_k P_l = P_l P_k$ for every $k, l \geq 1$, so $(I + P_k)$ and $(I + P_l)$ commute. Therefore, we write

$$\begin{aligned} e^{S_k} &= e^{\sum_{n=1}^{k}\log(I+P_n)} = e^{\log(I+P_1)} \cdots e^{\log(I+P_k)} \\ &= (I+P_1)(I+P_2)\cdots(I+P_k) \\ &= \prod_{n=1}^{k}(I+P_n) = U_k. \end{aligned} \tag{4.43}$$

Taking the limit as $k \to \infty$, one gets

$$\lim_{k\to\infty} e^{S_k} = e^{\lim_{k\to\infty} S_k} = \lim_{k\to\infty} U_k.$$

Therefore, U_k has a limit if and only if S_k has a limit. Thus, $\prod_{k=1}^{\infty}(I+P_k)$ and $\sum_{k=1}^{\infty}\log(I+P_k)$ have the same character. It completes the proof. □

In the next theorem, we prove the absolute convergence of an infinite matrix product.

Theorem 85. *Let $\{P_k\}_{k=1}^{\infty}$ be a sequence of matrices in $\mathbb{C}^{p\times p}$ such that $P_k P_l = P_l P_k$ for every $k, l \geq 1$. Then the product $\prod_{k=1}^{\infty}(I+P_k)$ is absolutely convergent if and only if the sum $\sum_{k=1}^{\infty} P_k$ converges absolutely.*

Proof. Taking into account Theorem 84, we can conclude that the infinite matrix product $\prod_{k=1}^{\infty}(I+P_k)$ is absolutely convergent if and only if the infinite matrix series $\sum_{k=1}^{\infty}\log(I+P_k)$ is absolutely convergent. So, it is enough to prove that $\sum_{k=1}^{\infty}\log(I+P_k)$ is absolutely convergent if and only if $\sum_{k=1}^{\infty} P_k$ is absolutely convergent, i. e.,

$$\sum_{k=1}^{\infty}\|\log(I+P_k)\| < \infty \iff \sum_{k=1}^{\infty}\|P_k\| < \infty.$$

Consider the case that $\sum_{k=1}^{\infty} \|P_k\|$ is convergent, then there exists $k_0 \in \mathbb{N}$ such that $\|P_k\| < \frac{1}{2}$ for all $k > k_0$. The logarithmic matrix function is defined by

$$\log(I + P_k) = \sum_{n=1}^{\infty} \frac{(-1)^{n-1}}{n} (P_k)^n, \quad \|P_k\| < \frac{1}{2}, \forall k > k_0. \tag{4.44}$$

Applying the triangle inequality and submultiplicative property of matrix norm, we get

$$\begin{aligned} \big|\|\log(I + P_k)\| \, \|P_k\|^{-1} - 1\big| &= \left| \left\| \sum_{n=1}^{\infty} \frac{(-1)^{n-1}}{n} (P_k)^n \right\| \|P_k\|^{-1} - 1 \right| \\ &\leq \left| \sum_{n=1}^{\infty} \frac{\|(P_k)\|^n}{n} \|P_k\|^{-1} - 1 \right| \leq \left| \sum_{n=1}^{\infty} \frac{\|(P_k)\|^{n-1}}{n} - 1 \right| \\ &\leq \left| \sum_{n=2}^{\infty} \frac{\|(P_k)\|^{n-1}}{n} \right| = \left| \sum_{n=1}^{\infty} \frac{\|(P_k)\|^n}{n+1} \right| \\ &\leq \sum_{n=1}^{\infty} \frac{1}{n+1} \left(\frac{1}{2}\right)^n < \sum_{n=1}^{\infty} \left(\frac{1}{2}\right)^{n+1} = \frac{\frac{1}{4}}{1 - \frac{1}{2}} = \frac{1}{2}. \end{aligned} \tag{4.45}$$

Therefore, we have proven that

$$\big|\|\log(I + P_k)\| \, \|P_k\|^{-1} - 1\big| < \frac{1}{2}, \quad \forall k > k_0. \tag{4.46}$$

That is,

$$\frac{1}{2} < \frac{\|\log(I + P_k)\|}{\|P_k\|} < \frac{3}{2}, \quad \forall k > k_0.$$

It leads to

$$\|\log(I + P_k)\| < \frac{3\|P_k\|}{2}, \quad \forall k > k_0.$$

By the comparison test for positive numerical series, the $\sum_{k=1}^{\infty} \log(I + P_k)$ converges as the series $\sum_{k=1}^{\infty} \|P_k\|$ is convergent. On the other hand, we have proven that

$$\frac{1}{2} < \frac{\|\log(I + P_k)\|}{\|P_k\|}, \quad \forall k > k_0,$$

which yields

$$\frac{\|P_k\|}{\|\log(I + P_k)\|} < 2, \quad \forall k > k_0;$$

i. e.,

$$\|P_k\| < 2\,\|\log(I + P_k)\|, \quad \forall k > k_0.$$

If we assume that $\sum_{k=1}^{\infty}\|\log(I+P_k)\|$ converges, then again by the comparison test for the positive numerical series $\sum_{k=1}^{\infty}\|P_k\|$ converges. It completes the proof. □

Note that due to Theorems 84 and 85, an infinite product, which is absolutely convergent, is also convergent.

Example 86. Show that the infinite matrix product $\prod_{k=1}^{\infty}(I+\frac{1}{(k+1)(k+3)}I)$ is absolutely convergent.

Proof. The given matrix product is in the form $\prod_{k=1}^{\infty}(I+P_k)$, where $P_k=\frac{1}{(k+1)(k+3)}I$ satisfying that $P_kP_l=P_lP_k$, $\forall k,l\geq 1$. Also, there is no singular factor in the product because

$$-1\notin\sigma(P_k)=\left\{\lambda_k \,\middle|\, \lambda_k=\frac{1}{(k+1)(k+3)}\neq -1,\ \forall k\geq 1\right\}.$$

Using Theorem 85, it is sufficient to show that the matrix series $\sum_{k=1}^{\infty}\frac{1}{(k+1)(k+3)}I$ is absolutely convergent. In fact,

$$\sum_{k=1}^{\infty}\left\|\frac{1}{(k+1)(k+3)}I\right\|=\sum_{k=1}^{\infty}\frac{1}{(k+1)(k+3)}<\sum_{k=1}^{\infty}\frac{1}{k^2},$$

which is convergent. Thus, the infinite matrix product $\prod_{k=1}^{\infty}(I+\frac{1}{(k+1)(k+3)}I)$ is absolutely convergent. Moreover, it is easy to compute the value of infinite product. The partial product U_k is given by

$$\begin{aligned}U_k&=\prod_{n=1}^{k}\left(I+\frac{1}{(n+1)(n+3)}I\right)\\&=\prod_{n=1}^{k}\left(\frac{(n+1)(n+3)+1}{(n+1)(n+3)}I\right)=\prod_{n=1}^{k}\left(\frac{(n+2)^2}{(n+1)(n+3)}I\right)\\&=\frac{[3\cdot 4\cdots(k+2)]^2}{[2\cdot 3\cdots(k+1)]\,[4\cdot 5\cdots(k+3)]}I\\&=\frac{k+2}{2}\frac{3}{k+3}I.\end{aligned}\tag{4.47}$$

Then

$$\prod_{k=1}^{\infty}\left(I+\frac{1}{(k+1)(k+3)}I\right)=\lim_{k\to\infty}U_k=\frac{3}{2}I.$$

Let us observe the necessary condition of convergence for the infinite matrix product:

$$\lim_{k\to\infty}P_k=\lim_{k\to\infty}\frac{1}{(k+1)(k+3)}I=0\in\mathbb{C}^{p\times p}.$$

It completes the proof. □

4.4 Gamma and beta matrix functions

If A is a matrix in $\mathbb{C}^{p\times p}$ such that the real part of all eigenvalues of A are positive, i. e., $\Re(z) > 0$, $\forall z \in \sigma(A)$, then the gamma matrix function is defined by

$$\Gamma(A) = \int_0^\infty e^{-u}\, u^{A-I}\, du, \tag{4.48}$$

where u^{A-I} is defined as $\exp((A-I)\ln u)$ and ln is the principal branch of the logarithmic function. Manually, one can evaluate the value of gamma matrix function in the following way.

Let $A = \left[\begin{smallmatrix} 1 & 0 \\ 0 & 2 \end{smallmatrix}\right]$. Then the matrix gamma function is obtained by

$$\begin{aligned}
\Gamma(A) &= \int_0^\infty e^{-u}\, u^{A-I}\, du \\
&= \int_0^\infty e^{-u} \exp((A-I)\ln u)\, du \\
&= \int_0^\infty e^{-u} \exp\left(\begin{bmatrix} 0 & 0 \\ 0 & 1 \end{bmatrix} \ln u\right) du \\
&= \int_0^\infty e^{-u} \exp\left(\begin{bmatrix} 0 & 0 \\ 0 & \ln u \end{bmatrix}\right) du \\
&= \int_0^\infty e^{-u} \left(I + \begin{bmatrix} 0 & 0 \\ 0 & \ln u \end{bmatrix} + \frac{1}{2!}\begin{bmatrix} 0 & 0 \\ 0 & (\ln u)^2 \end{bmatrix} + \cdots\right) du \\
&= \int_0^\infty e^{-u} \begin{bmatrix} 1 & 0 \\ 0 & 1 + \ln u + \frac{1}{2!}(\ln u)^2 + \cdots \end{bmatrix} du \\
&= \begin{bmatrix} \int_0^\infty e^{-u}\, du & 0 \\ 0 & \int_0^\infty e^{-u}\, u\, du \end{bmatrix} \\
&= \begin{bmatrix} \Gamma(1) & 0 \\ 0 & \Gamma(2) \end{bmatrix} = \begin{bmatrix} 1 & 0 \\ 0 & 1 \end{bmatrix} = I.
\end{aligned} \tag{4.49}$$

Thus, we obtain $\Gamma(A) = I$, when $A = \left[\begin{smallmatrix} 1 & 0 \\ 0 & 2 \end{smallmatrix}\right]$. Remember, it is not possible to get the value of the gamma matrix function manually for each matrix $A \in \mathbb{C}^{p\times p}$.

Let us recall the definition of the Pochhammer symbol in terms of the gamma function. According to the definition, the Pochhammer symbol for a complex number u is given by

$$(u)_k = u\,(u+1)\,\ldots\,(u+k-1) = \frac{\Gamma(u+k)}{\Gamma(u)}. \tag{4.50}$$

By application of the matrix functional calculus, this gives

$$(A)_k = A\,(A+I)\cdots(A+(k-1)I) = \Gamma^{-1}(A)\,\Gamma(A+kI), \quad n \geq 1. \tag{4.51}$$

Also, we have

$$u\,(u+1)\cdots(u+k-1) = \Gamma(u+k)\,\Gamma^{-1}(u). \tag{4.52}$$

Since $\Gamma^{-1}(u)$ is an entire function of u, the Riesz–Dunford calculus allows to define the image of $\Gamma^{-1}(u)$ with respect to a matrix $A \in \mathbb{C}^{p\times p}$. It will be denoted by $\Gamma^{-1}(A)$, called the inverse gamma matrix function, and defined by

$$A\,(A+I)\cdots(A+(k-1)I) = \Gamma(A+kI)\,\Gamma^{-1}(A). \tag{4.53}$$

With the help of these findings, one can easily verify the following important relations in terms of gamma and inverse gamma matrix functions:
1. $\Gamma(A+I) = A\,\Gamma(A)$;
2. $(A)_{m+k} = (A)_m\,(A+mI)_k$.

Whenever we have a matrix $A \in \mathbb{C}^{p\times p}$ such that $A, A+I, \ldots$ are invertible, then the matrix analogue of the Weierstrass gamma function is given by

$$\Gamma^{-1}(A) = A\,e^{\gamma A} \prod_{k=1}^{\infty} \left[\left(I + \frac{1}{k}A\right) e^{-\frac{1}{k}A}\right], \tag{4.54}$$

where γ is the Euler–Mescheroni constant defined in Chapter 1. Let us show that this function is well-defined by showing that such infinite product is absolutely convergent. Rather than dealing with an infinite matrix product (4.54), we will prove a more general statement, namely we will prove that the matrix infinite product

$$\prod_{k=1}^{\infty}\left(I - \frac{1}{c+k}M\right) e^{\frac{1}{k}M}$$

is absolutely convergent with $M \in \mathbb{C}^{p\times p}$ and $c \notin \mathbb{Z}\setminus\mathbb{N}$. Denoting $B_k = (I - \frac{1}{c+k}M)\,e^{\frac{1}{k}M} - I$, it is obvious that $B_k\,B_l = B_l\,B_k$ for every $k, l \geq 1$. Therefore, by Theorem 85 it is sufficient to show that

$$\sum_{k=1}^{\infty}\left[\left(I - \frac{1}{c+k}M\right) e^{\frac{1}{k}M} - I\right]$$

is absolutely convergent; i. e., we have to show that the numerical series

$$\sum_{k=1}^{\infty}\left\|\left[\left(I - \frac{1}{c+k}M\right) e^{\frac{1}{k}M} - I\right]\right\|$$

is convergent. For this, we will compare it with the convergent series $\sum_{k=1}^{\infty}(\frac{1}{k^2})$ and, therefore, it will be sufficient to show, by a comparison test, that

$$\lim_{k\to\infty} k^2 \left\| \left[\left(I - \frac{1}{c+k} M \right) e^{\frac{1}{k} M} - I \right] \right\| \\ = \left\| \lim_{k\to\infty} k^2 \left[\left(I - \frac{1}{c+k} M \right) e^{\frac{1}{k} M} - I \right] \right\| = a < \infty. \tag{4.55}$$

To compute the limit on the left-hand side, we make the change of the variable $\beta = \frac{1}{c+k}$, so $k = \frac{1-\beta c}{\beta}$ and $\frac{1}{k} = \frac{\beta}{1-\beta c}$, then limit transforms into

$$\left\| \lim_{\beta\to 0} \frac{(1-\beta c)^2}{\beta^2} [(I - \beta M)\, e^{\frac{\beta}{1-\beta c} M} - I] \right\| = \frac{1}{2} \| M(2cI - M) \| = a < \infty,$$

where the L'Hopital rule has been applied twice in the last step. Therefore, the infinite matrix series $\sum_{k=1}^{\infty}[(I - \frac{1}{c+k} M)\, e^{\frac{1}{k} M} - I]$ converges absolutely.

In particular, for $c = 0$, we get that the matrix product

$$\prod_{k=1}^{\infty} \left(I - \frac{1}{k} M \right) e^{\frac{1}{k} M}$$

converges absolutely. Also, the result applies to $A = -M \in \mathbb{C}^{p\times p}$ and gives

$$\prod_{k=1}^{\infty} \left(I + \frac{1}{k} A \right) e^{-\frac{1}{k} A}$$

an absolutely convergent infinite matrix product, and hence

$$\Gamma^{-1}(A) = A\, e^{\gamma A} \prod_{k=1}^{\infty} \left[\left(I + \frac{1}{k} A \right) e^{-\frac{1}{k} A} \right]$$

is absolutely convergent. One of the most useful applications from our result is to prove the existence of certain matrix limits.

Lemma 87. *If $A \in \mathbb{C}^{p\times p}$ such that $-k \notin \sigma(A)$ for all $k \geq 1$, then the following limit exists:*

$$\lim_{k\to\infty} (k-1)!\, [(A+I)\cdots(A+(k-1)I)]^{-1} k^A.$$

Proof. To verify this, we consider an infinite product whose kth partial product P_k coincides with the limit, i. e.,

$$P_k = (k-1)!\, [(A+I)\cdots(A+(k-1)I)]^{-1} k^A.$$

If we prove that the infinite product converges, we conclude the existence of $\lim_{k\to\infty} P_k$. Let us observe that

$$\begin{aligned}
P_{k+1} &= k!\left[(A+I)(A+kI)\right]^{-1}(k+1)^A \\
&= k!\,(A+I)^{-1}(A+kI)^{-1}(k+1)^A \\
&= \left[1(A+I)^{-1}2^A 1^{-A}\right]\left[2(A+2I)^{-1}3^A 2^{-A}\right]\cdots\left[k(A+kI)^{-1}(k+1)^A k^{-A}\right] \\
&= \prod_{n=1}^{k}\left[n(A+nI)^{-1}(n+1)^A n^{-A}\right] \\
&= \prod_{n=1}^{k}\left[\left(\frac{1}{n}A+I\right)^{-1}\left(1+\frac{1}{n}\right)^A\right].
\end{aligned} \tag{4.56}$$

Consider the product $\prod_{n=1}^{\infty} B_n$ being $B_n = (\frac{1}{n}A+I)^{-1}(1+\frac{1}{n})^A$, since $B_m B_l = B_l B_m$ for every $m, l \geq 1$. To show that it converges absolutely, by Theorem 85, it is enough to show that the matrix series sum

$$\sum_{n=1}^{\infty}\left[\left(\frac{1}{n}A+I\right)^{-1}\left(1+\frac{1}{n}\right)^A - I\right]$$

absolutely converges, i. e., the numerical series

$$\sum_{n=1}^{\infty}\left\|\left(\frac{1}{n}A+I\right)^{-1}\left(1+\frac{1}{n}\right)^A - I\right\|$$

converges. This is shown by comparison with the convergent series $\sum_{n=1}^{\infty}\frac{1}{n^2}$. In fact, notice that

$$\lim_{n\to\infty} n^2\left\|\left(\frac{1}{n}A+I\right)^{-1}\left(1+\frac{1}{n}\right)^A - I\right\| = \left\|\lim_{n\to\infty} n^2\left[\left(\frac{1}{n}A+I\right)^{-1}\left(1+\frac{1}{n}\right)^A - I\right]\right\|$$

and making the change of variable $\beta = \frac{1}{k}$, we get $k = \frac{1}{\beta}$ and

$$\lim_{\beta\to 0}\frac{(A\beta+I)^{-1}(1+\beta)^A - I}{\beta^2} = \frac{1}{2}\left\|A(A-I)\right\|,$$

where the L'Hopital rule is applied twice. Hence, the matrix infinite product

$$\prod_{n=1}^{\infty}\left[\left(\frac{1}{n}A+I\right)^{-1}\left(1+\frac{1}{n}\right)^A\right]$$

converges. Therefore,

$$\lim_{n\to\infty} P_k = \lim_{k\to\infty}(k-1)!\left[(A+I)\cdots(A+(k-1)I)\right]^{-1}k^A$$

exists. Notice that the assumption $-k \notin \sigma(A)$ for all $k \geq 1$ guarantees that the product has no singular matrices as a factor. This completes the proof. □

As a consequence, we have the following limiting definition of the gamma matrix function.

Theorem 88. *If $A \in \mathbb{C}^{p\times p}$ such that $-k \notin \sigma(A)$ for all $k \geq 1$ and $(A)_k$ denotes the Pochhammer's symbol, then the gamma matrix function obeys the following limiting definition:*

$$\Gamma(A) = \lim_{k\to\infty} (k-1)!\,(A)_k^{-1}\,k^A. \tag{4.57}$$

Proof. We know that

$$\Gamma^{-1}(A) = A\,e^{\gamma A} \prod_{k=1}^{\infty}\left[\left(I + \frac{1}{k}A\right)e^{-\frac{1}{k}A}\right], \tag{4.58}$$

and by assumption $-k \notin \sigma(A)$, $A + kI$ are invertible matrices for all $k = 1, 2, \ldots$ and so are $I + \frac{1}{k}A$. Since the exponential matrix is invertible and that $(e^A)^{-1} = e^{-A}$, we get

$$A\,\Gamma(A) = e^{-\gamma A} \prod_{k=1}^{\infty}\left[\left(I + \frac{1}{k}A\right)^{-1} e^{\frac{1}{k}A}\right]$$

or

$$A\,\Gamma(A) = e^{-\gamma A} \lim_{k\to\infty}\prod_{n=1}^{k}\left[\left(I + \frac{1}{n}A\right)^{-1} e^{\frac{1}{n}A}\right].$$

On the other hand, we know that

$$\gamma = \lim_{k\to\infty}[H_k - \ln(k+1)] = \lim_{k\to\infty}\left[H_k - \sum_{n=1}^{k}\ln\left(\frac{n+1}{n}\right)\right], \quad H_k = \sum_{n=1}^{k}\frac{1}{n},$$

and thus

$$\begin{aligned}
e^{-\gamma A} &= e^{-\lim_{k\to\infty}[H_k - \sum_{n=1}^{k}\ln(\frac{n+1}{n})]A} \\
&= \lim_{k\to\infty} e^{-H_k A + [\sum_{n=1}^{k}\ln(\frac{n+1}{n})]A} \\
&= \lim_{k\to\infty} e^{-(\sum_{n=1}^{k}\frac{1}{n})A + [\sum_{n=1}^{k}\ln(\frac{n+1}{n})]A} \\
&= \lim_{k\to\infty} e^{-(\sum_{n=1}^{k}\frac{1}{n})A}\, e^{[\sum_{n=1}^{k}\ln(\frac{n+1}{n})]A} \\
&= \lim_{k\to\infty}\left\{\left[\prod_{n=1}^{k} e^{-(\frac{1}{n})A}\right]\left[\prod_{n=1}^{k}\left(\frac{n+1}{n}\right)^{A}\right]\right\} \\
&= \lim_{k\to\infty}\left\{\left[\prod_{n=1}^{k} e^{-(\frac{1}{n})A}\left(\frac{n+1}{n}\right)^{A}\right]\right\}.
\end{aligned} \tag{4.59}$$

Thus, we have

$$A\,\Gamma(A) = \lim_{k\to\infty}\left\{\left[\prod_{n=1}^{k} e^{-(\frac{1}{n})A}\left(\frac{n+1}{n}\right)^{A}\left(I+\frac{1}{n}A\right)^{-1} e^{(\frac{1}{n})A}\right]\right\}$$
$$= \lim_{k\to\infty}\prod_{n=1}^{k}\left[\left(\frac{n+1}{n}\right)^{A}\left(I+\frac{1}{n}A\right)^{-1}\right]. \tag{4.60}$$

We have already proven that

$$\lim_{k\to\infty} P_k = \lim_{k\to\infty}\prod_{n=1}^{k}\left[\left(\frac{n+1}{n}\right)^{A}\left(I+\frac{1}{n}A\right)^{-1}\right]$$
$$= \lim_{k\to\infty}(k-1)!\,\left[(A+I)\cdots(A+(k-1)I)\right]^{-1}k^{A}. \tag{4.61}$$

Therefore,

$$A\,\Gamma(A) = \lim_{k\to\infty}(k-1)!\,\left[(A+I)\cdots(A+(k-1)I)\right]^{-1}k^{A}.$$

Taking into account the definition of Pochhammer symbol for a matrix, we have

$$\Gamma(A) = \lim_{k\to\infty}(k-1)!\,\left[A\,(A+I)\cdots(A+(k-1)I)\right]^{-1}k^{A}$$
$$= \lim_{k\to\infty}(k-1)!\,\left((A)_k\right)^{-1}k^{A}. \tag{4.62}$$

It completes the proof. □

Next, we give the matrix version of a beta function. For any two p-square matrices A and B such that $\beta(A), \beta(B) > 0$, the beta matrix function is denoted by $\beta(A,B)$ and defined in terms of an integral as

$$\beta(A,B) = \int_0^1 u^{A-I}(1-u)^{B-I}\,du. \tag{4.63}$$

Let us prove the convergence of the integral given above for beta matrix function. Using (4.35)–(4.37) and $\ln u < u$, $\ln(1-u) < (1-u)$, for $0 < u < 1$, it follows that

$$\int_0^1 \|u^{A-I}\|\,\|(1-u)^{B-I}\|\,du$$
$$\le \sum_{j=0}^{p-1}\sum_{k=0}^{p-1}\frac{(\|A\|+1)^j\,(\|B\|+1)^k\,(\sqrt{p})^{j+k}}{j!\,k!}\int_0^1 u^{\alpha(A)-1}(1-u)^{\alpha(B)-1}\ln^j(u)\ln^k(1-u)\,du$$
$$\le \sum_{j=0}^{p-1}\sum_{k=0}^{p-1}\frac{(\|A\|+1)^j\,(\|B\|+1)^k\,(\sqrt{p})^{j+k}}{j!\,k!}\int_0^1 u^{\alpha(A)+j-1}(1-u)^{\alpha(B)+k-1}\,du$$

$$\leq \sum_{j=0}^{p-1}\sum_{k=0}^{p-1} \frac{(\|A\|+1)^j\,(\|B\|+1)^k\,(\sqrt{p})^{j+k}\beta(\alpha(A)+j,\alpha(B)+k)}{j!\,k!}$$
$$\leq +\infty. \tag{4.64}$$

This shows that the integral converges absolutely, and thus the beta matrix function exists fairly.

If we allow the matrices A and B to commute, i. e., $AB = BA$, then the substitution $u = \frac{t}{1+t}$ and $du = \frac{1}{(1+t)^2}\,dt$ lead to get the another integral form of the beta matrix function as

$$\beta(A,B) = \int_0^\infty t^{A-I}\,(1+t)^{-(A+B)}\,dt. \tag{4.65}$$

If A and B are matrices in $\mathbb{C}^{p\times p}$ such that $\beta(A), \beta(B) > 0$ and A or B are scalar multiples of the identity matrix, then $\beta(A,B) = \beta(B,A)$. Commutativity is necessary for symmetry of a beta matrix function, one can verify this with the following example.

Example 89. Let $A = \begin{bmatrix} 1 & 0 \\ 1 & 2 \end{bmatrix}$, $B = \begin{bmatrix} 1 & 1 \\ 0 & 2 \end{bmatrix}$ and be matrices in $\mathbb{C}^{2\times 2}$ with $\sigma(A) = \sigma(B) = \{1,2\}$. Then, they are diagonalizable and

$$AB = \begin{bmatrix} 1 & 1 \\ 1 & 5 \end{bmatrix} \neq \begin{bmatrix} 2 & 2 \\ 2 & 4 \end{bmatrix} = BA.$$

Note that $\begin{bmatrix} 0 & 0 \\ 1 & 1 \end{bmatrix}^n = \begin{bmatrix} 0 & 0 \\ 1 & 1 \end{bmatrix}$ and $\begin{bmatrix} 0 & 1 \\ 0 & 1 \end{bmatrix}^n = \begin{bmatrix} 0 & 1 \\ 0 & 1 \end{bmatrix}$ for all $n \geq 1$. Hence, for $0 < t < 1$, one gets

$$t^{A-I} = t^{\begin{bmatrix} 0 & 0 \\ 1 & 1 \end{bmatrix}} = \begin{bmatrix} 1 & 0 \\ t-1 & t \end{bmatrix}, \quad (1-t)^{A-I} = (1-t)^{\begin{bmatrix} 0 & 0 \\ 1 & 1 \end{bmatrix}} = \begin{bmatrix} 1 & 0 \\ -t & 1-t \end{bmatrix},$$
$$t^{B-I} = t^{\begin{bmatrix} 0 & 1 \\ 0 & 1 \end{bmatrix}} = \begin{bmatrix} 1 & t-1 \\ 0 & t \end{bmatrix}, \quad (1-t)^{B-I} = (1-t)^{\begin{bmatrix} 0 & 1 \\ 0 & 1 \end{bmatrix}} = \begin{bmatrix} 1 & -t \\ 0 & 1-t \end{bmatrix}.$$

Thus, we get

$$\begin{aligned}\beta(A,B) &= \int_0^1 t^{A-I}\,(1-t)^{B-I}\,dt \\ &= \int_0^1 \begin{bmatrix} 1 & 0 \\ t-1 & t \end{bmatrix}\begin{bmatrix} 1 & -t \\ 0 & 1-t \end{bmatrix} dt \\ &= \int_0^1 \begin{bmatrix} 1 & -t \\ t-1 & 2t(1-t) \end{bmatrix} dt = \begin{bmatrix} 1 & \frac{-1}{2} \\ \frac{-1}{2} & \frac{1}{3} \end{bmatrix},\end{aligned} \tag{4.66}$$

$$\beta(B,A) = \int_0^1 t^{B-I}(1-t)^{A-I}\,dt$$
$$= \int_0^1 \begin{bmatrix} 1 & t-1 \\ 0 & t \end{bmatrix} \begin{bmatrix} 1 & 0 \\ -t & 1-t \end{bmatrix} dt$$
$$= \int_0^1 \begin{bmatrix} -t^2+t+1 & -(1-t)^2 \\ -t^2 & t(1-t) \end{bmatrix} dt = \begin{bmatrix} \frac{7}{6} & \frac{-1}{3} \\ \frac{-1}{3} & \frac{1}{6} \end{bmatrix}. \tag{4.67}$$

Hence, $\beta(A,B) \neq \beta(B,A)$.

In the following lemma, we prove the symmetric property of a beta matrix function.

Lemma 90. *Let A and B be commuting matrices in $\mathbb{C}^{p\times p}$ such that $\beta(A)$, $\beta(B) > 0$. Then $\beta(A,B) = \beta(B,A)$.*

Proof. Since $AB = BA$, it follows that $(A-I)(\ln u)(B-I)(\ln(1-u)) = (B-I)(\ln(1-u))(A-I)\ln u$, for $0 < u < 1$. Hence, we can write

$$\beta(A,B) = \int_0^1 u^{A-I}(1-u)^{B-I}\,du$$
$$= \int_0^1 e^{(A-I)\ln u}\, e^{(B-I)\ln(1-u)}\,du$$
$$= \int_0^1 e^{(B-I)\ln(1-u)}\, e^{(A-I)\ln u}\,du$$
$$= \int_0^1 e^{(B-I)\ln(t)}\, e^{(A-I)\ln(1-t)}\,dt = \beta(B,A). \tag{4.68}$$

It completes the proof. □

In the following theorems, we prove the distinct matrix versions of the fundamental relation between gamma and beta functions.

Theorem 91. *Let A, B be diagonalizable matrices in $\mathbb{C}^{p\times p}$ such that $AB = BA$ and $\beta(A)$, $\beta(B) > 0$. Then*

$$\beta(A,B) = \Gamma(A)\,\Gamma(B)\,\Gamma^{-1}(A+B).$$

Proof. Since A, B are diagonalizable and commute, they are simultaneously diagonalizable. Let S be an invertible matrix in $\mathbb{C}^{p\times p}$ such that

$$S^{-1}AS = D, \quad S^{-1}BS = E,$$

where D and E are a diagonal matrices. If $\sigma(A) = \{\lambda_1, \dots, \lambda_p\}$ and $\sigma(B) = \{\mu_1, \dots, \mu_p\}$, then $\sigma(A+B) = \{\lambda_i + \mu_i\}_{i=1}^{p}$ for some permutation $i_1, \dots, i_p$ of $1, 2, \dots, p$. Since matrices A and B satisfy $\beta(A), \beta(B) > 0$, it follows that

$$\Re(z) > 0, \quad \text{for all } z \in \sigma(A+B).$$

Since A and B are simultaneously diagonalizable and so is $A + B$, it follows $A + B = S\,(D+E)\,S^{-1}$ and

$$\Gamma(A+B) = S\left[\int_0^\infty e^{-u} u^{D+E-I}\,du\right] S^{-1} = S\,\Gamma(D+E)\,S^{-1},$$
$$\Gamma(A) = S\,\Gamma(D) S^{-1}, \quad \Gamma(B) = S\,\Gamma(E)\,S^{-1}.$$

Thus,

$$\begin{aligned}\beta(A,B) &= S\,\beta(D,E)\,S^{-1}\\ &= S\,[\Gamma(D)\,\Gamma(E)\,\Gamma^{-1}(D+E)]\,S^{-1}\\ &= [S\,\Gamma(D)\,S^{-1}]\,[S\,\Gamma(E)\,S^{-1}]\,[S\,\Gamma^{-1}(D+E)\,S^{-1}]\\ &= \Gamma(A)\,\Gamma(B)\,\Gamma^{-1}(A+B).\end{aligned} \tag{4.69}$$

It completes the proof. □

Apart from the commutative hypothesis, the diagonalizability condition of matrices guarantees that every eigenvalue z of the matrix $A + B$ lies in the right half-plane. The following example shows that, in general, if A, B are diagonalizable matrices such that $\beta(A), \beta(B) > 0$, its sum $A + B$ does not satisfy the condition.

Example 92. Let a, b be positive numbers such that $ab > 1$. Then matrices $A = \begin{bmatrix} \frac{1}{2} & 0 \\ a & \frac{1}{2} \end{bmatrix}$, $B = \begin{bmatrix} \frac{1}{2} & b \\ 0 & \frac{1}{2} \end{bmatrix}$ satisfy $\sigma(A) = \sigma(B) = \{\frac{1}{2}\}$, but $A + B = \left[\begin{smallmatrix} 1 & b \\ a & 1 \end{smallmatrix}\right]$ and $\sigma(A+B) = \{1 - \sqrt{ab}, 1 + \sqrt{ab}\}$ with $1 - \sqrt{ab} < 0$.

Now, we give some more relations between the gamma and beta matrix functions.

Lemma 93. *Let A, B be two positive stable matrices in $\mathbb{C}^{p\times p}$ such that $AB = BA$ and $A + B + nI$ is invertible for all integers $n \geq 0$. If $k \geq 0$, then the following relations hold:*
1. $\beta(A, B + kI) = (A+B)_k^{-1}\,(B)_k\,\beta(A,B)$.
2. $\beta(A + kI, B + kI) = (A)_k\,(B)_k\,(A+B)_{2k}^{-1}\,\beta(A,B)$.

Proof. When $k = 0$, the proof is obvious. Using $AB = BA$ and the integral representation of a beta matrix function given by (4.63), for $k \geq 1$, it follows that

$$\begin{aligned}\beta(A,B+kI) &= \int_0^1 t^{A-I}(1-t)^{B+(k-1)I}\,dt \\ &= \lim_{\delta\to 0}\int_\delta^{1-\delta} t^{A-I}(1-t)^{B+(k-1)I}\,dt \\ &= \lim_{\delta\to 0}\int_\delta^{1-\delta} t^{A+B+(k-2)I}(1-t)^{B+(k-1)I}t^{-(B+(k-1)I)}\,dt \\ &= \lim_{\delta\to 0}\int_\delta^{1-\delta} u(t)\,v'(t)\,dt, \end{aligned} \tag{4.70}$$

where

$$u(t) = (1-t)^{B+(k-1)I}t^{-(B+(k-1)I)}; \quad v'(t) = t^{A+B+(k-2)I}.$$

Integrating by parts in the last integral, we get

$$\begin{aligned}\beta(A,B+kI) &= \lim_{\delta\to 0}\left[(A+B+(k-1)I)^{-1}(1-t)^{B+(k-1)I}\right]_{t=\delta}^{t=1-\delta} \\ &\quad + \lim_{\delta\to 0}(A+B+(k-1)I)^{-1}\int_\delta^{1-\delta}\{(B+(k-1)I)(1-t)^{B+(k-2)I} \\ &\quad + (B+(k-1)I)(1-t)^{B+(k-1)I}\}dt \\ &= (A+B+(k-1)I)^{-1}(B+(k-1)I)\int_0^1 t^{A-I}(1-t)^{B+(k-2)I}\,dt \\ &= (A+B+(k-1)I)^{-1}(B+(k-1)I)\,\beta(A,B+(k-1)I). \end{aligned} \tag{4.71}$$

From (4.71), we have

$$\begin{aligned}&\beta(A,B+(k-1)I) \\ &\quad = (A+B+(k-2)I)^{-1}(B+(k-2)I)\,\beta(A,B+(k-2)I). \end{aligned} \tag{4.72}$$

Using (4.72) in (4.71) and repeating this process k-times yield

$$\beta(A,B+kI) = (A+B)_k^{-1}(B)_k\,\beta(A,B). \tag{4.73}$$

It completes the proof of the first part. To prove the second part, apply (4.73) by taking $\tilde{A} = A + kI$, $k \geq 1$. We get

$$\beta(\tilde{A},B+kI) = (\tilde{A}+B)_k^{-1}(B)_k\,\beta(\tilde{A},B). \tag{4.74}$$

Since $AB = BA$, it follows that $\tilde{A}B = B\tilde{A}$ and $\beta(\tilde{A}, B) = \beta(B, \tilde{A})$. The equation (4.74) can be rewritten as

$$\beta(\tilde{A}, B + kI) = (\tilde{A} + B)_k^{-1} (B)_k \, \beta(B, \tilde{A}). \tag{4.75}$$

Again using (4.73), we have

$$\beta(B, A + kI) = (A + B)_k^{-1} (A)_k \, \beta(A, B). \tag{4.76}$$

Equations (4.75) and (4.76) together yield

$$\beta(A + kI, B + kI) = (A + B + kI)_k^{-1} (B)_k \, (A + B)_k^{-1} (A)_k \, \beta(A, B). \tag{4.77}$$

By definition $(A + B + kI)_k \, (A + B)_k = (A + B)_{2k}$, one can write

$$\beta(A + kI, B + kI) = (A)_k \, (B)_k \, (A + B)_{2k}^{-1} \, \beta(A, B). \tag{4.78}$$

This completes the proof. □

Part (ii) of the above theorem permits to extend the definition of the beta matrix function of two and not necessarily positive stable matrix arguments.

Definition 94. Let A, B be commuting matrices in $\mathbb{C}^{p\times p}$, such that for all integers $k \geq 0$, one satisfies the condition

$$A + kI, \quad B + kI \quad \text{and} \quad A + B + kI \quad \text{are invertible.}$$

Also, let $\alpha(A, B) = \min\{\alpha(A), \alpha(B), \alpha(A + B)\}$ and $k_0 = k_0(A, B) = [|\alpha(A, B)|] + 1$, where $[\cdot]$ denotes the entire part function. Then the beta matrix function $\beta(A, B)$ is defined by the formula

$$\beta(A, B) = (A)_{k_0}^{-1} (B)_{k_0}^{-1} (A + B)_{2k_0} \, \beta(A + k_0 I, B + k_0 I).$$

Lemma 95. *Let $\tilde{A}$, $\tilde{B}$ be commuting matrices in $\mathbb{C}^{p\times p}$ such that $\tilde{A}$, $\tilde{B}$, $\tilde{A} + \tilde{B}$ are positive stable. Then*

$$\beta(\tilde{A}, \tilde{B}) = \Gamma(\tilde{A})\, \Gamma(\tilde{B})\, \Gamma^{-1}(\tilde{A} + \tilde{B}).$$

Proof. By hypothesis of positive stability and the commutativity of $\tilde{A}$, $\tilde{B}$, we can write

$$\begin{aligned} \Gamma(\tilde{A})\, \Gamma(\tilde{B}) &= \left(\int_0^\infty e^{-u} u^{\tilde{A}-I} \, du \right) \left(\int_0^\infty e^{-v} v^{\tilde{B}-I} \, dv \right) \\ &= \int_0^\infty \int_0^\infty e^{-(u+v)} u^{\tilde{A}-I} v^{\tilde{B}-I} \, du \, dv. \end{aligned} \tag{4.79}$$

Considering the change of variables $x = \frac{u}{u+v}, y = u + v$ in the above integral, the positive stability of $\tilde{A} + \tilde{B}$, one gets Jacobian $J(x,y) = y$ and

$$
\begin{aligned}
\Gamma(\tilde{A})\,\Gamma(\tilde{B}) &= \int_0^\infty \int_0^1 e^{-y}\,(xy)^{\tilde{A}-I}\,\big(y\,(1-x)\big)^{\tilde{B}-I}\,y\,dx\,dy \\
&= \left(\int_0^\infty e^{-y}\,y^{\tilde{A}+\tilde{B}-I}\,dy\right)\left(\int_0^1 x^{\tilde{A}-I}\,(1-x)^{\tilde{B}-I}\,dy\right) \\
&= \Gamma(\tilde{A}+\tilde{B})\,\beta(\tilde{A},\tilde{B}).
\end{aligned}
\tag{4.80}
$$

It completes the proof. □

Theorem 96. *Let A, B be commuting matrices in $\mathbb{C}^{p\times p}$ such that $A+kI$, $B+kI$ and $A+B+kI$ are invertible for all integers $k \geq 0$. Then*

$$\beta(A,B) = \Gamma(A)\,\Gamma(B)\,\Gamma^{-1}(A+B).$$

Proof. Let $k_0 = k_0(A,B)$ as defined in Definition 94 so that

$$\beta(A,B) = (A)_{k_0}^{-1}\,(B)_{k_0}^{-1}\,(A+B)_{2k_0}\,\beta(A+k_0I, B+k_0I),$$

where $A+k_0I, B+k_0I$ are positive stable. By the definition of an inverse gamma function, one gets

$$
\begin{aligned}
\Gamma(A) &= \Gamma(A+k_0I)\,\big(A+(k_0-1)I\big)^{-1}\cdots(A+I)^{-1}A^{-1}, \\
\Gamma(B) &= \Gamma(B+k_0I)\,\big(B+(k_0-1)I\big)^{-1}\cdots(B+I)^{-1}B^{-1}, \\
\Gamma(A+B) &= \Gamma(A+B+2k_0I)\,\big(A+B+(2k_0-1)I\big)^{-1}\cdots(A+B+I)^{-1}\,(A+B)^{-1}.
\end{aligned}
\tag{4.81}
$$

Since $AB = BA$, it follows from the last equations:

$$
\begin{aligned}
&\Gamma(A)\,\Gamma(B)\,\Gamma^{-1}(A+B) \\
&\quad = \Gamma(A+k_0I)\,\Gamma(B+k_0I)\,\Gamma^{-1}(A+B+2k_0I)\,\big(A+(k_0-1)I\big)^{-1}\cdots(A+I)^{-1}A^{-1} \\
&\qquad \times \big(B+(k_0-1)I\big)^{-1}\cdots(B+I)^{-1}B^{-1}\,\big(A+B+(2k_0-1)I\big)\cdots(A+B+I)\,(A+B) \\
&\quad = \Gamma(A+k_0I)\,\Gamma(B+k_0I)\,\Gamma^{-1}(A+B+2k_0I)\,(A)_{k_0}^{-1}\,(B)_{k_0}^{-1}\,(A+B)_{2k_0}.
\end{aligned}
\tag{4.82}
$$

Further, if $A+k_0I$, $B+k_0I$ and $A+B+2k_0I$ are positive stable matrices, then by Lemma 95, we get

$$\Gamma(A+k_0I)\,\Gamma(B+k_0I)\,\Gamma^{-1}(A+B+2k_0I) = \beta(A+k_0I,\, B+k_0I)$$

and by Lemma 93

$$\beta(A + k_0 I\, B + k_0 I) = (A)_{k_0}\, (B)_{k_0}\, (A + B)^{-1}_{2k_0}\, \beta(A, B).$$

Thus, one can conclude using the above three relations that

$$\beta(A, B) = \Gamma(A)\, \Gamma(B)\, \Gamma^{-1}(A + B).$$

It completes the proof. □

Now, with the integral of the gamma and beta matrix functions, we can derive the integrals for Pochhammer matrix function as

$$\begin{aligned}(A)_k &= \Gamma(A + kI)\, \Gamma^{-1}(A)\\ &= \Gamma^{-1}(A) \int_0^{\infty} e^{-u}\, u^{A+(k-1)I}\, du.\end{aligned} \tag{4.83}$$

Also, for $A, B \in \mathbb{C}^{p\times p}$ and $A\,B = B\,A$, we have

$$\begin{aligned}(A)_k\, (B)_k^{-1} &= \Gamma(A + kI)\, \Gamma^{-1}(A)\, \Gamma^{-1}(B + kI)\, \Gamma(B)\\ &= \Gamma(B)\, \Gamma^{-1}(A)\, \Gamma^{-1}(B - A)\, \Gamma(A + kI)\, \Gamma(B - A)\, \Gamma^{-1}(B + kI)\\ &= \Gamma(B)\, \Gamma^{-1}(A)\, \Gamma^{-1}(B - A)\, \beta(A + kI, B - A)\\ &= \Gamma(B)\, \Gamma^{-1}(A)\, \Gamma^{-1}(B - A) \int_0^1 u^{A+(k-1)I}\, (1 - u)^{B-A-I}\, du\end{aligned} \tag{4.84}$$

or

$$(A)_k\, (B)_k^{-1} = \Gamma(B)\, \Gamma^{-1}(A)\, \Gamma^{-1}(B - A) \int_0^{\infty} u^{A+(k-1)I}\, (1 + u)^{-(B+kI)}\, du. \tag{4.85}$$

4.5 On bilateral-type matrix differential equations

In this section, we summarize some results on bilateral-type linear matrix differential equations and their solutions.

Theorem 97. *Let $B(z_0, p)$ be an open disc in the complex plane of radius p centered at z_0. Let E be the Banach space of all $\mathbb{C}^{p\times p}$ matrices endowed with the two-norm. Let $f : B(z_0, p) \times E \to E$ be a continuous function such that*

$$\|f(z, X) - f(z, Y)\| \leq K(|z - z_0|)\, \|X - Y\|,$$

where $z \in B(z_0, p)$; $X, Y \in E$ and $\chi \to K(\chi)$ is a real valued continuous function on $[0, p]$. Then for each $X_0 \in E$, there exist one and only one solution U of

$$X' = f(z, X),$$

defined in $B(z_0, p)$, *such that* $U(z_0) = X_0$.

In particular, if we consider a bilateral linear matrix differential equation of the form

$$X' = P_1(z)X + X\,Q_1(z) + P_2(z)X\,Q_2(z) + R(z), \tag{4.86}$$

where $P_i, Q_i, i = 1, 2$ and R are continuous functions from $B(z_0, p)$ into E, then

$$f(z, X) = P_1(z)X + X\,Q_1(z) + P_2(z)X\,Q_2(z) + R(z)$$

and f satisfies the Lipschitz condition given in Theorem 97. By Theorem 97, for each X_0 in E, there exists one and only one solution U of (4.86), for $z \in B(z_0, p)$ such that $U(z_0) = X_0$.

Definition 98. Let $f_j : B(z_0, p) \to E$ be bounded continuous functions in $B(z_0, p)$ for $1 \le j \le 4$, and let U_1, U_2 be two solutions of the second-order differential equation

$$X'' = f_1(z)X' + f_2(z)X f_3(z) + X' f_4(z). \tag{4.87}$$

We say that $\{U_1, U_2\}$ is a fundamental set of solutions of (4.87) in $B(z_0, p)$, if for any solution U of (4.87) admits a unique representation of the form

$$U(z) = U_1(z)\,P + U_2(z)\,Q, \quad z \in B(z_0, p),$$

where P, Q are matrices in $\mathbb{C}^{p\times p}$, uniquely determined by U.

Theorem 99. *If* $\{U_1, U_2\}$ *is a pair of solutions of equation* (4.87) *in* $B(z_0, p)$ *such that the matrix in* $\mathbb{C}^{2p\times 2p}$,

$$W(U_1, U_2, z_0) = \begin{bmatrix} U_1(z_0) & U_2(z_0) \\ U_1'(z_0) & U_2'(z_0) \end{bmatrix} \quad \text{is invertible},$$

then $\{U_1, U_2\}$ *is a fundamental set of solutions of* (4.87) *in* $B(z_0, p)$.

Proof. Note that for any pair of matrices A, B in $\mathbb{C}^{p\times p}$, the function $V(z) = U_1(z)A + U_2(z)B$ is a solution of (4.87). Given a solution U of (4.87) with initial conditions $C_0 = U(z_0)$, $C_1 = U'(z_0)$. Let A_0, B_0 be the matrices defined by

$$\begin{bmatrix} A_0 \\ B_0 \end{bmatrix} = [W(U_1, U_2, z_0)]^{-1} \begin{bmatrix} C_0 \\ C_1 \end{bmatrix}.$$

Then $V(z) = U_1(z)\,A_0 + U_2(z)\,B_0$ is a solution of (4.87) in $B(z_0, p)$, satisfying the same initial conditions as U because

$$V'(z) = U_1'(z)\,A_0 + U_2'(z)\,B_0,$$

$$\begin{bmatrix} V(z_0) \\ V'(z_0) \end{bmatrix} = \begin{bmatrix} U_1(z_0) & U_2(z_0) \\ U_1'(z_0) & U_2'(z_0) \end{bmatrix} \begin{bmatrix} A_0 \\ B_0 \end{bmatrix}$$
$$= W(U_1, U_2, z_0)\,[W(U_1, U_2, z_0)]^{-1} \begin{bmatrix} C_0 \\ C_1 \end{bmatrix},$$
$$= \begin{bmatrix} C_0 \\ C_1 \end{bmatrix} = \begin{bmatrix} U(z_0) \\ U'(z_0) \end{bmatrix}. \tag{4.88}$$

Now, we prove that an initial value problem for equation (4.87) admits a unique solution. In fact, note that considering the change $Y = [X, X']^T$, then equation (4.87) is equivalent to

$$Y' = \begin{bmatrix} 0 & I \\ 0 & 0 \end{bmatrix} Y + \begin{bmatrix} 0 & 0 \\ 0 & f_1(z) \end{bmatrix} Y + \begin{bmatrix} 0 & 0 \\ 0 & I \end{bmatrix} Y f_4(z) + \begin{bmatrix} 0 & 0 \\ f_2(z) & 0 \end{bmatrix} Y f_3(z). \tag{4.89}$$

By Theorem 97, an initial value problem for (4.89) admits a unique solution. As the relationship between solutions of (4.87) and (4.89) is given by

$$X = [I, 0]\,Y,$$

then an initial value problem for (4.87) admits a unique solution. By the uniqueness, $V(z) = U(z)$ for all $z \in B(z_0, p)$. Hence, the result is established. □

Example 100. Let z_0 be a complex number with $0 < |z_0| < 1$ and let P, Q, R be complex matrices in $\mathbb{C}^{p\times p}$. Let $0 < \delta_0 < |z_0| < 1 - \delta_1$, with $\delta_1 < 1$ and $p = \min(\delta_0, \delta_1)$. Then in $B(z_0, p)$, the matrix differential equation

$$W'' = \frac{P}{1-z} W' + \frac{P}{z(1-z)} W Q - W' \left[\frac{R - z(Q+I)}{z(1-z)} \right] \tag{4.90}$$

is of type (4.87) with

$$f_1(z) = \frac{P}{1-z}; \quad f_2(z) = \frac{P}{z(1-z)}; \quad f_3(z) = Q; \quad f_4(z) = \frac{-R + z(Q+I)}{z(1-z)}.$$

4.6 Gauss hypergeometric matrix function

We begin this section with the definition of the Gauss hypergeometric matrix function.

Definition 101. Let P, Q, R be matrices in $\mathbb{C}^{p\times p}$ such that $R+kI$ is invertible for all positive integer k. Then the Gauss hypergeometric matrix function is denoted by ${}_2F_1(P, Q; R; z)$ or ${}_2F_1\left[{P,Q \atop R}; z\right]$ and defined as

$${}_2F_1(P, Q; R; z) = \sum_{k=0}^{\infty} (P)_k\,(Q)_k\,[(R)_k]^{-1} \frac{z^k}{k!}. \tag{4.91}$$

Now, we prove that the matrix power series (4.91) converges for all z with $|z| < 1$. Note that if k is large enough so that $k > \|R\|$, then by the perturbation lemma, we can write

$$\left\|\left(\frac{R}{k}+I\right)^{-1}\right\| \le \frac{1}{1-\frac{\|R\|}{k}} = \frac{k}{k-\|R\|}$$

and

$$\|(R+kI)^{-1}\| = \left\|\frac{1}{k}\left(\frac{R}{k}+I\right)^{-1}\right\| = \frac{1}{k}\left\|\left(\frac{R}{k}+I\right)^{-1}\right\| = \frac{1}{k-\|R\|}, \quad k \ge \|R\|.$$

Let us denote

$$\lambda(k) = \|R^{-1}\|\,\|(R+I)^{-1}\|\cdots\|(R+(k-I)I)^{-1}\|, \quad k \ge 0.$$

Also, note that

$$\|(P)_k\| \le (\|P\|)_k, \quad \|(Q)_k\| \le (\|Q\|)_k.$$

With these assumptions, one will get

$$\begin{aligned}\left\|(P)_k\,(Q)_k\,[(R)_k]^{-1}\frac{z^k}{k!}\right\| &\le \|(P)_k\|\,\|(Q)_k\|\,\lambda(k)\,\frac{|z|^k}{k!}\\ &\le (\|P\|)_k\,(\|Q\|)_k\,\lambda(k)\,\frac{|z|^k}{k!}.\end{aligned} \tag{4.92}$$

Now, we prove that

$$\sum_{k=0}^{\infty}(\|P\|)_k\,(\|Q\|)_k\,\lambda(k)\,\frac{|z|^k}{k!}$$

converges for $|z| < 1$ using the ratio test, and we have

$$\begin{aligned}\lim_{n\to\infty}\left|\frac{(\|P\|)_{k+1}\,(\|Q\|)_{k+1}\,\lambda(k+1)\,|z|^{k+1}}{(\|P\|)_k\,(\|Q\|)_k\,\lambda(k)\,|z|^k(k+1)}\right| &= \lim_{n\to\infty}\left|\frac{(\|P\|+k)\,(\|Q\|+k)\,\|(R+kI)^{-1}\|}{k+1}\,z\right|\\ &\le \lim_{n\to\infty}\frac{(\|P\|+k)\,(\|Q\|+k)}{(k+1)\,(k-\|R\|)}\,|z| = |z|.\end{aligned} \tag{4.93}$$

Thus, the power series converges absolutely for $|z| < 1$. Now, we study the conditions so that the Gauss matrix function ${}_2F_1(P,Q;R;z)$ converges when $|z| = 1$.

Theorem 102. *Let P, Q, R be positive stable matrices in $\mathbb{C}^{p\times p}$ such that*

$$\beta(R) > \alpha(P) + \alpha(Q).$$

Then the series (4.91) is absolutely convergent for $|z| = 1$.

Proof. By the assumption, there exists a positive number ϵ such that

$$\beta(R) - \alpha(P) - \alpha(Q) = 2\,\epsilon. \tag{4.94}$$

Let us write

$$\begin{aligned} & k^{1+\epsilon}\Big(\frac{1}{k!}(P)_k\,(Q)_k\,\big[(R)_k\big]^{-1}\Big) \\ &= \frac{k^{1+\epsilon}}{k!}\cdot\frac{(k-1)!\,k^P\,k^{-P}\,(P)_k}{(k-1)!}\cdot\frac{(k-1)!\,k^Q\,k^{-Q}\,(Q)_k}{(k-1)!}\cdot k^R\,k^{-R}\,(R)_k^{-1} \\ &= \frac{k^{1+\epsilon}}{k}\cdot\Big(\frac{k^{-P}\,(P)_k}{(k-1)!}\Big)k^P\cdot\Big(\frac{k^{-Q}\,(Q)_k}{(k-1)!}\Big)k^Q\cdot\big((k-1)!\,k^R\,(R)_k^{-1}\big)\,k^{-R}, \end{aligned} \tag{4.95}$$

or

$$\begin{aligned} & k^{1+\epsilon}\Big(\frac{1}{k!}(P)_k\,(Q)_k\,\big[(R)_k\big]^{-1}\Big) \\ &= k^{\epsilon}\cdot\Big(\frac{k^{-P}\,(P)_k}{(k-1)!}\Big)k^P\cdot\Big(\frac{k^{-Q}\,(Q)_k}{(k-1)!}\Big)k^Q\cdot\big((k-1)!\,k^R\,(R)_k^{-1}\big)\,k^{-R}. \end{aligned} \tag{4.96}$$

The Schur decomposition (4.35) and $\alpha(-R) = -\beta(R)$ together yield

$$\begin{aligned} & \|k^P\|\,\|k^Q\|\,\|k^{-R}\| \\ &\le k^{\alpha(P)+\alpha(Q)-\beta(R)}\left\{\sum_{j=0}^{p-1}\frac{(\|P\|\,p^{1/2}\,\ln k)^j}{j!}\right\}\left\{\sum_{j=0}^{p-1}\frac{(\|Q\|\,p^{1/2}\,\ln k)^j}{j!}\right\} \\ &\quad\times\left\{\sum_{j=0}^{p-1}\frac{(\|R\|\,p^{1/2}\,\ln k)^j}{j!}\right\} \\ &\le k^{-2\epsilon}\left\{\sum_{j=0}^{p-1}\frac{(\max\{\|P\|,\|Q\|,\|R\|\}\,p^{1/2}\,\ln k)^j}{j!}\right\}^3. \end{aligned} \tag{4.97}$$

Taking the norm on both sides in (4.96) gives

$$\begin{aligned} & \lim_{k\to\infty} k^{1+\epsilon}\Big\|\frac{1}{k!}(P)_k\,(Q)_k\,\big[(R)_k\big]^{-1}\,z^k\Big\| \\ &\le \lim_{k\to\infty} k^{\epsilon}\Big\|\frac{k^{-P}\,(P)_k}{(k-1)!}\Big\|\,\|k^P\|\,\Big\|\frac{k^{-Q}\,(Q)_k}{(k-1)!}\Big\|\,\|k^Q\|\,\|((k-1)!\,k^R\,(R)_k^{-1})\|\,\|k^{-R}\|. \end{aligned} \tag{4.98}$$

Now using the limiting definition of the gamma matrix function and (4.97), we get

$$\begin{aligned} & \lim_{k\to\infty} k^{1+\epsilon}\Big\|\frac{1}{k!}(P)_k\,(Q)_k\,\big[(R)_k\big]^{-1}\,z^k\Big\| \\ &\le \|\Gamma^{-1}(P)\|\,\|\Gamma^{-1}(Q)\|\,\|\Gamma^{-1}(R)\|\,\lim_{k\to\infty} k^{-\epsilon}\left\{\sum_{j=0}^{p-1}\frac{(\max\{\|P\|,\|Q\|,\|R\|\}\,p^{1/2}\,\ln k)^j}{j!}\right\}^3 \end{aligned}$$

$$= \|\Gamma^{-1}(P)\| \, \|\Gamma^{-1}(Q)\| \, \|\Gamma^{-1}(R)\| \cdot 0 = 0, \tag{4.99}$$

because

$$\lim_{k\to\infty} k^{-\epsilon} (\ln k)^k = 0, \quad \text{for all integers } k \geq 0.$$

Since

$$\lim_{k\to\infty} k^{1+\epsilon} \left\| \frac{1}{k!} (P)_k \, (Q)_k \left[(R)_k\right]^{-1} z^k \right\| = 0, \quad |z| = 1,$$

by the comparison theorem of the numerical series of positive numbers, one concludes the absolute convergence of series (4.91) at $|z| = 1$. □

Now, we are interested in the determination of a domain where $U_1(z) = {}_2F_1(P, Q; R; z)$ is invertible. Let $0 < \gamma < 1$ and note that

$$\lim_{k\to\infty} \frac{(\|P\| + k)\,(\|Q\| + k)}{(k+1)\,(k - \|R\|)} = 1.$$

Let $k_0 \geq \max(1, \|R\|)$ be a positive integer such that

$$\frac{(\|P\| + k)\,(\|Q\| + k)}{(k+1)\,(k - \|R\|)} < 1 + \gamma, \quad k \geq k_0.$$

Also

$$K = \sum_{k=1}^{k_0-1} \frac{(\|P\|)_k \, (\|Q\|)_k \, \lambda(k)}{k!}, \quad L = \frac{(\|P\|)_{k_0} \, (\|Q\|)_{k_0} \, \lambda(k_0)}{k_0!}, \tag{4.100}$$

where $\lambda(k)$ is defined by

$$\lambda(k) = \|R^{-1}\| \, \|(R+I)^{-1}\| \cdots \|(R + (k - I)I)^{-1}\|, \quad k \geq 0. \tag{4.101}$$

Since the function

$$f(x) = K\,x + L\,\frac{x}{1-x}, \quad 0 \leq x < 1 \tag{4.102}$$

increases, $f(0) = 0$ and $\lim_{x\to 1^-} f(x) = +\infty$, so there exists only one value ρ_0 such that

$$f(\rho_o) = 1 - \gamma.$$

Let us take ρ_1 with $0 < \rho_1 < \rho_0$, such that $(1+\gamma)\,\rho_1 < \rho_0$. Then we can write

$$\frac{k!\,(\|P\|)_{k+1}\,(\|Q\|)_{k+1}\,\lambda(k+1)}{(k+1)!\,(\|P\|)_k\,(\|Q\|)_k\,\lambda(k)} \leq \frac{(\|P\| + k)\,(\|Q\| + k)}{(k+1)\,(k - \|R\|)} < 1 + \gamma, \quad k \geq k_0, \tag{4.103}$$

or

$$\frac{(\|P\|)_{k+1}\,(\|Q\|)_{k+1}\,\lambda(k+1)}{(k+1)!}\,\rho_1 \le \rho_0\,\frac{1}{k!}\,(\|P\|)_k\,(\|Q\|)_k\,\lambda(k), \quad k \ge k_0. \tag{4.104}$$

Hence,

$$\begin{aligned}\|U_1(\rho_1) - U_1(0)\| &= \|U_1(\rho_1) - I\| \\ &\le \sum_{k=1}^{\infty} \frac{1}{k!}\,(\|P\|)_k\,(\|Q\|)_k\,\lambda(k)\,\rho_1^k \\ &\le \left\{\sum_{k=1}^{k_0-1} \frac{1}{k!}\,(\|P\|)_k\,(\|Q\|)_k\,\lambda(k)\right\}\rho_1 + \sum_{k=k_0}^{\infty} \frac{1}{k!}\,(\|P\|)_k\,(\|Q\|)_k\,\lambda(k)\,\rho_1^k \\ &\le K\rho_0 + \sum_{k=k_0}^{\infty} \frac{1}{k!}\,(\|P\|)_k\,(\|Q\|)_k\,\lambda(k)\,\rho_1^k. \end{aligned} \tag{4.105}$$

But

$$\begin{aligned}\sum_{k=k_0}^{\infty} \frac{1}{k!}\,(\|P\|)_k\,(\|Q\|)_k\,\lambda(k)\,\rho_1^k &\le \left[\frac{(\|P\|)_{k_0}\,(\|Q\|)_{k_0}\,\lambda(k_0)}{k_0!}\right]\sum_{k=k_0}^{\infty} \rho_0^k \\ &= \frac{(\|P\|)_{k_0}\,(\|Q\|)_{k_0}\,\lambda(k_0)}{k_0!}\,\frac{\rho_0^{k_0}}{1-\rho_0} \\ &\le L\,\frac{\rho_0}{1-\rho_0}. \end{aligned} \tag{4.106}$$

Therefore, (4.105) and (4.106) together yield

$$\|U_1(\rho_1) - I\| \le K\rho_0 + L\,\frac{\rho_0}{1-\rho_0} = f(\rho_0) = 1-\gamma. \tag{4.107}$$

Thus, $\|U_1(\rho) - I\| \le 1-\gamma$, for $0 \le \rho \le \rho_1$ and by the perturbation lemma, one gets that $U_1(z) = {}_2F_1(P,Q;R;z)$ is invertible and $\|U_1(z) - I\| \le 1-\gamma$, for $|z| < \rho_1$.

Further, by the perturbation lemma, one can obtain

$$\begin{aligned}\|(U_1(z))^{-1} - I\| &\le \frac{\|U_1(z) - I\|}{1 - \|U_1(z) - I\|} < \frac{1-\gamma}{\gamma}, \quad |z| < \rho_1, \\ \|(U_1(z))^{-1}\| &\le 1 + \frac{1-\gamma}{\gamma} = \frac{1}{\gamma}, \quad |z| < \rho_1. \end{aligned}$$

Furthermore, by (4.106) for $k \ge k_0$ and $|z| \le \rho_1$, it follows that

$$\left\| U_1(z) - \sum_{j=1}^{k-1} \frac{1}{j!} (P)_j \, (Q)_j \, \left[(R)_j\right]^{-1} z^j \right\| \leq \sum_{m \geq k} \frac{1}{m!} (\|P\|)_m \, (\|Q\|)_m \, \lambda(m) \, \rho_1^m$$
$$\leq L \sum_{m \geq k} \rho_0^m = L \frac{\rho_0^k}{1 - \rho_0}. \tag{4.108}$$

Hence, given $\epsilon > 0$, taking $k \geq k_0$ so that $\rho_0^k < \epsilon \frac{1-\rho_0}{L}$, or equivalently

$$k \geq \max\left\{ \frac{\ln[\frac{1-\rho_0}{L}]}{\ln(\rho_0)}, k_o \right\} = k_\epsilon. \tag{4.109}$$

If the truncation of order k of $F(P, Q; R; z)$ is defined by

$$F_k(P, Q; R; z) = \sum_{j=0}^{k} \frac{1}{j!} (P)_j \, (Q)_j \, \left[(R)_j\right]^{-1} z^j,$$

then it follows that

$$\left\| U_1(z) - F_{k-1}(P, Q; R; z) \right\| \leq \epsilon, \quad |z| \leq \rho_1, k \geq k_\epsilon.$$

We summarize our findings in the following theorem.

Theorem 103. *Let $0 < \gamma < 1$ and $\epsilon > 0$. Let k_0 and k_ϵ be a positive number satisfying (4.104) and (4.109), respectively. Let ρ_0 be the solution of equation $f(\rho_0) = 1 - \gamma$, where f is defined by*

$$f(x) = K\, x + L \frac{x}{1 - x}, \quad 0 \leq x < 1.$$

Let ρ_1 so that $\rho_1 < \frac{\rho_0}{1+\gamma}$, $U_1(z) = F(P, Q; R; z)$ and $F_k(P, Q; R; z) = \sum_{j=0}^{k} \frac{1}{j!} (P)_j \, (Q)_j \, [(R)_j]^{-1} z^j$. Then:

1. *$U_1(z)$ is invertible and $\|(U_1(z))^{-1}\| \leq \frac{1}{\gamma}$, $|z| < \rho_1$;*
2. *$\|U_1(z) - F_k(P, Q; R; z)\| < \epsilon$, $|z| < \rho_1$, $k > k_\epsilon$.*

Let us now consider the boundedness of $U'(z)$ for $|z| \leq \rho_1$ where ρ_1 is given by Theorem 103. Note that for $|z| \leq \rho_1$ one gets

$$\left\| U_1'(z) \right\| \leq \sum_{k \geq 1} \frac{1}{(k-1)!} (\|P\|)_k \, (\|Q\|)_k \, \lambda(k) \, \rho_1^{k-1}.$$

Given γ with $0 < \gamma < 1$, let $k \geq k_0$ such that

$$\frac{(\|P\| + k)\,(\|Q\| + k)}{k\,(k - \|R\|)} < 1 + \gamma, \quad k \geq k_1 > \max(1, \|R\|). \tag{4.110}$$

Let ρ_2 be so that $(1 + \gamma)\,\rho_2 < \rho_1$ and

$$K_1 = \sum_{k=1}^{k_1-1} \frac{1}{(k-1)!}(\|P\|)_k (\|Q\|)_k \lambda(k), \quad L_1 = \frac{1}{(k_1-1)!}(\|P\|)_{k_1} (\|Q\|)_{k_1} \lambda(k_1), \tag{4.111}$$

where $\lambda(k)$ is defined by (4.101). Note that for $|z| \leq \rho_2$, we can write

$$\begin{aligned} \|U_1'(z)\| \leq & \sum_{k=1}^{k_1-1} \frac{1}{(k-1)!}(\|P\|)_k (\|Q\|)_k \lambda(k) \rho_2^{k-1} \\ & + \sum_{k \geq k_1} \frac{1}{(k-1)!}(\|P\|)_k (\|Q\|)_k \lambda(k) \rho_2^{k-1}, \end{aligned} \tag{4.112}$$

and by (4.110),

$$\frac{1}{k!}(\|P\|)_{k+1} (\|Q\|)_{k+1} \lambda(k+1) < (1+\gamma)\frac{1}{(k-1)!}(\|P\|)_k (\|Q\|)_k \lambda(k), \quad k \geq k_1, \tag{4.113}$$

it follows

$$\begin{aligned} & \sum_{k \geq k_1} \frac{1}{(k-1)!}(\|P\|)_k (\|Q\|)_k \lambda(k) \rho_2^{k-1} \\ & \leq \frac{1}{(k_1-1)!}(\|P\|)_{k_1} (\|Q\|)_{k_1} \lambda(k_1) \sum_{k \geq k_1} (\rho_2(1+\gamma))^{k-1} \\ & \leq L_1 \sum_{k \geq k_1} \rho_1^{k-1} = L_1 \frac{\rho_1^{k-1}}{1-\rho_1} \leq \frac{L_1}{1-\rho_1}. \end{aligned} \tag{4.114}$$

If $F_k(P, Q; R; z)$ is defined as in Theorem 103, then we conclude

$$\|U_1'(z) - F_k'(P, Q; R; z)\| \leq L_1 \frac{\rho_1^{k-1}}{1-\rho_1}, \quad k \geq k_1, |z| \leq \rho_2, \tag{4.115}$$

$$\|U_1'(z)\| \leq K_1 + \frac{L_1}{1-\rho_1}, \quad |z| \leq \rho_2. \tag{4.116}$$

Thus, we summarize the result in the following theorem.

Theorem 104. *Let $\rho_2 > 0$ be chosen so that $\rho_2(1+\gamma) < \rho_1$. Let k_1 be a positive integer satisfying* (4.110) *and*

$$k_\epsilon' = 1 + \max\left\{\frac{\ln[\epsilon \frac{1-\rho_1}{L_1}]}{\ln(\rho_1)}\right\}. \tag{4.117}$$

Then:

1. $\|U_1'(z)\| \leq K_1 + \frac{L_1}{1-\rho_1}$, $|z| \leq \rho_2$,
2. $\|U_1'(z) - F_k'(P, Q; R; z)\| \leq \epsilon$, $|z| \leq \rho_2$, $k \geq k_\epsilon'$.

4.7 The hypergeometric matrix differential equation

Let us consider the hypergeometric matrix differential equation

$$z(1-z)\,W'' - z\,P\,W' + W'(R - z(Q+I)) - P\,W\,Q = 0, \quad 0 < |z| < 1, \tag{4.118}$$

where $QR = RQ$ and $R + kI$ is invertible for all positive integers $k \geq 0$. Note that equation (4.118) can be written in the form (4.87). Let us seek for a solution W of (4.118) of the form

$$W(z) = \sum_{k=0}^{\infty} W_k\, z^k, \quad |z| < 1.$$

Taking formal derivatives, we get

$$W'(z) = \sum_{k=0}^{\infty} k\, W_k\, z^{k-1}, \quad W''(z) = \sum_{k=0}^{\infty} k\,(k-1)\, W_k\, z^{k-2}.$$

Now substituting these values in (4.118), one gets

$$\begin{aligned}&\sum_{k=2}^{\infty} k\,(k-1)\, W_k\, z^{k-1} - \sum_{k=2}^{\infty} k\,(k-1)\, W_k\, z^k - P \sum_{k=1}^{\infty} k\, W_k\, z^k - \sum_{k=1}^{\infty} k\, W_k\, Q\, z^k \\ &\quad + \sum_{k=1}^{\infty} k\, W_k\, R\, z^{k-1} - \sum_{k=1}^{\infty} k\, W_k\, z^k - \sum_{k=0}^{\infty} P\, W_k\, Q\, z^k = 0.\end{aligned} \tag{4.119}$$

This can be rewritten as

$$\begin{aligned}&(W_1\, R - P\, W_0\, Q) + (2\,W_2 - P\,W_1 - W_1\,Q - W_1 + 2W_2\,R - P\,W_1\,Q)\,z \\ &\quad + \sum_{k=2}^{\infty} \{k\,(k+1)\, W_{k+1} - k(k-1)\, W_k - k\,P\,W_k - k\,W_k\,Q - k\,W_k + (k+1)\,W_{k+1}\,R \\ &\quad - P\,W_k\,Q\}\, z^k = 0.\end{aligned} \tag{4.120}$$

By equating to zero, the coefficients of each power of z^k, one gets

$$\begin{aligned}&W_1\,R - P\,W_0\,Q = 0, \\ &2\,W_2 - P\,W_1 - W_1\,Q - W_1 + 2W_2\,R - P\,W_1\,Q = 0, \\ &k\,(k+1)\,W_{k+1} - k(k-1)\,W_k - k\,P\,W_k - k\,W_k\,Q - k\,W_k + (k+1)\,W_{k+1}\,R \\ &\quad - P\,W_k\,Q = 0.\end{aligned} \tag{4.121}$$

Hence,

$$W_1 = P\,W_0\,Q\,R^{-1},$$

$$W_{k+1} = \frac{1}{k+1}(P + kI)\, W_k\, (Q + kI)\, (R + kI)^{-1}, \quad k \geq 0. \tag{4.122}$$

Using the commutativity of Q, R and equation (4.122) with $W_0 = I$, it follows that

$$\begin{aligned} W_{k+1} &= \frac{1}{(k+1)!} P\,(P+I) \cdots (P+kI)\, Q\,(Q+I) \cdots (Q+kI) \\ &\quad \times (R+kI)^{-1} \cdots (R+I)^{-1} R^{-1} \\ &= \frac{1}{(k+1)!}\, (P)_{k+1}\, (Q)_{k+1} \big[(R)_{k+1}\big]^{-1}. \end{aligned} \tag{4.123}$$

Hence, there exists a solution W_1 of (4.118) well-defined in $|z| < 1$ satisfying $W_1(0) = I$, given by

$$W_1(z) = {}_2F_1(P, Q; R; z). \tag{4.124}$$

Now, we seek a second solution W_2 of equation (4.118) under the assumption that $R + kI$ is invertible for all positive integers $k \geq 0$ and $QR = RQ$, $PR = RP$. Let D_0 be the complex plane cut along the negative real axis, and let us denote $z^{I-R} = \exp((I - R)\log z)$ where the log represents the principal logarithm. Let us seek a solution of the form

$$W_2(z) = V(z)\, z^{I-R}, \quad |z| < 1, z \in D_0,$$

where V is a function to be determined. Taking derivatives of W_2, one gets

$$\begin{aligned} W_2'(z) &= V'(z)\, z^{I-R} + V(z)\, z^{-R}(I - R), \\ W_2''(z) &= V''(z)\, z^{I-R} + 2\, V'(z)\, z^{-R}(I - R) - V(z)\, z^{-I-R}\, R\,(I - R). \end{aligned} \tag{4.125}$$

Substituting these values in (4.118), we get

$$\begin{aligned} &z(1-z)V''(z) + \{2\,V' - V'\,R - 3\,z\,V' + 2\,zV'R - PzV' - zV'Q\} \\ &\quad + \{V\,R - V\,R^2 - P\,V + P\,V\,R - V\,Q - V + V\,Q\,R + V\,R - P\,V\,Q\} = 0. \end{aligned} \tag{4.126}$$

Suppose for a moment that

$$V(z)\, R = R\, V(z), \quad V'(z)\, R = R\, V'(z).$$

Then (4.126) is equivalent to

$$\begin{aligned} &z(1-z)V'' - z(P + I - R)V' + V'\big[(2I - R) - z(Q + 2I - R)\big] \\ &\quad - (P + I - R)\, V\, (Q + I - R) = 0, \end{aligned} \tag{4.127}$$

which is a hypergeometric matrix differential equation of the form (4.118) with matrix parameters

$$\tilde{P} = P + I - R, \quad \tilde{Q} = Q + I - R, \quad \tilde{R} = 2I - R.$$

As we proved above,

$$V(z) = {}_2F_1(P + I - R, Q + I - R, 2I - R; z), \quad |z| < 1$$

is a solution of (4.127) because $QR = RQ$. Furthermore, note that under hypothesis $QR = RQ$ and $PR = RP$, in fact, ${}_2F_1(P + I - R, Q + I - R, 2I - R; z)$ and its derivative commutes with R because the matrix coefficients of ${}_2F_1(P + I - R, Q + I - R, 2I - R; z)$ are

$$\frac{1}{k!}(P + I - R)_k (Q + I - R)_k [(2I - R)_k]^{-1}, \quad k \geq 0.$$

Hence,

$$W_2(z) = {}_2F_1(P + I - R, Q + I - R, 2I - R; z)\, z^{I-R}$$

is also a solution of (4.118) for $z \in D_0$, $|z| < 1$.

In the following, we construct a closed-form general solution of the hypergeometric matrix equation (4.118) under the hypothesis:

$$R + kI \text{ is invertible for all positive integers } k \geq 0 \text{ and } PR = RP, QR = RQ.$$

Note that

$$W_1(z) = U_1(z), \quad W_2(z) = U_2(z)\, z^{I-R}, \quad 0 < |z| < 1, z \in D_0, \tag{4.128}$$

where $U_1(z) = {}_2F_1(P, Q; R; z)$ and $U_2(z) = {}_2F_1(P + I - R, Q + I - R, 2I - R; z)$ are solutions of (4.118). By Theorem 99, the pair $\{W_1, W_2\}$ is a fundamental set of solutions of (4.118) in the domain $\omega(\delta) = \{z \in D_0, 0 < |z| < \delta\}$, if the block matrix

$$S(z) = \begin{bmatrix} W_1(z) & W_2(z) \\ W_1'(z) & W_2'(z) \end{bmatrix}$$

is invertible in $\omega(\delta)$. By the properties of the Schur compliment of a matrix, the matrix $S(z)$ given above is invertible if and only if

$$M(z) = W_2'(z) - W_1'(z)[W_1(z)]^{-1} W_2(z) \quad \text{is invertible.}$$

Note that

$$\begin{aligned} M(z) &= U'(z)\, z^{I-R} + U_2(z)\, z^{-R}(I - R) - U_1'(z)\,[U_1(z)]^{-1}\, U_2(z)\, z^{I-R} \\ &= \{U'(z)\, z + U_2(z)(I - R) - U_1'(z)\,[U_1(z)]^{-1}\, U_2(z)\, z\}\, z^{-R}. \end{aligned} \tag{4.129}$$

Note that $M(z)$ is invertible if and only if

$$N(z) = z\left\{U'(z) - U_1'(z)\left[U_1(z)\right]^{-1}U_2(z)\right\} + U_2(z)(I-R) \quad \text{is invertible.} \tag{4.130}$$

We now determine the conditions where $N(z)$ is invertible. By the proofs of Theorems 103 and 104, we know that if $0 < \gamma < 1$ and k_2 is a positive integer such that

$$\frac{(\|P+I-R\|+k)(\|Q+I-R\|+k)}{(k+1)(k-\|2I-R\|)} < 1+\gamma, \quad k \geq k_2 \geq \max(1, \|2I-R\|), \tag{4.131}$$

with

$$\begin{aligned} L_2 &= \frac{(\|P+I-R\|)_{k_2}(\|Q+I-R\|)_{k_2}\Phi(k_2)}{k_2!}, \\ K_2 &= \sum_{k=1}^{k_2-1}\frac{(\|P+I-R\|)_k(\|Q+I-R\|)_k\Phi(k)}{k!}, \end{aligned} \tag{4.132}$$

where

$$\Phi(k) = \left\|(2I-R)^{-1}\right\|\left\|(I-R)^{-1}\right\|\cdots\left\|(-R+(k-1)I)^{-1}\right\|, \quad k \geq 0.$$

Let ρ_2 is the solution of

$$h_1(\rho_2) = 1-\gamma,$$

with

$$h_1(x) = K_2(x) + L_2\frac{x}{1-x}, \quad 0 \leq x < 1. \tag{4.133}$$

If $(1+\gamma)\rho_3 < \rho_2$, then $U_2(z) = {}_2F_1(P+I-R, Q+I-R; 2I-R; z)$ satisfies

$$\left\|U_2(z) - I\right\| < 1-\gamma \quad \text{and} \quad \left\|(U_2(z))^{-1}\right\| < \gamma^{-1}, \quad |z| \leq \rho_3.$$

Furthermore, if $k_3 \geq k_2$ is chosen so that

$$\frac{(\|P+I-R\|+k)(\|Q+I-R\|+k)}{k(k-\|2I-R\|)} < 1+\gamma, \quad k \geq k_3 > \max(1, \|2I-R\|), \tag{4.134}$$

with

$$\begin{aligned} L_3 &= \frac{(\|P+I-R\|)_{k_3}(\|Q+I-R\|)_{k_3}\Phi(k_3)}{(k_3-1)!}, \\ K_3 &= \sum_{k=1}^{k_3-1}\frac{(\|P+I-R\|)_k(\|Q+I-R\|)_k\Phi(k)}{(k-1)!}, \end{aligned} \tag{4.135}$$

From the previous comments and Theorem 104, one gets

$$\|U_2(z) - I\| < 1 - \gamma, \quad \|(U_2(z))^{-1}\| < \gamma^{-1},$$
$$\|U_2'(z)\| \le K_3 + \frac{L_3}{1-\rho_3}, \quad |z| \le \rho_3. \tag{4.136}$$

Thus, the following result has been proved.

Theorem 105. *Let P, Q, R be matrices in $\mathbb{C}^{p\times p}$ such that $R + kI$ is invertible for all positive integers $k \ge 0$, $PR = RP$ and $QR = RQ$. Also, let $U_2(z) = {}_2F_1(P + I - R, Q + I - R; 2I - R; z)$. Let $0 < \gamma < 1$ and k_3 be a positive integer satisfying $k \ge k_3 > \max(1, \|2I - R\|)$. Then* (4.136) *holds, if ρ_2, ρ_3, L_2, L_3, K_2 and K_3 are defined as above.*

Note that $N(z)$ given by (4.130) is well-defined in the disc $|z| \le \rho_1$ where $U_1(z) = {}_2F_1(P, Q; R; z)$ is invertible. Furthermore, $N(0) = I - R$ is invertible. By the perturbation lemma, $N(z)$ is invertible in the domain $|z| < p < \rho_1$ where

$$\|N(z) - (I - R)\| < \|(I - R)^{-1}\|^{-1}.$$

Let us write for $|z| \le \rho_1$,

$$\begin{aligned} N(z) - N(0) &= N(z) - (I - R) \\ &= z\left\{U'(z) - U_1'(z)\left[U_1(z)\right]^{-1} U_2(z)\right\} + (U_2(z) - I)(I - R). \end{aligned} \tag{4.137}$$

Note that given $0 < \epsilon < 1$, taking $\gamma = 1 - \epsilon$ in the proof of Theorem 105, considering $k_4 \ge \max(I, \|2I - R\|)$, so that

$$\frac{(\|P + I - R\| + k)(\|Q + I - R\| + k)}{k(k - \|2I - R\|)} < 2 - \epsilon, \quad k \ge k_4, \tag{4.138}$$

and taking

$$L_4 = \frac{(\|P + I - R\|)_{k_4} (\|Q + I - R\|)_{k_4} \Phi(k_4)}{(k_4 - 1)!},$$
$$K_4 = \sum_{k=1}^{k_4-1} \frac{(\|P + I - R\|)_k (\|Q + I - R\|)_k \Phi(k)}{(k-1)!}, \tag{4.139}$$
$$h_2(x) = K_4 x + L_4 \frac{x}{1-x}, \quad 0 \le x < 1. \tag{4.140}$$

If ρ_4 is the only solution of the above equation

$$h_2(\rho_4) = \epsilon,$$

taking ρ_5 such that

$$(1 + \gamma)\rho_5 = \rho_5(2 - \epsilon) < \rho_4,$$

then

$$\|U_2(z) - I\| < \epsilon, \quad |z| \le \rho_5.$$

By Theorems 103–105 and the above equations, taking $|z| < \min(\rho_1, \rho_2, \rho_3)$, it follows that

$$\begin{aligned}\|N(z) - N(0)\| &\le |z| \left[\|U_2'(z)\| + \|(U_1(z))^{-1}\| \|U_1'(z)\| \|U_2(z)\|\right] + \|U_2(z) - I\| \|I - R\| \\ &\le |z| \left[\left(K_3 + \frac{L_3}{1-\rho_3}\right) + \left(\frac{1+\epsilon}{1-\epsilon}\right)\left(K_1 + \frac{L_1}{1-\rho_1}\right)\right] + \epsilon \|I - R\|. \end{aligned} \tag{4.141}$$

Hence, taking

$$\epsilon = \frac{1}{2} \min\{\|(I-R)^{-1}\|^{-1} (\|I - R\|)^{-1}, 1\}$$

and $|z| < \min(\rho_1, \rho_3, \rho_5, \rho_6) = \rho^*$, where

$$\rho_6 = \frac{1}{2}\left\{\left[\left(K_3 + \frac{L_3}{1-\rho_3}\right) + \left(\frac{1+\epsilon}{1-\epsilon}\right)\left(K_1 + \frac{L_1}{1-\rho_1}\right)\right](\|(I-R)^{-1}\|)\right\}^{-1}. \tag{4.142}$$

By (4.141), one gets

$$\|N(z) - N(0)\| < \|(I-R)^{-1}\|^{-1}, \quad |z| < \rho^*,$$

and thus

$$N(z) \quad \text{is invertible for } |z| < \min(\rho_1, \rho_3, \rho_5, \rho_6) = \rho^*.$$

We can summarize the following results.

Theorem 106. *Let P, Q, R be matrices in $\mathbb{C}^{p\times p}$ such that $R + kI$ is invertible for all integers $k \ge 0$ and $PR = RP$, $QR = RQ$. Also, let $W_1(z)$ and $W_2(z)$ be defined by (4.128) for z in D_0 with $|z| < 1$. Then there exists a positive number $\rho^* < 1$ such that the general solution of equation (4.118) in $\omega(\rho^*) = \{z \in D_0; 0 < |z| < \rho^*\}$ is given by*

$$W(z) = W_1(z)A + W_2(z)B, \quad A, B \in \mathbb{C}^{p\times p}.$$

Furthermore, ρ^* can be determined according with the following procedure:
- Let $\epsilon = \frac{1}{2}\min\{\|(I-R)^{-1}\|^{-1} (\|I - R\|)^{-1}, 1\}$, $\gamma = 1 - \epsilon$.
- Let k_1 be defined as $k > k_1 > \max\{1, \|R\|\}$ and K_1 and L_1 are defined by (4.111).
- Let ρ_0 be the solution of $f(\rho_0) = 1 - \gamma$ where f is defined by (4.102) with K and L given by (4.100). Take ρ_1 such that $\rho_1 < \frac{\rho_0}{1+\gamma}$.
- Let k_2 be chosen so that (4.131) hold and let K_2 and L_2 be defined by (4.132). Let ρ_2 be the solution of $h_1(\rho_2) = 1 - \gamma$ with h_1 given by (4.133). Take ρ_3 with $\rho_3 < (1 + \gamma)^{-1}\rho_2$.
- Let k_3 be defined so that (4.134) holds and let K_3, L_3 be defined by (4.135).

- Let k_4 be chosen so that (4.138) holds and let K_4, L_4 be defined by (4.139). If ρ_4 is the solution of $h_2(\rho_4) = \epsilon$ where h_2 is defined by (4.140), take ρ_5 so that $\rho_5 = \frac{\rho_4}{1+\gamma}$.
- Let ρ_6 be defined by (4.142) and $\rho^* = \min\{\rho_1, \rho_3, \rho_5, \rho_6\}$.

4.8 Integral representation

We begin this section considering the binomial function of a matrix exponent. If x is a complex number with $|x| < 1$ and b is a complex number, then $(1-x)^b = \exp(b\log(1-x))$, where the log is the principal branch of the logarithm function. The Taylor expansion of $(1-x)^{-b}$ about $y = 0$ is given by

$$(1-x)^{-b} = \sum_{k=0}^{\infty} \frac{(b)_k}{k!} x^k, \quad |x| < 1, b \in \mathbb{C}. \tag{4.143}$$

Now, we consider the function of complex variable b defined by (4.143) as

$$f_k(b) = \frac{(b)_k}{k!} x^k = \frac{b\,(b+1)\cdots(b+k-1)}{k!} x^k, \quad b \in \mathbb{C}, \tag{4.144}$$

for a fixed complex number x with $|x| < 1$. It is clear that f_k is an holomorphic function of complex variable b defined in the complex plane. Given a closed bounded disc $D_R = \{b \in \mathbb{C} \mid |b| \le R\}$, one gets

$$|f_k(b)| \le \frac{(|b|)_k}{k!} |x|^k \le \frac{(R)_k}{k!} |x|^k, \quad k > 0, |b| \le R.$$

Since

$$\sum_{k=0}^{\infty} \frac{(R)_k}{k!} |x|^k < +\infty,$$

by the Weierstrass theorem for convergence of holomorphic functions, it follows that

$$g(b) = \sum_{k=0}^{\infty} \frac{(b)_k}{k!} x^k = (1-x)^{-b}$$

is holomorphic in $\mathbb{C}$. By the application of the matrix functional calculus, for any matrix $P \in \mathbb{C}^{p\times p}$, the image of g acting on P yields

$$(1-x)^{-P} = g(P) = \sum_{k=0}^{\infty} \frac{(P)_k}{k!} x^k, \quad |x| < 1, \tag{4.145}$$

where $(P)_k$ is given by

$$(P)_k = P\,(P+I)\,(P+(k-1)I), \quad \text{when } k \ge 1 \text{ and } P_k = I, \text{for } k = 0,$$

suppose that Q and R are matrices in $\mathbb{C}^{p\times p}$ such that

$$QR = RQ, \tag{4.146}$$

$$Q, R \text{ and } R - Q \text{ are positive stable.} \tag{4.147}$$

By the definition of inverse gamma function and shifted factorial for matrices with conditions (4.146) and (4.147), one get

$$\begin{aligned}(Q)_k\,(R)_k^{-1} &= \Gamma(Q+kI)\,\Gamma^{-1}(Q)\,\Gamma^{-1}(R+kI)\,\Gamma(R)\\ &= \Gamma(R)\,\Gamma^{-1}(Q)\,\Gamma^{-1}(R-Q)\,\Gamma(Q+kI)\,\Gamma(R-Q)\,\Gamma^{-1}(R+kI)\\ &= \Gamma(R)\,\Gamma^{-1}(Q)\,\Gamma^{-1}(R-Q)\,\beta(Q+kI, R-Q)\\ &= \Gamma(R)\,\Gamma^{-1}(Q)\,\Gamma^{-1}(R-Q)\int_0^1 u^{Q+(k-1)I}\,(1-u)^{R-Q-I}\,du.\end{aligned} \tag{4.148}$$

Hence, for $|z| < 1$, we can write

$$\begin{aligned}&{}_2F_1(P, Q; R; z)\\ &= \sum_{k=0}^{\infty} \frac{(P)_k\,(Q)_k\,(R)_k^{-1}}{k!}\,z^k\\ &= \sum_{k=0}^{\infty} \frac{(P)_k}{k!}\,\Gamma(R)\,\Gamma^{-1}(Q)\,\Gamma^{-1}(R-Q)\left(\int_0^1 u^{Q+(k-1)I}\,(1-u)^{R-Q-I}\,du\right)z^k\\ &= \sum_{k=0}^{\infty}\left(\int_0^1 \frac{(P)_k\,\Gamma(R)\,\Gamma^{-1}(Q)\,\Gamma^{-1}(R-Q)\,u^{Q+(k-1)I}\,(1-u)^{R-Q-I}\,z^k}{k!}\,du\right).\end{aligned} \tag{4.149}$$

Now, we are interested in the permutation of the series and the integral in the last expression. To legitimate this operation, let us consider the sequence of matrix functions defined by

$$S_k(u) = \frac{(P)_k\,\Gamma(R)\,\Gamma^{-1}(Q)\,\Gamma^{-1}(R-Q)\,u^{Q+(k-1)I}\,(1-u)^{R-Q-I}\,z^k}{k!}, \qquad 0 \le u \le 1, \tag{4.150}$$

and note that for $0 < u < 1$ and $k \ge 0$ one get

$$\|S_k(u)\| \le \frac{(\|P\|)_k\,\|\Gamma(R)\|\,\|\Gamma^{-1}(Q)\|\,\|\Gamma^{-1}(R-Q)\|\,\|u^{Q-I}\|\,\|(1-u)^{R-Q-I}\|\,|z|^k}{k!}. \tag{4.151}$$

By Schur decomposition (4.35)–(4.37), it follows that

$$\|u^{Q-I}\|\,\|(1-u)^{R-Q-I}\| \le u^{\alpha(Q)-1}\,(1-u)^{\alpha(R-Q)-1}\left\{\sum_{j=0}^{p-1}\frac{(\|Q-I\|\,p^{1/2}\,\ln u)^j}{j!}\right\}$$

$$\times \left\{ \sum_{j=0}^{p-1} \frac{(\|R-Q-I\|\, p^{1/2} \ln(1-u))^j}{j!} \right\}. \tag{4.152}$$

Taking into account that for $0 < u < 1$, one gets $\ln u < u < 1$ and $\ln(1-u) < 1-u < 1$, and from the above expression one gets

$$\|u^{Q-I}\|\, \|(1-u)^{R-Q-I}\| \le K\, u^{\alpha(Q)-1} (1-u)^{\alpha(R-Q)-1}, \quad 0 < u < 1, \tag{4.153}$$

where

$$K = \left\{ \sum_{j=0}^{p-1} \frac{(\|Q-I\|\, p^{1/2})^j}{j!} \right\} \left\{ \sum_{j=0}^{p-1} \frac{(\|R-Q-I\|\, p^{1/2})^j}{j!} \right\}. \tag{4.154}$$

Let S be the sum of the convergence numerical series

$$S = \sum_{k=0}^{\infty} \frac{(\|P\|)_k}{k!} |z|^k, \quad |z| < 1 \tag{4.155}$$

and note that by (4.151)–(4.155), we get

$$\sum_{k=0}^{\infty} \|S_k(u)\| \le \Phi(u) = L\, K\, S\, u^{\alpha(Q)-1} (1-u)^{\alpha(R-Q)-1}, \quad 0 < u < 1, \tag{4.156}$$

where

$$L = \|\Gamma(R)\|\, \|\Gamma^{-1}(Q)\|\, \|\Gamma^{-1}(R-Q)\|.$$

Since $\alpha(Q) > 0$, $\alpha(R-Q) > 0$, the function $\Phi(u) = L\, K\, S\, u^{\alpha(Q)-1} (1-u)^{\alpha(R-Q)-1}$ is integrable and

$$\int_0^1 \Phi(u)\, du - L\, K\, S\, \beta(u(Q), \alpha(R-Q)). \tag{4.157}$$

By the dominated convergence theorem, the series and integral can be permuted in (4.149) and using (4.146), we can write

$$\begin{aligned} {}_2F_1(P, Q; R; z) = & \int_0^1 \left\{ \sum_{k=0}^{\infty} \left(\frac{(P)_k}{k!} (uz)^k \right) u^{Q-I} (1-u)^{R-Q-I} \right\} du \\ & \times \Gamma(R)\, \Gamma^{-1}(Q)\, \Gamma^{-1}(R-Q). \end{aligned} \tag{4.158}$$

Now, by (4.145), one gets

$$\sum_{k=0}^{\infty} \frac{(P)_k}{k!} (uz)^k = (1-uz)^{-P}, \quad |z| < 1, 0 < u < 1. \tag{4.159}$$

Thus (4.158) takes the form

$$
{}_2F_1(P, Q; R; z) = \int_0^1 \{(1 - uz)^{-P} u^{Q-I} (1 - u)^{R-Q-I}\} \, du \, \Gamma(R) \Gamma^{-1}(Q) \Gamma^{-1}(R - Q). \tag{4.160}
$$

On summarizing, the following result has been established.

Theorem 107. *Let P, Q and R be matrices in $\mathbb{C}^{p\times p}$ such that $QR = RQ$ and Q, R, $R - Q$ are positive stable. Then for $|z| < 1$, it follows that*

$$
{}_2F_1(P, Q; R; z) = \int_0^1 \{(1 - uz)^{-P} u^{Q-I} (1 - u)^{R-Q-I}\} \, du \, \Gamma(R) \Gamma^{-1}(Q) \Gamma^{-1}(R - Q). \tag{4.161}
$$

The following corollary is a consequence of the above theorem.

Corollary 108. *Let P, Q and R be matrices in $\mathbb{C}^{p\times p}$ and let $\tilde{\alpha}(Q, R) = \min\{\alpha(Q), \alpha(R), \alpha(R - Q)\}$ and $k_1 = k_1(Q, R) = [|\tilde{\alpha}(Q, R)|] + 1$, where [] denotes the entire part function. Suppose that $QR = RQ$ and*

$$
\begin{aligned}
\sigma(Q) &\subset C \sim \{-k, k \geq k_1, k \text{ integer}\}, \\
\sigma(R - Q) &\subset C \sim \{-k, k \geq k_1, k \text{ integer}\}, \\
\sigma(R) &\subset C \sim \{-2k, k \geq k_1, k \text{ integer}\}.
\end{aligned} \tag{4.162}
$$

Then for $|z| < 1$, one gets

$$
\begin{aligned}
{}_2F_1(P, Q + k_1 I; R + 2k_1 I; z) = \int_0^1 &\{(1 - uz)^{-P} u^{Q+(k_1-1)I} (1 - u)^{R-Q+(k_1-1)I}\} \, du \\
&\times \Gamma(R + 2k_1 I) \Gamma^{-1}(Q + k_1 I) \Gamma^{-1}(R - Q + k_1 I).
\end{aligned} \tag{4.163}
$$

Proof. Consider the matrices P, $\tilde{Q} = Q + k_1 I$, $\tilde{R} = R + 2k_1 I$ and note that $\tilde{R}$, $\tilde{Q}$ and $\tilde{R} - \tilde{Q} = R - Q + k_1 I$ are positive stable. The result can be proved now in the same manner as in Theorem 107. □

4.9 Matrix valued Gauss hypergeometric function

In this section, we elaborate on the matrix valued Gauss hypergeometric function introduced by Tirao. This is another way to study the matrix version of the Gauss hypergeometric function whose parameters are taken from the set of linear maps defined on a vector space. We start with the definition of $\text{End}(V)$ on a vector space V.

Definition 109. Let V be a vector space over a field F. Then an endomorphism of vector space is a linear map defined from vector space V to itself. The set of all such linear maps of vector space V is denoted by $\mathrm{End}(V)$.

Given $P, Q, R \in \mathrm{End}(V)$, let us consider the following hypergeometric equation:

$$z(1-z)F'' + \left[R - (1+P+Q)\,z\right] F' - P\,QF = 0, \tag{4.164}$$

where F denotes a function on $\mathbb{C}$ with values in V.

Let us look for solutions of the form $F = z^a\,G$, with G analytic at $z = 0$ and $G(0) \neq 0$. We have

$$\begin{aligned} F' &= a\,z^{a-1}\,G + z^a\,G', \\ F'' &= a\,(a-1)\,z^{a-2}\,G + 2\,a\,z^{a-1}\,G' + z^a\,G''. \end{aligned}$$

From equation (4.164), the following differential equation for G follows:

$$\begin{aligned} G'' &+ \left[\frac{2a}{z} + \frac{R-(1+P+Q)\,z}{z\,(1-z)}\right] G' \\ &+ \left[\frac{a\,(a-1)}{z^2} + \frac{aR - a(1+P+Q)\,z}{z^2\,(1-z)} - \frac{P\,Q}{z(1-z)}\right] G = 0. \end{aligned}$$

If we put $G(z) = \sum_{k=0}^{\infty} z^k\,G_k$, we obtain the following recursion relation for the coefficients G_k. For all $n \geq -2$, we have

$$\begin{aligned} &(a+n+2)\,(R+a+n+1)\,G_{n+2} \\ &\quad + \sum_{m=0}^{n+1} [(a+m)\,(R-1-P-Q) - PQ]G_m = 0. \end{aligned}$$

For $n = -2$, we have $a\,(R + a - 1)\,G_0 = 0$ from which we derive the following indicial equation:

$$a^d\,\det(R + a - 1) = 0. \tag{4.165}$$

Let $\beta_1, \ldots, \beta_d$ be the eigenvalues of R; then the roots of the indicial equation are $a = 0, 1-\beta_1, \ldots, 1-\beta_d$.

For $a = 0$, we obtain the solutions F of (4.164) analytic at $z = 0$, the Taylor coefficients of which are given by the recursion relation

$$(n+2)\,(R+n+1)\,F_{n+2} + \sum_{m=0}^{n+1} [m\,(R-1-P-Q) - PQ]F_m = 0. \tag{4.166}$$

If the eigenvalues of R are not $0, -1, \ldots$, then the function F is characterized by its value at 0, because the matrix $(n+2)\,(R+n+1)$ is nonsingular for all $n \geq -1$.

Let us introduce the notation

$$(R,P,Q)_{m+1} = (R+m)^{-1}(P+m)(Q+m)(R+m-1)^{-1}(P+m-1) \\ \times (Q+m-1)\cdots R^{-1}PQ,$$

for all $m \geq 0$ and $(R,P,Q)_0 = 1$. One can observe that

$$(R,P,Q)_{m+1} = (R+m)^{-1}(P+m)(Q+m)(R,P,Q)_m.$$

Theorem 110. *If $\{F_k\}_{k\geq 0}$ is a sequence in V that satisfies the recursion relation* (4.166) *and $\sigma(R) \cap (-\mathbb{N}_0) = \phi$, then:*
1. $F_{k+1} = \frac{1}{k+1}(R+k)^{-1}(P+k)(Q+k)F_k$, *for all $k \geq 0$, and*
2. $F_k = \frac{1}{k!}(R,P,Q)_k F_0$, *for all $k \geq 0$.*

Proof. For (i), if we put $k = -1$ into equation (4.166), we get $F_1 = R^{-1}PQF_0$. Now, let $k \geq 1$ and set $n = k-1$ in (4.166). Then

$$(k+1)(R+k)F_{k+1} + \sum_{m=0}^{k}[m(R-1-P-Q) - PQ]F_m = 0.$$

For $n = k-2$, we obtain

$$k(R+k-1)F_k + \sum_{m=0}^{k-1}[m(R-1-P-Q) - PQ]F_m = 0.$$

Therefore,

$$\begin{aligned}(k+1)(R+k)F_{k+1} &= -\sum_{m=0}^{k}[m(R-1-P-Q) - PQ]F_m - [k(R-1-P-Q) - PQ]F_k \\ &= [k(R+k-1) - k(R-1-P-Q) + PQ]F_k \\ &= (k^2 + kP + kQ + PQ)F_k \\ &= (P+k)(Q+k)F_k,\end{aligned}$$

which proves (i). The statement in (ii) follows directly from (i) by induction on $k \geq 0$. □

Definition 111. If P, Q and $R \in \mathrm{End}(V)$ and no eigenvalues of R are in the set $\{0, -1, -2, \ldots\}$, we define

$${}_2F_1\binom{P,Q}{R};z\Big) = {}_2F_1(R,P;Q;z) = \sum_{k=0}^{\infty}\frac{z^k}{k!}(R,P,Q)_k. \tag{4.167}$$

In the next theorem, we summarize our results on the analytic solutions at $z = 0$ of differential equation (4.164).

Theorem 112. *If $\sigma(R) \cap (-\mathbb{N}_0)$ is empty, then:*
1. *the function ${}_2F_1\left({P,Q \atop R};z\right)$ is analytic on $|z| < 1$ with values in* $\mathrm{End}(V)$*, and*
2. *if $F_0 \in V$, then $F(z) = {}_2F_1\left({P,Q \atop R};z\right) F_0$ is a solution of the hypergeometric equation*

$$z(1-z)F'' + [R - (1+P+Q)z]F' - PQF = 0 \tag{4.168}$$

such that $F(0) = F_0$. Conversely, any solution F analytic at $z = 0$ is of this form.

Let $b \neq 1$ be an eigenvalue of R. Then $a = 1 - b$ is a root of indicial equation (4.165). Now, we want to study a sequence $\{G_k\}_{k\geq 0}$ satisfying

$$(a+n+2)(R+a+n+1)G_{n+2} \tag{4.169}$$

$$+\sum_{m=0}^{n+1}[(a+m)(R-1-P-Q)-PQ]G_m = 0. \tag{4.170}$$

For $n = -2$, we have $a(R+a-1)G_0 = 0$ if and only if G_0 is an eigenvector of R of eigenvalue b.

If $\beta \notin [\sigma(R) + \mathbb{N}] \cup (\mathbb{N}+1)$, then $\{G_k\}_{k\geq 0}$ is determined by G_0.

Theorem 113. *If $\{G_k\}_{k\geq 0}$ is a sequence in V that satisfies the recursion relation (4.170) and $b = 1 - a \notin [\sigma(R) + \mathbb{N}] \cup (\mathbb{N}+1)$ and $G_0 \in V$ is an eigenvector of R of eigenvalue b, then:*
1. $G_{k+1} = \frac{1}{(a+k+1)}(R+a+k)^{-1}(P+a+k)(Q+a+k)G_k$*, for all $k \geq 0$, and*
2. $G_k = \frac{1}{(a+1)_k}(R+a, P+a, Q+a)_k G_0$*, for all $k \geq 0$.*

Proof. For (i), if we put $k = -1$ into equation (4.170), we get

$$\begin{aligned}(a+1)(R+a)G_1 &= -[a(R-1-P-Q)-PQ]G_0\\ &= -[a(-a-P-Q)-PQ]G_0\\ &= [a^2 + a(P+Q) + PQ]G_0\\ &= (P+a)(Q+a)G_0,\end{aligned}$$

which proves (i) for $k = 0$. Now for $k \geq 1$, the proof continues in the same line as in Theorem 112. The statement in (ii) follows directly from (i) by induction on $k \geq 0$. □

Definition 114. If $P, Q, R \in \mathrm{End}(V)$ and $-a \notin [\sigma(R)+\mathbb{N}_0] \cup \mathbb{N}$, then we define the function

$${}_2F_1^a\left({P,Q \atop R};z\right) = {}_2F_1^a(R,P;Q;z) = \sum_{k=0}^{\infty}\frac{z^k}{(a+1)_k}(R+a, P+a, Q+a)_k. \tag{4.171}$$

Notice that for $a = 0$, we have

$${}_2F_1^a(R,P;Q;z) = {}_2F_1(R,P;Q;z).$$

Theorem 115. *If* $-a \notin [\sigma(R) + \mathbb{N}_0] \cup \mathbb{N}$, *then* ${}_2F_1^a\left({P,Q \atop R}; z\right)$ *is analytic on* $|z| < 1$ *with values in* $\operatorname{End}(V)$. *If* $b \notin [\sigma(R) + \mathbb{N}] \cup (\mathbb{N} + 1)$ *is an eigenvalue of* R *and* $G_0 \in V$ *is an eigenvector of* R *of eigenvalue* b, *then*

$$F(z) = z^{1-b}\, {}_2F_1^{1-b}\left({P,Q \atop R}; z\right) G_0$$

is a solution of the hypergeometric equation

$$z(1-z)F'' + [R - (1+P+Q)z]\,F' - P\,QF = 0. \tag{4.172}$$

Observe that when $V = \mathbb{C}$ *and* $R = w$, $P = u$, $Q = v$ *are complex numbers, then* $b = w$ *and the above function coincides with* $z^{1-w}\, {}_2F_1\left({u-w+1, v-w+1 \atop 2-w}; z\right)$ *for* $G_0 = 1$.

Corollary 116. *Let* R *be diagonalizable and let* $V(\beta_i)$ *be the eigenspace of* R *of eigenvalue* β_i. *Let* $\{F_j\}_{j=1}^d$ *be a basis of* V *and let* $\{G_{i,j}\}_j$ *be a basis of* $V(\beta_i)$ *for each* β_i. *If* $\sigma(R) \cap [(\sigma(R) + \mathbb{N}) \cup (\mathbb{Z} - \{1\})] = \phi$ *and* O *is a simply connected region in* $\mathbb{C} - \{0,1\}$ *with* $0 \in \overline{O}$, *then*

$$\left\{ {}_2F_1\left({P,Q \atop R}; z\right) F_j \right\}_{j=1}^{d}$$

is a basis of the space of all solutions of the hypergeometric equation

$$z(1-z)F'' + [R - (1+P+Q)z]\,F' - P\,QF = 0, \tag{4.173}$$

analytic at $z = 0$. *Moreover,*

$$\left\{ {}_2F_1\left({P,Q \atop R}; z\right) F_j \right\}_{j=1}^{d} \cup \left\{ z^{1-\beta_i}\, {}_2F_1^{1-\beta_i}\left({P,Q \atop R}; z\right) G_{i,j} \right\}$$

is a basis of the space of all analytic solutions on O.

When $V = \mathbb{C}$, a differential equation of the form

$$z(1-z)f'' + [w - u\,z]\,f' - vf = 0 \tag{4.174}$$

with $u, v, w \in \mathbb{C}$, after solving a quadratic equation becomes

$$z(1-z)f'' + [w - (1+u+v)z]\,f' - a\,bf = 0. \tag{4.175}$$

This is not necessarily the case when $\dim(V) > 1$. In other words, a differential equation of the form

$$z(1-z)F'' + [R - U\,z]\,F' - V\,F = 0, \tag{4.176}$$

with $P, Q, R \in \mathrm{End}(V)$ cannot always be reduced to one of the form of (4.164) because a quadratic equation in a noncommutative setting as $\mathrm{End}(V)$ may have no solutions. Thus, it is important to notice how to obtain the solutions of (4.176).

If the eigenvalues of R are not $0, -1, \ldots$. Let us introduce the sequence $[R, U, V]_k \in \mathrm{End}(V)$ by inductively

$$\begin{aligned}[R, U, V]_0 &= 1,\\ [R, U, V]_{k+1} &= (R + k)^{-1}\left[k^2 + k(U - 1) + V\right][R, U, V]_k,\end{aligned}$$

for all $k \geq 0$.

Definition 117. If $U, V, R \in \mathrm{End}(V)$ and no eigenvalues of R are in the set $\{0, -1, \ldots\}$, we define

$$ {}_2H_1\left(\begin{matrix}U, V\\ R\end{matrix}; z\right) = {}_2H_1(R, U; V; z) = \sum_{k=0}^{\infty} \frac{z^k}{k!}(R, U, V)_k. \tag{4.177}$$

If $-a \notin [\sigma(R) + \mathbb{N}_0] \cup \mathbb{N}$, then we define the function

$$ {}_2H_1^a\left(\begin{matrix}U, V\\ R\end{matrix}; z\right) = {}_2H_1^a(R, U; V; z) = \sum_{k=0}^{\infty} \frac{z^k}{(a+1)_k}(R + a, U + 2a, V + a^2 - a)_k. \tag{4.178}$$

Notice that for $a = 0$, we have

$$ {}_2H_1^a(R, U; V; z) = {}_2H_1(R, U; V; z),$$

and that if $U = 1 + P + Q$, $V = PQ$, then

$$ {}_2H_1^a(R, U; V; z) = {}_2F_1^a(R, P; Q; z).$$

Now, Theorem 115 and Corollary 116 can be generalized in the following way.

Theorem 118. *With the assumptions*

1. *If $-a \notin [\sigma(R) + \mathbb{N}_0] \cup \mathbb{N}$, the function ${}_2H_1^a\left(\begin{smallmatrix}U,V\\ R\end{smallmatrix}; z\right)$ is analytic on $|z| < 1$ with values in $\mathrm{End}(V)$.*
2. *If $b \notin [\sigma(R) + \mathbb{N}] \cup (\mathbb{N} + 1)$ is an eigenvalue of R and $G_0 \in V$ is an eigenvector of R of eigenvalue b.*
   The matrix valued function

$$F(z) = z^{1-b}\, {}_2H_1^{1-b}\left(\begin{matrix}U, V\\ R\end{matrix}; z\right) G_0$$

is a solution of the hypergeometric equation

$$z(1 - z)F'' + [R - U z]F' - VF = 0. \tag{4.179}$$

Corollary 119. *Let R be diagonalizable and let $V(\beta_i)$ be the eigenspace of R of eigenvalue β_i. Let $\{F_j\}_{j=1}^d$ be a basis of V and let $\{G_{i,j}\}_j$ be a basis of $V(\beta_i)$ for each β_i. If $\sigma(R) \cap [(\sigma(R) + \mathbb{N}) \cup (\mathbb{Z} - \{1\})] = \phi$ and O is a simply connected region in $\mathbb{C} - \{0,1\}$ with $0 \in \overline{O}$, then*

$$\left\{ {}_2H_1 \left(\begin{matrix} U, V \\ R \end{matrix} ; z \right) F_j \right\}_{j=1}^{d}$$

is a basis of the space of all solutions of the hypergeometric equation

$$z(1-z)F'' + [R - U\,z]\,F' - V\,F = 0 \tag{4.180}$$

analytic at $z = 0$. Moreover,

$$\left\{ {}_2H_1 \left(\begin{matrix} U, V \\ R \end{matrix} ; z \right) F_j \right\}_{j=1}^{d} \cup \left\{ z^{1-\beta_i}\, {}_2H_1^{1-\beta_i} \left(\begin{matrix} U, V \\ R \end{matrix} ; z \right) G_{i,j} \right\}$$

is a basis of the space of all analytic solutions on O.

Example 120. Following Grünbaum, let us consider the differential equation:

$$z(1-z)F'' + [X - U\,z]\,F' + (V\,F + F\,W) = 0, \tag{4.181}$$

where F denotes a function on $\mathbb{C}$ with values in $M(2,\mathbb{C})$, and X, U, V, W are the matrices given by

$$\begin{aligned}
X &= (b+2a+3)^{-1} \begin{pmatrix} (a+1)(b+2a+5) & 2(a+1) \\ b+1 & (a+2)\,b + 2a^2 + 5a + 4 \end{pmatrix}, \\
U &= \begin{pmatrix} b+a+3 & -1 \\ 0 & b+a+4 \end{pmatrix}, \\
V &= \begin{pmatrix} 0 & 0 \\ 0 & -\frac{1}{2}(b+2a+3) \end{pmatrix}, \\
W &= \begin{pmatrix} j\,(j+b+a+2) & 0 \\ 0 & j\,(j+b+a+3) + \frac{1}{2}(b+2a+3) \end{pmatrix}, \quad a, b \in \mathbb{C}, j = 0, 1, \ldots.
\end{aligned}$$

The term FW in equation (4.181) forces us to consider this equation as a differential equation on functions that take values on $\mathbb{C}^4$ and to consider the left and right multiplication by matrices in $M(2,\mathbb{C})$ as linear maps in $\mathbb{C}^4$. Thus, instead of (4.181), we shall consider the following equivalent differential equation:

$$z(1-z)\Phi'' + [R - \tilde{U}\,z]\,\Phi' - \tilde{T}\,\Phi = 0, \tag{4.182}$$

where

$$R = (b + 2a + 3)^{-1} \begin{pmatrix} (a+1)(b+2a+5)\,I & 2(a+1)\,I \\ (b+1)\,I & [(a+2)\,b + 2a^2 + 5a + 4]\,I \end{pmatrix},$$

$$\tilde{U} = \begin{pmatrix} (b+a+3)\,I & -I \\ 0 & (b+a+4)\,I \end{pmatrix},$$

$$\tilde{T} = \begin{pmatrix} t_1 & 0 & 0 & 0 \\ 0 & t_2 & 0 & 0 \\ 0 & 0 & t_3 & 0 \\ 0 & 0 & 0 & t_4 \end{pmatrix}.$$

Here, I denotes the 2×2 identity matrix and

$$\begin{aligned} t_1 &= -j\,(j + b + a + 2), \\ t_2 &= -j\,(j + b + a + 3) - \frac{1}{2}(b + 2a + 3), \\ t_3 &= -j\,(j + b + a + 2) + \frac{1}{2}(b + 2a + 3), \\ t_4 &= -j\,(j + b + a + 3). \end{aligned}$$

It is easy to verify that the matrices P and Q given below satisfy $\tilde{U} = 1 + P + Q$, $\tilde{T} = P\,Q$.

$$P = \begin{pmatrix} x_1 & 0 & \frac{-x_1}{x_1+x_3-(b+a+3)} & 0 \\ 0 & y_1 & 0 & \frac{-y_1}{y_1+y_3-(b+a+3)} \\ 0 & 0 & x_3 & 0 \\ 0 & 0 & 0 & y_3 \end{pmatrix},$$

$$Q = \tilde{U} - P - 1.$$

The parameters x_1, x_3, y_1 and y_3 are only subject to the following conditions:

$$\begin{aligned} &x_1 = -j \quad \text{or} \quad x_1 = (j + b + a + 2), \\ &y_3 = -j \quad \text{or} \quad y_3 = (j + b + a + 3), \\ &y_1^2 - (b + a + 2)\,y_1 - j\,(j + b + a + 3) - \frac{1}{2}(b + 2a + 3) = 0, \\ &x_3^2 - (b + a + 3)x_3 - j\,(j + b + a + 3) = 0. \end{aligned}$$

It is important to notice that $\sigma(R) = \{a+1, a+2\}$, where each eigenvalue has multiplicity 2. We also remark that $(P + k)\,(Q + k)$ is generically nonsingular for $k \neq j$, and that the kernel of $(P+j)\,(Q+j)$ is two-dimensional. Thus, $2F_1\left({P,Q \atop R} ; z\right)$ is not a polynomial function, as in the classical case, but nevertheless we have the following result

Corollary 121. *Differential equation* (4.181) *is equivalent to a hypergeometric equation of the form of* (4.164). *Therefore, the Jacobi polynomials* $P_j^{(a,b)}$ *introduced by Grünbaum are given in terms of hypergeometric functions, namely*

$$[P_j^{(a,b)}(z)]^T = {}_2F_1\left(\begin{matrix}P,Q\\R\end{matrix};z\right)[P_j^{(a,b)}(0)]^T.$$

This is an explicit example with in the theory of matrix valued orthogonal polynomials given by Krein.

4.10 The basic Gauss hypergeometric matrix function

In this section, we see the treatment of the matrix analogue of the basic Gauss hypergeometric function. In this matrix version, the parameters are taken in the q-exponent and the corresponding analysis has been done. We start with some basic definitions related to matrix q-functions.

4.10.1 Matrix analogue of q-numbers

Let P be a p-square complex matrix. Then the q-analogue of P can be defined as

$$[P]_q = \frac{I - q^P}{1-q}, \quad q \neq 1. \tag{4.183}$$

Since $[z]_q = \frac{1-z^q}{1-q}$ is an entire function of the complex variable z, for any complex square matrix P, the Riesz–Dunford functional calculus shows that the image of $[P]_q$ is a well-defined matrix for any matrix P.

Let us take into account the scalar q-shifted factorial function denoted by $(z,q)_k$ and defined by

$$(z,q)_0 = 1, \quad (z,q)_k = \prod_{n=0}^{k-1}(1 - zq^n), \quad k = 1,2,\dots. \tag{4.184}$$

Applying the matrix functional calculus to this function would yield for a complex square matrix P,

$$(P,q)_0 = I, \quad (P,q)_k = \prod_{n=0}^{k-1}(I - Pq^n), \quad k = 1,2,\dots. \tag{4.185}$$

Definition 122. An infinite product $\prod_{k=1}^{\infty} B_k$ of square matrices converges invertibly if there is an integer m such that B_k is invertible for $k \geq m$ and

$$B = \lim_{n\to\infty}\prod_{k=m}^{n} B_k$$

exists and is invertible.

The sufficient condition for convergence of infinite products of matrices is: if the series $\sum_{k=0}^{\infty} \|P_k\|$ converges, the infinite product $\prod_{k=0}^{\infty}(I + P_k)$ converges.

We have a notational convention used throughout in the book by Trefethen and Embree.

$$\text{If } z \in \sigma(P), \quad \text{then } \|(zI - P)^{-1}\| = \infty. \tag{4.186}$$

Using the previous condition and convention give us the opportunity to expand the q-shifted factorial matrix function to an invertibly convergent infinite product, which will be deduced in the following theorem.

Theorem 123. *For any complex square matrix P and $|q| < 1$, we have*

$$(P, q)_{\infty} = \lim_{n\to\infty} \prod_{k=0}^{n-1}(I - Pq^n) = \prod_{k=0}^{\infty}(I - Pq^k) \tag{4.187}$$

that converges. Furthermore, if $\|P\| < 1$ and $q^{-k} \notin \sigma(P)$, $k = 0, 1, \dots$, the infinite product (4.187) *converges invertibly.*

Proof. If $|q| < 1$, we have

$$\sum_{k=0}^{\infty} \|P q^k\| \le \|P\| \sum_{k=0}^{\infty} |q|^k = \frac{\|P\|}{1 - |q|},$$

which is the sufficient condition for convergence of an infinite product (4.187). Note that if $I - P q^k$ is singular for some $k \ge 0$, then the infinite product is singular, so we should choose $q^k \notin \sigma(P)$, $k = 0, 1, \dots$, to make, according to (4.186), the resolvent $(I - P q^k)^{-1}$ exists for all k. Now take $\|P\| < |q|^{-k}$ for all k (or equivalently $\|P\| < 1$) and take into account the Perturbation lemma, and we obtain

$$\begin{aligned} \|(I - P q^k)^{-1}\| &= \|(I - P q^k)^{-1} - I + I\| \le 1 + \|(I - P q^k)^{-1} - I\| \\ &= 1 + \frac{\|P\|\,|q|^k}{1 - \|P\|\,|q|^k} = \frac{1}{1 - \|P\|\,|q|^k}, \quad \|P\| < 1. \end{aligned} \tag{4.188}$$

This yields

$$\|(P, q)_{\infty}^{-1}\| \le \prod_{k=0}^{\infty} \|(I - P q^k)^{-1}\| \le \prod_{k=0}^{\infty} \frac{1}{1 - \|P\|\,|q|^k} < \infty, \quad \|P\| < 1.$$

It completes the proof. □

Theorem 124. *For any matrix $P \in \mathbb{C}^{p\times p}$ with numerical abscissa $w(P)$ and $\|\cdot\| = \|\cdot\|_2$,*

$$\|e^{tP}\| \le e^{tw(P)}, \quad \forall t \ge 0, \tag{4.189}$$

and

$$\|e^{tP}\| \le e^{tw(P)} + o(t), \quad \text{as } t \to 0. \tag{4.190}$$

In particular, $\|e^{tP}\| \le 1\ \forall t \ge 0$ *if and only if* $w(P) \le 0$.

Theorem 125. *For any matrix* $P \in \mathbb{C}^{p\times p}$ *with* $|q| < 1$, *then the infinite product of matrices*

$$(q^P, q)_\infty = \lim_{n\to\infty} \prod_{k=0}^{n-1} (I - q^{P+nI}) = \prod_{k=0}^{\infty} (I - q^{P+kI}) \tag{4.191}$$

converges invertibly if $\tilde{\mu}(P) > 0$ *and* $q^{-k} \notin \sigma(q^P)$, $k = 0, 1, \ldots$.

Theorem 126. *Let* P *be a complex square matrix,* $|q| < 1$ *and* $q^{-k} \notin \sigma(P)$, $k = 0, 1, \ldots$, *then we have*

$$\sum_{k=0}^{\infty} \frac{(a,q)_k}{(q,q)_k} P^k = (P,q)_\infty^{-1} (aP,q)_\infty, \quad \|P\| < 1. \tag{4.192}$$

Proof. The function $f_a(z) = \sum_{k=0}^{\infty} \frac{(a,q)_k}{(q,q)_k} z^k$ is analytic if $|z| < 1$, the matrix functional calculus tells us that the function $f_a(P)$ is convergent if $\|P\| < 1$. Now it is easy to prove that

$$f_a(P) = (I - aP) f_{aq}(P), \quad \text{and} \quad f_a(P) - f_a(qP) = (1-a) P f_{aq}(P).$$

These lead to

$$f_a(P) = (I-P)^{-1}(I - aP) f_a(qP).$$

Iterating this $(k-1)$ times and then letting $k \to \infty$, we obtain

$$\begin{aligned} f_a(P) &- (P,q)_k^{-1} (aP,q)_k f_a(q^k P) \\ &= (P,q)_\infty^{-1} (aP,q)_\infty f_a(\tilde{0}) \\ &= (P,q)_\infty^{-1} (aP,q)_\infty, \end{aligned} \tag{4.193}$$

where $q^k = 0$ as $k \to \infty$ and $f_a(\tilde{0}) = I$. This completes the proof. □

If we set $a = 0$ in (4.192), this gives

$$\sum_{k=0}^{\infty} \frac{P^k}{(q,q)_k} = (P,q)_\infty^{-1}, \quad \|P\| < 1, q^{-k} \notin \sigma(P), k = 0, 1, \ldots. \tag{4.194}$$

Another formula can be obtained from (4.192) by replacing P with $-P/a$ and then letting $a \to \infty$ to get

$$\sum_{k=0}^{\infty} \frac{q^{\frac{k(k-1)}{2}} P^k}{(q,q)_k} = (-P,q)_\infty. \tag{4.195}$$

Corollary 127. *For any two matrices P, Q in $\mathbb{C}^{p\times p}$ with $PQ = QP$, $q^{-k} \notin \sigma(P)$, $k = 0,1,\ldots$ and $|q| < 1$, we have*

$$\sum_{k=0}^{\infty} \frac{(Q,q)_k}{(q,q)_k} P^k = (P,q)_\infty^{-1} (PQ,q)_\infty, \quad \|P\| < 1. \tag{4.196}$$

Theorem 128. *Let $P \in \mathbb{C}^{p\times p}$ be a positive stable matrix or $\tilde{\mu} > 0$, then the matrix series*

$$\sum_{k=0}^{\infty} \frac{(Q,q)_k}{(q,q)_k} q^{kP}, \quad 0 < |q| < 1 \tag{4.197}$$

converges absolutely for any matrix Q.

Proof. It is easy to show that

$$\|(Q,q)_k\| \le (-\|Q\|,|q|)_k.$$

By using the Schur decomposition of matrices and the previous note, we have

$$\begin{aligned}\sum_{k=0}^{\infty} \frac{(Q,q)_k}{(q,q)_k} q^{kP} &\le \sum_{k=0}^{\infty} \frac{\|(Q,q)_k\|}{|(q,q)_k|} \|e^{kP(\log|q|+\imath\arg(q))}\| \\ &\le \sum_{k=0}^{\infty} \frac{(-\|Q\|,|q|)_k}{|(q,q)_k|} \|e^{kP\log|q|}\| \\ &\le \sum_{k=0}^{\infty} \frac{(-\|Q\|,|q|)_k\, |q|^{k\beta(P)}}{|(q,q)_k|} \sum_{j=0}^{p-1} \frac{(k\,\|P\|\,\sqrt{p}\log|q|^{-1})^j}{j!}.\end{aligned} \tag{4.198}$$

Using the ratio test would yield that the matrix series (4.197) converges absolutely for $\beta(P) > 0$ with $0 < |q| < 1$. To prove that the matrix series (4.197) converges absolutely in the case of $\tilde{\mu}(P) > 0$, we should use the logarithmic norm instead of Schur decomposition. This completes the proof. □

Lemma 129. *Let P be a square matrix in $\mathbb{C}^{p\times p}$ and $k \in \mathbb{N}_0$. Then we have the following three identities:*

1. $(P,q)_k = (P,q)_\infty (P\,q^k,q)_\infty^{-1} = \sum_{n=0}^{k} \binom{k}{n}_q q^{\binom{n}{2}} (-P)^n$, $\|P\| < 1$, *where* $q^{-m} \notin \sigma(P)$, $m = k, k+1, \ldots$.
2. $(P,q)_k^{-1} = (P,q)_\infty^{-1} (P\,q^k,q)_\infty = \sum_{n=0}^{\infty} \frac{(q^k,q)_n}{(q,q)_n} P^n$, $\|P\| < 1$, *where* $q^{-k} \notin \sigma(P)$, $k \in \mathbb{N}_0$.
3. $(P\,q^n,q)_{k-n} = (P,q)_n^{-1} (P,q)_k$, $n = 0,1,\ldots,k$, *holds for* $q^{-r} \notin \sigma(P)$, $r = 0,1,\ldots,n$.

Proof. Let $\|P\| < 1$ and $q^{-m} \notin \sigma(P)$, $m = k, k+1, \ldots$. Then we have

$$
\begin{aligned}
(P,q)_k &= \prod_{n=0}^{k-1}(I - Pq^n) \prod_{n=k}^{\infty}(I - Pq^n)(I - Pq^n)^{-1} \\
&= \prod_{n=0}^{\infty}(I - Pq^n) \prod_{n=k}^{\infty}(I - Pq^n)^{-1} \\
&= (P,q)_\infty \prod_{n=0}^{\infty}(I - Pq^{n+k})^{-1} = (P,q)_\infty (Pq^k, q)_\infty^{-1}. \qquad (4.199)
\end{aligned}
$$

Similarly, the other two identities can be proved. □

4.10.2 The *q*-gamma matrix function

Definition 130. For any complex square matrix P and $|q| < 1$, the q-analogues of the exponential matrix function can be defined as

$$E_q^P = \sum_{k=0}^{\infty} \frac{q^{\frac{k(k-1)}{2}} P^k}{[k]_q!} = (-(1-q)P, q)_\infty, \qquad (4.200)$$

$$e_q^P = \sum_{k=0}^{\infty} \frac{P^k}{[k]_q!} = (-(1-q)P, q)_\infty^{-1}. \qquad (4.201)$$

For convergence of the second form, we need $\|P\| < |1-q|^{-1}$ for the sum but for the infinite product we need $q^{-k}(1-q)^{-1} \notin \sigma(P)$, $k = 0, 1, \ldots$, in addition to the previous requirement.

Theorem 131. *For any matrix P in $\mathbb{C}^{p\times p}$ with $u^P = e^{P\log u}$, the matrix q-integral*

$$\int_0^{\frac{1}{1-q}} u^{P-I} E_q^{-qu}\, d_q u \qquad (4.202)$$

converges absolutely if P is a positive stable matrix or $\tilde{\mu}(P) > 0$.

Proof. By the Schur decomposition of matrices, the norm of the integral (4.202) yields

$$
\begin{aligned}
\left\| \int_0^{\frac{1}{1-q}} u^{P-I} E_q^{-qu}\, d_q u \right\| &= \left\| \sum_{k=0}^{\infty} q^k \left(\frac{q^k}{1-q}\right)^{P-I} E_q^{-\frac{q^{k+1}}{1-q}} \right\| \\
&\leq (1-q)(q,q)_\infty \sum_{k=0}^{\infty} \frac{1}{(q,q)_k} \|e^{P(k\log q - \log(1-q))}\|
\end{aligned}
$$

$$\begin{aligned}
&\leq (1-q)\,(q,q)_\infty \sum_{k=0}^{\lfloor \frac{\log(1-q)}{\log q} \rfloor} \frac{e^{\alpha(P)(k\log q-\log(1-q))}}{(q,q)_k} \\
&\times \left(\sum_{j=0}^{p-1} \frac{(\|P\|\,\sqrt{p}(k\log q-\log(1-q)))^j}{j!\,(1-q)^j} \right) \\
&+ (1-q)\,(q,q)_\infty \sum_{k=\lfloor \frac{\log(1-q)}{\log q} \rfloor+1}^{\infty} \frac{e^{\beta(P)(k\log q-\log(1-q))}}{(q,q)_k} \\
&\times \left(\sum_{j=0}^{p-1} \frac{(-\|P\|\,\sqrt{p}(k\log q-\log(1-q)))^j}{j!\,(1-q)^j} \right),
\end{aligned} \tag{4.203}$$

where $\lfloor\cdot\rfloor$ is the floor function. Using the ratio test would yield that the last summation converges absolutely for $\beta(P) > 0$. To prove the matrix q-integral (4.202) converges absolutely in the case of $\tilde{\mu}(P) > 0$, one can use the logarithmic norm instead of Schur decomposition. It completes the proof. □

Definition 132. Let P be a complex matrix such that P is positive stable or $\tilde{\mu}(P) > 0$, then a q-gamma matrix function can be defined as follows:

$$\Gamma_q(P) = \int_0^{\frac{1}{1-q}} u^{P-I}\, E_q^{-qu}\, d_q u. \tag{4.204}$$

Proposition 133. *A q-gamma matrix function satisfies the following identity:*

$$\Gamma_q(I) = I, \tag{4.205}$$

$$\Gamma_q(P+I) = [P]_q\, \Gamma_q(P). \tag{4.206}$$

Lemma 134. *Let $f(x) \geq g(x) \geq 0$, $\forall x \in [0,u]$. Then, we have*

$$\int_0^u f(x)\, d_q x \geq \int_0^u g(x)\, d_q x \geq 0. \tag{4.207}$$

Lemma 135. *Let $0 < u < [k]_q$. Then the following inequality holds for all positive integers k:*

$$0 < E_q^{-qu} - \left(\frac{qu}{[k]_q}, q \right)_k < \frac{q^{k+2}\, u^2}{(1+q)\,[k]_q}. \tag{4.208}$$

Proof. Applying the definition of the q-derivative, we get

$$D_q \left(\frac{qx}{[k]_q}, q \right)_k = -q \left(\frac{q^2 x}{[k]_q}, q \right)_{k-1} \quad \text{and} \quad D_q\, e_q^{qx} = q\, e_q^{qx}.$$

From the product rule for q-derivative,

$$D_q[f(z)\,g(z)] = g(z)\,D_q f(z) + f(qz)\,D_q g(z) = f(z)\,D_q g(z) + g(qz)\,D_q f(z),$$

we have

$$D_q\left(e_q^{qx}\left(\frac{qx}{[k]_q}, q\right)_k\right) = -q^{k+2}\, e_q^{qx}\left(\frac{q^2x}{[k]_q}, q\right)_{k-1}\frac{x}{[k]_q}.$$

On q-integrating both sides from 0 to u, we get

$$1 - e_q^{qu}\left(\frac{qu}{[k]_q}, q\right)_k = q^{k+2}\int_0^u e_q^{qx}\left(\frac{q^2x}{[k]_q}, q\right)_{k-1}\frac{x}{[k]_q}\, d_q x.$$

Using the Lemma 134 and the rule of q-integration by parts, we have for $0 < u < [k]_q$,

$$\begin{aligned} 0 &< \int_0^u e_q^{qx}\left(\frac{q^2x}{[k]_q}, q\right)_{k-1}\frac{x}{[k]_q}\, d_q x < \int_0^u e_q^{qx}\,\frac{x}{[k]_q}\, d_q x \\ &= \frac{u^2}{(1+q)\,[k]_q}\, e_q^{qu} - \frac{q^3}{(1+q)\,[k]_q}\int_0^u x^2 e_q^{qx}\, d_q x \\ &< \frac{u^2}{(1+q)\,[k]_q}\, e_q^{qu}. \end{aligned} \tag{4.209}$$

Recall that $E_q^{-x}\, e_q^x = 1$, $x < \frac{1}{1-q}$. This leads to the desired result. □

Theorem 136. *Let P be a p-square positive stable complex matrix or $\tilde{\mu}(P) > 0$. Then the q-gamma matrix function can be defined as follows:*

$$\Gamma_q(P) = (q,q)_\infty\,(q^P,q)_\infty^{-1}\,(1-q)^{I-P}, \quad q^{-k} \notin \sigma(q^P), k = 0,1,\ldots. \tag{4.210}$$

Proof. Consider the matrix function

$$f_k(P) = \int_0^{[k]_q} u^{P-I}\left(\frac{qu}{[k]_q}, q\right)_k d_q u, \quad k = 0,1,\ldots. \tag{4.211}$$

By changing the variable $u' = \frac{u}{[k]_q}$ and omitting the superscript dash, we have

$$\begin{aligned} f_k(P) &= \int_0^1 u^{P-I}\,(qu,q)_k\, d_q u \\ &= (1-q)\,[k]_q^P \sum_{n=0}^{\infty} q^{nP}\,(q^{k+1},q)_n \end{aligned}$$

$$
\begin{aligned}
&= (1-q)\,(q,q)_k\,[k]_q^P \sum_{n=0}^{\infty} \frac{(q^{k+1},q)_n}{(q,q)_n}\, q^{nP} \\
&= (1-q)\,(q,q)_k\,[k]_q^P\,(q^P,q)_\infty^{-1}\,(q^{P+(k+1)I},q)_\infty \\
&= (1-q)\,(q,q)_k\,[k]_q^P\,(q^P,q)_\infty^{-1}\,(q^P,q)_\infty\,(q^P,q)_{k+1}^{-1} \\
&= (1-q)\,(q,q)_k\,[k]_q^P\,(q^P,q)_{k+1}^{-1}.
\end{aligned}
\tag{4.212}
$$

Now, taking the limit to (4.211) and (4.212) as $k \to \infty$, we get

$$
\lim_{k\to\infty} f_k(P) = \int_0^{\frac{1}{1-q}} u^{P-I}\, E_q^{-qu}\, d_q u = (q,q)_\infty\,(q^P,q)_\infty^{-1}\,(1-q)^{I-P}. \tag{4.213}
$$

Now, we will prove that the interchange of the limit and the q-integral is justified. To do this, let

$$
\begin{aligned}
&\left\| \int_0^{\frac{1}{1-q}} u^{P-I}\, E_q^{-qu}\, d_q u - f_k(P) \right\| \\
&\quad \le \left\| \int_0^{[k]_q} u^{P-I} \left(E_q^{-qu} - \left(\frac{qu}{[k]_q}, q \right)_k \right) d_q u \right\| + \left\| \int_{[k]_q}^{\frac{1}{1-q}} u^{P-I}\, E_q^{-qu}\, d_q u \right\| \\
&\quad < \frac{q^{k+2}}{(1+q)\,[k]_q} \int_0^{[k]_q} \|u^{P+I}\|\, d_q u + \left\| \int_{[k]_q}^{\frac{1}{1-q}} u^{P-I}\, E_q^{-qu}\, d_q u \right\|.
\end{aligned}
\tag{4.214}
$$

The norm of the second q-integral approaches to zero as $k \to \infty$ and the q-integral in the first term is less than ∞. To verify this, we see

$$
\begin{aligned}
\int_0^{[k]_q} \|u^{P+I}\|\, d_q u &= \int_0^1 \|u^{P+I}\|\, d_q u + \int_1^{[k]_q} \|u^{P+I}\|\, d_q u \\
&\le \int_0^1 u^{\tilde{\mu}(P)+1} d_q u + \int_1^{[k]_q} u^{\mu(P)+1} d_q u < \infty.
\end{aligned}
\tag{4.215}
$$

The former relation and the fact that $q^{k+2} \to 0$ as $k \to \infty$ completes the proof. □

The reciprocal of q-gamma function denoted by $\Gamma_q^{-1}(z) = \frac{1}{\Gamma_q(z)}$ is an entire function of the complex variable z. Then the image of $\Gamma_q^{-1}(P)$ is a well-defined matrix. Furthermore, the previous theorem shows that $\Gamma_q^{-1}(P)$ is defined for all square matrices, which satisfies the condition of Theorem 136, $\tilde{\mu}(P) > 0$ or P is positive stable and we get the following formula:

$$[P]_q\,[P+I]_q\cdots[P+(k-1)I]_q\,\Gamma_q^{-1}(P+kI)=\Gamma_q^{-1}(P),\quad k=1,2,\dots. \tag{4.216}$$

The former equation can be rewritten in the form

$$[P]_q\,[P+I]_q\cdots[P+(k-1)I]_q=\Gamma_q^{-1}(P)\,\Gamma_q(P+kI),\quad k=1,2,\dots. \tag{4.217}$$

The q-shifted factorial matrix function can be defined in terms of q-gamma matrix function as

$$(q^P,q)_k=(1-q)^k\,\Gamma_q^{-1}(P)\,\Gamma_q(P+kI),\quad k=1,2,\dots. \tag{4.218}$$

Theorem 137. *Let P be a p-square positive stable matrix or $\tilde{\mu}(P) > 0$. Then the q-analogue of Legendre's duplication formula has the form*

$$\Gamma_q(2P)\,\Gamma_{q^2}\left(\frac{1}{2}\right)=\Gamma_{q^2}(P)\,\Gamma_{q^2}\left(P+\frac{1}{2}I\right)(1+q)^{2P-I} \tag{4.219}$$

such that $q^{-k}\notin\sigma(q^{2P})$, $q^{-2k}\notin\sigma(q^{2P})$ and $q^{-2k}\notin\sigma(q^{(P+\frac{1}{2}I)})$, $k=1,2,\dots$.

Proof. We have

$$\begin{aligned}\frac{\Gamma_{q^2}(P)\,\Gamma_{q^2}(P+\frac{1}{2}I)}{\Gamma_{q^2}(\frac{1}{2})}&=(q,q^2)_\infty\,(q^2,q^2)_\infty\,(q^{2P},q^2)_\infty^{-1}\,(q^{2P+I},q^2)_\infty^{-1}\,(1-q^2)^{I-2P}\\&=(q,q)_\infty\,(q^{2P},q)_\infty^{-1}\,(1-q)^{I-2P}\,(1+q)^{I-2P}\\&=\Gamma_q(2P)\,(1+q)^{I-2P}.\end{aligned} \tag{4.220}$$

It completes the proof. □

Similarly, it can be shown that the Gauss multiplication formula for a matrix $P\in\mathbb{C}^{p\times p}$ has the q-analogue in the form

$$\begin{aligned}&\Gamma_q(kP)\,\Gamma_r\left(\frac{1}{k}\right)\Gamma_r\left(\frac{2}{k}\right)\cdots\Gamma_r\left(\frac{k-1}{k}\right)\\&\quad=(1+q+\cdots+q^{k-1})^{kP-I}\,\Gamma_r(P)\,\Gamma_r\left(P+\frac{1}{k}I\right)\cdots\Gamma_r\left(P+\frac{k-1}{k}I\right),\end{aligned} \tag{4.221}$$

with $r=q^k$, $k=1,2,\dots$.

4.10.3 The q-beta matrix function

The q-beta function is defined as

$$\beta_q(x,y) = \int_0^1 u^{x-1}\frac{(qu,q)_\infty}{(uq^y,q)_\infty}\,d_qu, \quad \Re(x) > 0, y \neq 0,-1,\dots \tag{4.222}$$

$$= \frac{\Gamma_q(x)\,\Gamma_q(y)}{\Gamma_q(x+y)}, \quad x,y \neq 0,-1,\dots. \tag{4.223}$$

Theorem 138. *Let $P \in \mathbb{C}^{p\times p}$ be a positive stable matrix or $\tilde{\mu}(P) > 0$ and $Q \in \mathbb{C}^{p\times p}$ be a positive stable matrix or $\tilde{\mu}(Q) > 0$. Then the matrix q-integral*

$$\int_0^1 (uq,q)_\infty\,(uq^Q,q)_\infty^{-1}\,u^{P-I}\,d_qu \tag{4.224}$$

converges absolutely if $q^{-k} \notin \sigma(q^P) \cup \sigma(q^Q)$ for all $k = 0,1,\dots$.

Proof. It is easy to prove that $(q^k P,q)_\infty = (P,q)_k^{-1}\,(P,q)_\infty$, $k = 0,1,\dots$. Using this matrix identity yields

$$\int_0^1 (uq,q)_\infty\,(uq^Q,q)_\infty^{-1}\,u^{P-I}\,d_qu = (1-q)\sum_{k=0}^{\infty}(q^{k+1},q)_\infty\,(q^{Q+kI},q)_\infty^{-1}\,q^{kP}$$

$$= (1-q)\,(q,q)_\infty\,(q^Q,q)_\infty^{-1}\sum_{k=0}^{\infty}\frac{(q^Q,q)_k}{(q,q)_k}\,q^{kP}. \tag{4.225}$$

Taking the norm, we get

$$\left\|\int_0^1 (uq,q)_\infty\,(uq^Q,q)_\infty^{-1}\,u^{P-I}\,d_qu\right\|$$

$$\leq (1-q)\,(q,q)_\infty\,\|(q^Q,q)_\infty^{-1}\|\sum_{k=0}^{\infty}\frac{\|(q^Q,q)_k\|}{(q,q)_k}\,\|q^{kP}\| \tag{4.226}$$

converges absolutely, provided the summation converges absolutely and $(q^Q,q)_\infty^{-1}$ converges. Using the Schur decomposition, the summation converges absolutely if P ia a positive stable matrix or $\tilde{\mu}(P) > 0$ and the infinite product converges if Q is a positive stable matrix or $\tilde{\mu}(Q) > 0$ and $q^{-k} \notin \sigma(q^Q)$, $k = 0,1,\dots$. This completes the proof. □

Definition 139. Let $P,Q \in \mathbb{C}^{p\times p}$ be positive stable matrices or $\tilde{\mu}(P) > 0$ and $\tilde{\mu}(Q) > 0$. Then the q-beta matrix function can be defined as

$$\mathcal{B}_q(P,Q) = \int_0^1 (uq,q)_\infty \, (uq^Q,q)_\infty^{-1} \, u^{P-I} \, d_q u, \tag{4.227}$$

with $q^{-k} \notin \sigma(q^P) \cup \sigma(q^Q)$ for all $k = 0, 1, \ldots$.

Theorem 140. *Let P, Q be commuting p-square complex matrices such that* $\tilde{\mu}(P) > 0$ *and* $\tilde{\mu}(Q) > 0$. *Then the q-beta matrix function is related to the q-gamma matrix function in the following way:*

$$\mathcal{B}_q(P,Q) = \mathcal{B}_q(Q,P) = \Gamma_q(P)\,\Gamma_q(Q)\,\Gamma_q(P+Q)^{-1}, \tag{4.228}$$

with $q^{-k} \notin \sigma(q^P) \cup \sigma(q^Q)$ *for all* $k = 0, 1, \ldots$.

Proof. From the relations (4.196), (4.225) and (4.227), we have

$$\begin{aligned}
\mathcal{B}_q(P,Q) &= (1-q)\,(q,q)_\infty\,(q^Q,q)_\infty^{-1} \sum_{k=0}^{\infty} \frac{(q^Q,q)_k}{(q,q)_k}\, q^{kP} \\
&= (1-q)\,(q,q)_\infty\,(q^Q,q)_\infty^{-1}\,(q^P,q)_\infty^{-1}\,(q^{P+Q},q)_\infty \\
&= (1-q)\,(q,q)_\infty\,(q^P,q)_\infty^{-1}\,(1-q)\,(q,q)_\infty\,(q^Q,q)_\infty^{-1}\,(1-q)^{-1}\,(q,q)_\infty^{-1}\,(q^{P+Q},q)_\infty \\
&= \Gamma_q(P)\,\Gamma_q(Q)\,\Gamma_q(P+Q)^{-1}.
\end{aligned} \tag{4.229}$$

It completes the proof. □

Theorem 141. *Under the assumption in the previous theorem, a q-beta matrix function satisfies the following recurrence relations:*

$$\mathcal{B}_q(P,Q+kI) = (q^Q,q)_k\,(q^{P+Q},q)_k^{-1}\,\mathcal{B}_q(P,Q), \tag{4.230}$$

$$\mathcal{B}_q(P+kI,Q+kI) = (q^P,q)_k\,(q^Q,q)_k\,(q^{P+Q},q)_{2k}^{-1}\,\mathcal{B}_q(P,Q), \tag{4.231}$$

where k is a positive integer or zero.

Definition 142. For any matrices P, Q, R in $\mathbb{C}^{p\times p}$ such that $q^{-k} \notin \sigma(R)$ for all integers $k \geq 0$ and $|q| < 1$, the basic Gauss hypergeometric matrix function is defined by

$${}_2\phi_1(P,Q;R;q,z) = \sum_{k=0}^{\infty} \frac{(P,q)_k\,(Q,q)_k\,(R,q)_k^{-1}}{(q,q)_k}\, z^k, \quad |z| < 1. \tag{4.232}$$

In the following theorem, the absolute convergence of the matrix power series (4.232) will be proven for all complex variable z on the open unit disc $|z| < 1$.

Theorem 143. *For any matrices P, Q, R in* $\mathbb{C}^{p\times p}$ *such that* $q^{-k} \notin \sigma(R)$ *for all integers* $k \geq 0$ *and* $|q| < 1$, *the basic Gauss hypergeometric matrix function defined in* (4.232) *converges absolutely for all complex variable z on the open unit disc* $|z| < 1$.

Proof. The condition $q^{-k} \notin \sigma(R)$ guarantees that $I - q^k R$ is invertible for all integers $k \geq 0$. Now take k large enough so that $\|R\| < |q|^{-k}$. By the Perturbation lemma, one gets

$$\|(I - q^k R)^{-1}\| \leq \frac{1}{1 - |q|^k \|R\|}, \qquad k > -\frac{\ln \|R\|}{\ln |q|}. \tag{4.233}$$

It is easy to see that

$$\|(P, q)_k\| \leq (-\|P\|, |q|)_k, \quad \|(Q, q)_k\| \leq (-\|Q\|, |q|)_k. \tag{4.234}$$

If we take

$$G(k) = \begin{cases} 1, & k = 0, \\ \prod_{n=0}^{k-1} \|(I - q^n R)^{-1}\|, & k \geq 1, \end{cases} \tag{4.235}$$

then using (4.234), we get

$$\begin{aligned} \|{}_2\phi_1(P, Q; R; q, z)\| &\leq \sum_{k=0}^{\infty} \left\| \frac{(P, q)_k \, (Q, q)_k \, (R, q)_k^{-1}}{(q, q)_k} z^k \right\| \\ &\leq \sum_{k=0}^{\infty} \frac{(-\|P\|, |q|)_k \, (-\|Q\|, |q|)_k \, G(k)}{|(q, q)_k|} |z|^k. \end{aligned} \tag{4.236}$$

Using the ratio test and equation (4.233), one finds

$$\begin{aligned} &\lim_{k\to\infty} \left| \frac{(-\|P\|, |q|)_{k+1} \, (-\|Q\|, |q|)_{k+1} \, G(k+1) \, |(q, q)_k| \, |z|^{k+1}}{(-\|P\|, |q|)_k \, (-\|Q\|, |q|)_k \, G(k) \, |(q, q)_{k+1}| \, |z|^k} \right| \\ &= \lim_{k\to\infty} \left| \frac{(1 + |q|^k \|P\|) \, (1 + |q|^k \|Q\|) \, \|(I - q^k R)^{-1}\|}{(1 - |q|^{k+1})} |z| \right| \\ &\leq \lim_{k\to\infty} \left| \frac{(1 + |q|^k \|P\|) \, (1 + |q|^k \|Q\|)}{(1 - |q|^{k+1}) \, (1 - |q|^k \|R\|)} \right| |z| = |z|. \end{aligned} \tag{4.237}$$

Thus, the matrix power series (4.232) is absolutely convergent on the open unit disc $|z| < 1$. It completes the proof. □

Now, we are interested in identifying the domains in which the basic hypergeometric matrix function is invertible.

Let $0 < \gamma < 1$ and $k_1 \geq k_0 > \max\{1, -\frac{\ln \|R\|}{\ln |q|}\}$ be a positive integer such that

$$\frac{(1 + |q|^k \|P\|) \, (1 + |q|^k \|Q\|)}{(1 - |q|^{k+1-i}) \, (1 - |q|^k \|R\|)} < 1 + \gamma, \quad k \geq k_i, i = 0, 1. \tag{4.238}$$

Also

$$K_i = \frac{1}{1 - i|q|} \sum_{k=1}^{k_i - 1} \frac{(-\|P\|, |q|)_k \, (-\|Q\|, |q|)_k \, G(k)}{|(q, q)_{k-i}|}, \quad i = 0, 1; \tag{4.239}$$

$$L_i = \frac{1}{1-i|q|}\frac{(-\|P\|,|q|)_{k_i}\,(-\|Q\|,|q|)_{k_i}\,G(k_i)}{|(q,q)_{k_i-i}|}, \quad i=0,1 \tag{4.240}$$

and

$$f(x) = K_0\,x + L_0\,\frac{x}{1-x}, \quad 0 \le x < 1. \tag{4.241}$$

Since the function $f(x)$ is increasing in the interval $[0,1[$, $f(0)=0$ and $\lim_{x\to 1^-} f(x) = \infty$, we find there is only one value $0 < \delta < 1$ such that $f(\delta) = 1-\gamma$. Thus, we can summarize the finding in the following theorem.

Theorem 144. *Suppose that $|q| < 1$ and $0 < \gamma < 1$. Let k_0 be a positive integer and δ be the solution of the equation $f(x) = 1-\gamma$, and f is defined by*

$$f(x) = K_0\,x + L_0\,\frac{x}{1-x}, \quad 0 \le x < 1,$$

where K_0 and L_0 are given by the equations (4.239) and (4.240), respectively. Also let $\rho > 0$ so that $\rho(1+\gamma) < \delta$ and $\Phi(z) = {}_2\phi_1(P,Q;R;q,z)$. Then $\Phi(z)$ is invertible and norm value is given by

$$\|(\Phi(z))^{-1}\| < \gamma^{-1}, \quad \textit{for } |z| < \rho.$$

Theorem 145. *With the notation of Theorem 144, suppose $\epsilon > 0$ and k_ϵ be a positive integer such that*

$$k_\epsilon = \max\left\{k_0, \frac{\ln(\frac{\epsilon(1-\delta)}{L_0})}{\ln\delta}\right\}. \tag{4.242}$$

Also let

$$\Phi_k(P,Q;R;q,z) = \sum_{n=0}^{k}\frac{(P,q)_n\,(Q,q)_n\,(R,q)_n^{-1}}{(q,q)_n}\,z^n. \tag{4.243}$$

Then

$$\|\Phi(z) - \Phi_{k-1}(P,Q;R;q,z)\| < \epsilon, \quad |z| \le \rho, k \ge k_\epsilon.$$

The following theorem will be interested in the determination of boundedness of the function $D_q\Phi(z)$ for $|z| \le \lambda$, where λ will be defined later.

Theorem 146. *With the notation of Theorems 144 and 145, suppose $\lambda > 0$ so that $\lambda(1+\gamma) < \rho$, k_1 is a positive integer defined by (4.238) and k'_ϵ is a positive real number defined by*

$$k'_\epsilon = 1 + \max\left\{k_1, \frac{\ln(\frac{\epsilon(1-\rho)}{L_1})}{\ln\rho}\right\}. \tag{4.244}$$

Then:
1. $\|D_q\Phi(z)\| < K_1 + \frac{T_1}{1-\rho}$, $|z| \le \lambda$,
2. $\|D_q \Phi(z) - D_q \Phi_{k-1}(P, Q; R; q, z)\| < \epsilon$, $|z| \le \lambda$, $k \ge k'_\epsilon$.

It is well known that $\frac{d}{du} e^{uP} = P e^{uP}$, for some $P \in \mathbb{C}^{p\times p}$. Using this and the L'hopital rule, one can get

$$\lim_{q\to 1} \frac{I - q^P}{1-q} = \lim_{q\to 1} \frac{I - e^{P \log q}}{1-q} = P,$$

which reveals that

$$\lim_{q\to 1} \frac{(P, q)_k}{(1-q)^k} = \lim_{q\to 1} \prod_{n=0}^{k-1} \frac{I - q^{P+nI}}{1-q} = \prod_{n=0}^{k-1} (P + nI) = (P)_k.$$

Therefore, the basic Gauss hypergeometric matrix function ${}_2\phi_1(q^P, q^Q; q^R; q, z)$ approaches the Gauss hypergeometric matrix function (4.91) when letting $q \to 1$.

4.10.4 Hypergeometric matrix q-difference equation

James in 1975 introduced a hypergeometric type q-difference equation, which is a second-order homogeneous q-difference equation

$$\sigma(z) D_q D_{q^{-1}} u(z) + \tau(z) D_q U(z) - \lambda u(z) = 0, \tag{4.245}$$

where $\sigma(z)$ and $\tau(z)$ are polynomials in z of at most second and first degree, respectively, and λ is a constant (but all possibly depend on q). Some solutions of (4.245) are functions extensively used in mathematical physics such as classical q-orthogonal polynomials (big and little q-Jacobi, q-Laguerre and q-Hermite polynomials), basic Gauss hypergeometric and basic confluent hypergeometric functions. The basic Gauss hypergeometric function ${}_2\phi_1(q^a, q^b; q^c; q, z)$ is a solution of the second-order q-difference equation (4.245) when

$$\sigma(z) = (q - z)\, z, \quad \tau(z) = q[c]_q - [a + b + 1]_q\, z, \quad \lambda = q\, [a]_q\, [b]_q, \quad 0 < |z| < 1. \tag{4.246}$$

In the present section, we find two solutions for the matrix q-difference equation

$$\begin{aligned} &z(q - z) D_q D_{q^{-1}} U(z) - z[P]_q D_q U(z) + D_q U(z)(q\, [R]_q - z[Q + I]_q) \\ &\quad + (1 - q)\, z\, [P]_q D_q U(z)\, [Q + I]_q - q[P]_q U(z)\, [Q]_q = 0, \quad 0 < |z| < 1, \end{aligned} \tag{4.247}$$

where $P, Q, R \in \mathbb{C}^{p\times p}$, $QR = RQ$, $\tilde{\mu}(R) \ge 0$ and $q^{-k} \notin \sigma(q^R)$, $k = 0, 1, \ldots$.

Theorem 147. *Let $P, Q, R \in \mathbb{C}^{p\times p}$, $QR = RQ$, $\tilde{\mu}(R) \geq 0$, $q^{-k} \notin \sigma(q^R)$, $k = 0, 1, \ldots$ and $0 < |q| < 1$. Then the basic Gauss hypergeometric matrix function ${}_2\phi_1(q^P, q^Q; q^R; q, z)$ is a solution of the matrix difference equation* (4.247) *and satisfying* ${}_2\phi_1(q^P, q^Q; q^R; q, 0) = I$.

Proof. Let us seek a solution $U(z)$ of (4.247) in the form

$$U(z) = \sum_{k=0}^{\infty} U_k z^k, \quad |z| < 1. \tag{4.248}$$

Taking into account the q-derivatives of (4.248), given by

$$D_q U(z) = \sum_{k=1}^{\infty} U_k [k]_q z^{k-1}, \quad D_q D_{q^{-1}} U(z) = \sum_{k=2}^{\infty} U_k [k]_q [k-1]_q z^{k-2},$$

and substituting into (4.247), we get

$$\begin{aligned} &q(U_1[R]_q - [P]_q U_0 [Q]_q) + (q U_2 [2]_q [R]_q + U_2 [2]_q - q [P]_q U_1[Q]_q - [P]_q U_1 \\ &- U_1[Q+I]_q + (1-q) [P]_q U_1[Q+I]_q) z + \sum_{k=2}^{\infty} (q U_{k+1} [k+1]_q [R]_q \\ &+ qU_{k+1} [k+1]_q [k]_q q^{-k} - [P]_q U_k [k]_q - q [P]_q U_k [Q]_q - U_k [k]_q [Q+I]_q \\ &+ (1-q) [P]_q U_k [k]_q [Q+I]_q - U_k [k]_q [k-1]_q q^{1-k}) z^k = 0. \end{aligned} \tag{4.249}$$

In view of the previous equation, equating the coefficients of each power of z^k to zero and taking into account that the hypothesis $\tilde{\mu}(R) \geq 0$, $q^{-k} \notin \sigma(q^R)$, $k = 0, 1, \ldots$ yield that $[R + kI]_q$ are invertible for all integers $k \geq 0$, one gets

$$U_1 = [P]_q U_0 [Q]_q [R]_q^{-1}, \tag{4.250}$$

$$U_2 = \frac{[P+I]_q U_1 [Q+I]_q [R+I]_q^{-1}}{[2]_q}, \tag{4.251}$$

$$U_{k+1} = \frac{[P+kI]_q U_k [Q+kI]_q [R+kI]_q^{-1}}{[k+1]_q}, \quad k \geq 2. \tag{4.252}$$

It is easy to prove (4.250) and (4.251). To prove (4.252), equating to zero, the coefficient of z^k in (4.249),

$$\begin{aligned} U_{k+1} [k+1]_q q^{1-k} (q^k [R]_q + [kI]_q) = {} & [P]_q U_k ([kI]_q + q[Q]_q) + (1 - (1-q) [P]_q) \\ & \times U_k [k]_q [Q+I]_q + U_k [k]_q [k-1]_q q^{1-k} \end{aligned} \tag{4.253}$$

and taking into account the following identities:

$$q^k [P]_q + [kI]_q = [P]_q + q^P [k]_q = [P+kI]_q, \quad [kI]_q + q[Q]_q = [Q+I]_q + q [(k-1)I]_q. \tag{4.254}$$

Substituting into (4.253), we get

$$\begin{aligned} U_{k+1}[k+1]_q[R+kI]_q &= q^{k-1}[P]_q U_k([Q+I]_q + q[(k-1)I]_q) \\ &\quad + q^{P+(k-1)I} U_k [k]_q [Q+I]_q + U_k [k]_q [k-1]_q \\ &= q^{k-1}([P]_q + q^P [k]_q) U_k [Q+I]_q + (q^k [P]_q + [k]_q I) U_k [k-1]_q \\ &= q^{k-1}[P+kI]_q U_k [Q+I]_q + [P+kI]_q U_k [k-1]_q \\ &= [P+kI]_q U_k (q^{k-1}[Q+I]_q + [k-1]_q) \\ &= [P+kI]_q U_k [Q+I]_q, \end{aligned} \tag{4.255}$$

which transforms into (4.252). From (4.250)–(4.252), taking $U_0 = I$, we have

$$\begin{aligned} U_k &= \frac{[P+(k-1)I]_q U_{k-1} [Q+(k-1)I]_q [R+(k-1)I]_q^{-1}}{[k]_q} \\ &= \frac{(q^P,q)_k (q^Q,q)_k (q^R,q)_k^{-1}}{(q,q)_k}, \quad k \geq 0. \end{aligned} \tag{4.256}$$

It completes the proof. □

Recall that if k, n are positive integers, the identity $(q^{-k},q)_n = 0$ for all $k \leq n-1$. Hence, we have

$$_2\phi_1(q^P, q^{-kI}; q^R; q, z) = \sum_{n=0}^{k} \frac{(q^P,q)_n (q^{-kI},q)_n (q^R,q)_n^{-1}}{(q,q)_n} z^n. \tag{4.257}$$

From the above note, we can deduce the following corollary by putting $Q = -kI$.

Corollary 148. *Let P, R be matrices in $\mathbb{C}^{p\times p}$ such that $\tilde{\mu}(R) \geq 0$, $q^{-k} \notin \sigma(q^R)$, $0 < |q| < 1$ and k be a positive integer. Then the hypergeometric matrix q-difference equation*

$$z(q-z) D_q D_{q^{-1}} U(z) - z[P+(1-q)I]_q D_q U(z) + q D_q U(z) [R]_q + q^{1-k} [k]_q [P]_q U(z) = 0, \tag{4.258}$$

has a matrix polynomial solution of degree k.

Now, we will be interested to find the second solution of the difference equation (4.247) along the complex plane D_0 cut along the negative real axis.

Theorem 149. *Let $P, Q, R \in \mathbb{C}^{p\times p}$, $QR = RQ$, $PR = RP$, $\tilde{\mu}(R) \leq 2$, $q^{k+2} \notin \sigma(q^R)$, $k \geq 0$. Then the basic Gauss hypergeometric matrix function*

$$_2\phi_1(q^{P+I-R}, q^{Q+I-R}; q^{2I-R}; q, z) z^{I-R}, \quad |z| < 1, z \in D_0, |q| < 1 \tag{4.259}$$

is a solution of the matrix difference equation (4.247).

Proof. Suppose that the second solution of equation (4.247) has the form

$$U(z) = V(z)\,z^{I-R}, \quad z^{I-R} = \exp((I-R)\ln z), |z| < 1, z \in D_0, \tag{4.260}$$

where ln is the principal branch of the logarithm function and $V(z)$ is an unknown function to be determined. The q-derivatives of (4.260) are given by

$$D_{q^{-1}}(z^{I-R}) = q^R\,[I-R]_q\,z^{-R}, \tag{4.261}$$

$$D_qU(z) = V(z)\,[I-R]_q z^{-R} + D_qV(z)z^{I-R}, \tag{4.262}$$

$$D_{q^{-1}}U(z) = V(z)\,q^R\,[I-R]_q z^{-R} + D_{q^{-1}}V(z)q^{R-I}z^{I-R}, \tag{4.263}$$

$$\begin{aligned} D_qD_{q^{-1}}U(z) &= D_qD_{q^{-1}}V(z)\,q^{R-I}\,z^{I-R} + D_qV(z)\,[I-R]_q\,z^{-R} \\ &\quad + D_qV(z)\,[I-R]_q\,q^{R-I}\,z^{-R} + V(z)\,[R]_q\,[I-R]_q\,z^{-R-I}. \end{aligned} \tag{4.264}$$

Substituting these values in (4.247) yield

$$\begin{aligned} &z(q-z)\,D_q\,D_{q^{-1}}\,V(z) - z[P+I-R]_q\,D_q\,V(z) + D_q\,V(z)(q\,[2I-R]_q \\ &\quad - z[Q+2I-R]_q) + (1-q)\,z\,[P+I-R]_q\,D_q\,V(z)\,[Q+2I-R]_q \\ &\quad - q[P+I-R]_q\,V(z)\,[Q+I-R]_q = 0, \end{aligned} \tag{4.265}$$

which is a matrix q-difference equation of the form (4.247) with

$$\tilde{P} = P+I-R, \quad \tilde{Q} = Q+I-R, \quad \tilde{R} = 2I-R.$$

Hence, the matrix q-difference equation (4.247) has a solution

$${}_2\phi_1(q^{P+I-R}, q^{Q+I-R}; q^{2I-R}; q, z)\,z^{I-R}, \quad |z| < 1, |q| < 1. \tag{4.266}$$

Noto that the hypothesis $\mu(R) \le 2$, which is precisely equivalent $\tilde{\mu}(2I-R) \ge 0$ ensures that $[(2+k)I-R]_q$ is invertible for all integers $k \ge 0$. Furthermore, under the hypothesis of this theorem, $V(z)$ and its q-derivatives commutes with a matrix R. □

4.10.5 A q-integral representation

The basic hypergeometric function ${}_2\phi_1(q^a, q^b; q^c; q, z)$ has a q-integral representation for $|z| < 1$, $0 < q < 1$ and $\Re(b) > 0$ in the form

$${}_2\phi_1(q^a, q^b; q^c; q, z) = \frac{\Gamma_q(c)}{\Gamma_q(b)\,\Gamma_q(c-b)} \int_0^1 t^{b-1} \frac{(uzq^a, uq, q)_\infty}{(uz, uq^{c-b}, q)_\infty}\,d_qu.$$

This section is devoted to establish a q-integral representation for the hypergeometric matrix function ${}_2\phi_1(q^P, q^Q; q^R; q, z)$ via the definitions of the q-beta matrix function. To reach the goal first, we give few definitions and theorems as per the requirements.

Let a_{ij} be a complex valued function of two subscripts i, j defined for $i, j = 0, 1, \ldots$. The formal series $\sum_{i,j=0}^{\infty} a_{ij}$ is called a double series. If $\sum_{j=0}^{\infty} a_{ij}$ converges for all i, then $\sum_{i=0}^{\infty}(\sum_{j=0}^{\infty} a_{ij})$ is called an iterated series by rows. If $\sum_{i=0}^{\infty} a_{ij}$ converges for all j, then $\sum_{j=0}^{\infty}(\sum_{i=0}^{\infty} a_{ij})$ is called an iterated series by columns. Even if two iterated series by rows and columns are convergent, if two sums are not always identical, then the double series is not always convergent. If any one of the three series $\sum_{i,j=0}^{\infty} |a_{ij}|$, $\sum_{i=0}^{\infty}(\sum_{j=0}^{\infty} |a_{ij}|)$ and $\sum_{j=0}^{\infty}(\sum_{i=0}^{\infty} |a_{ij}|)$ is convergent, the double series $\sum_{i,j=0}^{\infty} a_{ij}$ is called absolutely convergent. If $\sum_{i,j=0}^{\infty} a_{ij}$ is absolutely convergent, then any series obtained from $\sum_{i,j=0}^{\infty} a_{ij}$ by arranging the term in arbitrary order converges to the same sum and this yields the following.

Theorem 150. *Let $0 < q < 1$ and f_k be a complex valued function defined for $k = 0, 1, \ldots$. If any one of the $\int_0^a \sum_{k=0}^{\infty} |f_k(u)|\, d_q u$, $\int_0^a (\sum_{k=0}^{\infty} |f_k(u)|)\, d_q u$ and $\sum_{k=0}^{\infty}(\int_0^a |f_k(u)|\, d_q u)$ is convergent, then other two converge to the same sum.*

Definition 151. Let $0 < q < 1$ and f_k be a complex valued function defined for $k = 0, 1, \ldots$. Then the q-integral $\int_0^a \sum_{k=0}^{\infty} |f_k(u)|\, d_q u$ is said to be absolutely convergent if the double series $\sum_{m,k=0}^{\infty} q^m\, |f_k(q^m\, a)|$ is convergent.

Theorem 152. *Let $0 < q < 1$ and f_k be a complex valued function defined for $k = 0, 1, \ldots$. If the q-integral $\int_0^a \sum_{k=0}^{\infty} f_k(u)\, d_q u$ is absolutely convergent, then both $\int_0^a (\sum_{k=0}^{\infty} f_k(u))\, d_q u$ and $\sum_{k=0}^{\infty}(\int_0^a f_k(u)\, d_q u)$ are convergent and identical.*

A more general version of the codomain of the functions $f_k, k = 0, 1, \ldots$ is any Banach space. In such case, the statement $|f_k|$ may be replaced by $\|f_k\|$ where $\|\cdot\|$ is the norm on the Banach space.

Theorem 153. *For any matrices P, Q, $R \in \mathbb{C}^{p\times p}$ such that Q, R are commuting matrices, $\tilde{\mu}(Q) \geq 0$, $\tilde{\mu}(R) \geq 0$, $\tilde{\mu}(R-Q) \geq 0$ and $q^{\;k} \notin \sigma(q^R) \cup \sigma(q^Q) \cup \sigma(q^{R-Q})$, $k = 0, 1, \ldots$. Then the basic hypergeometric matrix function ${}_2\phi_1(q^P, q^Q; q^R; q, z)$ for $0 < q < 1$, $|z| < 1$ can be written in the q-integral form as*

$$
\begin{aligned}
{}_2\phi_1(q^P, q^Q; q^R; q, z) = \int_0^1 & \frac{(uq, q)_\infty\, (uzq^P, q)_\infty\, (uq^{R-Q}, q)_\infty^{-1}}{(uz, q)_\infty}\, u^{Q-I}\, d_q u \\
& \times \Gamma_q^{-1}(Q)\, \Gamma_q^{-1}(R-Q)\, \Gamma_q(R).
\end{aligned} \tag{4.267}
$$

Proof. From the definitions of q-shifted factorial for matrices and q-beta matrix function, we have

$$
\begin{aligned}
(q^Q, q)_k\, (q^R, q)_k^{-1} &= \Gamma_q^{-1}(Q)\, \Gamma_q(Q+kI)\, \Gamma_q^{-1}(R+kI)\, \Gamma_q(R) \\
&= \beta_q(R-Q, Q+kI)\, \Gamma_q^{-1}(Q)\, \Gamma_q^{-1}(R-Q)\, \Gamma_q(R)
\end{aligned}
$$

$$= \int_0^1 (uq,q)_\infty \,(uq^{R-Q},q)_\infty^{-1}\, u^{Q+(k-1)I}\, d_q u\, \Gamma_q^{-1}(Q)\,\Gamma_q^{-1}(R-Q)\,\Gamma_q(R). \quad (4.268)$$

Substituting this value in (4.232), for $|z| < 1$, we obtain

$$\begin{aligned}
{}_2\phi_1(P,Q;R;q,z) &= \sum_{k=0}^{\infty} \frac{(P,q)_k\,(Q,q)_k\,(R,q)_k^{-1}}{(q,q)_k} z^k \\
&= \sum_{k=0}^{\infty} \frac{(P,q)_k}{(q,q)_k} z^k\,(Q,q)_k\,(R,q)_k^{-1} \\
&= \sum_{k=0}^{\infty} \frac{(P,q)_k}{(q,q)_k} z^k \int_0^1 (uq,q)_\infty\,(uq^{R-Q},q)_\infty^{-1}\, u^{Q+(k-1)I}\, d_q u \\
&\quad \times \Gamma_q^{-1}(Q)\,\Gamma_q^{-1}(R-Q)\,\Gamma_q(R) \\
&= \sum_{k=0}^{\infty} \int_0^1 \frac{(uq,q)_\infty\,(q^P,q)_k\,(uq^{R-Q},q)_\infty^{-1}}{(q,q)_k} z^k\, u^{Q+(k-1)I}\, d_q u \\
&\quad \times \Gamma_q^{-1}(Q)\,\Gamma_q^{-1}(R-Q)\,\Gamma_q(R).
\end{aligned} \quad (4.269)$$

Suppose that we are allowed to interchange the q-integral and summation on the right-hand side. This would give

$$\begin{aligned}
{}_2\phi_1(P,Q;R;q,z) &= \int_0^1 (uq,q)_\infty \left(\sum_{k=0}^{\infty} \frac{(q^P,q)_k}{(q,q)_k} (zu)^k \right) (uq^{R-Q},q)_\infty^{-1}\, u^{Q-I}\, d_q u \\
&\quad \times \Gamma_q^{-1}(Q)\,\Gamma_q^{-1}(R-Q)\,\Gamma_q(R) \\
&= \int_0^1 \frac{(uq,q)_\infty\,(uzq^P,q)_\infty\,(uq^{R-Q},q)_\infty^{-1}}{(uz,q)_\infty} u^{Q-I}\, d_q u \\
&\quad \times \Gamma_q^{-1}(Q)\,\Gamma_q^{-1}(R-Q)\,\Gamma_q(R).
\end{aligned} \quad (4.270)$$

In order to justify this interchange of the q-integral and sum, we claim that for $0 < t < 1$, $0 < q < 1$, the sequence of matrix functions

$$F_k(u) = \frac{(uq,q)_\infty\,(q^P,q)_k\,(uq^{R-Q},q)_\infty^{-1}}{(q,q)_k} z^k\, u^{Q+(k-1)I}\, \Gamma_q^{-1}(Q)\,\Gamma_q^{-1}(R-Q)\,\Gamma_q(R) \quad (4.271)$$

converges. For $0 < q < 1$, $k \geq 0$ and $N = \|\Gamma_q^{-1}(Q)\|\,\|\Gamma_q^{-1}(R-Q)\|\,\|\Gamma_q(R)\|$, the above matrix sequence yields

$$\|F_k(u)\| \leq N \frac{(uq,q)_\infty\,(-\|q^P\|,q)_k\,\|(uq^{R-Q},q)_\infty^{-1}\|}{(q,q)_k} |z|^k\,\|u^{Q-I}\|, \quad 0 < u < 1. \quad (4.272)$$

Since $\tilde{\mu}(R-Q) > 0$, the matrix $R-Q$ is positive stable and so $I - q^{R-Q+kI}$ is invertible for all integers $k \geq 0$. By the perturbation lemma, we have

$$\|I - u\,q^{R-Q+kI}\| \leq \frac{1}{1 - u\,q^{k}\,\|q^{R-Q}\|} \leq \frac{1}{1 - u\,q^{\tilde{\mu}(R-Q)+k}}, \quad 0 < u < 1, k \geq 0. \tag{4.273}$$

This would give

$$\begin{aligned} \|(uq^{R-Q}, q)_{\infty}^{-1}\| &\leq \prod_{k=0}^{\infty} \|I - u\,q^{R-Q+kI}\|^{-1} \\ &\leq \prod_{k=0}^{\infty} \frac{1}{1 - u\,q^{\tilde{\mu}(R-Q)+k}} = \frac{1}{(u\,q^{\tilde{\mu}(R-Q)+k}, q)_{\infty}}. \end{aligned} \tag{4.274}$$

Also, by the logarithm norm, we have

$$\|u^{Q-I}\| \leq u^{\tilde{\mu}(Q)-1}, \quad 0 < u < 1. \tag{4.275}$$

Suppose that F denotes the sum of convergent series

$$F = \sum_{k=0}^{\infty} \frac{(-\|q^{P}\|, q)_{k}}{(q, q)_{k}} |z|^{k} = \frac{(-|z|\|q^{P}\|, q)_{\infty}}{(|z|, q)_{\infty}}, \quad |z| < 1. \tag{4.276}$$

From (4.271)–(4.276), we obtain

$$\sum_{k=0}^{\infty} \|F_{k}(u)\| = F\,N \frac{(uq, q)_{\infty}}{(u\,q^{\tilde{\mu}(R-Q)+k}, q)_{\infty}} u^{\tilde{\mu}(Q)-1}. \tag{4.277}$$

Since $\tilde{\mu}(Q) > 0$, $\tilde{\mu}(R) > 0$ and $\tilde{\mu}(R-Q) > 0$, the matrices $Q, R, R-Q$ are positive stable, by the dominated convergence theorem

$$\int_{0}^{1} \sum_{k=0}^{\infty} \|F_{k}(u)\| = F\,N\,\beta_{q}(Q, R-Q) < \infty. \tag{4.278}$$

This yields the absolute convergence of (4.271), and hence the q-integral and summation can be interchanged. This competes the proof. □

This chapter consists of three different matrix versions of the Gaussian hypergeometric function and its basic properties. Basically, the content of this chapter is inspired by the work of L. Jódar, J. C. Cortés, Fransico J. Solís, Roberto Ku-Carrillo, Juan A. Tirao and Ahmed Salem. One can visit [1, 2, 4, 48–52] to see the further devlopment of special matrix functions.

Exercises

1. Find the spectrum and check whether the given matrices are positive stable or not.

$$\begin{bmatrix} 2 & -1 \\ -1 & 2 \end{bmatrix}, \quad \begin{bmatrix} 1 & 0 & 1 \\ 0 & 1 & 2 \\ 0 & 2 & 3 \end{bmatrix}, \quad \begin{bmatrix} 1 & 1 & 2 & 0 \\ 0 & 2 & 1 & 0 \\ 3 & 1 & 0 & 2 \\ 0 & 0 & 1 & 1 \end{bmatrix}.$$

2. Find the Euclidean norm, 1-norm, 2-norm and ∞-norm of the following matrices.

$$\begin{bmatrix} 1 & 2 \\ -2 & 1 \end{bmatrix}, \quad \begin{bmatrix} 1 & 0 \\ -1 & 2 \end{bmatrix}.$$

3. Show that the following infinite matrix product are absolutely convergent:

$$\text{(i)} \prod_{k=1}^{\infty}\left(I + \frac{6}{(k+1)\,(2k+9)}\, I\right) \qquad \text{(ii)} \prod_{k=0}^{\infty}(I + P^{2k}), \quad P \in \mathbb{C}^{p\times p}.$$

4. If $P \in \mathbb{C}^{p\times p}$, then prove that $(P)_{n-k} = (-1)^k\,(P)_k\,(I - P - nI)_k$.
5. If P is a p-square complex matrix, then show that the matrix analogue of Legendre duplication formula is given by

$$\Gamma(2P) = \pi^{\frac{-1}{2}}\, 2^{2P-I}\, \Gamma(P)\, \Gamma\!\left(P + \frac{1}{2} I\right).$$

Further, it can be generalized to Gauss multiplication theorem:

$$\Gamma(kP) = (2\pi)^{\frac{-1}{2}(k-1)}\, k^{kP-\frac{1}{2}I} \prod_{s=1}^{k} \Gamma\!\left(P + \frac{s-1}{k} I\right).$$

6. If P and Q are positive stable matrices in $\mathbb{C}^{p\times p}$ such that $PQ = QP$, then show that

$$\int_{-1}^{1} (1+x)^{P-I}\,(1-x)^{Q-I}\, dx = 2^{P+Q-1}\, \beta(P, Q).$$

7. If A, B, C and D are commuting matrices in $\mathbb{C}^{p\times p}$, then the beta matrix function satisfies the following functional relations:
   (a) $\beta(A, B+I) = B\,\beta(A+I, B)\,A^{-1} + B\,\beta(A, B)\,(A+B)^{-1}$;
   (b) $\beta(A, B)\,\beta(A+B, C) = \beta(B, C)\,\beta(B+C, A) = \beta(C, A)\,\beta(A+C, B)$;
   (c) $\beta(A, B)\,\beta(A+B, C)\,\beta(A+B+C, D) = \Gamma(A)\,\Gamma(B)\,\Gamma(C)\,\Gamma(D)\,\Gamma^{-1}(A+B+C+D)$.
8. If A and B are diagonal matrices in $\mathbb{C}^{p\times p}$ such that $\beta(A), \beta(B) > 0$, then show that

$$\beta(A, B) = \Gamma(A)\,\Gamma(B)\,\Gamma^{-1}(A+B).$$

9. Prove that

$$\frac{d}{dz}\,{}_2F_1(P,Q;R;z) = P\,{}_2F_1(P+I,Q+I;R+I;z)\,Q\,R^{-1}, \quad QR = RQ.$$

In general,

$$\frac{d^r}{dz^r}\,{}_2F_1(P,Q;R;z) = (P)_r\,{}_2F_1(P+rI,Q+rI;R+rI;z)\,(Q)_r\,(R)_r^{-1}, \quad QR = RQ.$$

10. Let P, Q, R be commuting matrices and $F, F(P\pm), F(Q\pm), F(R\pm)$ denote ${}_2F_1(P,Q;R;z)$, ${}_2F_1(P\pm I,Q;R;z)$, ${}_2F_1(P,Q\pm I;R;z)$ and ${}_2F_1(P,Q;R\pm I;z)$, respectively. Then verify the following matrix contiguous formulas:

$$\begin{aligned}
PF - FQ &= PF(P+) - F(Q+)\,Q;\\
(P+I)\,F - FR &= PF(P+) - F(R-)\,(R-I);\\
PF + zF\,(Q-R) &= (1-z)PF(P+) - z\,F(R+)\,(R-P)\,(R-Q)\,R^{-1};\\
(1-z)F &= F(P-) - zF(R+)\,(R-Q)\,R^{-1};\\
(1-z)F &= F(Q-) - z\,(R-Q)\,R^{-1}\,F(R+).
\end{aligned}$$

11. If P, Q, R be commuting matrices in $\mathbb{C}^{p\times p}$ such that Q, R, $R-Q$ are positive stable. Then prove that

$${}_2F_1(P,Q;R;1) = \Gamma(R)\,\Gamma(R-P-Q)\,\Gamma^{-1}(R-P)\,\Gamma^{-1}(R-Q).$$

12. Show that

$${}_2F_1(P,Q;R;z) = (1-z)^{-P}\,{}_2F_1\left(P,R-Q;R;\frac{z}{z-1}\right), \quad QR = RQ.$$

13. If ${}_1F_1(P;R;z)$ denotes the matrix analogue of the confluent hypergeometric function, then verify that

$$\lim_{\varepsilon\to 0}\,{}_2F_1\left(P,\frac{1}{\varepsilon}I;R;\varepsilon\,z\right) = {}_1F_1(P;R;z).$$

14. Verify the following identities for q-shifted factorial matrix functions:
(a) $(P;q)_{m+k} = (P;q)_m\,(Pq^m;q)_k$;
(b) $(P;q)_m = (P;q)_\infty\,(Pq^m;q)_\infty^{-1}$;
(c) $(P;q)_{2m} = (P;q^2)_m\,(Pq;q^2)_m$;
(d) $(P^2;q^2)_m = (P;q)_m\,(-P;q)_m$.

15. If P, Q, R are commuting matrices in $\mathbb{C}^{p\times p}$ and

$$\begin{aligned}
\phi &= {}_2\phi_1(q^P,q^Q;q^R;q;z);\\
\phi(P\pm) &= {}_2\phi_1(q^{P\pm I},q^Q;q^R;q;z);
\end{aligned}$$

$$\phi(Q\pm) = {}_2\phi_1(q^P, q^{Q\pm I}; q^R; q; z);$$
$$\phi(R\pm) = {}_2\phi_1(q^P, q^Q; q^{R\pm I}; q; z).$$

Then prove that

$$([P]_q - [Q]_q)\,\phi = q^Q\,[P]_q\,\phi(P+) - q^P\,[Q]_q\,\phi(Q+),$$
$$([P]_q - [R-I]_q)\,\phi = q^{R-I}\,[P]_q\,\phi(P+) - q^P\,[R-I]_q\,\phi(R-).$$

5 Finite field analogue of the Gauss hypergeometric function

In this chapter, we look up the finite field version of Gauss hypergeometric function, which was introduced by Greene in 1987. After reading this chapter, the reader will be able to understand the working for finite field analogue of special functions. The content of this paper is motivated with the work of J. Greene, D. Stanton, R. Evans, A. Helversen-Pasotto, K. Ireland, M. Rosen, N. Koblitz, W. Li and J. Soto-Andrade. For proper understanding, we suggest that the reader review Appendix C. To elaborate more on this topic, see [54, 66–68, 83, 109].

5.1 Quadratic residue and Legendre symbol

Let u and m be any two co-prime integers. Then u is called a quadratic residue if there exists a number v such that

$$v^2 \equiv u \pmod{m}.$$

Otherwise it is called quadratic nonresidue modulo m. For example, $3^2 = 9$ is congruent to 2 modulo 7. Hence 2 is a quadratic residue modulo 7. Similarly, one can see, 4 is a quadratic residue modulo 7 but 3 is a quadratic nonresidue modulo 7 because there does not exist any numnber v such that

$$v^2 \equiv 3 \pmod{7}.$$

The following proposition brief some conditions which can be used for deciding when a given integer is quadratic residue modulo m.

Proposition 154. *Let $m = 2^c p_1^{c_1} \cdots p_k^{c_k}$ be the prime decomposition of m, also assume u and m are co-prime. Then $v^2 \equiv u \pmod{m}$ has a solution if and only if following conditions are satisfied:*
1. *If $c = 2$, then $u \equiv 1 \pmod{4}$;*
2. *If $c \geq 3$, then $u \equiv 1 \pmod{8}$;*
3. *$u^{\frac{p_i-1}{2}} \equiv 1 \pmod{p_i}$, for each $i = 1, \ldots, k$.*

The Legendre symbol is a function denoted by (u/p), where p is a prime and defined as

$$(u/p) = \begin{cases} 0, & \text{if } p \text{ divides } u, \\ 1, & \text{if } u \text{ is a quadratic residue modulo } p, \\ -1, & \text{if } u \text{ is a quadratic nonresidue modulo } p. \end{cases}$$

Lemma 155. *The Legendre symbol satisfies the following properties and identities:*

https://doi.org/10.1515/9783111324586-005

1. $(u\,v/p) = (u/p)\,(v/p)$;
2. $(-1/p) = (-1)^{\frac{p-1}{2}}$,
   $(2/p) = (-1)^{\frac{p^2-1}{8}}$,
   $$(-3/p) = \begin{cases} 1, & \text{if } p \equiv 1 \pmod 6, \\ -1, & \text{if } p \equiv 5 \pmod 6; \end{cases}$$
3. $(u/p) = (v/p)$, *if* $u \equiv v \pmod p$;
4. $(u^2/p) = 1$, *if* $u \not\equiv 0 \pmod 6$;
5. $(u^2\,v/p) = (v/p)$, *if* $u \not\equiv 0 \pmod 6$, $v \in \mathbb{Z}$.

Lemma 156. *If* (u/p) *denotes the Legendre symbol, then*

$$\sum_{u=0}^{p-1} (u/p) = 0.$$

Proof. By the definition of the Legendre symbol, $(0/p) = 0$. Now, for remaining $p-1$ terms, half of the terms having value +1 and remaining have −1. On adding all of these together yields $\sum_{u=0}^{p-1}(u/p) = 0$. □

5.2 Multiplicative character

A multiplicative character on a finite field is a map ϕ from $\mathbb{F}_p^*$ to the set of nonzero complex numbers such that

$$\phi(u\,v) = \phi(u)\,\phi(v), \quad \text{for all } u, v \in \mathbb{F}_p^*. \tag{5.1}$$

As an example, the Legendre symbol (u/p) is of such character if it is considered as a function of the coset of u modulo p.

The trivial multiplicative character ϵ is a map whose image is 1 for every element u of $\mathbb{F}_p^*$, i. e.,

$$\epsilon(u) = 1, \quad \text{for all } u \in \mathbb{F}_p^*.$$

The above definition of multiplicative character can be extended for all $u \in \mathbb{F}_p$ by defining

$$\phi(0) = \begin{cases} 0, & \text{if } \phi \neq \epsilon, \\ 1, & \text{if } \phi = \epsilon. \end{cases}$$

In the following proposition, we present several properties of a multiplicative character.

Proposition 157. *Let* ϕ *be a multiplicative character and* $u \in \mathbb{F}_p^*$. *Then:*
1. $\phi(1) = 1$;

2. *$\phi(u)$ is a $(p-1)^{st}$ root of unity;*
3. $\phi(u^{-1}) = [\phi(u)]^{-1} = \overline{\phi(u)}$.

Proof. By the definition of multiplicative character,

$$\phi(1) = \phi(1 \cdot 1) = \phi(1)\,\phi(1) \implies \phi(1) = 1, \quad \text{because } \phi(1) \neq 0.$$

Again, using the definition of multiplicative character and the fact that $u^{p-1} = 1$, for all $u \in \mathbb{F}_p^*$, we get

$$1 = \phi(1) = \phi(u^{p-1}) = (\phi(u))^{p-1}.$$

Thus, $\phi(u)$ is a $(p-1)^{st}$ root of unity.

The set $\mathbb{F}_p^*$ is a collection of nonzero elements, which are invertible. Hence, there exists an element $u \in \mathbb{F}_p^*$ such that

$$1 = \phi(u\,u^{-1}) = \phi(u)\,\phi(u^{-1}),$$

which gives

$$\phi(u^{-1}) = \frac{1}{\phi(u)} = [\phi(u)]^{-1}. \tag{5.2}$$

Also by part two, $\phi(u)$ is a complex number such that $[\phi(u)]^{p-1} = 1$. It follows that the absolute value $|\phi(u)| = 1$. We say

$$|\phi(u)|^2 = 1 \implies \phi(u)\overline{\phi(u)} = 1 \implies \overline{\phi(u)} = [\phi(u)]^{-1}. \tag{5.3}$$

From (5.2) and (5.3), we have

$$\phi(u^{-1}) = [\phi(u)]^{-1} = \overline{\phi(u)}.$$

It completes the proof. □

Proposition 158. *Let ϕ be a multiplicative character. Then*

$$\sum_{u \in \mathbb{F}_p} \phi(u) = \begin{cases} 0, & \text{if } \phi \neq \epsilon, \\ p, & \text{if } \phi = \epsilon. \end{cases} \tag{5.4}$$

Proof. If $\phi = \epsilon$, then by definition of trivial character $\epsilon(u) = 1$, for all $u \in \mathbb{F}_p^*$, the proof is obvious. If $\phi \neq \epsilon$, then there exists an element $b \in \mathbb{F}_p^*$ such that $\phi(b) \neq 1$. Consider $U = \sum_{u \in \mathbb{F}_p^*} \phi(u)$. Then

$$\phi(b)\,U = \sum_{u \in \mathbb{F}_p^*} \phi(b)\,\phi(u) = \sum_{u \in \mathbb{F}_p^*} \phi(b\,u) = U.$$

Thus, we obtain $\phi(b)\,U = U$ and $\phi(b) \neq 1$. So,

$$U = \sum_{u \in \mathbb{F}_p^*} \phi(u) = 0.$$

It completes the proof. □

The set of multiplicative characters forms a group with respect to the following definitions:

1. If ϕ and ψ are multiplicative characters, then $\phi \cdot \psi$ is the map that takes $u \in \mathbb{F}_p^*$ to $\phi(u) \cdot \psi(u)$.
2. If ϕ is a multiplicative character, then ϕ^{-1} is the map that takes $u \in \mathbb{F}_p^*$ to $[\phi(u)]^{-1}$.

Further, the group of characters is a cyclic group of order $p-1$. We list several other elementary results related to multiplicative character that is needed in the chapter.

1. If $u \in \mathbb{F}_p^*$ and $u \neq 1$, then there is a character ϕ such that $\phi(u) \neq 1$.
2. If $u \in \mathbb{F}_p^*$ and $u \neq 1$, then $\sum_{\phi \in \mathbb{F}_p^*} \phi(u) = 0$, where the summation is over all multiplicative characters.
3. If $u \in \mathbb{F}_p^*$, $k/p-1$ and $x^k = u$ is not solvable, then there is a multiplicative character ϕ such that

$$\text{(i) } \phi^k = \epsilon, \quad \text{(ii) } \phi(u) \neq 1.$$

4. For $u \in \mathbb{F}_p$, let $K(x^k = u)$ be the number of solutions of the equation $x^k = u$, then $K = \sum_{\phi^k = \epsilon} \phi(u)$.

5.3 Gauss sum

Definition 159. If ζ is the primitive p^{th} root of unity, then quadratic Gauss sum is defined by

$$g_b = \sum_{u=0}^{p-1} (u/p)\,\zeta^{bu}. \tag{5.5}$$

Proposition 160. *The quadratic Gauss sum g_b obeys the following relation:*

$$g_b = (b/p) \cdot g_1. \tag{5.6}$$

Proof. Suppose $b \equiv 0 \pmod{p}$, then $\zeta^{bu} = 1$ for all u. It follows by Lemma 156 that

$$g_b = \sum_{u=0}^{p-1} (u/p) = 0 \tag{5.7}$$

and the proof of (5.6) is obvious. Now, consider that $b \not\equiv 0 \pmod{p}$. Then

$$(b/p)\, g_b = \sum_{u=0}^{p-1} (b/p)\,(u/p)\zeta^{bu} = \sum_{u=0}^{p-1} (b\,u/p)\zeta^{bu} = \sum_{t} (t/p)\zeta^{t} = g_1. \tag{5.8}$$

Now, multiplying the equation by (b/p) and using $(b^2/p) = 1$, $b \not\equiv 0 \pmod{p}$, we get the required result. □

Squaring both sides of the equation (5.6) yield

$$(g_b)^2 = (b/p)^2\, g_1^2 = g_1^2, \quad b \not\equiv 0 \pmod{p}. \tag{5.9}$$

Thus, we have the following proposition.

Proposition 161. *The following result holds:*

$$g_1^2 = g^2 = (-1)^{\frac{p-1}{2}}\, p. \tag{5.10}$$

Proposition 162. *Let ζ be the primitive p^{th} root of unity, i. e., $\zeta = e^{\frac{2\pi\iota}{p}}$. Then*

$$\sum_{u=0}^{p-1} \zeta^{b\,u} = \begin{cases} p, & \textit{if } b \equiv 0 \pmod{p}, \\ 0, & \textit{otherwise}. \end{cases} \tag{5.11}$$

Proof. The proof is simple when $b \equiv 0 \pmod{p}$, then $\zeta^b = 1$, and thus $\sum_{u=0}^{p-1} \zeta^{b\,u} = p$. If $b \not\equiv 0 \pmod{p}$, then $\zeta^b \neq 1$ and

$$\sum_{u=0}^{p-1} \zeta^{b\,u} = \frac{(\zeta^{b\,p} - 1)}{(\zeta^b - 1)} = 0, \quad \because \zeta^p = 1. \tag{5.12}$$

It completes the proof. □

An immediate corollary can be obtained as the following.

Corollary 163. *The following relation follows:*

$$p^{-1} \sum_{u=0}^{p-1} \zeta^{(a-b)\,u} = \delta(a,b), \tag{5.13}$$

where

$$\delta(a,b) = \begin{cases} 1, & \textit{if } a \equiv b \pmod{p}, \\ 0, & \textit{if } a \not\equiv b \pmod{p}. \end{cases} \tag{5.14}$$

Proof. The definition of $\delta(a,b)$ is equivalent to

$$\delta(a,b) = \begin{cases} 1, & \text{if } a-b \equiv 0 \pmod p, \\ 0, & \text{if } a-b \not\equiv 0 \pmod p. \end{cases}$$

Let $a-b=c$ and rewrite the problem

$$\sum_{u=0}^{p-1} \zeta^{(a-b)\,u} = p\,\delta(c) = \begin{cases} p, & \text{if } c \equiv 0 \pmod p, \\ 0, & \text{if } c \not\equiv 0 \pmod p), \end{cases}$$

which coincide with the problem given in Proposition 162 and can be proved in the similar manner. □

Gauss sum generalizes the Gauss quadratic sum.

Definition 164. Let ϕ be a multiplicative character on a finite field $\mathbb{F}_p$ and $b \in \mathbb{F}_p$. Then $\sum_{u\in\mathbb{F}_p} \phi(u)\,\zeta^{b\,u}$ is called the Gauss sum on $\mathbb{F}_p$ with respect to multiplicative character ϕ. It is denoted by $g_b(\phi)$.

Let us see some fundamental properties obeyed by the Gauss sum.

Proposition 165. *If b is nonzero quantity and ϕ is nontrivial, then we have $g_b(\phi) = \phi(b^{-1})\,g_1(\phi)$. In case b is nonzero but ϕ is trivial, then $g_b(\epsilon) = 0$. In another case, if $b = 0$ and $\phi \neq \epsilon$, then $g_0(\phi) = 0$. If b is zero and ϕ is trivial, then $g_0(\epsilon) = 0$.*

Proof. To prove the first assertion, we have assumed that $b \neq 0, \phi \neq \epsilon$, so we have

$$\phi(b)\,g_b(\phi) = \phi(b) \sum_{u\in\mathbb{F}_p} \phi(u)\,\zeta^{b\,u} = \sum_{u\in\mathbb{F}_p} \phi(b\,u)\,\zeta^{b\,u} = g_1(\phi). \tag{5.15}$$

This implies

$$g_b(\phi) = [\phi(b)]^{-1}\,g_1(\phi) = \phi(b^{-1})\,g_1(\phi). \tag{5.16}$$

If $b \neq 0$ and $\phi = \epsilon$, then

$$g_b(\epsilon) = \sum_{u\in\mathbb{F}_p} \epsilon(u)\,\zeta^{b\,u} = \sum_{u\in\mathbb{F}_p} \zeta^{b\,u} = 0. \tag{5.17}$$

If $b = 0$ and $\phi \neq \epsilon$, then

$$g_0(\phi) = \sum_{u\in\mathbb{F}_p} \phi(u) = 0, \quad \text{by Proposition 158.} \tag{5.18}$$

If $b = 0$ and $\phi = \epsilon$, then again from Proposition 158,

$$g_0(\epsilon) = \sum_{u\in\mathbb{F}_p} \epsilon(u) = p. \tag{5.19}$$

It completes the proof. □

Proposition 166. *If $\phi \neq \epsilon$, then $|g| = |g_1(\phi)| = \sqrt{p}$.*

Proof. The proof is based on the idea that we evaluate the sum $\sum_{b\in\mathbb{F}_p}^{p-1} g_b(\phi)\,\overline{g_b(\phi)}$ in two ways and equate them. If b is nonzero, then

$$g_b(\phi) = \phi(b^{-1})\,g(\phi) \quad \text{and} \quad \overline{g_b(\phi)} = \overline{\phi(b^{-1})\,g(\phi)} = \phi(b)\,\overline{g(\phi)}. \tag{5.20}$$

Thus,

$$\begin{aligned} g_b(\phi)\,\overline{g_b(\phi)} &= \phi(b^{-1})\,g(\phi)\,\phi(b)\,\overline{g(\phi)} \\ &= [\phi(b)]^{-1}\,\phi(b)\,g(\phi)\,\overline{g(\phi)} \\ &= |g(\phi)|^2. \end{aligned} \tag{5.21}$$

Since $g_0(\phi) = 0$, so we get

$$\sum_{b\in\mathbb{F}_p}^{p-1} g_b(\phi)\,\overline{g_b(\phi)} = (p-1)\,|g(\phi)|^2. \tag{5.22}$$

On the other hand,

$$g_b(\phi)\,\overline{g_b(\phi)} = \sum_{u\in\mathbb{F}_p}^{p-1}\sum_{v\in\mathbb{F}_p}^{p-1} \phi(u)\,\overline{\phi(v)}\,\zeta^{bu-bv}. \tag{5.23}$$

This yields

$$\begin{aligned} \sum_{b\in\mathbb{F}_p} g_b(\phi)\,\overline{g_b(\phi)} &= \sum_{u\in\mathbb{F}_p}^{p-1}\sum_{v\in\mathbb{F}_p}^{p-1} \phi(u)\,\overline{\phi(v)} \sum_{b\in\mathbb{F}_p} \zeta^{bu-bv} \\ &= \sum_{u\in\mathbb{F}_p}^{p-1}\sum_{v\in\mathbb{F}_p}^{p-1} \phi(u)\,\overline{\phi(v)}\,\delta(u,v)\,p \\ &= (p-1)\,p. \end{aligned} \tag{5.24}$$

Because $\delta(u,v) = 1$, when $u = v$, otherwise it is zero. Equating (5.22) and (5.24) give $|g(\phi)| = \sqrt{p}$. It completes the proof. □

This proposition and fact that $\overline{\phi(-1)} = \phi(-1)$ together yield

$$\overline{g(\phi)} = \sum_{u\in\mathbb{F}_p} \overline{\phi(u)}\,\zeta^{-u} = \phi(-1)\sum_u \overline{\phi(-u)}\,\zeta^{-1} = \phi(-1)\,g(\overline{\phi}). \tag{5.25}$$

Thus, $|g(\phi)|^2 = p \implies g(\phi)\,g(\overline{\phi}) = \phi(-1)\,p$.

5.4 Jacobi sum

For multiplicative characters ϕ and ψ defined on $\mathbb{F}_p^*$, the Jacobi sum is given by

$$J(\phi,\psi) = \sum_{x\in\mathbb{F}_p} \phi(x)\,\psi(1-x) = \sum_{x+y=1} \phi(x)\,\psi(y). \tag{5.26}$$

Proposition 167. *Let ϕ and ψ be nontrivial multiplicative character. Then the Jacobi sum satisfies the following properties:*

1. $J(\epsilon,\epsilon) = p$;
2. $J(\phi,\psi) = J(\psi,\phi)$;
3. $J(\epsilon,\phi) = 0$;
4. $J(\phi,\phi^{-1}) = -\phi(-1)$;
5. *Further, if* $\phi\,\psi \neq \epsilon$, *then* $J(\phi,\psi) = \frac{G(\phi)\,G(\psi)}{G(\phi,\psi)}$.

Proof. By definition,

$$J(\epsilon,\epsilon) = \sum_{x\in\mathbb{F}_p} \epsilon(x)\,\epsilon(1-x) = \sum_{x\in\mathbb{F}_p} (1) = p.$$

This is our first assertion. The second is followed by replacing x *b* y $(1-y)$ in the definition $J(\phi,\psi) = \sum_{x\in\mathbb{F}_p} \phi(x)\,\psi(1-x)$. Thus,

$$J(\phi,\psi) = \sum_y \phi(1-y)\,\psi(y) = J(\psi,\phi).$$

For third one, we have

$$J(\epsilon,\phi) = \sum_{x\in\mathbb{F}_p} \epsilon(x)\,\phi(1-x) = \sum_y \phi(y) = 0.$$

To prove the fourth assertion, we write

$$J(\phi,\phi^{-1}) = \sum_{x\neq 1} \phi(x)\,\phi^{-1}(1-x) = \sum_{x\neq 1} \phi(x)\,\phi\left(\frac{1}{1-x}\right) = \sum_{x\neq 1} \phi\left(\frac{x}{1-x}\right).$$

Let $\frac{x}{1-x} = y \implies x = \frac{y}{1+y}$. Then

$$J(\phi,\phi^{-1}) = \sum_{y\neq -1} \phi(y) = -\phi(-1).$$

For the fifth assertion, notice that

$$\begin{aligned} g(\phi)\,g(\psi) &= \Big(\sum_x \phi(x)\,\zeta^x\Big)\Big(\sum_y \psi(y)\,\zeta^y\Big) \\ &= \sum_{x,y} \phi(x)\,\psi(y)\,\zeta^{x+y} \\ &= \sum_u \Big(\sum_{x+y=u} \phi(x)\,\psi(y)\Big)\,\zeta^u. \end{aligned} \tag{5.27}$$

If $u = 0$, then

$$\sum_{x+y=0} \phi(x)\,\psi(y) = \sum_x \phi(x)\,\psi(-x) = \psi(-1)\sum_x \phi(x)\,\psi(x) = \psi(-1)\sum_x \phi\psi(x).$$

Since $\phi\,\psi$ is nontrivial, so by Proposition 158,

$$\sum_{x+y=0} \phi(x)\,\psi(y) = 0.$$

If $u \neq 0$, then replacing x by $u\,x'$ and y by $u\,y'$ implies

$$\begin{aligned} \sum_{x+y=u} \phi(x)\,\psi(y) &= \sum_{x'+y'=1} \phi(u\,x')\,\psi(u\,y') \\ &= \sum_{x'+y'=1} \phi\,\psi(u)\,\phi(x')\,\phi(y') \\ &= \phi\,\psi(u) \sum_{x'+y'=1} \phi(x')\,\phi(y') = \phi\,\psi(u)\,J(\phi,\psi). \end{aligned}$$

Substituting this value in (5.27) leads

$$\begin{aligned} g(\phi)\,g(\psi) &= \sum_u \phi\,\psi(u)\,J(\phi,\psi)\,\zeta^u \\ &= J(\phi,\psi)\sum_u \phi\,\psi(u)\,\zeta^u = J(\phi,\psi)\,g(\phi\,\psi). \end{aligned}$$

It completes the proof. □

5.5 Gaussian analogue of binomial theorem

The binomial theorem states that

$$(1+t)^a = \sum_{k=0}^{\infty} \binom{a}{k} t^k,$$

where $\binom{a}{k}$ is the "binomial coefficient" defined by

$$\binom{a}{k} = \frac{a(a-1)\cdots(a-k+1)}{k!} = \frac{\Gamma(a+1)}{\Gamma(k+1)\,\Gamma(a-k+1)}.$$

A character sum analogue of the binomial theorem can be obtained easily.

Theorem 168. *For any character A of $\mathbb{F}_p$ and $u \in \mathbb{F}_p$,*

$$A(1+u) = \delta(u) + \frac{1}{p-1}\sum_{\phi\in\mathbb{F}_p} J(A,\overline{\phi})\,\phi(-u). \tag{5.28}$$

Proof. Let f be a function defined by $f : \mathbb{F}_p \to \mathbb{C}$. Then the finite field analogue for the power series expansion of f has a unique representation,

$$f(t) = f_\delta\,\delta(t) + \sum_{\phi\in\mathbb{F}_p} f_\phi\,\phi(t), \tag{5.29}$$

where the constants f_δ, f_ϕ are given by

$$f_\delta = f(0), \quad f_\phi = \frac{1}{p-1}\sum_{t\in\mathbb{F}_p} f(t)\,\overline{\phi}(t)$$

and δ is the function

$$\delta(t) = \begin{cases} 1, & \text{if } t = 0, \\ 0, & \text{if } t \neq 0. \end{cases}$$

Consider $f(t) = A(1+t)$, for a character A on $\mathbb{F}_p$. Then we have

$$f_\phi = \frac{1}{p-1}\sum_{t\in\mathbb{F}_p} A(1+t)\,\overline{\phi}(t). \tag{5.30}$$

Changing variable from $1+u \to t$ and using the definition of Jacobi sum, we get

$$A(1+u) = \delta(u) + \frac{1}{p-1}\sum_{\phi\in\mathbb{F}_p} J(A,\overline{\phi})\,\phi(-u). \tag{5.31}$$

Now, substituting this value in (5.30) and using (5.29) gives the required identity. It completes the proof. □

From this theorem, we conclude that the finite field analogue of binomial coefficient is the Jacobi sum. For characters ϕ and ψ of $\mathbb{F}_p$, the binomial coefficient is denoted by $\binom{\phi}{\psi}$ and defined as

$$\binom{\phi}{\psi} = \frac{\psi(-1)}{q} J(\phi,\overline{\psi}).$$

The finite field analogue of the binomial theorem can be redefined now as

$$A(1+u) = \delta(u) + \frac{q}{q-1}\sum_{\phi}\binom{A}{\phi}\phi(u). \tag{5.32}$$

Some useful properties of the binomial coefficient in a finite field case are as follows.

Theorem 169.

$$\binom{\phi}{\psi} = \binom{\phi}{\phi\overline{\psi}}; \tag{5.33}$$

$$\binom{\phi}{\psi} = \binom{\psi\overline{\phi}}{\psi}\psi(-1); \tag{5.34}$$

$$\binom{\phi}{\psi} = \binom{\overline{\psi}}{\overline{\phi}}\phi\,\psi(-1); \tag{5.35}$$

$$\binom{\phi}{\psi} = \frac{G(\phi)\,G(\overline{\psi})}{G(\phi\,\overline{\psi})\,q} + \frac{(q-1)}{q}\psi(-1)\,\delta(\phi\,\overline{\psi}) \tag{5.36}$$

$$= \frac{G(\phi)}{G(\phi)\,G(\phi\,\overline{\psi})} + \frac{(q-1)}{q}\delta(\psi) + \frac{(q-1)}{q}\delta(\phi\,\overline{\psi}). \tag{5.37}$$

With the help of this binomial coefficient, other variants of the binomial theorem are as follows.

Theorem 170.

$$\overline{A}(1-t) = \delta(t) + \frac{q}{q-1}\sum_{\phi}\binom{A\,\phi}{\phi}\phi(t) \tag{5.38}$$

and

$$\overline{B}(t)\,\overline{AB}(1-t) = \frac{q}{q-1}\sum_{\phi}\binom{A\,\phi}{B\,\phi}\phi(t), \quad t \neq 0. \tag{5.39}$$

Also, a very important formula in terms of the binomial coefficient is

$$\binom{\phi^2}{\phi\,\psi}\binom{\chi}{\psi} = \binom{\phi}{\psi}\binom{\chi\,\phi}{\phi\,\psi}\phi(4). \tag{5.40}$$

5.6 Hypergeometric series over finite field

Throughout this section, $GF(q)$ will be denoting the field with q elements, where $q = p^s$ for some odd prime p. A multiplicative character ϕ on $GF(q)^*$ is defined by

$$\phi : GF(q)^* \to \mathbb{C}^*.$$

The collection of all such characters will be denoted by $\widehat{GF(q)^*}$ and the inverse of a multiplicative character ϕ by $\overline{\phi}$. The following lemma is a consequence of Proposition 158.

Lemma 171. *Let ϕ be a multiplicative character. Then*

$$\sum_{u\in GF(q)} \phi(u) = \begin{cases} q-1, & \text{if } \phi = \epsilon, \\ 0, & \text{if } \phi \neq \epsilon. \end{cases} \tag{5.41}$$

Further,

$$\sum_{\phi\in \widehat{GF(q)^*}} \phi(u) = \begin{cases} q-1, & \text{if } u = 1, \\ 0, & \text{if } u \neq 1. \end{cases} \tag{5.42}$$

Let $\xi_p = e^{2\pi i/p}$ be the fixed primitive p^{th} root and map $\mathrm{Tr} : GF(q) \to GF(p)$ is defined by

$$\mathrm{Tr}(x) = x + x^p + \cdots + x^{p^{r-1}}, \quad x \in GF(q).$$

Then the Gauss sum of character ϕ on $GF(q)$ is defined as follows:

$$G(\phi) = \sum_{x\in GF(q)} \phi(x)\xi_p^{\mathrm{tr}(x)}.$$

More general, if τ is the additive character $\tau : GF(q) \to \mathbb{C}$ and defined by

$$\tau(x) = \xi_p^{\mathrm{tr}(x)}.$$

Then the Gauss sum is defined by

$$G(\phi;\tau) = \sum_{x\in GF(q)} \phi(x)\tau(ax), \quad a \neq 0 \in GF(q).$$

In particular,

$$G(\phi;\tau) = \sum_{x\in GF(q)} \phi(x)\tau(x).$$

Recall the integral definition of gamma and beta function as

$$\Gamma(u) = \int_0^\infty t^{u-1}e^{-t}dt, \quad \Re(u) > 0$$

and

$$\beta(u,v) = \int_0^1 t^{u-1}(1-t)^{v-1}\,dt,$$

respectively. The relation between these two functions is given by

$$\beta(u,v) = \frac{\Gamma(u)\,\Gamma(v)}{\Gamma(u+v)}.$$

Now consider the integral $\int dt$ as the sum in the interval $(0,1)$. The multiplicative characters ϕ and ψ are defined by $\phi : t \to t^{u-1}$, $\psi : t \to t^{v-1}$ and visualize ϕ in place of u; ψ in place of v. Also, consider e^{-t} as the additive character from $[0,\infty)$ to $[0,1]$. Then one can clearly visualize that the Gauss sum

$$G(\phi) = \sum_{t\in GF(q)} \phi(t)\,\xi_p^{\mathrm{tr}(t)}$$

is the character sum analogue of the gamma function and Jacobi sum, and

$$J(\phi,\psi) = \sum_{t\in GF(q)} \phi(t)\,\psi(1-t)$$

is the Gauss sum analogue of the beta function.

The Gauss hypergeometric series is defined by

$$_2F_1(u,v;w;z) = \sum_{k=0}^{\infty} \frac{(u)_k\,(v)_k}{(w)_k\,k!} z^k, \tag{5.43}$$

where

$$(u)_k = \frac{\Gamma(u+k)}{\Gamma(u)} = u(u+1)\cdots(u+k-1)$$

The most elementary attempt to get a character sum analogue of the Gauss hypergeometric series is using $\frac{G(\phi\chi)}{G(\phi)}$ as an analogue of $\frac{\Gamma(u+k)}{\Gamma(u)}$ is

$$\frac{G(\xi)}{G(\phi)\,G(\psi)} \sum_{\chi} \frac{G(\phi\chi)\,G(\psi\chi)}{G(\xi\chi)} \chi(z). \tag{5.44}$$

Unfortunately, this approach leads to rather poor results. An alternative way for a better result is to find the analogue of the integral representation for the hypergeometric series,

$$_2F_1(a,b;c;z) = \frac{\Gamma(c)}{\Gamma(b)\Gamma(c-b)} \int_0^1 t^{b-1}(1-t)^{c-b-1}\,(1-zt)^{-a}\,dt. \tag{5.45}$$

Definition 172. For characters ϕ, ψ and ξ of $GF(q)$ and $t \in GF(q)$, the character sum analogue of (5.45) is given by

$$ {}_2F_1(\phi, \psi; \xi; t) = \epsilon(t) \frac{\psi\xi(-1)}{q} \sum_u \psi(u) \overline{\psi}\xi(1-u) \overline{\phi}(1-tu). \tag{5.46} $$

To obtain simpler results, the normalizing factor $\frac{\Gamma(c)}{\Gamma(b)\Gamma(c-b)}$ is dropped and another factor $\epsilon(t) \frac{\psi\xi(-1)}{q}$ is chosen.

Theorem 173. *For characters ϕ, ψ and ξ of $GF(q)$ and $t \in GF(q)$, the following identity holds:*

$$ {}_2F_1(\phi, \psi; \xi; t) = \frac{q}{q-1} \sum_\chi \binom{\phi\chi}{\chi} \binom{\psi\chi}{\xi\chi} \chi(t). \tag{5.47} $$

Proof. Using equation (5.38), we have

$$ \overline{A}(1-tu) = \delta(tu) + \frac{q}{q-1} \sum_\chi \binom{A\chi}{\chi} \chi(tu). \tag{5.48} $$

Since $\epsilon(t)\,\psi(u)\,\delta(tu) = 0$ for all t, u, we have

$$ \begin{aligned} {}_2F_1(\phi, \psi; \xi; t) &= \frac{\psi\xi(-1)}{q-1} \sum_{u,\chi} \binom{\phi\chi}{\chi} \chi(t)\, \psi\chi(u)\, \overline{\psi}\xi(1-u)\, \overline{\phi}(1-tu) \\ &= \frac{\psi\xi(-1)}{q-1} \sum_{u,\chi} \binom{\phi\chi}{\chi} J(\psi\chi, \overline{\psi}\xi)\, \chi(t) \\ &= \frac{q}{q-1} \sum_\chi \binom{\phi\chi}{\chi} \binom{\psi\chi}{\psi\overline{\xi}} \chi(t). \end{aligned} \tag{5.49} $$

Since

$$ \binom{\psi\chi}{\psi\overline{\xi}} = \binom{\psi\chi}{\xi\chi}. $$

Thus, we have

$$ {}_2F_1(\phi, \psi; \xi; t) = \frac{q}{q-1} \sum_\chi \binom{\phi\chi}{\chi} \binom{\psi\chi}{\xi\chi} \chi(t). \tag{5.50} $$

It completes the proof. □

This is the analogue of identity, where the classical hypergeometric series is expressed in terms of binomial coefficients as

$$ {}_2F_1(u, v; w; z) = \binom{v-1}{w-1}^{-1} \sum_{k=0}^{\infty} \binom{u+k-1}{k} \binom{v+k-1}{w+k-1} z^k. $$

In the following theorem, we give the particular forms of Gauss function in terms of binomial coefficients without proofs.

Theorem 174. *For characters ϕ, ψ and ξ of $GF(q)$ and $t \in GF(q)$, we have*

$$ {}_2F_1(\epsilon, \psi; \xi; t) = \binom{\psi}{\xi} \epsilon(t) - \frac{1}{q} \overline{\xi}(t) \overline{\psi} \xi(1-t), \tag{5.51}$$

$$\begin{aligned} {}_2F_1(\phi, \epsilon; \xi; t) = {} & \binom{\xi}{\phi} \phi(-1) \overline{\xi}(t) \overline{\phi} \xi(1-t) \epsilon(t) - \frac{1}{q} \xi(-1) \epsilon(t) \\ & + \frac{q-1}{q} \phi(-1) \delta(1-t) \delta(\overline{\phi} \xi), \end{aligned} \tag{5.52}$$

$$ {}_2F_1(\phi, \psi; \phi; t) = \binom{\psi}{\phi} \epsilon(t) \overline{\psi}(1-t) - \frac{1}{q} \overline{\phi}(-t) + \frac{q-1}{q} \phi(-1) \delta(1-t) \delta(\psi), \tag{5.53}$$

$$ {}_2F_1(\phi, \psi; \psi; t) = \binom{\psi}{\phi} \phi(-1) \overline{\psi}(t) - \frac{1}{q} \epsilon(t) \overline{\phi}(1-t). \tag{5.54}$$

Theorem 175. *For characters ϕ, ψ, ξ and ϵ of $GF(q)$ and $t \in GF(q)$, we have*

$$ {}_2F_1(\phi, \psi; \xi; t) = \phi(-1)\, {}_2F_1(\phi, \psi; \phi \psi \overline{\xi}; 1-t) + \phi(-1) \binom{\psi}{\overline{\phi} \xi} \delta(1-t) - \binom{\psi}{\xi} \delta(t), \tag{5.55}$$

$$ {}_2F_1(\phi, \psi; \xi; t) = \xi(-1) \overline{\phi}(1-t)\, {}_2F_1\left(\phi, \xi \overline{\psi}; \xi; \frac{t}{t-1} \right) + \phi(-1) \binom{\psi}{\overline{\phi} \xi} \delta(1-t), \tag{5.56}$$

$$ {}_2F_1(\phi, \psi; \xi; t) = \overline{\phi}(1-t)\, {}_2F_1\left(\xi \overline{\phi}, \psi; \xi; \frac{t}{t-1} \right) + \phi(-1) \binom{\psi}{\overline{\phi} \xi} \delta(1-t), \tag{5.57}$$

$$ {}_2F_1(\phi, \psi; \xi; t) = \xi(-1) \xi \overline{\phi \psi}(1-t)\, {}_2F_1(\xi \overline{\phi}, \xi \overline{\psi}; \xi; t) + \phi(-1) \binom{\psi}{\overline{\phi} \xi} \delta(1-t). \tag{5.58}$$

Proof. We change variables in the definition,

$$ {}_2F_1(\phi, \psi; \xi; t) = \epsilon(t) \frac{\psi \xi(-1)}{q} \sum_u \psi(u) \overline{\psi} \xi(1-u) \overline{\phi}(1-tu). \tag{5.59}$$

For (5.55), we put $u = \frac{y}{y-1}$. Then we have $1-u = \frac{1}{1-y}$ and $1-tu = \frac{(1-(1-t)y)}{1-y}$. Thus,

$$\begin{aligned} {}_2F_1(\phi, \psi; \xi; t) &= \epsilon(t) \frac{\psi \xi(-1)}{q} \sum_y \psi(u) \overline{\psi}(y-1) \psi \overline{\xi}(1-y) \overline{\phi}(1-y) \overline{\phi}(1-(1-t)y) \\ &= \epsilon(t) \phi(-1) \frac{\phi \xi(-1)}{q} \sum_y \psi(u) \phi \overline{\xi}(1-y) \overline{\phi}(1-(-t)y). \end{aligned} \tag{5.60}$$

The result now follows from the observation that $\epsilon(t) = \epsilon(1-t) + \delta(1-t) - \delta(t)$. For (5.56)–(5.58), make the changes of variables

$$u = 1 - y, \quad u = \frac{y}{1 - t + ty} \quad \text{and} \quad u = \frac{1 - y}{1 - ty},$$

respectively. □

In the above theorem, equations (5.56) and (5.57) are analogues of Pfaff's transformation

$${}_2F_1(u, v; w; z) = (1 - z)^{-u} {}_2F_1\left(u, w - v; w; \frac{z}{z - 1}\right).$$

Equation (5.58) is an analogue of Euler's transformation

$${}_2F_1(u, v; w; z) = (1 - z)^{w-u-v} {}_2F_1(w - u, w - v; w; z).$$

Equation (5.55) is an analogue of the transformation

$$\begin{aligned}{}_2F_1(u, v; 1 + u + v - w; 1 - z) = {} & \frac{\Gamma(1 + u + v - w)\,\Gamma(1 - w)}{\Gamma(1 + u - w)\,\Gamma(1 + v - w)} {}_2F_1(u, v; w; z) \\ & + \frac{\Gamma(1 + u + v - w)\,\Gamma(w - 1)}{\Gamma(u)\,\Gamma(v)} z^{1-w} \\ & \times {}_2F_1(1 + u - w, 1 + v - w; 2 - w; z).\end{aligned}$$

It may be pointed out that if one includes the trivial $u \to y$ then there are 24 changes of variables of the type used in the proof of Theorem 175. This gives rise to 23 transformations (24 if one includes the identity transformation). These 24 changes of variables and the resulting correspond to Kummer's 24 solutions. We listed a few; the others may be obtained by various compositions of the transformations.

Theorem 176. *For characters ϕ, ψ and ξ of $GF(q)$,*

$${}_2F_1(\phi, \psi; \xi; 1) = \phi(-1)\left(\frac{\psi}{\overline{\phi}\,\xi}\right). \tag{5.61}$$

Proof. The set $t = 1$ in (5.55) gives

$$\begin{aligned}{}_2F_1(\phi, \psi; \xi; 1) &= \phi(-1)\, {}_2F_1(\phi, \psi; \phi\,\psi\,\overline{\xi}; 0) + \phi(-1)\left(\frac{\psi}{\overline{\phi}\,\xi}\right)\delta(0) - \left(\frac{\psi}{\xi}\right)\delta(1) \\ &= \phi(-1)\left(\frac{\psi}{\overline{\phi}\,\xi}\right).\end{aligned} \tag{5.62}$$

It completes the proof. □

Alternatively, if we put $t = 1$ in the definition

$${}_2F_1(\phi, \psi; \xi; t) = \epsilon(t)\,\frac{\psi\,\xi(-1)}{q} \sum_u \psi(u)\,\overline{\psi}\,\xi(1 - u)\,\overline{\phi}(1 - tu), \tag{5.63}$$

we get

$$
\begin{aligned}
{}_2F_1(\phi,\psi;\xi;1) &= \frac{\psi\,\xi(-1)}{q}\sum_u \psi(u)\,\overline{\psi}\,\xi(1-u)\,\overline{\phi}(1-u)\\
&= \frac{\psi\,\xi(-1)}{q}J(\psi,\overline{\phi\,\psi\,\xi}) = \phi(-1)\begin{pmatrix}\psi\\ \overline{\phi}\,\xi\end{pmatrix}.
\end{aligned}\tag{5.64}
$$

The above relation is an analogue of Gauss's evaluation

$$
{}_2F_1(u,v;w;1) = \frac{\Gamma(w)\,\Gamma(w-u-v)}{\Gamma(w-u)\,\Gamma(w-v)}.
$$

From the definition of the Gauss hypergeometric series for a finite field version, the other summation theorem can be deduced. For example, if we set $t=-1$ we have

$$
{}_2F_1(\phi,\psi;\xi;-1) = \frac{\psi\,\xi(-1)}{q}\sum_u \psi(u)\,\overline{\psi}\,\xi(1-u)\,\overline{\phi}(1+u).
$$

If $\xi = \overline{\phi}\,\psi$, this becomes

$$
{}_2F_1(\phi,\psi;\overline{\phi}\,\psi;-1) = \frac{\phi(-1)}{q}\sum_u \psi(u)\,\overline{\phi}(1-u^2) \tag{5.65}
$$

$$
= \frac{\phi(-1)}{q}\sum_u \psi(u)\,\frac{q}{q-1}\sum_\chi \begin{pmatrix}\phi\chi\\ \chi\end{pmatrix}\chi(u^2) \tag{5.66}
$$

$$
= \frac{\phi(-1)}{q-1}\sum_\chi \begin{pmatrix}\phi\chi\\ \chi\end{pmatrix}\sum_u \psi\chi^2(u). \tag{5.67}
$$

This result is the analogue of Kummer's theorem,

$$
{}_2F_1(u,v;1+v-u;-1) = \frac{\Gamma(1+v-u)\,\Gamma(1+\frac{1}{2}v)}{\Gamma(1+v)\,\Gamma(1+\frac{1}{2}v-u)},
$$

The inner sum is zero unless $\psi\chi^2$ is trivial, so we are lead to

$$
{}_2F_1(\phi,\psi;\overline{\phi}\,\psi;-1) = \begin{cases} 0, & \text{if } \psi \text{ is not a square},\\ \binom{\xi}{\phi} + \binom{\Phi\xi}{\phi}, & \text{if } \psi = \xi^2. \end{cases}
$$

Using Theorem 175, we are lead to

$$
{}_2F_1(\phi,\psi;\phi^2;2) = \phi(-1)\begin{cases} 0, & \text{if } \psi \text{ is not a square},\\ \binom{\xi}{\phi} + \binom{\Phi\xi}{\phi}, & \text{if } \psi = \xi^2. \end{cases}
$$

$$
{}_2F_1\left(\phi,\overline{\phi};\overline{\phi}\psi;\frac{1}{2}\right) = \phi(-2)\begin{cases} 0, & \text{if } \psi \text{ is not a square},\\ \binom{\xi}{\phi} + \binom{\Phi\xi}{\phi}, & \text{if } \psi = \xi^2. \end{cases}
$$

A large number of transformations exist for the classical hypergeometric series. Similar results hold for hypergeometric series over $GF(q)$. For example, we have the following.

Theorem 177. *For characters ϕ and ψ of $GF(q)$ and $t \in GF(q)$, we have*

$$
\begin{aligned}
&{}_2F_1(\phi, \psi; \overline{\phi}\,\psi; t)\\
&\quad = \delta(1+t) \begin{cases} 0, & \text{if } \psi \text{ is not a square} \\ \binom{\xi}{\phi} + \binom{\Phi\xi}{\phi}, & \text{if } \psi = \xi^2 \end{cases}\\
&\qquad + \frac{\phi(-1)}{q}\,\psi\!\left(\frac{1+t}{t}\right) + \overline{\psi}(1+t)\,\frac{q}{q-1}\sum_{\chi}\binom{\psi\chi^2}{\chi}\binom{\psi\chi}{\overline{\phi}\,\psi\chi}\chi\!\left(\frac{t}{(1+t)^2}\right).
\end{aligned} \tag{5.68}
$$

Proof. Using the definition for ${}_2F_1$, we have

$$
\begin{aligned}
{}_2F_1(\phi, \psi; \overline{\phi}\,\psi; t) &= \epsilon(t)\,\frac{\phi(-1)}{q}\sum_{u}\psi(u)\,\overline{\phi}(1-u)\,\overline{\phi}(1-tu) && (5.69)\\
&= \epsilon(t)\,\frac{\phi(-1)}{q}\sum_{u}\psi(u)\,\overline{\phi}(1-u(1+t-tu)). && (5.70)
\end{aligned}
$$

By the binomial theorem,

$$
\overline{\phi}(1-u(1+t-tu)) = \delta(u(1+t-tu)) + \frac{q}{q-1}\sum_{\chi}\binom{\phi\chi}{\chi}\chi(u)\chi(1+t-tu).
$$

Using this in (5.70) and changing $u \to \frac{u}{t}$ gives

$$
{}_2F_1(\phi, \psi; \overline{\phi}\,\psi; t) = \frac{\phi(-1)}{q}\,\psi\!\left(\frac{1+t}{t}\right) + \frac{\phi(-1)}{q-1}\sum_{\chi,u}\binom{\phi\chi}{\chi}\overline{\psi\chi}(t)\,\psi\chi(u)\,\chi(1+t-u). \tag{5.71}
$$

Since $1 = \epsilon(1+t) + \delta(1+t)$, then we have

$$
\begin{aligned}
{}_2F_1(\phi, \psi; \overline{\phi}\,\psi; t) = {}& \frac{\phi(-1)}{q}\,\psi\!\left(\frac{1+t}{t}\right) + \delta(1+t)\,\frac{\phi(-1)}{q-1}\sum_{\chi,u}\binom{\phi\chi}{\chi}\psi(-1)\,\psi\chi^2(u)\\
&+ \epsilon(1+t)\,\frac{\phi(-1)}{q-1}\sum_{\chi,u}\binom{\phi\chi}{\chi}\overline{\psi\chi}(t)\,\psi\chi(u)\,\chi(1+t-u).
\end{aligned} \tag{5.72}
$$

The second term on the right-hand side was evaluated in (5.67). To evaluate the third term, make the change of variable $u \to (1+t)\,u$ to obtain

$$
\begin{aligned}
{}_2F_1(\phi, \psi; \overline{\phi}\,\psi; t) = {}& \frac{\phi(-1)}{q}\,\psi\!\left(\frac{1+t}{t}\right) + \delta(1+t)\begin{cases} 0, & \text{if } \psi \text{ is not a square} \\ \binom{\xi}{\phi} + \binom{\Phi\xi}{\phi}, & \text{if } \psi = \xi^2 \end{cases}\\
&+ \psi\!\left(\frac{1+t}{t}\right)\frac{\phi(-1)}{q-1}\sum_{\chi}\binom{\phi\chi}{\chi}\chi\!\left(\frac{(1+t)^2}{t}\right)\sum_{u}\psi\chi(u)\,\chi(1-u).
\end{aligned} \tag{5.73}
$$

The inner sum on this last term is

$$J(\psi\chi,\chi) = q\chi(-1)\binom{\psi\chi}{\overline{\chi}}.$$

Using this and making the change $\chi \to \overline{\psi\chi}$ gives the desired result. □

Theorem 178. *For characters ϕ and ψ of $GF(q)$ and $t \in GF(q)$, we have*

$$\begin{aligned}
&{}_2F_1(\phi,\psi^2;\overline{\phi}\psi^2;t) \\
&= \delta(1+t)\left[\binom{\psi}{\phi}+\binom{\Phi\psi}{\phi}\right] + \phi(-1)\frac{q-1}{q}\epsilon(t)\epsilon(1+t)\delta(\Phi\psi) \\
&+ \frac{q-1}{q^2}\binom{\Phi}{\Phi\psi}^{-1}\phi(-1)\psi^2\left(\frac{2}{1-t}\right)\epsilon(t)\epsilon(1+t)\delta(\phi\overline{\psi}) \\
&+ \binom{\psi}{\phi}\binom{\Phi}{\Phi\psi}^{-1}\phi(-1)\psi^2\left(\frac{2}{1+t}\right){}_2F_1\left(\Phi\psi,\psi;\overline{\phi}\psi^2;\frac{4t}{(1+t)^2}\right).
\end{aligned} \tag{5.74}$$

Proof. It follows from (5.40) that

$$\binom{\psi^2\chi^2}{\chi} = \binom{\Phi\psi\chi}{\chi}\binom{\psi\chi}{\psi^2\chi}\binom{\Phi}{\Phi\psi}^{-1}\psi\chi(4). \tag{5.75}$$

Consequently, if ψ is replaced by ψ^2 in Theorem 177, we obtain

$$\begin{aligned}
&{}_2F_1(\phi,\psi^2;\overline{\phi}\psi^2;t) \\
&= \delta(1+t)\left[\binom{\psi}{\phi}+\binom{\Phi\psi}{\phi}\right] + \frac{\phi(-1)}{q}\psi^2\left(\frac{1+t}{t}\right) + \overline{\psi}^2(1+t)\psi(4)\binom{\Phi}{\Phi\psi}^{-1} \\
&\times \frac{q}{q-1}\sum_{\chi}\binom{\Phi\psi\chi}{\chi}\binom{\psi\chi}{\psi^2\chi}\binom{\psi^2\chi}{\overline{\phi}\psi^2\chi}\chi\left(\frac{4t}{(1+t)^2}\right).
\end{aligned}$$

The character sum on the right-hand side is a degenerate ${}_3F_2$, which when reduced to ${}_2F_1$ gives rise to the desired result. □

5.7 Generalized hypergeometric series over a finite field

The generalized hypergeometric function is defined by

$${}_rF_s\left[\begin{matrix}a_1,a_2,\ldots,a_r\\ b_1,b_2,\ldots,b_s\end{matrix};t\right] = \sum_{k=0}^{\infty}\frac{(a_1)_k\cdots(a_r)_k}{(b_1)_k\cdots(b_s)_k}\frac{t^k}{k!}. \tag{5.76}$$

The first elementary attempt to define a character sum analogue of generalized hypergeometric function would be

$$\frac{G(B_1)\,G(B_2)\cdots G(B_s)}{G(A_1)\,G(A_2)\cdots G(A_r)}\sum_{\chi}\frac{G(A_1\chi)\,G(A_2\chi)\cdots G(A_r\chi)}{G(B_1\chi)\,G(B_2\chi)\cdots G(B_s\chi)}\chi(t). \tag{5.77}$$

As we have seen for a Gauss hypergeometric series that the present form is not appropriate for further investigations, and hence we give the definition of generalized hypergeometric function analogous to the Gauss hypergeometric function over the finite field.

Definition 179. For characters $A_0, A_1, \ldots, A_s, B_1, \ldots, B_s$ of $GF(q)$ and $t \in GF(q)$, the $GF(q)$ version for generalized hypergeometric function in a special case when $r = s+1$ is given by

$$ {}_{s+1}F_s\left[\begin{matrix}A_0, A_1, \ldots, A_s\\ B_1, B_2, \ldots, B_s\end{matrix}; t\right] = \frac{q}{q-1}\sum_{\chi}\binom{A_0\chi}{\chi}\binom{A_1\chi}{B_1\chi}\cdots\binom{A_s\chi}{B_s\chi}\chi(t). \tag{5.78}$$

The above definition is the analogue of the following expression of a generalized hypergeometric series in terms of binomial coefficients:

$$ {}_{s+1}F_s\left[\begin{matrix}a_1, a_2, \ldots, a_r\\ b_1, b_2, \ldots, b_s\end{matrix}; t\right] = C\sum_{k=0}^{\infty}\binom{a_0+k-1}{k}\binom{a_1+k-1}{b_1+k-1}\cdots\binom{a_s+k-1}{b_s+k-1}t^k,$$

where

$$C = \left\{\binom{a_1-1}{b_1-1}\cdots\binom{a_s-1}{b_s-1}\right\}^{-1}.$$

Again, we drop the factor C from Definition (5.78) to obtain a simpler result. The following inductive integral representation is held by the generalized hypergeometric function

$$\begin{aligned} &{}_{s+1}F_s\left[\begin{matrix}a_0, a_1, a_2, \ldots, a_s\\ b_1, b_2, \ldots, b_s\end{matrix}; t\right] \\ &= \frac{\Gamma(b_s)}{\Gamma(a_s)\Gamma(b_s-a_s)}\int_0^1 u^{a_s-1}(1-u)^{b_s-a_s-1}\,{}_sF_{s-1}\left[\begin{matrix}a_0, a_1, \ldots, a_{s-1}\\ b_1, \ldots, b_{s-1}\end{matrix}; ut\right]du. \end{aligned} \tag{5.79}$$

The analogue for this result also holds.

Theorem 180. *For characters $A_0, A_1, \ldots, A_s, B_1, \ldots, B_s$ of $GF(q)$ and $t \in GF(q)$, the following summation formula is true:*

$$\begin{aligned} &{}_{s+1}F_s\left[\begin{matrix}A_0, A_1, \ldots, A_s\\ B_1, B_2, \ldots, B_s\end{matrix}; t\right] \\ &= \frac{A_sB_s(-1)}{q}\sum_{u}{}_sF_{s-1}\left[\begin{matrix}A_0, A_1, \ldots, A_{s-1}\\ B_1, B_2, \ldots, B_{s-1}\end{matrix}; tu\right]A_s(u)\overline{A}_sB_s(1-u). \end{aligned} \tag{5.80}$$

Proof. We use

$$\binom{A_s\chi}{B_s\chi}=\binom{A_s\chi}{A_s\overline{B}_s}=\frac{A_sB_s(-1)}{q}J(A_s\chi,A_s\overline{B}_s)$$
$$=\frac{A_sB_s(-1)}{q}\sum_u A_s(u)\overline{A}_s\chi(u)\overline{A}_sB_s(1-u),$$

and we get

$$\begin{aligned}&{}_{s+1}F_s\left[\begin{matrix}A_0,A_1,\dots,A_s\\B_1,B_2,\dots,B_s\end{matrix};t\right]\\&=\frac{A_sB_s(-1)}{q}\sum_u\frac{q}{q-1}\sum_\chi\binom{A_0\chi}{\chi}\binom{A_1\chi}{B_1\chi}\cdots\binom{A_{s-1}\chi}{B_{s-1}\chi}\chi(tu)A_s(u)\overline{A}_sB_s(1-u)\\&=\frac{A_sB_s(-1)}{q}\sum_u{}_sF_{s-1}\left[\begin{matrix}A_0,A_1,\dots,A_{s-1}\\B_1,B_2,\dots,B_{s-1}\end{matrix};tu\right]A_s(u)\overline{A}_sB_s(1-u).\end{aligned}\tag{5.81}$$

It completes the proof. ☐

With the conventional notation,

$$\binom{\vec{A}}{\vec{B}}=\prod_{k=1}^{s}\binom{A_k}{B_k}$$

and

$$F\left(\begin{matrix}C,\vec{A}\\\vec{B}\end{matrix};t\right)=F\left(\begin{matrix}C,A_1,\dots,A_s\\B_1,\dots,B_s\end{matrix};t\right).$$

We have the following identities to be considered.

Theorem 181. *The $G\Gamma(q)$ version of generalized hypergeometric function satisfy the following identities:*

$$F\left(\begin{matrix}\epsilon,A,\vec{B}\\C,\vec{D}\end{matrix};t\right)=-\frac{1}{q}\overline{C}(t)F\left(\begin{matrix}A\overline{C},\vec{B}\overline{C}\\\vec{D}\overline{C}\end{matrix};t\right)+\epsilon(t)\binom{A}{C}\binom{\vec{B}}{\vec{D}};\tag{5.82}$$

$$F\left(\begin{matrix}A,B,\vec{C}\\B,\vec{D}\end{matrix};t\right)=-\frac{1}{q}F\left(\begin{matrix}A,\vec{C}\\\vec{D}\end{matrix};t\right)+\overline{B}(t)\binom{A\overline{B}}{\overline{B}}\binom{\vec{C}\overline{B}}{\vec{D}\overline{B}};\tag{5.83}$$

$$\begin{aligned}F\left(\begin{matrix}A,\epsilon,B,\vec{C}\\D,E,\vec{F}\end{matrix};t\right)&=A(-1)\overline{D}(t)\binom{D}{A}F\left(\begin{matrix}A\overline{D},B\overline{D},\vec{C}\overline{D}\\E\overline{D},\vec{F}\overline{D}\end{matrix};t\right)\\&\quad-\frac{1}{q}D(-1)\epsilon(t)\binom{B}{E}\binom{\vec{C}}{\vec{F}}+\frac{q-1}{q^2}D(-1)\overline{E}(t)\\&\quad\cdot F\left(\begin{matrix}B\overline{E},\vec{C}\overline{E}\\\vec{F}\overline{E}\end{matrix};t\right)\delta(A\overline{D});\end{aligned}\tag{5.84}$$

$$F\left(\begin{matrix}A,B,C,\vec{D}\\ A,E,\vec{F}\end{matrix};t\right)=\binom{B}{A}F\left(\begin{matrix}B,C,\vec{D}\\ E,\vec{F}\end{matrix};t\right)-\frac{1}{q}\overline{A}(-t)\binom{C\overline{A}}{E\overline{A}}\binom{\vec{D}\overline{A}}{\vec{F}\overline{A}}$$
$$+\frac{q-1}{q^2}A(-1)\,\overline{E}(t)\,F\left(\begin{matrix}C\overline{E},\vec{D}\overline{E}\\ \vec{F}\overline{E}\end{matrix};t\right)\delta(B); \tag{5.85}$$

$$F\left(\begin{matrix}A,B,C,\vec{D}\\ E,B,\vec{F}\end{matrix};t\right)=\binom{C\overline{E}}{B\overline{E}}F\left(\begin{matrix}A,C,\vec{D}\\ E,\vec{F}\end{matrix};t\right)-\frac{1}{q}BE(-1)\,\overline{B}(t)\binom{A\overline{B}}{\overline{B}}\binom{\vec{D}\overline{B}}{\vec{F}\overline{B}}$$
$$+\frac{q-1}{q^2}BE(-1)\,F\left(\begin{matrix}A,\vec{D}\\ \vec{F}\end{matrix};t\right)\delta(C\overline{E}). \tag{5.86}$$

Theorem 182. *With the notation used in the previous theorem, the following identities hold:*

$$\binom{B\overline{D}}{A\overline{D}}F\left(\begin{matrix}A,B,\vec{C}\\ D,\vec{E}\end{matrix};t\right)-\frac{q-1}{q^2}\binom{\vec{C}}{\vec{E}}D(-1)\,\epsilon(t)\,\delta(A)$$
$$+\frac{q-1}{q^2}AD(-1)\,F\left(\begin{matrix}A,\vec{C}\\ \vec{E}\end{matrix};t\right)\delta(B\overline{D})$$
$$=\binom{B}{A}F\left(\begin{matrix}B,A,\vec{C}\\ D,\vec{E}\end{matrix};t\right)-\frac{q-1}{q^2}\binom{\vec{C}\overline{A}}{\vec{E}\overline{A}}\overline{A}(-t)\,\delta(D\overline{A})$$
$$+\frac{q-1}{q^2}A(-1)\,\overline{D}(t)\,F\left(\begin{matrix}A\overline{D},\vec{C}\overline{D}\\ \vec{E}\overline{D}\end{matrix};t\right)\delta(B), \tag{5.87}$$

$$\binom{C\overline{F}}{B\overline{F}}F\left(\begin{matrix}A,B,C,\vec{D}\\ E,F,\vec{G}\end{matrix};t\right)-\frac{q-1}{q^2}\binom{A\overline{B}}{\overline{B}}\binom{\vec{D}\overline{B}}{\vec{G}\overline{B}}BF(-1)\,\overline{B}(t)\,\delta(B\overline{E})$$
$$+\frac{q-1}{q^2}BF(-1)\,F\left(\begin{matrix}A,B,\vec{D}\\ E,\vec{G}\end{matrix};t\right)\delta(C\overline{F})$$
$$=\binom{C\overline{E}}{B\overline{E}}F\left(\begin{matrix}A,C,B,\vec{D}\\ E,F,\vec{G}\end{matrix};t\right)-\frac{q-1}{q^2}\binom{A\overline{B}}{\overline{B}}\binom{\vec{D}\overline{B}}{\vec{G}\overline{B}}\overline{B}E(-1)\,\overline{B}(t)\,\delta(F\overline{B})$$
$$+\frac{q-1}{q^2}BF(-1)\,F\left(\begin{matrix}A,B\overline{D}\\ \vec{F},\overline{G}\end{matrix};t\right)\delta(C\overline{E}). \tag{5.88}$$

Proof. We use the reduction formulas of Theorem 181 as follows. For (5.87), consider the series:

$$F\left(\begin{matrix}A,B,A,\vec{C}\\ A,D,\vec{E}\end{matrix};t\right).$$ ☐

This series can be reduced into either (5.85) or (5.86). Setting the two expressions equal results (5.87). For (5.88), apply (5.86) in two different ways on the series:

$$F\left(\begin{matrix}A,B,C,B,\vec{D}\\ E,B,F,\vec{G}\end{matrix};t\right).$$

Corollary 183. *If the series is nondegenerate, then*

$$\binom{\vec{B}}{\vec{C}}^{-1} F\left(\begin{matrix}A,\vec{B}\\ \vec{C}\end{matrix};t\right)$$

is invariant under any permutation of the numerator or denominator parameters.

Proof. Without loss of generality, we may restrict our attention to transpositions of numerator parameters. Again as in Theorem 182, there are two cases; we will consider the second. By (5.88), since the series is nondegenerate,

$$\binom{C\overline{F}}{B\overline{F}} F\left(\begin{matrix}A,B,C,\vec{D}\\ E,F,\vec{G}\end{matrix};t\right) = \binom{C\overline{E}}{B\overline{E}} F\left(\begin{matrix}A,C,B,\vec{D}\\ E,F,\vec{G}\end{matrix};t\right).$$

Thus,

$$\begin{aligned}&\left\{\binom{B}{E}\binom{C}{F}\binom{\vec{D}}{\vec{G}}\right\}^{-1} F\left(\begin{matrix}A,B,C,\vec{D}\\ E,F,\vec{G}\end{matrix};t\right)\\ &= \left\{\binom{C\overline{F}}{B\overline{F}}\binom{B}{E}\binom{C}{F}\binom{\vec{D}}{\vec{G}}\right\}^{-1} \binom{C\overline{E}}{B\overline{E}} F\left(\begin{matrix}A,B,C,\vec{D}\\ E,F,\vec{G}\end{matrix};t\right)\end{aligned}$$

and so, we must show that

$$\binom{C\overline{F}}{B\overline{F}}\binom{B}{E}\binom{C}{F}\binom{C\overline{E}}{B\overline{E}}^{-1} = \binom{C}{E}\binom{B}{F},$$

or

$$\binom{C\overline{F}}{B\overline{F}}\binom{B}{E}\binom{C}{F} = \binom{C\overline{E}}{B\overline{E}}\binom{C}{E}\binom{B}{F}.$$

Since none of the characters involved are trivial, by (5.36) we have

$$\binom{C\overline{F}}{B\overline{F}}\binom{B}{E}\binom{C}{F} = \frac{G(B)\,G(C)}{G(E)\,G(F)\,G(B\,\overline{E})\,C(B\,\overline{F})\,G(C\,\overline{B})} = \binom{C\overline{E}}{B\overline{E}}\binom{C}{E}\binom{B}{F}.$$

It completes the proof. □

Theorem 184. *For characters A, B, D, $\vec{C}$, $\vec{E}$ of $GF(q)$ and $t \in GF(q)$,*

$$F\left(\begin{matrix}A,B,\vec{C}\\ D,\vec{E}\end{matrix};t\right) = \overline{D}(t)\,F\left(\begin{matrix}B\overline{D},A\overline{D},\vec{C}\,\overline{D}\\ \overline{D},E\overline{D}\end{matrix};t\right), \tag{5.89}$$

$$F\left(\begin{matrix}A,B,\vec{C}\\ D,\vec{E}\end{matrix};t\right) = A\,B\,D\,\vec{C}\,\vec{E}(-1)\,\overline{A}(t)\,F\left(\begin{matrix}A,A\,\overline{D},A\,\overline{\vec{E}}\\ A\overline{B},A\,\overline{\vec{C}}\end{matrix};t\right). \tag{5.90}$$

Proof. We change variables in the definition,

$$F\begin{pmatrix} A, B, \vec{C} \\ D, \vec{E} \end{pmatrix}; t = \frac{q}{q-1} \sum_{\chi} \binom{A\chi}{\chi} \binom{B\chi}{D\chi} \binom{\vec{C}\chi}{\vec{E}\chi} \chi(t).$$

For, the first identity replace χ by $\overline{D}\chi$ and for second replace χ by $\overline{A}\chi$ and appeal to the property (5.35) for binomial coefficients. □

Exercises

1. For $\phi \in \widehat{GF(q)^*}$, prove that

$$G(\phi)\, G(\overline{\phi}) = q\, \phi(-1) - (q-1)\, \delta(\phi);$$

and

$$\frac{1}{G(\overline{\phi})} = \frac{\phi(-1)\, G(\phi)}{q} - \frac{(q-1)}{q}\, \delta(\phi).$$

2. For multiplicative characters ϕ and ψ of $GF(q)$, prove that

$$J(\phi, \psi) = \frac{G(\phi)\, G(\psi)}{G(\phi, \psi)} + (q-1)\, \psi(-1)\, \delta(\phi\, \psi).$$

Also, if $\phi, \psi, \phi\,\psi \neq \epsilon$, then show that

$$|J(\phi, \psi)| = \sqrt{q}.$$

3. Verify the following for the finite field analogue of the binomial coefficient:

$$\binom{\phi}{\epsilon} = \binom{\phi}{\phi} = -\frac{1}{q} + \frac{q-1}{q}\, \delta(\phi),$$
$$\binom{\epsilon}{\phi} = -\frac{\phi(-1)}{q} + \frac{q-1}{q}\, \delta(\phi),$$
$$\binom{\epsilon}{\epsilon} = \frac{q-2}{q}, \quad \phi, \epsilon \in GF(q).$$

4. Show that, for $\phi, \epsilon \in GF(q)$, the analogue of the well-known classical identity for the binomial coefficient

$$\binom{n}{k}\binom{m}{n} = \binom{m}{k}\binom{m-k}{n-k}$$

is

$$\binom{\phi}{\psi}\binom{\chi}{\phi} = \binom{\chi}{\psi}\binom{\chi\overline{\psi}}{\phi\overline{\psi}} - \frac{q-1}{q^2}\, \psi(-1)\, \delta(\phi) + \frac{q-1}{q^2}\, \phi\, \psi(-1)\, \delta(\psi\overline{\chi}).$$

5. For multiplicative characters A, B, C, D, E prove that

$$ {}_3F_2\left(\begin{matrix}A, B, D\\ C, E\end{matrix}; 1\right) = A\,E(-1)\,{}_3F_2\left(\begin{matrix}A, B, \overline{D}\,E\\ A\,B\,\overline{C}, E\end{matrix}; 1\right). $$

6. Establish the following identities:

$$ {}_3F_2\left(\begin{matrix}A, B, D\\ C, E\end{matrix}; 1\right) = A\,C(-1)\,{}_3F_2\left(\begin{matrix}A, C\,\overline{B}, D\\ C, A\,D\,\overline{E}\end{matrix}; 1\right), $$

$$ {}_3F_2\left(\begin{matrix}A, B, D\\ C, E\end{matrix}; 1\right) = B(-1)\,{}_3F_2\left(\begin{matrix}C\overline{A}, B, D\\ C, B\,D\,\overline{E}\end{matrix}; 1\right), $$

$$ {}_3F_2\left(\begin{matrix}A, B, D\\ C, E\end{matrix}; 1\right) = A\,B(-1)\,{}_3F_2\left(\begin{matrix}C\,\overline{A}, C\,\overline{B}, D\\ C, \overline{AB}\,CE\end{matrix}; 1\right). $$

7. For characters A, B and C of $GF(q)$, obtain the following relation:

$$ \begin{aligned} &{}_3F_2\left(\begin{matrix}A, B, C\\ C\,\overline{A}\,C\,\overline{B}\end{matrix}; 1\right) \\ &= \frac{q-1}{q^2}\,BC(-1)\,\delta(A) - \frac{q-1}{q^2} AC(-1)\delta(B) - \frac{q-1}{q^2} A(-1)\delta(C\,\overline{AB}) \\ &\quad + AB(-1)\begin{cases} 0, & \text{if } C \text{ is not a square,}\\ \binom{D}{A}\binom{B\overline{D}}{AB\,\overline{D}} + \binom{\Phi D}{A}\binom{\Phi B\overline{D}}{\Phi AB\overline{D}}, & \text{if } C = D^2. \end{cases} \end{aligned} $$

8. If the series involved are nondegenerate, then show that

$$ \begin{aligned} &{}_4F_3\left(\begin{matrix}A, B, C, \overline{M}\\ C, \overline{A}, C\,\overline{B}, W\end{matrix}; 1\right) \\ &= \frac{A\,B\,C\,W(-1)}{q^2}\,M\,W(2) + \frac{B\,C\,W(-1)}{q^2}\binom{B}{A\,B}\binom{M\,W\,C}{W} \\ &\quad + B(-1)\,\frac{q}{q-1}\sum_{\chi}\binom{C\chi^2}{\chi}\binom{C\chi}{\overline{A}\,C\chi}\binom{\overline{AB}\,C\chi}{\overline{B}\,C\chi}\binom{\overline{M}\chi}{\overline{MW}\,C\chi^2}\chi(-1). \end{aligned} $$

9. Prove that

$$ \begin{aligned} &{}_4F_3\left(\begin{matrix}A, B, C, \overline{M}\\ C, \overline{A}, C\,\overline{B}, W\end{matrix}; 1\right) \\ &= \frac{A\,D\,G(-1)}{q}\binom{B}{\overline{A}\,E}\binom{C\overline{G}}{\overline{D}\,F} - \frac{A\,D\,G(-1)}{q}\binom{E\,\overline{B}}{A}\binom{\overline{D}\,G}{C\,\overline{F}} \\ &\quad + C\,D\,E(-1)\,{}_4F_3\left(\begin{matrix}E\,\overline{A}, E\,\overline{B}, C, D\\ E, C\,D\,\overline{F}, C\,D\,\overline{G}\end{matrix}; 1\right), \end{aligned} $$

provided $ABCD = EFG$.

6 Extended Gauss hypergeometric functions

The present paper is influenced by the work of Chaudhary et al. [22, 23], Lee et al. [108], Özergin et al. [126] and Choi et al. [31]. We elaborated four extended versions of Gauss hypergeometric function together with the Gamma, beta and confluent hypergeometric functions. Euler generalized the factorial function from the domain of natural numbers to the gamma function as

$$\Gamma(u) = \int_0^\infty u^{u-1} e^{-u}\, du, \quad \Re(u) > 0, \tag{6.1}$$

which is defined by the right half of the complex plane. This led Legendre to decompose the gamma function into the incomplete gamma functions $\gamma(u,x)$ and $\Gamma(u,x)$, which are obtained from (6.1) by replacing the upper and lower bounds by x. These functions develop singularities for negative integers.

Chaudhry et al. extended the domain of these functions to the entire complex plane by inserting a regularization factor $e^{-\frac{p}{u}}$ in the integrand of (6.1). It turns out that this generalized gamma function, denoted by $\Gamma_p(u)$, and defined as

$$\Gamma_p(u) = \int_0^\infty u^{u-1} e^{-u-\frac{p}{u}}\, du, \quad \Re(u) > 0, \Re(p) > 0. \tag{6.2}$$

For $\Re(p) > 0$, this factor clearly removes the singularity coming from the $u = 0$, and for $p = 0$ reduces to the original gamma function. It satisfies the recursion relation and reflection formula:

$$\begin{aligned}\Gamma_p(u+1) &= u\,\Gamma_p(u) + p\,\Gamma_p(u-1),\\ \Gamma_p(-u) &= p^{-u}\,\Gamma_p(u).\end{aligned}$$

In view of the effectiveness of the above regularized gamma function, it seems worthwhile to look into the possibility that the domains of other special functions could be usefully extended in the similar manner. This motivation results the study of special functions from various extensions. In this chapter, we list few of them.

6.1 First extension of Gauss hypergeometric function

In this section, we deal with the expansion of Gauss and the confluent hypergeometric function studied by Chaudhry et al. We start with the extension of the beta function. Based on the generalization of the gamma function discussed above, the beta function has an extended version as

https://doi.org/10.1515/9783111324586-006

$$\beta(u,v,p) = \int_0^1 u^{u-1}(1-u)^{v-1} e^{-\frac{p}{u(1-u)}}\,du, \quad \Re(p) > 0. \tag{6.3}$$

This extension will prove extremely useful because most of the properties of the beta function can be naturally and easily transferred to it. It provides connections with the error function and the Whittaker functions and even new representations for special cases of these functions. When $p = 0$, it clearly reduces to the original beta function. The following relation in the form of an improper integral shows the connection between the extended beta function (6.3) and the classical beta function (1.51).

Theorem 185. *The following relations hold true:*

$$\int_0^\infty p^{r-1}\beta(u,v,p)\,dp \tag{6.4}$$

$$= \Gamma(r)\,\beta(u+r, v+r), \quad \Re(r) > 0, \Re(u+r) > 0, \Re(v+r) > 0. \tag{6.5}$$

Proof. Multiplying (6.3) by p^{r-1} and integrating with respect to p from $p = 0$ to $p = \infty$, we get

$$\int_0^\infty p^{r-1}\beta(u,v,p)\,dp = \int_0^\infty p^{r-1}\left(\int_0^1 u^{u-1}(1-u)^{v-1} e^{-\frac{p}{u(1-u)}}\,du\right)dp. \tag{6.6}$$

The order of integration in (6.6) can be interchanged because of the uniform convergence of the integral. Therefore,

$$\int_0^\infty p^{r-1}\beta(u,v,p)\,dp = \int_0^1 u^{u-1}(1-u)^{v-1}\left(\int_0^\infty p^{r-1} e^{-\frac{p}{u(1-u)}}\,dp\right)du. \tag{6.7}$$

However, we have

$$\int_0^\infty p^{r-1} e^{-\frac{p}{u(1-u)}}\,dp = u^r(1-u)^r\,\Gamma(r), \quad \Re(r) > 0, 0 < u < 1.$$

It follows that

$$\begin{aligned}\int_0^\infty p^{r-1}\beta(u,v,p)\,dp &= \Gamma(r)\int_0^1 u^{u+r-1}(1-u)^{v+r-1}\,du \\ &= \Gamma(r)\,\beta(u+r, v+r).\end{aligned} \tag{6.8}$$

It completes the proof. □

By putting $r = 1$, we get the interesting result between original and extended beta function

$$\int_0^\infty \beta(u,v,p)\,dp = \beta(u+1,v+1), \quad \Re(u) > -1, \Re(v) > -1.$$

It is interesting to note that all the derivativse of the extended beta function with respect to the parameter p can be expressed in terms of the function to give

$$\frac{\partial^k}{\partial p^k}\beta(u,v,p) = (-1)^k \beta(u-k,v-k,p), \quad k = 0,1,\dots. \tag{6.9}$$

Theorem 186. *The extended beta function can be presented in the various integral forms as given below:*

$$\beta(u,v,p) = 2\int_0^{\frac{\pi}{2}} (\cos\theta)^{2u-1}(\sin\theta)^{2v-1} e^{-p\sec^2\theta\csc^2\theta}\,d\theta; \tag{6.10}$$

$$= e^{-2p}\int_0^\infty \frac{t^{u-1}}{(1-t)^{u+v}} e^{-p(t+t^{-1})}\,dt; \tag{6.11}$$

$$= 2^{1-u-v}\int_{-1}^{1} (1+t)^{u-1}(1-t)^{v-1} e^{\frac{-4p}{(1-t^2)}}\,dt; \tag{6.12}$$

$$= \frac{1}{2}e^{-2p}\int_0^\infty \frac{t^{u-1}+t^{v-1}}{(1-t)^{u+v}} e^{-p(t+t^{-1})}\,dt; \tag{6.13}$$

$$= (\gamma-\alpha)^{1-u-v}\int_\alpha^\gamma (t-\alpha)^{u-1}(\gamma-t)^{v-1}\exp\left[-p\frac{(\gamma-\alpha)^2}{(t-\alpha)(\gamma-t)}\right]dt. \tag{6.14}$$

The above integrals are valid when $\Re(p) > 0$. For $p = 0$, $\Re(u) > 0$, $\Re(v) > 0$ will be the required condition of validity.

Proof. The proofs of (6.10)–(6.14) are straightforward. In particular, (6.10) follows when we use the transformation $u = \cos^2\theta$ in (6.3) and (6.11) follows from (6.3) when we use the transformation $u = \frac{t}{1+t}$. Similarly, (6.14) follows from (6.3) when we use the transformation $u = \frac{t-\alpha}{\gamma-\alpha}$. □

Proposition 187. *A useful inequality*

$$|\beta(u,v,p)| \le e^{-4p}\beta(u,v), \quad u > 0, v > 0, p \ge 0,$$

follows from the integral representation (6.11).

Proof. Since the function $e^{-p(t+t^{-1})}$ attains the maximum value at $t = 1$,

$$|\beta(u,v,p)| \le e^{-4p} \int_0^\infty \frac{t^{u-1}}{(1+t)^{u+v}}\, dt$$
$$\le e^{-4p}\, \beta(u,v).$$

Hence, we have the inequality. □

Theorem 188 (Functional relation). *The extension of beta function* (6.3) *obeys the following functional relation:*

$$\beta(u,v+1,p) + \beta(u+1,v,p) = \beta(u,v,p). \tag{6.15}$$

Proof. The left-hand side and the integral definition (6.3) together yield

$$\beta(u,v+1,p) + \beta(u+1,v,p) = \int_0^1 \{u^{u-1}(1-u)^{v} + u^{u}(1-u)^{v-1}\}\, e^{-\frac{p}{u(1-u)}}\, du$$
$$= \int_0^1 u^{u-1}(1-u)^{v-1}\, e^{-\frac{p}{u(1-u)}}\, du$$
$$= \beta(u,v,p). \tag{6.16}$$

It completes the proof. □

As a direct corollary, on putting $p = 0$ in (6.15), we get the usual relation for the beta function. Since the generalized gamma function extends the Euler function in a natural way, it seems worthwhile to extend the product of the gamma function proved by Euler.

Theorem 189. *Let $\Gamma_p(u)$ be the generalized gamma function as defined in* (6.2). *Then*

$$\Gamma_p(u)\,\Gamma_p(v) = 2\int_0^\infty r^{2(u+v)-1}\, e^{-r^2}\, \beta\left(u, v, \frac{p}{r^2}\right) dr,$$
$$\Re(p) > 0.\ For\ p = 0,\ \Re(u) > 0,\ \Re(v) > 0. \tag{6.17}$$

Proof. Substituting $u = t_1^2$ in (6.2), we get

$$\Gamma_p(u) = 2\int_0^\infty t_1^{2u-1}\, e^{-t_1^2 - p t_1^{-2}}\, dt_1.$$

Therefore,

$$\Gamma_p(u)\,\Gamma_p(v) = 4\int_0^\infty\int_0^\infty t_1^{2u-1}\,t_2^{2v-1}\,e^{-(t_1^2+t_2^2)}\,e^{-p(t_1^{-2}+t_2^{-1})}\,dt_1\,dt_2.$$

The substitution $t_1 = r\cos\theta$ and $t_2 = r\sin\theta$ yield

$$\begin{aligned}&\Gamma_p(u)\,\Gamma_p(v)\\&= 4\int_0^{\frac{\pi}{2}}\int_0^\infty r^{2(u+v)-1}\,e^{-r^2}\,(\cos\theta)^{2u-1}\,(\sin\theta)^{2v-1}\,\exp\left[\frac{p}{r^2\sin^2\theta\cos^2\theta}\right]dr\,d\theta.\end{aligned}$$

Interchanging the order of integration on the right side, we get

$$\begin{aligned}&\Gamma_p(u)\,\Gamma_p(v)\\&= 2\int_0^\infty r^{2(u+v)-1}\,e^{-r^2}\left(2\int_0^{\frac{\pi}{2}}(\cos\theta)^{2u-1}\,(\sin\theta)^{2v-1}\,\exp\left[\frac{p/r^2}{\sin^2\theta\cos^2\theta}\right]\right)dr\,d\theta.\end{aligned}$$

Using (6.10), we have

$$\begin{aligned}&\Gamma_p(u)\,\Gamma_p(v)\\&= 2\int_0^\infty r^{2(u+v)-1}\,e^{-r^2}\,\beta\!\left(u, v, \frac{p}{r^2}\right)dr, \quad \Re(p) > 0. \text{ For } p = 0, \Re(u), \Re(v) > 0.\end{aligned}$$

It completes the proof. □

In the following theorem, we obtain the infinite summation formula for the extension of the beta function.

Theorem 190. *The infinite summation formula holds:*

$$\beta(u, 1-v, p) = \sum_{k=0}^{\infty}\frac{(v)_k}{k!}\,\beta(u+k, 1, p), \quad \Re(p) > 0. \tag{6.18}$$

Proof. Using the identity,

$$(1-u)^{-b} = \sum_{k=0}^{\infty}(b)_k\,\frac{u^k}{k!} \tag{6.19}$$

in (6.3), we obtain

$$\beta(u, 1-v, p) = \int_0^1 \sum_{k=0}^{\infty}\frac{(v)_k}{k!}\,u^{u+k-1}\,e^{-\frac{p}{u(1-u)}}\,du. \tag{6.20}$$

Now the summation could be divergent at the end points of integration for particular values of u and v. However, for $\Re(p) > 0$, the regularizing factor dumps the singularity arising there. Thus, we can interchange the order of integration and summation to obtain

$$\begin{aligned}\beta(u, 1-v, p) &= \sum_{k=0}^{\infty} \frac{(v)_k}{k!} \int_0^1 u^{u+k-1} e^{-\frac{p}{u(1-u)}} du \\ &= \sum_{k=0}^{\infty} \frac{(v)_k}{k!} \beta(u+k, 1, p), \quad \Re(p) > 0. \end{aligned} \tag{6.21}$$

It completes the proof. □

In Chapter 2, we discussed the Gauss and confluent hypergeometric function in detail. Recalling the integral definition of Gauss and confluent hypergeometric functions, given by

$${}_2F_1(u, v; w; z) = \frac{\Gamma(w)}{\Gamma(v)\Gamma(w-v)} \int_0^1 u^{v-1}(1-u)^{w-v-1}(1-zu)^{-u} du \tag{6.22}$$

and

$${}_1F_1(u; w; z) = \frac{\Gamma(w)}{\Gamma(u)\Gamma(w-u)} \int_0^1 u^{u-1}(1-u)^{w-u-1} e^{-zu} du, \tag{6.23}$$

respectively. Using the series representations $(1-zu)^{-u} = \sum_{k=0}^{\infty} \frac{(u)_k}{k!}(zu)^k$, $e^{-zu} = \sum_{k=0}^{\infty} \frac{(zu)^k}{k!}$ and the integral representation of the beta function

$$\beta(u, v) = \int_0^1 u^{u-1}(1-u)^{w-u-1} du, \tag{6.24}$$

together yield the alternative definition of Gauss and confluent hypergeometric functions in terms of the beta function as

$$\begin{aligned}&{}_2F_1(u, v; w; z) \\ &= \frac{\Gamma(w)}{\Gamma(v)\Gamma(w-v)} \sum_{k=0}^{\infty} \frac{(u)_k}{k!} z^k \beta(v+k, w-v), \quad |z| < 1; \Re(w) > \Re(v) > 0, \end{aligned} \tag{6.25}$$

and

$${}_1F_1(u; w; z) = \frac{\Gamma(w)}{\Gamma(u)\Gamma(w-u)} \sum_{k=0}^{\infty} \frac{z^k}{k!} \beta(u+k, w-u), \quad \Re(w) > \Re(u) > 0. \tag{6.26}$$

The proposed extension of the Gauss hypergeometric and confluent hypergeometric functions can be obtained by simply taking the extended beta function at a place of simple beta function in the above definitions as follows:

$$F_p(u, v; w; z) = \frac{\Gamma(w)}{\Gamma(v)\Gamma(w-v)} \sum_{k=0}^{\infty} \frac{(u)_k}{k!} z^k \beta(v+k, w-v, p), \quad \Re(p) \geq 0; |z| < 1; \Re(w) > \Re(v) > 0, \tag{6.27}$$

which can be rewritten in the form

$$F_p(u, v; w; z) = \sum_{k=0}^{\infty} \frac{(u)_k}{k!} z^k \frac{\beta(v+k, w-v, p)}{\beta(v, w-v)}, \quad \Re(p) \geq 0; |z| < 1; \Re(w) > \Re(v) > 0. \tag{6.28}$$

Using the formula,

$$\beta(u, v, p) \sim \frac{\pi^{1/2} p^{1/4}}{u^{3/4}} \left(\frac{p}{u}\right)^{(v-1)/2} \exp\left[-2(pu)^{1/2} - \frac{3p}{2}\right] \quad (u \to \infty),$$

one can verify that the series (6.27) is convergent when $|z| < 1$, provided that w is not a negative integer or zero. Also, for $p = 0$, it reduces to the usual Gauss hypergeometric function.

Similarly, the extended confluent hypergeometric function is defined by

$$\Phi_p(u; w; z) = \sum_{k=0}^{\infty} \frac{\beta(u+k, w+k, p)}{\beta(u, w-u)} \frac{z^k}{k!}, \quad \Re(p) \geq 0; \Re(w) > \Re(u) > 0. \tag{6.29}$$

For $p > 0$, the series converges for all z, provided that w is not a negative integer or zero. For $p = 0$, it obviously reduces to the usual confluent hypergeometric function.

6.1.1 Integral representations of the extended Gauss and confluent hypergeometric function

The extended Gauss hypergeometric function can be provided with an integral representation by using (6.3) in (6.28). Thus, we get

$$F_p(u, v; w; z) = \frac{1}{\beta(v, w-v)} \int_0^1 u^{v-1}(1-u)^{w-v-1} \exp\left[-\frac{p}{u(1-u)}\right] \sum_{k=0}^{\infty} (u)_k \frac{(zu)^k}{k!} du,$$

$$\Re(p) \geq 0; |z| < 1; \Re(w) > \Re(v) > 0. \tag{6.30}$$

Since the series in (6.30) represents $(1-zu)^{-u}$, we can rewrite this equation as follows:

$$F_p(u, v; w; z) = \frac{1}{\beta(v, w-v)} \int_0^1 \mathrm{u}^{v-1}(1-\mathrm{u})^{w-v-1} \exp\left[-\frac{p}{\mathrm{u}(1-\mathrm{u})}\right] (1-z\mathrm{u})^{-u}\, d\mathrm{u},$$
$$\Re(p) \geq 0; |z| < 1; \Re(w) > \Re(v) > 0. \tag{6.31}$$

An alternative form is obtained by setting $t = \frac{\mathrm{u}}{1-\mathrm{u}}$ in (6.31), and we have

$$F_p(u, v; w; z) = \frac{\exp(-2p)}{\beta(v, w-v)} \int_0^\infty t^{v-1}(1+t)^{u-w} \exp\left[-p\left(t + \frac{1}{t}\right)\right] [1 + t(1-z)]^{-u}\, dt,$$
$$\Re(p) > 0; p = 0 \text{ and } |\arg(1-z)| < \pi; \Re(w) > \Re(v) > 0. \tag{6.32}$$

Again by the substitution $\mathrm{u} = \sin^2\theta$ in (6.31), we get

$$F_p(u, v; w; z) = \frac{2}{\beta(v, w-v)} \int_0^{\frac{\pi}{2}} \frac{(\sin\theta)^{2v-1}(\cos\theta)^{2w-2v-1}}{(1 - z\sin^2\theta)^u} \exp(-p\sec^2\theta\csc^2\theta)\, d\theta,$$
$$\Re(p) > 0; p = 0 \text{ and } |\arg(1-z)| < \pi; \Re(w) > \Re(v) > 0. \tag{6.33}$$

Yet another integral representation is obtained by means of the substitution $\mathrm{u} = \tanh^2\mathfrak{v}$ in (6.31) and get

$$F_p(u, v; w; z) = \frac{2}{\beta(v, w-v)} \int_0^\infty \frac{(\sinh\mathfrak{v})^{2v-1}(\cosh\mathfrak{v})^{2u-2w-1}}{(\cosh^2\mathfrak{v} - z\sin^2\mathfrak{v})^u} \exp(-p\cosh^2\mathfrak{v}\coth^2\mathfrak{v})\, d\mathfrak{v},$$
$$\Re(p) > 0; p = 0 \text{ and } |\arg(1-z)| < \pi; \Re(w) > \Re(v) > 0. \tag{6.34}$$

A similar procedure yields integral representations of the extended confluent hypergeometric function by using (6.29) in place of (6.28). We thus obtain

$$\Phi_p(u; w; z) = \frac{1}{\beta(u, w-u)} \int_0^1 \mathrm{u}^{u-1}(1-\mathrm{u})^{w-u-1} \exp\left[z\mathrm{u} - \frac{p}{\mathrm{u}(1-\mathrm{u})}\right] d\mathrm{u},$$
$$\Re(p) > 0; p = 0 \text{ and } \Re(w) > \Re(u) > 0. \tag{6.35}$$

The corresponding expression for the above substitution can easily be obtained as well. In particular,

$$\Phi_p(u; w; z) = \frac{\exp(z)}{\beta(u, w-u)} \int_0^1 \mathrm{u}^{w-u-1}(1-\mathrm{u})^{u-1} \exp\left[-z\mathrm{u} - \frac{p}{\mathrm{u}(\mathrm{u}-1)}\right] d\mathrm{u},$$
$$\Re(p) \geq 0; |z| < 1; \Re(w) > \Re(u) > 0. \tag{6.36}$$

The substitution $p = 0$ in the above integral representations lead for usual integrals of Gauss and confluent hypergeometric functions.

The Gauss summation formula,

$$_2F_1(u, v; w; 1) = \frac{\Gamma(w)\,\Gamma(w-u-v)}{\Gamma(w-u)\,\Gamma(w-v)} = \frac{\beta(v, w-u-v)}{\beta(v, w-v)}, \tag{6.37}$$

can be extended more elegantly for the extended Gauss hypergeometric function. Setting $z = 1$ in (6.31), we have

$$F_p(u, v; w; 1) = \frac{1}{\beta(v, w-v)} \int_0^1 u^{v-1}(1-u)^{w-u-v-1} \exp\left[-\frac{p}{u\,(1-u)}\right] du. \tag{6.38}$$

Now, the integral on the right-hand side is simply the extended beta function. Hence,

$$F_p(u, v; w; 1) = \frac{\beta(v, w-u-v, p)}{\beta(v, w-v)}, \quad \Re(p) \geq 0; p = 0 \text{ and } \Re(w-u-v) > 0. \tag{6.39}$$

6.1.2 Differential formulas

We can obtain a formula for the derivative of the extended Gauss hypergeometric functions with respect to the variable z in terms of a shift of the parameters by using the identities

$$\beta(v, w-v) = \frac{w}{v}\,\beta(v+1, w-v), \tag{6.40}$$

and

$$(u)_{k+1} = u\,(u+1)_k. \tag{6.41}$$

For the extended Gauss hypergeometric function, differentiation of (6.28) with respect to z yields

$$\frac{d}{dz}\{F_p(u, v; w; z)\} = \sum_{k=1}^{\infty} \frac{(u)_k}{(k-1)!}\, z^{k-1}\, \frac{\beta(v+k, w-v, p)}{\beta(v, w-v)}.$$

Changing k to $k+1$, the derivative becomes

$$\frac{d}{dz}\{F_p(u, v; w; z)\} = \frac{u\,v}{w} F_p(u+1, v+1; w+1; z). \tag{6.42}$$

Recursive application of this procedure finally gives

$$\frac{d^k}{dz^k}\{F_p(u, v; w; z)\} = \frac{(u)_k\,(v)_k}{(w)_k} F_p(u+k, v+k; w+k; z).$$

6.1.3 Transformation formulas

By writing,

$$[1 - z(1 - u)]^{-u} = (1 - z)^{-u}\left(1 + \frac{z}{1 - z}u\right)^{-u},$$

and replacing u by $1 - u$ in (6.31), we find that

$$F_p(u, v; w; z) = \frac{(1 - z)^{-u}}{\beta(v, w - v)} \int_0^1 u^{v-1}(1 - u)^{w-v-1} \exp\left[-\frac{p}{u(1 - u)}\right] \times \left(1 + \frac{z}{1 - z}u\right)^{-u} du, \quad \Re(p) \geq 0; |z| < 1; \Re(w) > \Re(v) > 0. \tag{6.43}$$

Since the beta function is symmetric in its arguments, the above equation leads to

$$F_p(u, v; w; z) = (1 - z)^{-u} F_p\left(u, w - v; w; \frac{z}{1 - z}\right), \quad |\arg(1 - z)| < \pi. \tag{6.44}$$

On putting $p = 0$, this transformation formula for an extended Gauss hypergeometric function reduces to the standard formula for the usual hypergeometric function.

For the extended confluent hypergeometric function, from (6.35) and (6.36), we find the following transformation formula:

$$\Phi_p(u; w; z) = \exp(z)\, \Phi_p(w - u; w; -z), \tag{6.45}$$

which reduces to Kummer's first formula when $p = 0$.

6.1.4 Recurrence relations

From the integral representation (6.31) of extended Gauss hypergeometric function and the difference operator $\Delta_u f(u) = f(u + 1) - f(u)$, we find that

$$\begin{aligned}\Delta_u F_p(u, v; w; z) &= F_p(u + 1, v; w; z) - F_p(u, v; w; z) \\ &= \frac{z}{\beta(v, w - v)} \int_0^1 u^{v-1}(1 - u)^{w-v-1} (1 - z u)^{-u-1} \exp\left[-\frac{p}{u(1 - u)}\right] du.\end{aligned} \tag{6.46}$$

Now, changing the arguments in the extended Gauss hypergeometric function by raising each parameter by one, we find that

$$F_p(u+1, v+1; w+1; z) = \frac{1}{\beta(v+1, w-v)} \int_0^1 u^v (1-u)^{w-v-1} (1-zu)^{-u-1} \times \exp\left[-\frac{p}{u(1-u)}\right] du. \tag{6.47}$$

Finally, by comparing (6.46) and (6.47), we get

$$\Delta_u F_p(u, v; w; z) = \frac{vz}{w} F_p(u+1, v+1; w+1; z), \tag{6.48}$$

which is our first recurrence relation for the extended Gauss hypergeometric function. Further, by using the differential formula (6.42), we obtain another differential recurrence relation as follows:

$$\frac{d}{dz}\{F_p(u, v; w; z)\} = \frac{u}{z} \Delta_u F_p(u, v; w; z). \tag{6.49}$$

Thus, the operators $z\frac{d}{dz}$ and $u\,\Delta_u$ have equal actions on the extended Gauss hypergeometric function.

6.2 Second extension of Gauss hypergeometric function

Lee et al. in 2011, presented another extension of the Gauss hypergeometric function with the help of an extended beta hypergeometric function. The following generalization of a beta function is considered by them:

$$\beta(u, v, p, m) = \int_0^1 u^{u-1} (1-u)^{v-1} \exp\left[-\frac{p}{u^m (1-u)^m}\right] du, \tag{6.50}$$

where $\Re(p) > 0$, $m > 0$. Clearly, when $m = 1$, equation (6.50) reduces to the extended beta function (6.3) and $p = 0$; it reduces to the usual Euler's beta function.

6.2.1 Integral representation of new generalized beta function

Theorem 191. *For $\Re(p) > 0$, $m > 0$, $\Re(s) > 0$, $\Re(u+ms) > 0$ and $\Re(v+ms) > 0$, we have*

$$\int_0^\infty p^{s-1} \beta(u, v, p, m)\, dp = \Gamma(s)\, \beta(u+ms, v+ms). \tag{6.51}$$

Proof. Multiplying (6.50) by p^{s-1} and integrating with respect to p from $p = 0$ to $p = \infty$, we get

$$\int_0^\infty p^{s-1}\,\beta(u,v,p,m)\,dp = \int_0^\infty p^{s-1}\left(\int_0^1 u^{u-1}(1-u)^{v-1}\exp\left[-\frac{p}{u^m(1-u)^m}\right]du\right)dp. \tag{6.52}$$

The order of the integral in (6.52) can be interchanged because of the uniform convergence of the integral. Therefore, we have

$$\int_0^\infty p^{s-1}\,\beta(u,v,p,m)\,dp = \int_0^1 u^{u-1}(1-u)^{v-1}\left(\int_0^1 p^{s-1}\exp\left[-\frac{p}{u^m(1-u)^m}\right]dp\right)du. \tag{6.53}$$

Using the definition of the gamma function in (6.53), we have

$$\begin{aligned}\int_0^\infty p^{s-1}\,\beta(u,v,p,m)\,dp &= \int_0^1 u^{u-1}(1-u)^{v-1}\left(u^{ms}(1-u)^{ms}\,\Gamma(s)\right)du\\ &= \Gamma(s)\int_0^1 u^{u+ms-1}(1-u)^{v+ms-1}\,du\\ &= \Gamma(s)\,\beta(u+ms, v+ms).\end{aligned} \tag{6.54}$$

It completes the proof. □

Remark that when $s = 1$, it reduces to the integral

$$\int_0^\infty \beta(u,v,p,m)\,dp = \beta(u+m, v+m) \tag{6.55}$$

shows relation between the classical and new generalized beta functions, where $\mathbb{R}(u+m) > 0$, $\mathbb{R}(v+m) > 0$. Further, when $m = 1$ it coincides with the relation (6.5).

The generalization of Euler's beta function is presented by the various integral representations as follows.

Theorem 192.

$$\beta(u,v,p,m) = 2\int_0^{\frac{\pi}{2}} \cos^{2u-1}\theta\,\sin^{2v-1}\theta\,\exp[-p\sec^{2m}\theta\csc^{2m}\theta]\,d\theta \tag{6.56}$$

$$= \int_0^\infty \frac{t^{u-1}}{(1-t)^{u+v}}\exp\left[-p\left(2+t+\frac{1}{t}\right)^m\right]dt \tag{6.57}$$

$$= 2^{1-u-v}\int_{-1}^1 (1+t)^{u-1}(1-t)^{v-1}\exp\left[-\frac{4^m p}{(1-t^2)^m}\right]dt \tag{6.58}$$

$$= (w-u)^{1-u-v} \int_u^w (t-u)^{u-1} (w-t)^{v-1} \exp\left[-p \frac{(w-u)^{2m}}{(t-u)^m (w-t)^m}\right] dt,$$

$$\Re(p) > 0 \textit{ and } m > 0; \textit{ For } p = 0, \Re(u) > 0, \Re(v) > 0 \tag{6.59}$$

$$= 2^{1-u-v} \int_{-\infty}^{\infty} \exp[(u-v)\,t - 4^m p \cosh^{2m} t] \frac{dt}{(\cosh t)^{u+v}},$$

$$\Re(p) > 0 \textit{ and } m > 0; \textit{ For } p = 0, \Re(u) > 0, \Re(v) > 0. \tag{6.60}$$

Proof. Equations (6.56)–(6.60) can be obtained by using the transformations $\mathrm{u} = \cos^2\theta$, $\mathrm{u} = \frac{t}{1+t}$, $\mathrm{u} = \frac{1+t}{2}$, $\mathrm{u} = \frac{t-u}{w-u}$ and $\mathrm{u} = \tanh t$ in equation (6.50), respectively. □

6.2.2 Properties of the new generalized beta function

Theorem 193.

$$\beta(u, v+1, p, m) + \beta(u+1, v, p, m) = \beta(u, v, p, m). \tag{6.61}$$

Proof. The left-hand side can be written as

$$\int_0^1 \{\mathrm{u}^{u-1} (1-\mathrm{u})^v + \mathrm{u}^u (1-\mathrm{u})^{v-1}\} \exp\left[-\frac{p}{\mathrm{u}^m (1-\mathrm{u})^m}\right] d\mathrm{u}, \tag{6.62}$$

which after simple algebraic manipulation yields

$$\int_0^1 \mathrm{u}^{u-1} (1-\mathrm{u})^{v-1} \exp\left[-\frac{p}{\mathrm{u}^m (1-\mathrm{u})^m}\right] d\mathrm{u}. \tag{6.63}$$

This is equal to the right-hand side of (6.61). □

Theorem 194. *For* $\Re(p) > 0$, $\Re(u) > 0$, $\Re(v) < 1$ *and* $m > 0$,

$$\beta(u, 1-v, p, m) = \sum_{k=0}^{\infty} \frac{(v)_k}{k!} \beta(u+k, 1, p, m). \tag{6.64}$$

Proof. By the definition of the new generalized beta function, we have

$$\beta(u, 1-v, p, m) = \int_0^1 \mathrm{u}^{u-1} (1-\mathrm{u})^{-v} \exp\left[-\frac{p}{\mathrm{u}^m (1-\mathrm{u})^m}\right] d\mathrm{u}.$$

Using the following binomial series expansion:

$$(1-u)^{-v} = \sum_{k=0}^{\infty} \frac{(v)_k}{k!} u^k, \quad |u| < 1,$$

we obtain

$$\beta(u, 1-v, p, m) = \int_0^1 \sum_{k=0}^{\infty} \frac{(v)_k}{k!} u^{u+k-1} \exp\left[-\frac{p}{u^m (1-u)^m}\right] du. \tag{6.65}$$

Therefore, interchanging the order of integration and summation in (6.65) together with (6.50), we have

$$\begin{aligned}\beta(u, 1-v, p, m) &= \sum_{k=0}^{\infty} \frac{(v)_k}{k!} \int_0^1 u^{u+k-1} \exp\left[-\frac{p}{u^m (1-u)^m}\right] du \\ &= \sum_{k=0}^{\infty} \frac{(v)_k}{k!} \beta(u+k, 1, p, m).\end{aligned}$$

It completes the proof. □

Theorem 195. *For* $\Re(p) > 0$ *and* $m > 0$,

$$\beta(u, v, p, m) = \sum_{k=0}^{\infty} \beta(u+k, v+1, p, m).$$

Proof. Replacing $(1-u)^{v-1}$ in (6.50) by its series representation,

$$(1-u)^{v-1} = (1-u)^v \sum_{k=0}^{\infty} u^k,$$

and thus, we obtain

$$\beta(u, v, p, m) = \int_0^1 (1-u)^v \sum_{k=0}^{\infty} u^{u+k-1} \exp\left[-\frac{p}{u^m (1-u)^m}\right] du.$$

Interchanging the order of integration and summation, we have

$$\begin{aligned}\beta(u, v, p, m) &= \sum_{k=0}^{\infty} \int_0^1 (1-u)^v u^{u+k-1} \exp\left[-\frac{p}{u^m (1-u)^m}\right] du \\ &= \sum_{k=0}^{\infty} \beta(u+k, v+1, p, m).\end{aligned}$$

It completes the proof. □

6.2.3 Generalized extended Gauss and confluent hypergeometric functions

Using the new generalized beta function (6.50), we are capable of defining the generalized extended Gauss and confluent hypergeometric functions by

$$\begin{aligned} F_p(u, v; w; z; m) \\ = \sum_{k=0}^{\infty} \frac{(u)_k}{k!} z^k \frac{\beta(v+k, w-v, p, m)}{\beta(v, w-v)}, \quad \Re(p) \geq 0, m > 0; |z| < 1; \Re(w) > \Re(v) > 0, \end{aligned} \tag{6.66}$$

and

$$\Phi_p(u; w; z; m) = \sum_{k=0}^{\infty} \frac{\beta(u+k, w+k, p, m)}{\beta(u, w-u)} \frac{z^k}{k!}, \quad \Re(p) \geq 0, m > 0; \Re(w) > \Re(u) > 0, \tag{6.67}$$

respectively. The integral representation of a generalized extended Gauss hypergeometric function for $m > 0$ is given by

$$F_p(u, v; w; z; m) = \frac{1}{\beta(v, w-v)} \int_0^1 u^{v-1} (1-u)^{w-v-1} \exp\left[-\frac{p}{u^m (1-u)^m}\right] \sum_{k=0}^{\infty} \frac{(u)_k}{k!} (zu)^k \, du,$$
$$\Re(p) \geq 0, m > 0; |z| < 1; \Re(w) > \Re(v) > 0. \tag{6.68}$$

Since $(1-zu)^{-u} = \sum_{k=0}^{\infty} \frac{(u)_k}{k!} (zu)^k$, so we can write

$$F_p(u, v; w; z; m) = \frac{1}{\beta(v, w-v)} \int_0^1 u^{v-1} (1-u)^{w-v-1} (1-zu)^{-u} \exp\left[-\frac{p}{u^m (1-u)^m}\right] du,$$
$$\Re(p) \geq 0; p = 0, m > 0 \text{ and } |\arg(z)| < \pi; \Re(w) > \Re(v) > 0. \tag{6.69}$$

The another form is obtained by putting $u = \frac{t}{1+t}$ in (6.69), and we get

$$\begin{aligned} F_p(u, v; w; z; m) \\ = \frac{1}{\beta(v, w-v)} \int_0^{\infty} t^{v-1} (1+t)^{u-w} (1+t(1-z))^{-u} \exp\left[-p\left(2+t+\frac{1}{t}\right)^m\right] dt, \\ \Re(p) \geq 0; p = 0, m > 0 \text{ and } |\arg(z)| < \pi; \Re(w) > \Re(v) > 0. \end{aligned} \tag{6.70}$$

Also, by substitution $u = \sin^2 t$ in (6.69), we get

$$F_p(u,v;w;z;m)$$
$$= \frac{2}{\beta(v,w-v)} \int_0^{\frac{\pi}{2}} \frac{(\sin t)^{2v-1}(\cos t)^{2w-2v-1}}{(1-z\sin^2 t)^u} \exp[-p\sec^{2m} t \csc^{2m} t]\, dt. \tag{6.71}$$

Further, if we substitute $u = \tanh^2 t$ in (6.69), the another integral representation is obtained as

$$F_p(u,v;w;z;m)$$
$$= \frac{2}{\beta(v,w-v)} \int_0^{\infty} \frac{(\sinh t)^{2v-1}(\cosh t)^{2u-2w-1}}{(\cosh^2 t - z\sinh^2 t)^u} \exp[-p\cosh^{2m} t \coth^{2m} t]\, dt. \tag{6.72}$$

Similarly, the integral representation of a generalized extended confluent hypergeometric function is defined by using (6.67) as

$$\Phi_p(u;w;z;m) = \frac{1}{\beta(u,w-u)} \int_0^1 u^{u-1}(1-u)^{w-u-1} \exp\left[zu - \frac{p}{u^m(1-u)^m}\right] du,$$
$$\Re(p) \geq 0, m > 0 \text{ and } \Re(w) > \Re(u) > 0. \tag{6.73}$$

Also, if we replace u by $1-u$, we get another integral form for an extended generalized confluent hypergeometric function of (6.73), and so we have

$$\Phi_p(u;w;z;m) = \frac{\exp(z)}{\beta(u,w-u)} \int_0^1 u^{w-u-1}(1-u)^{u-1} \exp\left[-zu - \frac{p}{u^m(1-u)^m}\right] du,$$
$$\Re(p) \geq 0, p = 0 \text{ and } \Re(w) > \Re(u) > 0. \tag{6.74}$$

Clearly, for $m = 1$ in (6.68)–(6.74) leads to the integral representations for extended Gauss and confluent hypergeometric functions, and for $p = 0$, it leads to the integral representation of the usual Gauss and confluent hypergeometric functions.

Putting $z = 1$ in (6.69), we get

$$F_p(u,v;w;1;m) = \frac{1}{\beta(v,w-v)} \int_0^1 u^{v-1}(1-u)^{w-u-v-1} \exp\left[-\frac{p}{u^m(1-u)^m}\right] du. \tag{6.75}$$

Now, the integral on the right-hand side of (6.75) can be written in terms of the new generalized beta function. So, we obtain

$$F_p(u,v;w;1;m) = \frac{\beta(v,w-u-v,p,m)}{\beta(v,w-v)}, \quad \Re(p) \geq 0, p = 0 \text{ and } \Re(w-u-v) > 0. \tag{6.76}$$

The differentiation formulas for the generalized extended Gauss and confluent hypergeometric functions can be obtained by differentiating (6.66) and (6.67) with respect to the variable z and the identity (6.40). Differentiating (6.66) yields

$$\frac{d}{dz}F_p(u,v;w;z;m)=\sum_{k=1}^{\infty}\frac{(u)_k}{(k-1)!}z^{k-1}\frac{\beta(v+k,w-v,p,m)}{\beta(v,w-v)}. \tag{6.77}$$

Changing k into $k+1$ and using (6.66), we have

$$\frac{d}{dz}F_p(u,v;w;z;m)=\frac{uv}{w}F_p(u+1,v+1;w+1;z;m). \tag{6.78}$$

Recursive application of this procedure gives the general form,

$$\frac{d^k}{dz^k}F_p(u,v;w;z;m)=\frac{(u)_k\,(v)_k}{(w)_k}F_p(u+k,v+k;w+k;z;m). \tag{6.79}$$

A similar procedure can be used to get the differential formula for generalized extended confluent hypergeometric functions as

$$\frac{d^k}{dz^k}\Phi_p(u;w;z;m)=\frac{(u)_k}{(w)_k}\Phi_p(u+k;w+k;z;m). \tag{6.80}$$

Now, we list the transformation formulas for generalized extended Gauss and confluent hypergeometric functions. The following is the transformation formula obeyed by the generalized extended Gauss hypergeometric function.

Theorem 196. *The following formula holds true:*

$$F_p(u,v;w;z;m)=(1-z)^{-u}F_p\left(u,w-v;w;\frac{z}{z-1};m\right),\quad |\arg(1-z)|<\pi. \tag{6.81}$$

Proof. By writing,

$$[1-z(1-\mathrm{u})]^{-u}=(1-z)^{-u}\left(1+\frac{z}{1-z}\mathrm{u}\right)^{-u},$$

and replacing u by $1-\mathrm{u}$ in (6.69), we have

$$\begin{aligned}&F_p(u,v;w;z;m)\\&=\frac{(1-z)^{-u}}{\beta(v,w-v)}\int_0^1\mathrm{u}^{w-v-1}(1-\mathrm{u})^{v-1}\left(1+\frac{z}{1-z}\mathrm{u}\right)^{-u}\exp\left[-\frac{p}{\mathrm{u}^m(1-\mathrm{u})^m}\right]d\mathrm{u},\\&\mathbb{R}(p)\geq 0;p=0,m>0\text{ and }|\arg(z)|<\pi;\mathbb{R}(w)>\mathbb{R}(v)>0,\end{aligned} \tag{6.82}$$

which can be rewritten as

$$F_p(u,v;w;z;m) = (1-z)^{-u} F_p\left(u, w-v; w; \frac{z}{z-1}; m\right), \quad |\arg(1-z)| < \pi. \tag{6.83}$$

It competes the proof. □

Clearly, when $m = 1$, the transformation formulas lead to the transformation formula satisfied by the extended Gauss hypergeometric function given in the previous section. For $p = 0$, they lead to the transformation formulas for the usual Gauss hypergeometric function.

In the similar way, we can obtain the transformation formulas for the generalized extended confluent hypergeometric function as follows.

Theorem 197. *The following identity is the extended version of Kummer's formula:*

$$\Phi_p(u;w;z;m) = \exp(z)\,\Phi_p(w-u;w;-z;m). \tag{6.84}$$

Replacing z by $1-\frac{1}{z}$ in (6.83), we get the following transformation formula:

$$F_p\left(u,v;w;1-\frac{1}{z};m\right) = z^u F_p(u, w-v; w; 1-z; m), \quad |\arg(1-z)| < \pi. \tag{6.85}$$

Replacing z by $\frac{z}{1+z}$ in (6.83), we get the following transformation formula:

$$F_p\left(u,v;w;\frac{z}{1+z};m\right) = z^u F_p(u, w-v; w; 1-z; m), \quad |\arg(1-z)| < \pi. \tag{6.86}$$

Now, we list some recurrence relations obeyed by the generalized extended Gauss hypergeometric and confluent hypergeometric function.

Theorem 198. *The following formula holds true:*

$$\Delta_u F_p(u,v;w;z;m) = \frac{v\,z}{w} F_p(u+1, v+1; w+1; z; m). \tag{6.87}$$

Proof. By the integral representation (6.69) of a generalized extended Gauss hypergeometric function and the difference operator Δ_u, we see that

$$\begin{aligned}\Delta_u F_p(u,v;w;z;m) &= F_p(u+1,v;w;z;m) - F_p(u,v;w;z;m)\\ &= \frac{1}{\beta(v,w-v)} \int_0^1 u^{v-1}(1-u)^{w-v-1}(1-zu)^{-u}\{(1-z\,u)^{-1}-1\}\\ &\quad \times \exp\left[-\frac{p}{u^m(1-u)^m}\right] du\end{aligned}$$

$$= \frac{z}{\beta(v, w-v)} \int_0^1 u^v (1-u)^{w-v-1} (1-zu)^{-u-1}$$
$$\times \exp\left[-\frac{p}{u^m (1-u)^m}\right] du. \tag{6.88}$$

It completes the proof. □

Changing u, v and w to $u+1$, $v+1$ and $w+1$ in (6.69), respectively, yield

$$F_p(u+1, v+1; w+1; z; m) = \frac{1}{\beta(v+1, w-v)} \int_0^1 u^v (1-u)^{w-v-1} (1-zu)^{-u-1}$$
$$\times \exp\left[-\frac{p}{u^m (1-u)^m}\right] du. \tag{6.89}$$

Using the result (6.40) and (6.89) in (6.88), we have

$$\Delta_u F_p(u, v; w; z; m) = \frac{vz}{w} F_p(u+1, v+1; w+1; z; m), \tag{6.90}$$

which is our first recurrence formula for the generalized extended Gauss hypergeometric function. For the generalized extended confluent hypergeometric function, we have the following recurrence relation.

Theorem 199.

$$u \Delta_u \Phi_p(u; w+1; z; m) + w \Delta_w \Phi_p(u; w; z; m) = 0. \tag{6.91}$$

Now replacing u by $u+1$ and w by $w+1$ in (6.73) and using (6.40), we have

$$u \Delta_u \Phi_p(u; w+1; z; m) + w \Delta_w \Phi_p(u; w; z; m) = 0, \tag{6.92}$$

which is a recurrence relation for the generalized extended confluent hypergeometric function.

Clearly, for $m = 1$, (6.92) leads to the recurrence relation for an extended confluent hypergeometric function and for $p = 0$. It reduces to

$${}_1F_1(u+1; w+1; z) = \frac{u}{w} {}_1F_1(u; w; z) + \left(1 - \frac{w}{u}\right) {}_1F_1(u; w+1; z).$$

Again, using the differentiation formula (6.80) with $k = 1$, we have the following difference-differential formula as

$$\frac{d}{dz} \Phi_p(u; w; z; m) = \frac{u}{w} \Phi_p(u; w+1; z; m) - \Delta_w \Phi_p(u; w; z; m).$$

6.3 Third extension of the Gauss hypergeometric function

Özergin et al. considered the following generalizations of gamma and beta functions:

$$\Gamma_p^{a_1,a_2}(u) = \int_0^\infty \mathrm{u}^{u-1}\,{}_1F_1\left(a_1; a_2; -\mathrm{u} - \frac{p}{\mathrm{u}}\right) d\mathrm{u},$$
$$\Re(a_1) > 0, \Re(a_2) > 0, \Re(p) > 0, \Re(u) > 0, \tag{6.93}$$

and

$$\beta_p^{a_1,a_2}(u,v) = \int_0^1 \mathrm{u}^{a-1}(1-\mathrm{u})^{v-1}\,{}_1F_1\left(a_1; a_2; -\frac{p}{\mathrm{u}(1-\mathrm{u})}\right) du,$$
$$\Re(a_1) > 0, \Re(a_2) > 0, \Re(p) > 0, \Re(u) > 0, \Re(v) > 0, \tag{6.94}$$

respectively. It is obvious that $\Gamma_p^{a_1,a_1}(u) = \Gamma_p(u)$, $\Gamma_0^{a_1,a_1}(u) = \Gamma(u)$, $\beta_p^{a_1,a_1}(u,v) = \beta_p(u,v)$, $\beta_0^{a_1,a_2}(u,v) = \beta(u,v)$.

It is important to see some useful integral representations of the new generalized gamma and beta functions

Theorem 200. *For the new generalized gamma function, we have*

$$\Gamma_p^{a_1,a_2}(u) = \frac{\Gamma(a_2)}{\Gamma(a_1)\,\Gamma(a_2-a_1)} \int_0^1 \Gamma_{p\,\mathrm{u}^2}(u)\,\mathrm{u}^{a_1-u-1}(1-\mathrm{u})^{a_2-a_1-1}\,d\mathrm{u}. \tag{6.95}$$

Proof. The integral representation of the confluent hypergeometric function is given by

$${}_1F_1(a_1; a_2; z) = \frac{\Gamma(a_2)}{\Gamma(a_1)\,\Gamma(a_2-a_1)} \int_0^1 \mu^{a_1-1}(1-\mu)^{a_2-a_1-1}\exp(z\mu)\,d\mu.$$

Using this, we have

$$\Gamma_p^{a_1,a_2}(u) = \frac{\Gamma(a_2)}{\Gamma(a_1)\,\Gamma(a_2-a_1)} \int_0^\infty\int_0^1 \mathfrak{w}^{u-1}\exp\left(-\mathfrak{w}\,\mu - \frac{p\,\mu}{\mathfrak{w}}\right)\mu^{a_1-1}$$
$$\times (1-\mu)^{a_2-a_1-1}\exp(z\mu)\,d\mathfrak{w}\,d\mu.$$

Now, using a one-to-one transformation $\mathfrak{v} = \mathfrak{w}\,\mu$, $\mathrm{u} = \mu$ and Jacobian of the transformation $J = \frac{1}{\mathrm{u}}$ in the above inequality, we get

$$\Gamma_p^{a_1,a_2}(u) = \frac{\Gamma(a_2)}{\Gamma(a_1)\,\Gamma(a_2-a_1)} \int_0^\infty\int_0^1 \mathfrak{v}^{u-1}\exp\left(-\mathfrak{v} - \frac{p\,\mathrm{u}^2}{\mathfrak{v}}\right) d\mathfrak{v}\,\mathrm{u}^{a_1-u-1}(1-\mathrm{u})^{a_2-a_1-1}\,d\mathrm{u}.$$

From the uniform convergence of the integrals, the order of integration can be interchanged to yield that

$$\Gamma_p^{a_1,a_2}(u) = \frac{\Gamma(a_2)}{\Gamma(a_1)\,\Gamma(a_2-a_1)} \int_0^1 \int_0^\infty \left[\mathfrak{v}^{u-1} \exp\left(-\mathfrak{v} - \frac{p\,\mathfrak{u}^2}{\mathfrak{v}}\right) d\mathfrak{v} \right] \times \mathfrak{u}^{a_1-u-1}\,(1-\mathfrak{u})^{a_2-a_1-1}\,d\mathfrak{u}$$
$$= \frac{\Gamma(a_2)}{\Gamma(a_1)\,\Gamma(a_2-a_1)} \int_0^1 \Gamma_{p\,\mathfrak{u}^2}(u)\,\mathfrak{u}^{a_1-u-1}\,(1-\mathfrak{u})^{a_2-a_1-1}\,d\mathfrak{u}.$$

It completes the proof. □

In the above theorem, for $p = 0$, we have

$$\Gamma^{a_1,a_2}(u) = \frac{\Gamma(a_2)}{\Gamma(a_1)\,\Gamma(a_2-a_1)} \int_0^1 \Gamma(u)\,\mathfrak{u}^{a_1-u-1}\,(1-\mathfrak{u})^{a_2-a_1-1}\,d\mathfrak{u} = \frac{\Gamma(a_2)\,\Gamma(a_1-u)\,\Gamma(u)}{\Gamma(a_1)\,\Gamma(a_2-u)}. \tag{6.96}$$

The next theorem gives the Mellin transform representation of the function $\beta_p^{a_1,a_2}(u,v)$ in terms of the ordinary beta function and $\Gamma^{a_1,a_2}(u)$.

Theorem 201. *The Mellin transform representation of the new generalized beta function is given by*

$$\int_0^\infty p^{x-1}\,\beta_p^{a_1,a_2}(u,v)\,dp = \beta(x+u,x+v)\,\Gamma^{a_1,a_2}(x), \quad \mathfrak{R}(x) > 0, \mathfrak{R}(x+u) > 0,$$
$$\mathfrak{R}(x+v) > 0, \mathfrak{R}(p) > 0, \mathfrak{R}(a_1) > 0, \mathfrak{R}(a_2) > 0. \tag{6.97}$$

Proof. Multiplying (6.94) by p^{x-1} and integrating with respect to p from $p = 0$ to $p = \infty$, we get

$$\int_0^\infty p^{x-1}\,\beta_p^{a_1,a_2}(u,v)\,dp = \int_0^\infty p^{x-1} \int_0^1 \mathfrak{u}^{u-1}\,(1-\mathfrak{u})^{v-1}\,{}_1F_1\left(a_1;a_2;-\frac{p}{\mathfrak{u}(1-\mathfrak{u})}\right) d\mathfrak{u}\,dp. \tag{6.98}$$

From the uniform convergence of the integral, the order of integration can be interchanged, therefore we have

$$\int_0^\infty p^{x-1}\,\beta_p^{a_1,a_2}(u,v)\,dp = \int_0^1 \mathfrak{u}^{u-1}\,(1-\mathfrak{u})^{v-1} \int_0^\infty p^{x-1}\,{}_1F_1\left(a_1;a_2;-\frac{p}{\mathfrak{u}(1-\mathfrak{u})}\right) dp\,d\mathfrak{u}. \tag{6.99}$$

Now, using the one-to-one transformation $\mathfrak{v} = \frac{p}{\mathfrak{u}(1-\mathfrak{u})}$, $\mu = \mathfrak{u}$, we get

$$\int_0^\infty p^{x-1}\,\beta_p^{a_1,a_2}(u,v)\,dp = \int_0^1 \mu^{u+x-1}(1-\mu)^{v+x-1}\,d\mu \int_0^\infty \upsilon^{x-1}{}_1F_1(a_1;a_2;-\upsilon)\,d\upsilon. \tag{6.100}$$

Therefore, using (6.96), we have

$$\int_0^\infty p^{x-1}\,\beta_p^{a_1,a_2}(u,v)\,dp = \beta(x+u,x+v)\,\Gamma^{a_1,a_2}(x). \tag{6.101}$$

It completes the proof. □

Remark 202. Putting $x = 1$ and considering that $\Gamma^{a_1,a_2}(1) = \frac{\Gamma(a_2)\Gamma(a_1-1)}{\Gamma(a_1)\Gamma(a_2-1)}$ in (6.97), we get

$$\int_0^\infty \beta_p^{a_1,a_2}(u,v)\,dp = \beta(u+1,v+1)\,\frac{\Gamma(a_2)\,\Gamma(a_1-1)}{\Gamma(a_1)\,\Gamma(a_2-1)}. \tag{6.102}$$

Letting $\beta_p^{a_1,a_1}(u,v) = \beta_p(u,v)$, it reduces to the identity

$$\int_0^\infty \beta_p(u,v)\,dp = \beta(u+1,v+1), \qquad \Re(u) > -1, \Re(v) > -1, \tag{6.103}$$

in terms of classical and extended beta function.

Theorem 203. *For the new generalized beta function, we have the following integral representations:*

$$\beta_p^{a_1,a_2}(u,v) = 2\int_0^{\frac{\pi}{2}} \cos^{2u-1}\theta\,\sin^{2v-1}\theta\,{}_1F_1(a_1;a_2;-p\sec^2\theta\,\csc^2\theta)\,d\theta, \tag{6.104}$$

$$\beta_p^{a_1,a_2}(u,v) = \int_0^\infty \frac{\mathfrak{u}^{u-1}}{(1+\mathfrak{u})^{u+v}}\,{}_1F_1\left(a_1;a_2;-2p-p\left(\mathfrak{u}+\frac{1}{\mathfrak{u}}\right)\right)d\mathfrak{u}. \tag{6.105}$$

Proof. Letting $\mathfrak{u} = \cos^2\theta$ in (6.94), we get

$$\begin{aligned}\beta_p^{a_1,a_2}(u,v) &= \int_0^1 \mathfrak{u}^{u-1}(1-\mathfrak{u})^{v-1}{}_1F_1\left(a_1;a_2;-\frac{p}{\mathfrak{u}(1-\mathfrak{u})}\right)d\mathfrak{u}\\ &= 2\int_0^{\frac{\pi}{2}} \cos^{2u-1}\theta\,\sin^{2v-1}\theta\,{}_1F_1(a_1;a_2;-p\sec^2\theta\,\csc^2\theta)\,d\theta.\end{aligned} \tag{6.106}$$

On the other hand, putting $\mathfrak{u} = \frac{t}{1+t}$ in (6.94), we get

$$B_p^{a_1,a_2}(u,v) = \int_0^\infty \frac{t^{u-1}}{(1+t)^{u+v}}\, {}_1F_1\left(a_1; a_2; -2p - p\left(t + \frac{1}{t}\right)\right) dt. \tag{6.107}$$

It completes the proof. □

Theorem 204. *For the new generalized beta function, we have following functional relation:*

$$B_p^{a_1,a_2}(u,v+1) + B_p^{a_1,a_2}(u+1,v) = B_p^{a_1,a_2}(u,v). \tag{6.108}$$

Proof. Direct calculation yields

$$\begin{aligned}
&B_p^{a_1,a_2}(u,v+1) + B_p^{a_1,a_2}(u+1,v) \\
&\quad= \int_0^1 u^u (1-u)^{v-1} {}_1F_1\left(a_1; a_2; -\frac{p}{u(1-u)}\right) du \\
&\qquad+ \int_0^1 u^{u-1} (1-u)^v {}_1F_1\left(a_1; a_2; -\frac{p}{u(1-u)}\right) du \\
&\quad= \int_0^1 [u^u (1-u)^{v-1} + u^{u-1} (1-u)^v]\, {}_1F_1\left(a_1; a_2; -\frac{p}{u(1-u)}\right) du \\
&\quad= \int_0^1 u^{u-1} (1-u)^{v-1} {}_1F_1\left(a_1; a_2; -\frac{p}{u(1-u)}\right) du = B_p^{a_1,a_2}(u,v).
\end{aligned} \tag{6.109}$$

It completes the proof. □

Theorem 205. *For the product of two new generalized gamma function, we have the following integral representation:*

$$\begin{aligned}
&\Gamma_p^{a_1,a_2}(u)\,\Gamma_p^{a_1,a_2}(v) \\
&\quad= 4 \int_0^{\frac{\pi}{2}} \int_0^\infty r^{2(u+v)-1} \cos^{2u-1}\theta \sin^{2v-1}\theta \\
&\qquad\times {}_1F_1\left(a_1; a_2; -r^2\cos^2\theta - \frac{p}{r^2\cos^2\theta}\right) {}_1F_1\left(a_1; a_2; -r^2\sin^2\theta - \frac{p}{r^2\sin^2\theta}\right) dr\, d\theta.
\end{aligned} \tag{6.110}$$

Proof. Substituting $u = \eta^2$ in (6.93), we get

$$\Gamma_p^{a_1,a_2}(u) = 2\int_0^\infty \eta^{2u-1}\, {}_1F_1\left(a_1; a_2; -\eta^2 - \frac{p}{\eta^2}\right) d\eta.$$

Therefore,

$$\Gamma_p^{a_1,a_2}(u)\,\Gamma_p^{a_1,a_2}(v)$$
$$= 4\int_0^\infty\int_0^\infty \eta^{2u-1}\chi^{2v-1}\,{}_1F_1\Big(a_1;a_2;-\eta^2-\frac{p}{\eta^2}\Big)\,{}_1F_1\Big(a_1;a_2;-\chi^2-\frac{p}{\chi^2}\Big)\,d\eta\,d\chi.$$

Letting $\eta = r\cos\theta, \chi = r\sin\theta$ in the above inequality give

$$\Gamma_p^{a_1,a_2}(u)\,\Gamma_p^{a_1,a_2}(v)$$
$$= 4\int_0^{\frac{\pi}{2}}\int_0^\infty r^{2(u+v)-1}\,\cos^{2u-1}\theta\,\sin^{2v-1}\theta$$
$$\times\,{}_1F_1\Big(a_1;a_2;-r^2\cos^2\theta-\frac{p}{r^2\cos^2\theta}\Big)\,{}_1F_1\Big(a_1;a_2;-r^2\sin^2\theta-\frac{p}{r^2\sin^2\theta}\Big)dr\,d\theta.$$

It completes the proof. □

Remark 206. Putting $p = 0$ and $a_1 = a_2$ in (6.110), we get the classical relation between the gamma and beta functions:

$$\beta(u,v) = \frac{\Gamma(u)\,\Gamma(v)}{\Gamma(u+v)}.$$

Theorem 207. *For new generalized beta function, we have the following infinite summation relation:*

$$\beta_p^{a_1,a_2}(u,1-v) = \sum_{k=0}^{\infty}\frac{(v)_k}{k!}\,\beta_p^{a_1,a_2}(u+k,1),\qquad \Re(p) > 0.$$

Proof. From the definition of a new generalized beta function, we get

$$\beta_p^{a_1,a_2}(u,1-v) = \int_0^1 u^{u-1}\,(1-u)^{-v}\,{}_1F_1\Big(a_1;a_2;-\frac{p}{u(1-u)}\Big)\,du.$$

Using the following binomial series expansion,

$$(1-u)^{-v} = \sum_{k=0}^{\infty}(v)_k\,\frac{u^k}{k!},\qquad |u| < 1,$$

we obtain

$$\beta_p^{a_1,a_2}(u,1-v) = \int_0^1\sum_{k=0}^{\infty}\frac{(v)_k}{k!}u^{u+k-1}\,(1-u)^{-v}\,{}_1F_1\Big(a_1;a_2;-\frac{p}{u(1-u)}\Big)\,du.$$

Therefore, interchanging the order of integration and summation, using (6.94), we obtain

$$\begin{aligned} \beta_p^{a_1,a_2}(u,1-v) &= \sum_{k=0}^{\infty} \frac{(v)_k}{k!} \int_0^1 u^{u+k-1}(1-u)^{-v} {}_1F_1\left(a_1;a_2;-\frac{p}{u(1-u)}\right) du \\ &= \sum_{k=0}^{\infty} \frac{(v)_k}{k!} \beta_p^{a_1,a_2}(u+k,1). \end{aligned}$$

It completes the proof. □

Now, we give differential recurrence relations for the generalized gamma and generalized beta functions defined by (6.93) and (6.94), respectively.

Theorem 208. *For the generalized gamma function, we have the following recurrence relation:*

$$\begin{aligned} &\frac{d^2(\Gamma_p^{a_1,a_2}(u+5))}{dp^2} + p\frac{d^2(\Gamma_p^{a_1,a_2}(u+3))}{dp^2} - a_2\frac{d(\Gamma_p^{a_1,a_2}(u+2))}{dp} - \frac{d(\Gamma_p^{a_1,a_2}(u+3))}{dp} \\ &- p\frac{d(\Gamma_p^{a_1,a_2}(u+1))}{dp} + a_1\Gamma_p^{a_1,a_2}(u) = 0. \end{aligned}$$

Proof. Taking the derivative under the integral symbol by using the Leibnitz rule, we get

$$\begin{aligned} &\frac{d^2(\Gamma_p^{a_1,a_2}(u+5))}{dp^2} + p\frac{d^2(\Gamma_p^{a_1,a_2}(u+3))}{dp^2} - a_2\frac{d(\Gamma_p^{a_1,a_2}(u+2))}{dp} - \frac{d(\Gamma_p^{a_1,a_2}(u+3))}{dp} \\ &- p\frac{d(\Gamma_p^{a_1,a_2}(u+1))}{dp} + a_1\Gamma_p^{a_1,a_2}(u) \\ &= \int_0^{\infty} u^{u-1}\left[(u^3+pu)\frac{d^2\mathfrak{w}}{dp^2} + (u^2+a_2u+p)\frac{d\mathfrak{w}}{dp} + a_1\mathfrak{w}\right] du, \end{aligned}$$

where $\mathfrak{w} = {}_1F_1(a_1;a_2;-u-\frac{p}{u})$. On the other hand, since $\mathfrak{w} = {}_1F_1(a_1;a_2;-u-\frac{p}{u})$ is a solution of the equation,

$$(u^3+pu)\frac{d^2\mathfrak{w}}{dp^2} + (u^2+a_2u+p)\frac{d\mathfrak{w}}{dp} + a_1\mathfrak{w} = 0,$$

we get the result. □

Theorem 209. *For the generalized beta function, we have the following recurrence relation:*

$$p\frac{d^2(\beta_p^{a_1,a_2}(u+3,v+3))}{dp^2}+a_2\frac{d(\beta_p^{a_1,a_2}(u+2,v+2))}{dp}+p\frac{d(\beta_p^{a_1,a_2}(u+1,v+1))}{dp}$$
$$+a_1\,\beta_p^{a_1,a_2}(u,v)=0.$$

Proof. Taking the derivative by using the Leibnitz rule, we get

$$p\frac{d^2(\beta_p^{a_1,a_2}(u+3,v+3))}{dp^2}+a_2\frac{d(\beta_p^{a_1,a_2}(u+2,v+2))}{dp}+p\frac{d(\beta_p^{a_1,a_2}(u+1,v+1))}{dp}$$
$$+a_1\,\beta_p^{a_1,a_2}(u,v)$$
$$=\int_0^1 \mathfrak{u}^{u-1}(1-\mathfrak{u})^{v-1}\left[p\mathfrak{u}(1-\mathfrak{u})\frac{d^2\mathfrak{w}}{dp^2}+(a_2\mathfrak{u}(1-\mathfrak{u})+p)\frac{d\mathfrak{w}}{dp}+a_1\,\mathfrak{w}\right]d\mathfrak{u},$$

where $\mathfrak{w}={}_1F_1(a_1;a_2;-\frac{p}{\mathfrak{u}(1-\mathfrak{u})})$. Since $\mathfrak{w}={}_1F_1(a_1;a_2;-\frac{p}{\mathfrak{u}(1-\mathfrak{u})})$ is a solution of the equation

$$p\mathfrak{u}(1-\mathfrak{u})\frac{d^2\mathfrak{w}}{dp^2}+(a_2\mathfrak{u}(1-\mathfrak{u})+p)\frac{d\mathfrak{w}}{dp}+a_1\,\mathfrak{w}=0,$$

we get the result. □

6.3.1 Generalized Gauss hypergeometric and confluent hypergeometric functions

Here, we use the new generalization (6.94) of the beta function to generalize the Gauss hypergeometric and confluent hypergeometric functions defined as

$$F_p^{a_1,a_2}(u,v;w;z)=\sum_{k=0}^{\infty}(u)_k\,\frac{\beta_p^{a_1,a_2}(v+k,w-v)}{\beta(v,w-v)}\,\frac{z^k}{k!}$$

and

$${}_1F_1^{a_1,a_2;p}(u;w;z)=\sum_{k=0}^{\infty}(u)_k\,\frac{\beta_p^{a_1,a_2}(u+k,w-u)}{\beta(u,w-u)}\,\frac{z^k}{k!},$$

respectively. We call $F_p^{a_1,a_2}(u,v;w;z)$ by the generalized Gauss hypergeometric function and ${}_1F_1^{a_1,a_2;p}(u;w;z)$ by the generalized confluent hypergeometric function. Observe that

$$F_p^{a_1,a_1}(u,v;w;z)=F_p(u,v;w;z),\quad F_0^{a_1,a_2}(u,v;w;z)={}_2F_1(u,v;w;z)$$

and

$${}_1F_1^{a_1,a_1;p}(u;w;z)=\Phi_p(u;w;z),\quad {}_1F_1^{a_1,a_2;0}(u;w;z)={}_1F_1(u;w;z).$$

The generalized Gauss hypergeometric function can be provided with an integral representation by using the definition of the new generalized beta function (6.94). We get the following theorem.

Theorem 210. *For the generalized Gauss hypergeometric function, we have the following integral representations:*

$$F_p^{a_1,a_2}(u,v;w;z)$$

$$= \frac{1}{B(v,w-v)} \int_0^1 u^{v-1}(1-u)^{w-v-1} {}_1F_1\left(a_1;a_2;-\frac{p}{u(1-u)}\right)(1-zu)^{-u}\,du,$$

$$\Re(p) \geq 0; p = 0, |\arg(1-z)| < \pi; \Re(w) > \Re(v) > 0 \tag{6.111}$$

$$= \frac{1}{B(v,w-v)} \int_0^1 t^{v-1}(1+t)^{u-w}[1+t(1-z)]^{-u} {}_1F_1\left(a_1;a_2;-2p-p\left(t+\frac{1}{t}\right)\right)dt \tag{6.112}$$

$$= \frac{1}{B(v,w-v)} \int_0^{\frac{\pi}{2}} 2\cos^{2w-2v-1}\theta\,\sin^{2v-1}\theta\,(1-z\sin^2\theta)^{-u}$$

$$\times {}_1F_1(a_1;a_2;-p\sec^2\theta\,\csc^2\theta)\,d\theta. \tag{6.113}$$

Proof. The direct calculation yields

$$F_p^{a_1,a_2}(u,v;w;z)$$

$$= \sum_{k=0}^{\infty}(u)_k \frac{B_p^{a_1,a_2}(v+k,w-v)}{B(v,w-v)}\frac{z^k}{k!}$$

$$= \frac{1}{B(v,w-v)}\sum_{k=0}^{\infty}(u)_k \int_0^1 u^{v+k-1}(1-u)^{w-v-1} {}_1F_1\left(a_1;a_2;-\frac{p}{u(1-u)}\right)\frac{z^k}{k!}\,du$$

$$= \frac{1}{B(v,w-v)} \int_0^1 u^{v-1}(1-u)^{w-v-1} {}_1F_1\left(a_1;a_2;-\frac{p}{u(1-u)}\right)\sum_{k=0}^{\infty}(u)_k\frac{(zu)^k}{k!}\,du$$

$$= \frac{1}{B(v,w-v)} \int_0^1 u^{v-1}(1-u)^{w-v-1} {}_1F_1\left(a_1;a_2;-\frac{p}{u(1-u)}\right)(1-zu)^{-u}\,du.$$

Setting $t = \frac{u}{1-u}$ in (6.111), we get

$$F_p^{a_1,a_2}(u,v;w;z) = \frac{1}{B(v,w-v)} \int_0^1 t^{v-1}(1+t)^{u-w}[1+t(1-z)]^{-u}$$

$$\times {}_1F_1\left(a_1;a_2;-2p-p\left(t+\frac{1}{t}\right)\right)dt.$$

On the other hand, substituting $u = \sin^2\theta$ in (6.111), we have

$$F_p^{a_1,a_2}(u,v;w;z) = \frac{1}{\beta(v,w-v)} \int_0^{\frac{\pi}{2}} 2\cos^{2w-2v-1}\theta \, \sin^{2v-1}\theta \, (1-z\sin^2\theta)^{-u} \times {}_1F_1(a_1;a_2;-p\sec^2\theta\csc^2\theta)\, d\theta.$$

It completes the proof. □

A similar procedure yields an integral representation of the generalized confluent hypergeometric function by using the definition of new generalized beta function.

Theorem 211. *For the generalized confluent hypergeometric function, we have the following integral representations:*

$$ {}_1F_1^{a_1,a_2;p}(u;w;z) = \frac{1}{\beta(u,w-u)} \int_0^1 u^{u-1}(1-u)^{w-u-1} e^{zu}\, {}_1F_1\left(a_1;a_2;-\frac{p}{u(1-u)}\right) du, \quad (6.114)$$

$$ {}_1F_1^{a_1,a_2;p}(u;w;z) = \frac{1}{\beta(u,w-u)} \int_0^1 u^{w-u-1}(1-u)^{u-1} e^{z(1-u)}\, {}_1F_1\left(a_1;a_2;-\frac{p}{u(1-u)}\right) du,$$

$\Re(p) \geq 0$ *and* $\Re(w) > \Re(u) > 0$.

Remark 212. Putting $p = 0$ in (6.111) and (6.114), we get the integral representations of the classical Gauss hypergeometric function and confluent hypergeometric function.

By using the identity (6.40), we obtain new formulas including derivatives of generalized Gauss hypergeometric and confluent hypergeometric functions with respect to variable z.

Theorem 213. *For a generalized Gauss hypergeometric function, we have the following differentiation formula:*

$$\frac{d^k}{dz^k} F_p^{a_1,a_2}(u,v;w;z) = \frac{(v)_k\,(u)_k}{(w)_k} F_p^{a_1,a_2}(u+k,v+k;w+k;z). \quad (6.115)$$

Proof. Taking the derivative of $F_p^{a_1,a_2}(u,v;w;z)$ with respect to z, we obtain

$$\frac{d}{dz} F_p^{a_1,a_2}(u,v;w;z) = \frac{d}{dz}\left\{\sum_{k=0}^{\infty}(u)_k \frac{\beta_p^{a_1,a_2}(v+k,w-v)}{\beta(v,w-v)} \frac{z^k}{k!}\right\}$$
$$= \sum_{k=1}^{\infty}(u)_k \frac{\beta_p^{a_1,a_2}(v+k,w-v)}{\beta(v,w-v)} \frac{z^{k-1}}{(k-1)!}.$$

Replacing $k \to k+1$, we get

$$\frac{d}{dz}F_p^{a_1,a_2}(u,v;w;z) = \frac{u\,v}{w}\sum_{k=0}^{\infty}(u+1)_k\,\frac{B_p^{a_1,a_2}(v+k+1,w-v)}{B(v+1,w-v)}\,\frac{z^k}{k!}$$
$$= \frac{u\,v}{w}\,F_p^{a_1,a_2}(u+1,v+1;w+1;z).$$

A recursive application of this procedure gives us the general form:

$$\frac{d^k}{dz^k}F_p^{a_1,a_2}(u,v;w;z) = \frac{(v)_k\,(u)_k}{(w)_k}\,F_p^{a_1,a_2}(u+k,v+k;w+k;z).$$

It completes the proof. □

Theorem 214. *For a generalized confluent hypergeometric function, we have the following differentiation formula:*

$$\frac{d^k}{dz^k}\,{}_1F_1^{a_1,a_2;p}(u;w;z) = \frac{(u)_k}{(w)_k}\,{}_1F_1^{a_1,a_2;p}(u+k;w+k;z).$$

Now, we obtain the Mellin transform representation of the generalized Gauss and confluent hypergeometric function.

Theorem 215. *For the generalized Gauss hypergeometric function, we have the following Mellin transform representation:*

$$\int_0^{\infty} p^{\chi-1}\,F_p^{a_1,a_2}(u,v;w;z)\,dp = \frac{\Gamma^{a_1,a_2}(\chi)\,B_p^{a_1,a_2}(v+\chi,w+\chi-v)}{B(v,w-v)}$$
$$\times\,{}_2F_1(u,v+\chi;w+2\chi;z). \tag{6.116}$$

Proof. To obtain the Mellin transform, we multiply both sides of (6.111) by $p^{\chi-1}$ and integrate it with respect to p from $0 \to \infty$. Thus, we get

$$\int_0^{\infty} p^{\chi-1}\,F_p^{a_1,a_2}(u,v;w;z)\,dp = \frac{1}{B(v,w-v)}\int_0^1 \mathrm{u}^{v-1}\,(1-\mathrm{u})^{w-v-1}(1-z\mathrm{u})^{-u}$$
$$\times\left[\int_0^{\infty} p^{\chi-1}\,{}_1F_1\!\left(a_1;a_2;-\frac{p}{\mathrm{u}(1-\mathrm{u})}\right)dp\right]d\mathrm{u}. \tag{6.117}$$

Substituting $t = \frac{p}{\mathrm{u}(1-\mathrm{u})}$ in (6.117) gives

$$\int_0^{\infty} p^{\chi-1}\,{}_1F_1\!\left(a_1;a_2;-\frac{p}{\mathrm{u}(1-\mathrm{u})}\right)dp = \int_0^{\infty} t^{\chi-1}\,\mathrm{u}^{\chi}\,(1-\mathrm{u})^{\chi}\,{}_1F_1(a_1;a_2;-t)\,dt$$
$$= \mathrm{u}^{\chi}\,(1-\mathrm{u})^{\chi}\int_0^{\infty} t^{\chi-1}\,{}_1F_1(a_1;a_2;-t)\,dt$$
$$= \mathrm{u}^{\chi}\,(1-\mathrm{u})^{\chi}\,\Gamma^{a_1,a_2}(\chi).$$

Thus, we get

$$\begin{aligned}
&\int_0^\infty p^{x-1} F_p^{a_1,a_2}(u, v; w; z)\, dp \\
&= \frac{1}{\beta(v, w-v)} \int_0^1 u^{v+x-1} (1-u)^{w+x-v-1} (1-zu)^{-u}\, \Gamma^{a_1,a_2}(x)\, du \\
&= \frac{\Gamma^{a_1,a_2}(x)\, \beta_p^{a_1,a_2}(v+x, w+x-v)}{\beta(v, w-v)} \frac{1}{\beta(v+x, w+x-v)} \\
&\quad \times \int_0^1 u^{v+x-1} (1-u)^{w+2x-(v+x)-1} (1-zu)^{-u}\, du \\
&= \frac{\Gamma^{a_1,a_2}(x)\, \beta_p^{a_1,a_2}(v+x, w+x-v)}{\beta(v, w-v)}\, {}_2F_1(u, v+x; w+2x; z).
\end{aligned}$$

It completes the proof. □

Proceeding in the similar way, one can determine the Mellin transform representation for generalized confluent hypergeometric function.

Theorem 216. *For generalized confluent hypergeometric function, we have the following Mellin transform representation:*

$$\begin{aligned}
\int_0^\infty p^{x-1}\, {}_1F_1^{a_1,a_2;p}(u; w; z)\, dp &= \frac{\Gamma^{a_1,a_2}(x)\, \beta_p^{a_1,a_2}(u+x, w+x-u)}{\beta(u, w-u)} \\
&\quad \times {}_1F_1^{a_1,a_2;p}(u+x; w+2x; z).
\end{aligned} \tag{6.118}$$

Theorem 217. *For the new generalized Gauss hypergeometric function, we have the following transformation formula:*

$$F_p^{a_1,a_2}(u, v; w; z) = (1-z)^{-u} F_p^{a_1,a_2}\left(u, w-v; v; \frac{z}{z-1}\right), \quad |\arg(1-z)| < \pi.$$

Proof. By writing

$$[1 - z(1-u)]^{-u} = (1-z)^{-u} \left(1 + \frac{z}{1-z} u\right)^{-u}$$

and replacing $u \to 1-u$ in (6.114), we obtain

$$\begin{aligned}
&F_p^{a_1,a_2}(u, v; w; z) \\
&= \frac{(1-z)^{-u}}{\beta(v, w-v)} \int_0^1 u^{w-v-1} (1-u)^{v-1} \left(1 - \frac{z}{z-1} u\right)^{-u} {}_1F_1\left(a_1; a_2; -\frac{p}{u(1-u)}\right) du,
\end{aligned}$$

$$\Re(p) \geq 0; p = 0, \text{and } |\arg(1-z)| < \pi; \Re(w) > \Re(v) > 0.$$

Hence,

$$F_p^{a_1,a_2}(u, v; w; z) = (1-z)^{-u} F_p^{a_1,a_2}\left(u, w-v; v; \frac{z}{z-1}\right).$$

It completes the proof. □

Remark 218. Note that replacing z by $1 - \frac{1}{z}$ in Theorem 217, one can easily obtain the following transformation formula:

$$F_p^{a_1,a_2}\left(u, v; w; 1 - \frac{1}{z}\right) = z^u F_p^{a_1,a_2}(u, w-v; v; 1-z), \quad |\arg z| < \pi.$$

Furthermore, replacing z by $\frac{z}{1+z}$ in Theorem 217, we get the following transformation formula:

$$F_p^{a_1,a_2}\left(u, v; w; \frac{z}{1+z}\right) = (1+z)^u F_p^{a_1,a_2}(u, w-v; v; z), \quad |\arg(1+z)| < \pi.$$

Remark 219. Setting $z = 1$, in (6.111), we have the following relation between the generalized hypergeometric and beta functions:

$$\begin{aligned} F_p^{a_1,a_2}(u, v; w; 1) &= \frac{1}{\beta(v, w-v)} \int_0^1 \mathrm{u}^{v-1} (1-\mathrm{u})^{w-u-v-1} {}_1F_1\left(a_1; a_2; -\frac{p}{\mathrm{u}(1-\mathrm{u})}\right) du, \\ &= \frac{\beta_p^{a_1,a_2}(v, w-u-v)}{\beta(v, w-v)}. \end{aligned}$$

We now obtain some differential recurrence relations for generalized Gauss and confluent hypergeometric functions. We start with the following theorem.

Theorem 220. *For a generalized Gauss hypergeometric function, we have the following recurrence relation:*

$$\begin{aligned} &p\,\beta(v+3, w-v+3) \frac{d^2 F_p^{a_1,a_2}(u, v+3; w+6; z)}{dp^2} - a_2\,\beta(v+2, w-v+2) \\ &\times \frac{d F_p^{a_1,a_2}(u, v+2; w+4; z)}{dp} - p\,\beta(v+1, w-v+1) \frac{d F_p^{a_1,a_2}(u, v+1; w+2; z)}{dp} \\ &+ a_1 F_p^{a_1,a_2}(u, v; w; z) = 0. \end{aligned}$$

Proof. Taking derivatives under the integral in (6.111) by using the Leibnitz rule, we get

$$p\,\beta(v+3,w-v+3)\,\frac{d^2\,F_p^{a_1,a_2}(u,v+3;w+6;z)}{dp^2}-a_2\,\beta(v+2,w-v+2)$$
$$\times\,\frac{d\,F_p^{a_1,a_2}(u,v+2;w+4;z)}{dp}-p\,\beta(v+1,w-v+1)\,\frac{d\,F_p^{a_1,a_2}(u,v+1;w+2;z)}{dp}$$
$$+\,a_1\,F_p^{a_1,a_2}(u,v;w;z)$$
$$=\int_0^1 \mathrm{u}^{v-1}\,(1-\mathrm{u})^{w-v-1}\left[p\,\mathrm{u}(1-\mathrm{u})\frac{d^2w}{dp^2}+(a_2\,\mathrm{u}(1-\mathrm{u})+p)\,\frac{dw}{dp}+a_1\,w\right]du,$$

where $w={}_1F_1(a_1;a_2;-\frac{p}{\mathrm{u}(1-\mathrm{u})})$. Since $w={}_1F_1(a_1;a_2;-\frac{p}{\mathrm{u}(1-\mathrm{u})})$ is a solution of the equation

$$p\,\mathrm{u}(1-\mathrm{u})\frac{d^2w}{dp^2}+(a_2\,\mathrm{u}(1-\mathrm{u})+p)\,\frac{dw}{dp}+a_1\,w=0,$$

we get the result. □

In a similar manner, we have the following theorem.

Theorem 221.

$$p\,\beta(u+3,w-u+3)\,\frac{d^2\,{}_1F_1^{a_1,a_2;p}(u+3;w+6;z)}{dp^2}-a_2\,\beta(u+2,w-u+2)$$
$$\times\,\frac{d\,{}_1F_1^{a_1,a_2;p}(u+2;w+4;z)}{dp}-p\,\beta(u+1,w-u+1)\,\frac{d\,{}_1F_1^{a_1,a_2;p}(u+1;w+2;z)}{dp}$$
$$+\,a_1\,{}_1F_1^{a_1,a_2;p}(u;w;z)=0.$$

6.4 Fourth extension of Gauss hypergeometric functions

Here, we see the further extension of the extended beta function,

$$\beta(u,v,p)=\int_0^1 \mathrm{u}^{u-1}\,(1-\mathrm{u})^{v-1}\,\exp\left[-\frac{p}{\mathrm{u}\,(1-\mathrm{u})}\right]du,\quad \Re(p)>0,$$

by introducing an additional parameter q as follows:

$$\beta(u,v,p,q)=\int_0^1 \mathrm{u}^{u-1}\,(1-\mathrm{u})^{v-1}\,\exp\left[-\frac{p}{\mathrm{u}}-\frac{q}{(1-\mathrm{u})}\right]du,\quad \Re(p)>0,\Re(q)>0. \tag{6.119}$$

It can be easily seen that the cases $p=q$ and $p=0=q$ of (6.119) reduce it to the extended beta function given in (6.50) and the usual beta function respectively.

First, we investigate various integral representations of the function $\beta(u,v,p,q)$.

Theorem 222. *The following integral formula holds true:*

$$\int_0^\infty\int_0^\infty p^{r-1}q^{s-1}\beta(u,v,p,q)\,dp\,dq = \Gamma(r)\Gamma(s)\beta(u+r,v+s),$$
$$\Re(p)>0,\Re(q)>0,\Re(r)>0,\Re(s)>0,\Re(u+r)>0,\Re(v+s)>0. \tag{6.120}$$

Proof. Multiplying each side of (6.119) by $p^{r-1}q^{s-1}$ and integrating with respect to p and q, from 0 to ∞, we obtain

$$\begin{aligned}&\int_0^\infty\int_0^\infty p^{r-1}q^{s-1}\beta(u,v,p,q)\,dp\,dq\\&\quad=\int_0^\infty\int_0^\infty p^{r-1}q^{s-1}\left[\int_0^1 \mathrm{u}^{u-1}(1-\mathrm{u})^{v-1}\exp\left(-\frac{p}{\mathrm{u}}-\frac{q}{(1-\mathrm{u})}\right)du,\right]dp\,dq.\end{aligned}$$

The uniform convergence of the above integral guarantees that the order of the triple integrals can be interchanged. We therefore have

$$\begin{aligned}&\int_0^\infty\int_0^\infty p^{r-1}q^{s-1}\beta(u,v,p,q)\,dp\,dq\\&\quad=\int_0^1 \mathrm{u}^{u-1}(1-\mathrm{u})^{v-1}\left\{\int_0^\infty p^{r-1}\exp\left(-\frac{p}{\mathrm{u}}\right)dp\int_0^\infty q^{s-1}\left(-\frac{q}{(1-\mathrm{u})}\right)dq\right\}du,\end{aligned}$$

which can be simplified in terms of the gamma function, and thus we get

$$\begin{aligned}&\int_0^\infty\int_0^\infty p^{r-1}q^{s-1}\beta(u,v,p,q)\,dp\,dq\\&\quad=\Gamma(r)\Gamma(s)\int_0^1 \mathrm{u}^{u+r-1}(1-\mathrm{u})^{v+s-1}du=\Gamma(r)\Gamma(s)\beta(u+r,v+s).\end{aligned}$$

It completes the proof. □

Remark 223. The special case $r=1=s$ of the above integral gives an interesting result between the extension of the extended beta function and usual beta function:

$$\int_0^\infty\int_0^\infty \beta(u,v,p,q)\,dp\,dq=\beta(u+1,v+1,p,q),\quad \Re(u)>-1,\Re(v)>-1.$$

Theorem 224. *The following various integral representations for $\beta(u,v,p,q)$ hold true:*

$$\beta(u,v,p,q) = 2\int_0^{\frac{\pi}{2}} \cos^{2u-1}\theta \sin^{2v-1}\theta \exp\left[-\frac{p}{\cos^2\theta} - \frac{q}{\sin^2\theta}\right] d\theta; \tag{6.121}$$

$$\beta(u,v,p,q) = e^{-p-q}\int_0^{\infty} \frac{t^{u-1}}{(1+t)^{u+v}} \exp\left(-qt - \frac{p}{t}\right) dt; \tag{6.122}$$

$$\beta(u,v,p,q) = 2^{1-u-v}\int_{-1}^{1} (1+t)^{u-1}(1-t)^{v-1} \exp\left[-\frac{2(p+q)+2(q-p)t}{1-t^2}\right] dt; \tag{6.123}$$

$$\begin{aligned}\beta(u,v,p,q) &= (y-x)^{1-u-v}\int_x^y (t-x)^{u-1}(y-t)^{v-1}\\ &\times \exp\left[-\frac{(y-x)}{(t-x)(y-t)}\{(q-p)t + (py-qa)\}\right] dt,\\ &\Re(p) > 0, \Re(q) > 0; p = 0 = q; \Re(u) > 0, \Re(v) > 0.\end{aligned} \tag{6.124}$$

Proof. The proofs of the integrals from (6.121) to (6.124) can be obtained easily by employing the transformations $u = \cos^2\theta$, $u = \frac{t}{1+t}$, $u = \frac{1+t}{2}$ and $u = \frac{t-x}{y-x}$ in equation (6.119), respectively. □

Theorem 225. *The following functional relation holds true:*

$$\beta(u+1,v,p,q) + \beta(u,v+1,p,q) = \beta(u,v,p,q). \tag{6.125}$$

Proof. The left hand side together with (6.119) yield

$$\begin{aligned}&\beta(u+1,v,p,q) + \beta(u,v+1,p,q)\\ &= \int_0^1 [u^u(1-u)^{v-1} + u^{u-1}(1-u)^v] \exp\left[-\frac{p}{u} - \frac{q}{(1-u)}\right] du\\ &= \int_0^1 u^{u-1}(1-u)^{v-1} \exp\left[-\frac{p}{u} - \frac{q}{(1-u)}\right] du = \beta(u,v,p,q).\end{aligned} \tag{6.126}$$

It completes the proof. □

Theorem 226. *The following summation formula holds true:*

$$\beta(u,1-v,p,q) = \sum_{k=0}^{\infty} \frac{(v)_k}{k!}\beta(u+k,1,p,q), \quad \Re(p) > 0, \Re(q) > 0. \tag{6.127}$$

Proof. Applying the generalized binomial theorem,

$$(1-u)^{-v} = \sum_{k=0}^{\infty} \frac{(v)_k}{k!} u^k, \quad |u| < 1 \tag{6.128}$$

to the definition (6.119) of $\beta(u, v, p, q)$, we obtain

$$\beta(u, 1-v, p, q) = \int_0^1 \sum_{k=0}^{\infty} \frac{(v)_k}{k!} u^{u+k-1} \exp\left[-\frac{p}{u} - \frac{q}{(1-u)}\right] du. \tag{6.129}$$

Now, interchanging the order of integration and summation in the last expression and again using (6.119), we get

$$\beta(u, 1-v, p, q) = \sum_{k=0}^{\infty} \frac{(v)_k}{k!} \beta(u+k, 1, p, q). \tag{6.130}$$

It completes the proof. □

Theorem 227. *The following infinite summation formula holds true:*

$$\beta(u, v, p, q) = \sum_{k=0}^{\infty} \beta(u+k, v+1, p, q), \quad \Re(p) > 0, \Re(q) > 0. \tag{6.131}$$

Proof. Replacing $(1-u)^{-v}$ in (6.119) by its series representation,

$$(1-u)^{v-1} = (1-u)^{v} \sum_{k=0}^{\infty} u^k, \quad |u| < 1, \tag{6.132}$$

we obtain

$$\beta(u, v, p, q) = \int_0^1 (1-u)^{v} \sum_{k=0}^{\infty} u^{u+k-1} \exp\left[-\frac{p}{u} - \frac{q}{(1-u)}\right] du. \tag{6.133}$$

Now, interchanging the order of integration and summation in the last expression and using (6.119), we get

$$\beta(u, v, p, q) = \sum_{k=0}^{\infty} \beta(u+k, v+1, p, q), \quad \Re(p) > 0, \Re(q) > 0. \tag{6.134}$$

It completes the proof. □

Theorem 228. *For the product of two extended gamma functions, we have the following product formula:*

$$\Gamma_p(u)\,\Gamma_q(v) = 2\int_0^\infty r^{2(u+v)-1}\,\exp(-r^2)\,\beta\Big(u, v, \frac{p}{r^2}, \frac{q}{r^2}\Big)\,dr,$$

$$\Re(p) > 0, \Re(q) > 0; p = 0 = q; \Re(u) > 0, \Re(v) > 0. \tag{6.135}$$

Proof. Setting $u = \eta^2$ and $u = \chi^2$ sequentially in

$$\Gamma_p(u) = \int_0^\infty u^{u-1}\,\exp\Big(-u - \frac{p}{u}\Big)\,du, \quad \Re(p) > 0, \tag{6.136}$$

we get

$$\Gamma_p(u) = 2\int_0^\infty \eta^{2u-1}\,\exp\Big(-\eta^2 - \frac{p}{\eta^2}\Big)\,d\eta \tag{6.137}$$

and

$$\Gamma_q(v) = 2\int_0^\infty \chi^{2v-1}\,\exp\Big(-\chi^2 - \frac{q}{\chi^2}\Big)\,d\chi. \tag{6.138}$$

We therefore have

$$\Gamma_p(u)\,\Gamma_q(v) = 4\int_0^\infty\int_0^\infty \eta^{2u-1}\,r^{2(u+v)-1}\,\chi^{2v-1}\,\exp(-\eta^2 - \chi^2)\,\exp\Big(-\frac{p}{\eta^2} - \frac{q}{\chi^2}\Big)\,d\eta\,d\chi. \tag{6.139}$$

Substituting $\eta = r\cos\theta$ and $\chi = r\sin\theta$ in the above equality, we get

$$\Gamma_p(u)\,\Gamma_q(v) = 4\int_0^{\frac{\pi}{2}}\int_0^\infty r^{2(u+v)-1}\,\exp(-r^2)\,\cos^{2u-1}\theta\,\sin^{2v-1}\theta$$
$$\times \exp\Big(-\frac{p}{r^2\cos^2\theta} - \frac{q}{\sin^2\theta}\Big)\,dr\,d\theta. \tag{6.140}$$

Interchanging the order of integration, we have

$$\Gamma_p(u)\,\Gamma_q(v) = 2\int_0^\infty r^{2(u+v)-1}\,\exp(-r^2)$$
$$\times \left\{\int_0^{\frac{\pi}{2}} \cos^{2u-1}\theta\,\sin^{2v-1}\theta\,\exp\Big(-\frac{p}{r^2\cos^2\theta} - \frac{q}{\sin^2\theta}\Big)\,d\theta\right\}dr. \tag{6.141}$$

Using (6.121), we get

$$\Gamma_p(u)\,\Gamma_q(v) = 2\int_0^\infty r^{2(u+v)-1}\exp(-r^2)\,\beta\Big(u, v, \frac{p}{r^2}, \frac{q}{r^2}\Big)\,dr. \tag{6.142}$$

It completes the proof. □

Theorem 229. *The following identity holds true:*

$$\beta(\alpha, -\alpha - k, p, q) = \sum_{n=0}^{k} \binom{n}{k} \beta(\alpha + n, -\alpha - n, p, q), \quad n \in \mathbb{N}_0. \tag{6.143}$$

Proof. By the fundamental relation, we have

$$\beta(u + 1, v, p, q) + \beta(u, v + 1, p, q) = \beta(u, v, p, q).$$

Setting $u = \alpha$ and $v = -\alpha - k$ in the above formula, we get

$$\beta(\alpha, -\alpha - k, p, q) = \beta(\alpha, -\alpha - k + 1, p, q) + \beta(\alpha + 1, -\alpha - k, p, q).$$

Starting with $k = 1$, we can write this formula recursively to obtain

$$\begin{aligned}
\beta(\alpha, -\alpha - 1, p, q) &= \beta(\alpha, -\alpha, p, q) + \beta(\alpha + 1, -\alpha - 1, p, q);\\
\beta(\alpha, -\alpha - 2, p, q) &= \beta(\alpha, -\alpha, p, q) + 2\,\beta(\alpha + 1, -\alpha - 1, p, q) + \beta(\alpha + 2, -\alpha - 2, p, q);\\
\beta(\alpha, -\alpha - 3, p, q) &= \beta(\alpha, -\alpha, p, q) + 3\,\beta(\alpha + 1, -\alpha - 1, p, q) + 3\,\beta(\alpha + 2, -\alpha - 2, p, q)\\
&\quad + \beta(\alpha + 3, -\alpha - 3, p, q),
\end{aligned}$$

and so on. Thus, it can be seen that the series rises exactly as the expansion of the finite binomial series. Therefore, we get the desired identity. □

Theorem 230. *The following Mellin transformation formula holds true:*

$$\begin{aligned}
&\beta(u, v, p, q)\\
&= \frac{1}{(2\pi\iota)^2}\int_{-\iota\infty}^{\iota\infty}\int_{-\iota\infty}^{\iota\infty} \frac{\Gamma(r)\,\Gamma(s)\,\Gamma(u + r)\,\Gamma(v + s)}{\Gamma(u + v + r + s)}\, p^{-r}\, q^{-s}\, dr\, ds, \quad \Re(p) > 0, \Re(q) > 0.
\end{aligned} \tag{6.144}$$

Proof. We already proved that

$$\begin{gathered}
\int_0^\infty\int_0^\infty p^{r-1} q^{s-1}\,\beta(u, v, p, q)\,dp\,dq = \Gamma(r)\,\Gamma(s)\,\beta(u + r, v + s),\\
\Re(p) > 0, \Re(q) > 0, \Re(r) > 0, \Re(s) > 0, \Re(u + r) > 0, \Re(v + s) > 0.
\end{gathered} \tag{6.145}$$

Using the fundamental relation between the beta and gamma functions, we have

$$\int_0^\infty \int_0^\infty p^{r-1} q^{s-1} \beta(u,v,p,q)\, dp\, dq = \frac{\Gamma(r)\Gamma(s)\Gamma(u+r)\Gamma(v+s)}{\Gamma(u+v+r+s)}. \tag{6.146}$$

Finally, taking the inverse Mellin transform yields

$$\beta(u,v,p,q) = \frac{1}{(2\pi\iota)^2} \int_{-\iota\infty}^{\iota\infty} \int_{-\iota\infty}^{\iota\infty} \frac{\Gamma(r)\Gamma(s)\Gamma(u+r)\Gamma(v+s)}{\Gamma(u+v+r+s)} p^{-r} q^{-s}\, dr\, ds. \tag{6.147}$$

It completes the proof. □

6.4.1 Extension of Gauss and confluent hypergeometric functions

Here, we use $\beta(u,v,p,q)$ to extend the Gauss and confluent hypergeometric functions as

$$F_{p,q}(u,v;w;z) = \sum_{k=0}^{\infty} \frac{\beta(v+k, w-v, p, q)}{\beta(v, w-v)} (u)_k \frac{z^k}{k!}, \quad p \geq 0, q \geq 0, |z| < 1; \Re(w) > \Re(v) > 0 \tag{6.148}$$

and

$$\Phi_{p,q}(u;w;z) = \sum_{k=0}^{\infty} \frac{\beta(u+k, w-u, p, q)}{\beta(u, w-u)} (u)_k \frac{z^k}{k!}, \quad p \geq 0, q \geq 0, \Re(w) > \Re(u) > 0. \tag{6.149}$$

The functions $F_{p,q}(u,v;w;z)$ and $\Phi_{p,q}(u;w;z)$ will be called the further extension of the extended Gauss hypergeometric and confluent hypergeometric functions, respectively.

Theorem 231. *The following integral formulas hold true:*

$$F_{p,q}(u,v;w;z) = \frac{1}{\beta(v,w-v)} \int_0^1 u^{v-1} (1-u)^{w-v-1} \exp\left[-\frac{p}{u} - \frac{q}{(1-u)}\right] \sum_{k=0}^{\infty} (u)_k \frac{z^k}{k!}\, du,$$
$$p > 0, q > 0; p = 0, q = 0 \text{ and } |z| < 1; \Re(w) > \Re(v) > 0 \tag{6.150}$$

$$= \frac{1}{\beta(v,w-v)} \int_0^1 u^{v-1} (1-u)^{w-v-1} (1-zu)^{-u} \exp\left[-\frac{p}{u} - \frac{q}{(1-u)}\right] du,$$
$$p > 0, q > 0; p = 0, q = 0 \text{ and } |\arg(1-z)| < \pi; \Re(w) > \Re(v) > 0 \tag{6.151}$$

$$= \frac{e^{-p-q}}{\beta(v,w-v)} \int_0^1 t^{v-1} (1+t)^{u-w} (1+t(1-z))^{-u} \exp\left(-qt - \frac{p}{t}\right) dt,$$
$$p > 0, q > 0; p = 0, q = 0 \text{ and } |\arg(1-z)| < \pi; \Re(w) > \Re(v) > 0 \tag{6.152}$$

$$= \frac{2}{\beta(v, w-v)} \int_0^{\frac{\pi}{2}} \frac{\sin^{2v-1}\theta \cos^{2w-2v-1}\theta}{(1 - z\sin^2\theta)^u}$$

$$\times \exp[-p \sec^2\theta \csc^2\theta + (p-q)\sec^2\theta]\, d\theta, \quad p > 0, q > 0;$$

$$p = 0, q = 0 \; and \; |\arg(1-z)| < \pi; \Re(w) > \Re(v) > 0 \tag{6.153}$$

$$= \frac{2}{\beta(v, w-v)} \int_0^{\infty} \frac{\sinh^{2v-1}\theta \cosh^{2u-2w-1}\theta}{(\cosh^2\theta - z\sinh^2\theta)^u}$$

$$\times \exp[-p \cosh^2\theta \coth^2\theta + (p-q)\cosh^2\theta]\, d\theta, \quad p > 0, q > 0;$$

$$p = 0, q = 0 \; and \; |\arg(1-z)| < \pi; \Re(w) > \Re(v) > 0; \tag{6.154}$$

$$\Phi_{p,q}(u; w; z) = \frac{1}{\beta(u, w-u)} \int_0^1 u^{u-1}(1-u)^{w-u-1} \exp\left[zu - \frac{p}{u} - \frac{q}{(1-u)}\right]$$

$$\times \sum_{k=0}^{\infty} (u)_k \frac{z^k}{k!}\, du, \quad p > 0, q > 0; p = 0, q = 0; \Re(w) > \Re(u) > 0 \tag{6.155}$$

$$= \frac{e^z}{\beta(u, w-u)} \int_0^1 u^{w-u-1}(1-u)^{u-1}(1-zu)^{-u}$$

$$\times \exp\left[-zu - \frac{p}{u} - \frac{q}{(1-u)}\right] du,$$

$$p > 0, q > 0; p = 0, q = 0; \Re(w) > \Re(u) > 0. \tag{6.156}$$

Proof. The integral formula in (6.150) can be easily obtained by using the integral definition of $\beta(v+k, w-v, p, q)$ in (6.148). Employing the generalized binomial expansion,

$$(1-zu)^{-u} = \sum_{k=0}^{\infty} (u)_k \frac{(zu)^k}{k!}, \quad |zu| < 1, \tag{6.157}$$

in (6.150), we find the integral in (6.151). Similarly, the integral (6.152)–(6.154) can be obtained by using the transformations $u = \frac{t}{1+t}$, $u = \sin^2\theta$, $u = \tanh^2\theta$ in (6.151), respectively. A similar argument will be established in the integral (6.155) for further extension of extended confluent hypergeometric function. The integral in (6.156) is obtained by replacing t with $1-t$ in (6.155). □

Differentiation formulas for further extension of the extended Gauss hypergeometric and extended confluent hypergeometric functions can be found by differentiating (6.148) and (6.149) with respect to the z together with the formula (6.40).

Theorem 232. *The following differentiation formulas hold true:*

$$\frac{d}{dz} F_{p,q}(u, v; w; z) = \frac{u\,v}{w} F_{p,q}(u+1, v+1; w+1; z); \tag{6.158}$$

$$\frac{d^k}{dz^k} F_{p,q}(u, v; w; z) = \frac{(u)_k\,(v)_k}{(w)_k} F_{p,q}(u+k, v+k; w+k; z), \quad k \in \mathbb{N}_0; \tag{6.159}$$

$$\frac{d^k}{dz^k} \Phi_{p,q}(u; w; z) = \frac{(u)_k}{(w)_k} \Phi_{p,q}(u+k; w+k; z), \quad k \in \mathbb{N}_0. \tag{6.160}$$

Proof. Differentiating (6.148) with respect to z, we have

$$\begin{aligned} \frac{d}{dz} F_{p,q}(u, v; w; z) &= \sum_{k=1}^{\infty} \frac{\beta(v+k, w-v, p, q)}{\beta(v, w-v)} (u)_k \frac{z^{k-1}}{(k-1)!} \\ &= \sum_{k=0}^{\infty} \frac{\beta(v+k+1, w-v, p, q)}{\beta(v, w-v)} (u)_{k+1} \frac{z^k}{k!}. \end{aligned} \tag{6.161}$$

Using (6.40), we get

$$\frac{d}{dz} F_{p,q}(u, v; w; z) = \frac{u\,v}{w} F_{p,q}(u+1, v+1; w+1; z). \tag{6.162}$$

Repeating this process k-times yields

$$\frac{d^k}{dz^k} F_{p,q}(u, v; w; z) = \frac{(u)_k\,(v)_k}{(w)_k} F_{p,q}(u+k, v+k; w+k; z). \tag{6.163}$$

Again differentiating (6.149) and using the same argument as above, we get

$$\frac{d^k}{dz^k} \Phi_{p,q}(u; w; z) = \frac{(u)_k}{(w)_k} \Phi_{p,q}(u+k; w+k; z). \tag{6.164}$$

It completes the proof. □

Theorem 233. *The following Mellin–Barnes integral representations hold true:*

$$\begin{aligned} F_{p,q}(u, v; w; z) = {} & \frac{1}{(2\pi\iota)^2\,\beta(v, w-v)} \int_{-\iota\infty}^{\iota\infty}\int_{-\iota\infty}^{\iota\infty} \frac{\Gamma(r)\,\Gamma(s)\,\Gamma(v+r)\,\Gamma(w-v+s)}{\Gamma(w+r+s)} \\ & \times F_{p,q}(u, v+r; w+r+s; z)\, p^{-r}\, q^{-s}\, dr\, ds, \end{aligned} \tag{6.165}$$

$$\begin{aligned} \Phi_{p,q}(u; w; z) = {} & \frac{1}{(2\pi\iota)^2\,\beta(u, w-u)} \int_{-\iota\infty}^{\iota\infty}\int_{-\iota\infty}^{\iota\infty} \frac{\Gamma(r)\,\Gamma(s)\,\Gamma(v+r)\,\Gamma(w-v+s)}{\Gamma(w+r+s)} \\ & \times \Phi_{p,q}(u+r; w+r+s; z)\, p^{-r}\, q^{-s}\, dr\, ds. \end{aligned} \tag{6.166}$$

Proof. In order to find the Mellin transform of further extended Gauss hypergeometric function, we multiply both sides of (6.151) by $p^{-r}\,q^{-s}$ and integrating with respect to p from 0 to ∞ and q from 0 to ∞, thus we get

$$\int_0^\infty\int_0^\infty p^{-r}\,q^{-s}\,F_{p,q}(u,v;w;z)\,dp\,dq$$
$$=\frac{1}{\beta(v,w-v)}\int_0^1 u^{v-1}\,(1-u)^{w-v-1}\,(1-zu)^{-u}$$
$$\times\left\{\int_0^\infty p^{r-1}\,\exp\left(-\frac{p}{u}\right)dp\int_0^\infty q^{s-1}\left(-\frac{q}{(1-u)}\right)dq\right\}du. \tag{6.167}$$

Using the definition of the gamma function, we get

$$\int_0^\infty\int_0^\infty p^{-r}\,q^{-s}\,F_{p,q}(u,v;w;z)\,dp\,dq$$
$$=\frac{\Gamma(r)\,\Gamma(s)}{\beta(v,w-v)}\int_0^1 u^{v+r-1}\,(1-u)^{w-v+s-1}\,(1-zu)^{-u}\,du$$
$$=\frac{\Gamma(r)\,\Gamma(s)\,\beta(v+r,w-v+s)}{\beta(v,w-v)}\,F(u,v+r;w+r+s;z). \tag{6.168}$$

Now, taking the inverse Mellin transform and using the fundamental relation between classical gamma and beta functions, we get

$$F_{p,q}(u,v;w;z)=\frac{1}{(2\pi\iota)^2\,\beta(v,w-v)}\int_{-\iota\infty}^{\iota\infty}\int_{-\iota\infty}^{\iota\infty}\frac{\Gamma(r)\,\Gamma(s)\,\Gamma(v+r)\,\Gamma(w-v+s)}{\Gamma(w+r+s)}$$
$$\times F_{p,q}(u,v+r;w+r+s;z)\,p^{-r}\,q^{-s}\,dr\,ds. \tag{6.169}$$

Similarly, we can establish the Mellin transform of the further extended confluent hypergeometric function. □

Theorem 234. *The following transformation formulas hold true:*

$$F_{p,q}(u,v;w;z)=(1-z)^{-u}\,F_{p,q}\left(u,w-v;w;-\frac{z}{1-z}\right),\quad p>0,q>0;$$
$$p=0,q=0\ and\ |\arg(1-z)|<\pi;\ \Re(w)>\Re(v)>0; \tag{6.170}$$
$$F_{p,q}\left(u,v;w;1-\frac{1}{z}\right)=z^{u}\,F_{p,q}(u,w-v;w;1-z),$$
$$p>0,q>0;p=0,q=0\ and\ |\arg z|<\pi;\ \Re(w)>\Re(v)>0; \tag{6.171}$$
$$F_{p,q}\left(u,v;w;\frac{z}{1+z}\right)=(1+z)^{u}\,F_{p,q}(u,w-v;w;-z),\quad p>0,q>0;$$
$$p=0,q=0\ and\ |\arg(1+z)|<\pi;\ \Re(w)>\Re(v)>0; \tag{6.172}$$
$$\Phi_{p,q}(u;w;z)=e^{z}\,\Phi_{p,q}(w-u;w;-z). \tag{6.173}$$

Proof. By writing

$$[1 - z(1 - u)]^{-u} = (1 - z)^{-u}\left(1 + \frac{z}{1 - z}\right)^{-u}, \tag{6.174}$$

and replacing u by 1 − u in (6.151), we get

$$\begin{aligned} F_{p,q}(u, v; w; z) &= \frac{(1 - z)^{-u}}{\beta(v, w - v)} \int_0^1 u^{v-1} (1 - u)^{w-v-1} \left(1 + \frac{z}{1 - z}\right)^{-u} \\ &\quad \times \exp\left[-\frac{p}{u} - \frac{q}{(1 - u)}\right] du \\ &= (1 - z)^{-u} F_{p,q}\left(u, w - v; w; -\frac{z}{1 - z}\right), \quad p > 0, q > 0; \\ &\quad p = 0, q = 0 \text{ and } |\arg(1 - z)| < \pi; \Re(w) > \Re(v) > 0. \end{aligned} \tag{6.175}$$

Now, replacing z by $1 - \frac{z}{1+z}$ and $\frac{z}{1+z}$ in (6.170), gives the other transformations of $F_{p,q}(u, v; w; z)$. Similarly, the transformation formula for $\Phi_{p,q}(u; w; z)$ can be obtained. □

Theorem 235. *The following difference relation holds true:*

$$\Delta_u F_{p,q}(u, v; w; z) = \frac{vz}{w} F_{p,q}(u + 1, v + 1; w + 1; z). \tag{6.176}$$

Proof. The action of the difference operator Δ_u gives

$$\begin{aligned} &\Delta_u F_{p,q}(u, v; w; z) \\ &\quad = F_{p,q}(u + 1, v; w; z) - F_{p,q}(u, v; w; z) \\ &\quad = \frac{z}{\beta(v, w - v)} \int_0^1 u^{v} (1 - u)^{w-v-1} (1 - uz)^{-u-1} \exp\left[-\frac{p}{u} - \frac{q}{(1 - u)}\right] du. \end{aligned}$$

Replacing the parameter u, v and w by $u + 1$, $v + 1$ and $w + 1$, we get

$$\begin{aligned} &F_{p,q}(u + 1, v + 1; w + 1; z) \\ &\quad = \frac{1}{\beta(v + 1, w - v)} \int_0^1 u^{v} (1 - u)^{w-v-1} (1 - uz)^{-u-1} \exp\left[-\frac{p}{u} - \frac{q}{(1 - u)}\right] du. \end{aligned}$$

The above two equations together yield

$$\Delta_u F_{p,q}(u, v; w; z) = \frac{vz}{w} F_{p,q}(u + 1, v + 1; w + 1; z).$$

It completes the proof. □

Exercises

1. Prove that the Mellin transforms of the extended Gauss and confluent hypergeometric functions are given by

$$\int_0^\infty p^{s-1} F_p(u, v; w; z)\, dp = \Gamma(s) \frac{\beta(v+s, w+s-v)}{\beta(v, w-v)} \, {}_2F_1(u, v+s, w+2s; z);$$

$$\int_0^\infty p^{s-1} \Phi_p(v; w; z)\, dp = \Gamma(s) \frac{\beta(v+s, w+s-v)}{\beta(v, w-v)} \, {}_1F_1(v+s, w+2s; z).$$

2. Verify the following recurrence relations for the extended confluent hypergeometric function:

$$v \Delta_v \Phi_p(v; w+1; z) + w \Delta_w \Phi_p(v; w; z) = 0;$$

$$\frac{d}{dz}\{\Phi_p(v; w; z)\} = \frac{v}{w} \Phi_p(v; w+1; z) - \Delta_w \Phi_p(v; w; z).$$

3. Show that the extended version of Gauss summation formula

$${}_2F_1(u, v, w; 1) = \frac{\Gamma(w)\,\Gamma(w-u-v)}{\Gamma(w-u)\,\Gamma(w-v)} = \frac{\beta(v, w-u-v)}{\beta(v, w-v)}$$

is given by

$$F_p(u, v, w; 1) = \frac{\beta(v, w-u-v; p)}{\beta(v, w-v; p)}, \quad p > 0; p = 0 \text{ and } \Re(w-u-v) > 0.$$

4. Prove that the differential formula for the extended confluent hypergeometric function is

$$\frac{d^k}{dz^k}\{\Phi_p(u; w; z)\} = \frac{(u)_k}{(w)_k} \Phi_p(u+k; w+k; z).$$

5. Show that the following relation between new generalized beta function and classical beta function is true:

$$\int_0^\infty p^{s-1} \beta(u, v; p; m)\, dp = \beta(u+m, v+m).$$

6. Prove that the derivative of the new generalized beta function with respect to the parameter p, for nonnegative integer k, is given by

$$\frac{\partial^k}{\partial p^k}[\beta(u, v; p; m)] = (-1)^k \beta(u-km, v-km; p; m), \quad \Re(p) > 0, m > 0.$$

7. For $\Re(p) > 0, m > 0$ and $c > 0$, prove that

$$\beta(u, v; p; m) = \frac{1}{2\pi\iota} \int_{c-\iota\infty}^{c+\iota\infty} \frac{\Gamma(s)\,\Gamma(u + m s)\,(v + ms)}{\Gamma(u + v + 2 m s)}\, p^{-s}\, ds.$$

8. Obtain the following relations for generalized extended Gauss hypergeometric function and generalized extended confluent hypergeometric function:

$$\int_0^{\infty} p^{s-1} F_p(u, v; w; z; m)\, dp = \Gamma(s)\, \frac{\beta(v + ms, w + ms - v)}{\beta(v, w - v)} \times {}_2F_1(u, v + ms, w + 2ms; z);$$

$$\int_0^{\infty} p^{s-1} \Phi_p(v; w; z; m)\, dp = \Gamma(s)\, \frac{\beta(v + ms, w + ms - v)}{\beta(v, w - v)}\, {}_1F_1(v + ms, w + 2ms; z);$$

9. Show that the extended version of Gauss summation formula is

$$F_p(u, v, w; 1; m) = \frac{\beta(v, w - u - v; p; m)}{\beta(v, w - v; p)}, \quad p > 0; p = 0; m > 0 \text{ and } \Re(w - u - v) > 0.$$

10. By the Mellin inversion formula, prove that the following complex integral representation holds true:

$$\beta_p^{a_1,a_2}(u, v) = \frac{1}{2\pi\iota} \int_{-\iota\infty}^{\iota\infty} \beta(x + u, x + v)\, \Gamma^{a_1,a_2}(x)\, p^{-x}\, dx,$$

$$F_p^{a_1,a_2}(u, v; w; z) = \frac{1}{2\pi\iota} \int_{-\iota\infty}^{\iota\infty} \frac{\Gamma^{a_1,a_2}(x)\, \beta_p^{a_1,a_2}(v + x, w + x - v)}{\beta(v, w - v)} \times {}_2F_1(u, v + x; w + 2x; z)\, p^{-x}\, dx,$$

$${}_1F_1^{a_1,a_2;p}(v; w; z) = \frac{1}{2\pi\iota} \int_{-\iota\infty}^{\iota\infty} \frac{\Gamma^{a_1,a_2}(x)\, \beta_p^{a_1,a_2}(v + x, w + x - v)}{\beta(v, w - v)} \times {}_1F_1^{a_1,a_2;p}(v + x; w + 2x; z)\, p^{-x}\, dx.$$

11. For the generalized confluent hypergeometric function, verify the following transformation formula:

$${}_1F_1^{a_1,a_2;p}(u; w; z) = \exp(z)\, {}_1F_1^{a_1,a_2;p}(w - u; w; -z).$$

12. Verify the following recurrence type relation for extension of extended beta function:

$$\begin{aligned} &u\,\beta(u, v+1, p, q) - v\,\beta(u+1, v, p, q) \\ &\quad = q\,\beta(u+1, v-1, p, q) - p\,\beta(u-1, v+1, p, q), \quad \Re(p) > 0, \Re(q) > 0. \end{aligned}$$

13. Verify the following differential-difference relations for extension of extended Gauss and confluent hypergeometric function:

$$\begin{aligned} &\frac{d}{dz} F_{p,q}(u, v; w; z) = \frac{u}{z} \Delta_u F_{p,q}(u, v; w; z); \\ &u\,\Delta_u \Phi_{p,q}(u; w+1; z) + w\,\Delta_w \Phi_{p,q}(u; w; z) = 0; \\ &\frac{d}{dz} \Phi_{p,q}(u; w; z) = \frac{u}{w} \Phi_{p,q}(u; w+1; z) - \Delta_w \Phi_{p,q}(u; w; z); \\ &\Phi_{p,q}(u+1; w+1; z) = \frac{w}{u} \Phi_{p,q}(u; w; z) + \left(1 - \frac{w}{u}\right) \Phi_{p,q}(u; w+1; z). \end{aligned}$$

14. Prove the following summation formula:

$$\begin{aligned} &F_{p,q}(u, v; w; 1) \\ &\quad = \frac{\beta(v, w-u-v, p, q)}{\beta(v, w-v)}, \quad p > 0, q > 0; p = 0, q = 0 \text{ and } \Re(w-u-v) > 0. \end{aligned}$$

7 Other popular forms of a Gauss hypergeometric function

In this chapter, we discuss the other three forms of the Gaussian hypergeometric function that have been developed in recent years. In particular, we have given the (p, q) analogue, the incomplete forms and the discrete version of the Gaussian hypergeometric function. Properties such as integral representations, differential or difference equations, transformation and summation formulas in terms of these functions are also presented in the chapter. Following literature and references therein can be accessed to elaborate these three forms in detail [11, 12, 15, 24–27, 86, 100, 123–125, 131, 144, 150].

7.1 The (p, q) Gauss hypergeometric function

The (p, q)-analogue of a complex number u, for $p, q > 0$ and $pq \neq 1$, is defined by

$$[u]_{p,q} = \frac{p^{\frac{1}{2}u} - q^{-\frac{1}{2}u}}{p^{\frac{1}{2}} - q^{-\frac{1}{2}}}. \tag{7.1}$$

The (p, q)-shifted factorial function for a complex number u, denoted by $([u]_{p,q})_n$, is defined as [131],

$$([u]_{p,q})_n = \begin{cases} [u]_{p,q}[u+1]_{p,q} \cdots [u+n-1]_{p,q}, & \text{if } n = 1, 2, \ldots, \\ 1, & \text{if } n = 0. \end{cases} \tag{7.2}$$

We now list several results in the following theorem associated with (p, q)-shifted factorial function.

Theorem 236. *Let $u \in \mathbb{C}$ be such that $\mathbb{R}(u) \neq k$ for all integers k. Then the following identities for the (p, q)-shifted factorial function hold:*
1. $([u]_{p,q})_{m+n} = ([u]_{p,q})_m([u+m]_{p,q})_n = ([u]_{p,q})_n([u+n]_{p,q})_m$;
2. $([u-k]_{p,q})_k = (-1)^k(\frac{p}{q})^{\frac{1}{2}ku - \frac{1}{2}\binom{k+1}{2}}([-u+1]_{p,q})_k$;
3. $([u]_{p,q})_{n-k} = (-1)^k(\frac{q}{p})^{\frac{1}{2}k(u+n) - \frac{1}{2}\binom{k+1}{2}}\frac{([u]_{p,q})_n}{([1-(u+n)]_{p,q})}$;
4. $([u-k]_{p,q})_n = ([u]_{p,q})_k([1-u]_{p,q})_n([1-(u+k)]_{p,q})_n^{-1}(\frac{p}{q})^{-\frac{1}{2}nk}$;
5. $([u]_{p,q})_n = (-1)^n([1-(u+n)]_{p,q})_n(\frac{p}{q})^{\frac{1}{2}\binom{n}{2} + \frac{1}{2}nu}$.

Proof. The proof of the first identity follows directly from the definition of (p, q)-shifted factorial function. To prove the second, consider

$$\begin{aligned} &([u-k]_{p,q})_k \\ &\quad = [u-k]_{p,q}[u-k+1]_{p,q} \cdots [u-1]_{p,q} \end{aligned}$$

https://doi.org/10.1515/9783111324586-007

$$
\begin{aligned}
&= \frac{p^{\frac{1}{2}(u-k)} - q^{-\frac{1}{2}(u-k)}}{p^{\frac{1}{2}} - q^{-\frac{1}{2}}} \frac{p^{\frac{1}{2}(u-k+1)} - q^{-\frac{1}{2}(u-k+1)}}{p^{\frac{1}{2}} - q^{-\frac{1}{2}}} \cdots \frac{p^{\frac{1}{2}(u-1)} - q^{-\frac{1}{2}(u-1)}}{p^{\frac{1}{2}} - q^{-\frac{1}{2}}} \\
&= (-1)\left(\frac{p}{q}\right)^{\frac{1}{2}(u-k)} \frac{p^{\frac{1}{2}(-u+k)} - q^{-\frac{1}{2}(-u+k)}}{p^{\frac{1}{2}} - q^{-\frac{1}{2}}} (-1)\left(\frac{p}{q}\right)^{\frac{1}{2}(u-k+1)} \\
&\quad \times \frac{p^{\frac{1}{2}(-u+k-1)} - q^{-\frac{1}{2}(-u+k-1)}}{p^{\frac{1}{2}} - q^{-\frac{1}{2}}} \cdots (-1)\left(\frac{p}{q}\right)^{\frac{1}{2}(u-1)} \frac{p^{\frac{1}{2}(-u+1)} - q^{-\frac{1}{2}(-u+1)}}{p^{\frac{1}{2}} - q^{-\frac{1}{2}}} \\
&= (-1)^k \left(\frac{p}{q}\right)^{\frac{1}{2}ku - \frac{1}{2}\binom{k+1}{2}} ([-u+1]_{p,q})_k. \qquad (7.3)
\end{aligned}
$$

Similarly, other identities can be proved. □

We know that for every infinite product we can assign an infinite series whose terms are logarithms. We prove that both the infinite product and the series have the same character. Consequently, the character of an infinite product can be analyzed using an infinite series. This concept leads us to extend the (p,q)-shifted factorial function to an invertible convergent infinite product, which is described in the following theorem.

Theorem 237. *Let u be a complex number such that $(pq)^{-\frac{1}{2}k} \neq \mathbb{R}((pq)^{\frac{1}{2}u})$ for all integers $k \geq 0$. Then*

$$
\begin{aligned}
([u]_{p,q})_\infty &= \lim_{k\to\infty} \prod_{m=0}^{k-1} \frac{p^{\frac{1}{2}(u+m)} - q^{-\frac{1}{2}(u+m)}}{p^{\frac{1}{2}} - q^{-\frac{1}{2}}} \\
&= \prod_{m=0}^{\infty} \frac{p^{\frac{1}{2}(u+m)} - q^{-\frac{1}{2}(u+m)}}{p^{\frac{1}{2}} - q^{-\frac{1}{2}}}. \qquad (7.4)
\end{aligned}
$$

Proof. The infinite product defined in (7.4) converges if and only if the infinite product $\prod_{m=0}^{\infty}(p^{\frac{1}{2}(u+m)} - q^{-\frac{1}{2}(u+m)})$ converges. Since we are assuming that $(pq)^{-\frac{1}{2}k} \neq \mathbb{R}((pq)^{\frac{1}{2}u})$ for all integers $k \geq 0$, the term of the form $(p^{\frac{1}{2}(u+m)} - q^{-\frac{1}{2}(u+m)})$ is invertible and its logarithm exists. If the infinite product has (a finite number of) a singular factor, we remove them. Consider

$$
P_k = \prod_{m=0}^{k-1}(p^{\frac{1}{2}(u+m)} - q^{-\frac{1}{2}(u+m)}) \quad \text{and} \quad S_k = \sum_{m=0}^{k-1} \ln(p^{\frac{1}{2}(u+m)} - q^{-\frac{1}{2}(u+m)}). \qquad (7.5)
$$

Therefore, we have

$$
\begin{aligned}
e^{S_k} &= e^{\sum_{m=0}^{k-1} \ln(p^{\frac{1}{2}(u+m)} - q^{-\frac{1}{2}(u+m)})} \\
&= e^{\ln(p^{\frac{1}{2}u} - q^{-\frac{1}{2}u})} e^{\ln(p^{\frac{1}{2}(u+1)} - q^{-\frac{1}{2}(u+1)})} \cdots e^{\ln(p^{\frac{1}{2}(u+k-1)} - q^{-\frac{1}{2}(u+k-1)})} \\
&= (p^{\frac{1}{2}u} - q^{-\frac{1}{2}u})(p^{\frac{1}{2}(u+1)} - q^{-\frac{1}{2}(u+1)}) \cdots (p^{\frac{1}{2}(u+k-1)} - q^{-\frac{1}{2}(u+k-1)}) \\
&= P_k. \qquad (7.6)
\end{aligned}
$$

Taking the limit as $k \to \infty$, we get

$$\lim_{k\to\infty} e^{S_k} = e^{\lim_{k\to\infty} S_k} = \lim_{k\to\infty} P_k. \tag{7.7}$$

Hence, P_k has a limit if and only if S_k has a limit. Therefore, the infinite product $([u]_{p,q})_\infty$ and series $\sum_{m=0}^{\infty} \ln(p^{\frac{1}{2}(u+m)} - q^{-\frac{1}{2}(u+m)})$ have the same character. □

The (p,q)-analogues of exponential functions are given by

$$e_{p,q}(u) = \sum_{k=0}^{\infty} \frac{(-1)^k q^{-\frac{1}{2}\binom{k}{2}} u^k}{(p^{\frac{1}{2}} - q^{-\frac{1}{2}})^k [k]_{p,q}!}, \quad pq < 1, |uq^{\frac{1}{2}}| < 1, \tag{7.8}$$

$$E_{p,q}(u) = \sum_{k=0}^{\infty} \frac{(-1)^k p^{\frac{1}{2}\binom{k}{2}} u^k}{(p^{\frac{1}{2}} - q^{-\frac{1}{2}})^k [k]_{p,q}!}, \quad pq < 1, \text{for all } u. \tag{7.9}$$

These two exponential functions are important, and as such they satisfy

$$e_{p,q}(u)E_{p,q}(-u) = 1 \tag{7.10}$$

and with the simultaneous limit $q \to 1$ and $p \to q^2$ the two (p,q)-exponential functions reduce into the q-exponential functions,

$$e_q(u) = \sum_{k=0}^{\infty} \frac{u^k}{(q;q)_k}, \tag{7.11}$$

$$E_q(u) = \sum_{k=0}^{\infty} \frac{q^{\binom{k}{2}} u^k}{(q;q)_k}, \tag{7.12}$$

respectively. To find the convergence of these functions, we use the convergence criteria of the infinite q-shifted factorial. Using the q-calculus and some rearrangement, the (p,q)-exponential functions can be written in the form of an infinite q-shifted factorial as

$$e_{p,q}(u) = \sum_{k=0}^{\infty} \frac{(-1)^k q^{-\frac{1}{2}\binom{k}{2}} u^k}{(p^{\frac{1}{2}} - q^{-\frac{1}{2}})^k [k]_{p,q}!} = \frac{1}{(q^{\frac{1}{2}}u; (pq)^{\frac{1}{2}})_\infty}, \tag{7.13}$$

$$E_{p,q}(u) = \sum_{k=0}^{\infty} \frac{(-1)^k p^{\frac{1}{2}\binom{k}{2}} u^k}{(p^{\frac{1}{2}} - q^{-\frac{1}{2}})^k [k]_{p,q}!} = (-q^{\frac{1}{2}}u; (pq)^{\frac{1}{2}})_\infty. \tag{7.14}$$

Thus, from equations (7.13) and (7.14), we can conclude that the (p,q)-exponential function $e_{p,q}(u)$ converges for $pq < 1$ and $|q^{\frac{1}{2}}u| < 1$. Similarly, the function $E_{p,q}(u)$ converges for $pq < 1$ and for all complex numbers $u \in \mathbb{C}$.

Besides these, the other (p,q)-analogue of the exponential function exists in the literature:

$$\exp_{p,q}(u) = \sum_{k=0}^{\infty} \frac{u^k}{[k]_{p,q}!}. \tag{7.15}$$

The (p,q)-derivative operator $D_{p,q}$ is defined by

$$D_{p,q}f(u) = \frac{T_p^{\frac{1}{2}} - T_q^{-\frac{1}{2}}}{(p^{\frac{1}{2}} - q^{-\frac{1}{2}})u} f(u) = \frac{f(p^{\frac{1}{2}}u) - f(q^{-\frac{1}{2}}u)}{(p^{\frac{1}{2}} - q^{-\frac{1}{2}})u}, \tag{7.16}$$

where T_p^k is p-dilation operator defined by

$$T_p^k f(u) = f(p^k u). \tag{7.17}$$

For the product $f_1(u)f_2(u)$ of functions, the (p,q)-derivative is given by

$$D_{p,q}(f_1(u)f_2(u)) = f_1(q^{-\frac{1}{2}}u)\{D_{p,q}f_2(u)\} + \{D_{p,q}f_1(u)\}f_2(p^{\frac{1}{2}}u). \tag{7.18}$$

The inverse operation to (p,q)-differentiation is called (p,q)-integration. The (p,q)-integral for $|pq| < 1$ is defined as

$$\int_0^x f(t)d_{p,q}(t) = (q^{-\frac{1}{2}} - p^{\frac{1}{2}})x \sum_{k=0}^{\infty} p^{\frac{k}{2}} q^{\frac{(k+1)}{2}} f(p^{\frac{k}{2}} q^{\frac{(k+1)}{2}} x). \tag{7.19}$$

The formula for (p,q)-integration by parts is given by

$$\int_0^x f_1(q^{-\frac{1}{2}}u)\{D_{p,q}f_2(u)\}d_{p,q}(u) = [f_1(u)f_2(u)]_0^x - \int_0^x \{D_{p,q}f_1(u)\}f_2(p^{\frac{1}{2}}u)d_{p,q}(u). \tag{7.20}$$

7.1.1 The (p,q)-gamma function

In this section, we define the (p,q)-gamma function with the help of a (p,q)-integral. Some results related to the (p,q)-gamma function will be determined. We start with the convergence of the following (p,q)-integral.

Theorem 238. *For a complex number a, the (p,q)-integral*

$$\int_0^{\frac{1}{q^{-\frac{1}{2}} - p^{\frac{1}{2}}}} u^{a-1} E_{p,q}^{((q^{-\frac{1}{2}} - p^{\frac{1}{2}})q^{-1}u)} d_{p,q}u, \tag{7.21}$$

converges absolutely for $pq < 1$ and $\Re(a) > 0$.

Let a be a complex number. Then we define a (p,q)-gamma function in terms of the (p,q)-integral as

$$\Gamma_{p,q}(a) = q^{-\frac{1}{4}a(a+3)}p^{-\frac{1}{2}a} \int_0^{\frac{1}{q^{-\frac{1}{2}}-p^{\frac{1}{2}}}} u^{a-1}E_{p,q}^{((q^{-\frac{1}{2}}-p^{\frac{1}{2}})q^{-1}u)} d_{p,q}u. \tag{7.22}$$

Theorem 239. *Let a be a complex number. Then the (p,q)-gamma function satisfies the following functional identities:*

$$\Gamma_{p,q}(1) = 1, \tag{7.23}$$

$$\Gamma_{p,q}(a+1) = [a]_{p,q}\Gamma_{p,q}(a). \tag{7.24}$$

Proof. From the definition of (p,q)-gamma function given in (7.22), we have

$$\Gamma_{p,q}(1) = \left(q^{-1}p^{-\frac{1}{2}} \int_0^{\frac{1}{q^{-\frac{1}{2}}-p^{\frac{1}{2}}}} E_{p,q}^{((q^{-\frac{1}{2}}-p^{\frac{1}{2}})q^{-1}u)} d_{p,q}u \right). \tag{7.25}$$

Using the (p,q)-integral for $pq < 1$, the above equation yields

$$\begin{aligned}
\Gamma_{p,q}(1) &= \left(q^{-1}p^{-\frac{1}{2}} \sum_{k=0}^{\infty} p^{\frac{k}{2}} q^{\frac{k+1}{2}} E_{p,q}^{(p^{\frac{k}{2}}q^{\frac{k+1}{2}}q^{-1})} \right) \\
&= \left((pq)^{-\frac{1}{2}} \sum_{k=0}^{\infty} (pq)^{\frac{1}{2}k} ((pq)^{\frac{1}{2}k}; (pq)^{\frac{1}{2}})_{\infty} \right) \\
&= (((pq)^{\frac{1}{2}}; (pq)^{\frac{1}{2}})_{\infty}^{-1} ((pq)^{\frac{1}{2}}; (pq)^{\frac{1}{2}})_{\infty}) = 1.
\end{aligned} \tag{7.26}$$

This completes the proof of (7.23). To prove (7.24), we consider the inverse property of the (p,q)-differentiation and the product rule for (p,q)-differentiation. Consider the integral

$$\begin{aligned}
&\int_0^{\frac{1}{q^{-\frac{1}{2}}-p^{\frac{1}{2}}}} D_{p,q}(u^a E_{p,q}^{((q^{-\frac{1}{2}}-p^{\frac{1}{2}})q^{-1}p^{-\frac{1}{2}}u)}) d_{p,q}u \\
&\quad = \int_0^{\frac{1}{q^{-\frac{1}{2}}-p^{\frac{1}{2}}}} (uq^{-\frac{1}{2}})^a D_{p,q}(E_{p,q}^{((q^{-\frac{1}{2}}-p^{\frac{1}{2}})q^{-1}p^{-\frac{1}{2}}u)}) d_{p,q}u \\
&\quad\quad + \int_0^{\frac{1}{q^{-\frac{1}{2}}-p^{\frac{1}{2}}}} D_{p,q}(u^a) E_{p,q}^{((q^{-\frac{1}{2}}-p^{\frac{1}{2}})q^{-1}u)} d_{p,q}u.
\end{aligned} \tag{7.27}$$

Using the (p,q)-differentiation of the (p,q)-exponential function, we get

$$D_{p,q}\left(E_{p,q}^{((q^{-\frac{1}{2}}-p^{\frac{1}{2}})q^{-1}p^{-\frac{1}{2}}u)}\right) = -q^{-1}p^{-\frac{1}{2}}E_{p,q}^{((q^{-\frac{1}{2}}-p^{\frac{1}{2}})q^{-1}u)}. \tag{7.28}$$

Equations (7.19), (7.27) and (7.28) together yield

$$\begin{aligned} q^{-(\frac{1}{2}a+1)}p^{-\frac{1}{2}} & \int_0^{\frac{1}{q^{-\frac{1}{2}}-p^{\frac{1}{2}}}} u^a E_{p,q}^{((q^{-\frac{1}{2}}-p^{\frac{1}{2}})q^{-1}u)} d_{p,q}u \\ &= [a]_{p,q} \int_0^{\frac{1}{q^{-\frac{1}{2}}-p^{\frac{1}{2}}}} u^{a-1} E_{p,q}^{((q^{-\frac{1}{2}}-p^{\frac{1}{2}})q^{-1}u)} d_{p,q}u. \end{aligned} \tag{7.29}$$

Taking into account equations (7.22) and (7.29), the required result (7.24) is obtained. □

Theorem 240. *Let a be a complex number such that $(pq)^{-\frac{1}{2}k} \neq \mathbb{R}((pq)^{\frac{1}{2}a})$ for all integers $k \geq 0$. Then the (p,q)-gamma function is expressible in terms of infinite shifted factorials as*

$$\Gamma_{p,q}(a) = q^{\frac{a(1-a)}{4}}(q^{-\frac{1}{2}}-p^{\frac{1}{2}})^{1-a}((pq)^{\frac{1}{2}};(pq)^{\frac{1}{2}})_{\infty}((pq)^{\frac{1}{2}a};(pq)^{\frac{1}{2}})_{\infty}^{-1}. \tag{7.30}$$

Proof. Consider the function

$$f_k(a) = \int_0^{\frac{1-(pq)^{\frac{1}{2}k}}{q^{-\frac{1}{2}}-p^{\frac{1}{2}}}} u^{a-1}\left(\frac{q^{-\frac{1}{2}}(q^{-\frac{1}{2}}-p^{\frac{1}{2}})u}{1-(pq)^{\frac{1}{2}k}};(pq)^{\frac{1}{2}}\right)_k d_{p,q}u, \quad k=0,1,2,\dots. \tag{7.31}$$

Making the substitution $\frac{u}{1-(pq)^{\frac{1}{2}k}} = u'$, we get

$$\begin{aligned} f_k(a) &= (1-(pq)^{\frac{1}{2}k})^a \int_0^{\frac{1}{q^{-\frac{1}{2}}-p^{\frac{1}{2}}}} u^{a-1}(q^{-\frac{1}{2}}(q^{-\frac{1}{2}}-p^{\frac{1}{2}})u;(pq)^{\frac{1}{2}})_k d_{p,q}u \\ &= (1-(pq)^{\frac{1}{2}k})^a(q^{-\frac{1}{2}}-p^{\frac{1}{2}})^{1-a}\sum_{l=0}^{\infty}(p^{\frac{l}{2}}q^{\frac{l+1}{2}})^a((pq)^{\frac{1}{2}l};(pq)^{\frac{1}{2}})_k. \end{aligned} \tag{7.32}$$

Taking the limit $k \to \infty$ on both sides and using the definition of the (p,q)-gamma function, we get the desired equation (7.30). □

Replacing a by $a+1$ in (7.24) yields

$$\Gamma_{p,q}(a+2) = [a]_{p,q}[a+1]_{p,q}\Gamma_{p,q}(a). \tag{7.33}$$

Iterating this process $(k-1)$-times, we get

$$\Gamma_{p,q}(a+k) = [a]_{p,q}\,[a+1]_{p,q}\cdots[a+k-1]_{p,q}\,\Gamma_{p,q}(a). \tag{7.34}$$

Thus, for a complex number a such that $(pq)^{-\frac{1}{2}k} \neq \mathbb{R}(pq)^{\frac{1}{2}a}$ for all integers k, the inverse of the (p,q)-gamma function is given by the following formula:

$$\Gamma_{p,q}^{-1}(a) = [a]_{p,q}\,[a+1]_{p,q}\cdots[a+k-1]_{p,q}\,\Gamma_{p,q}^{-1}(a+k). \tag{7.35}$$

From equations (7.34) and (7.35), we are able to write the (p,q)-shifted factorial function in terms of the (p,q)-gamma function as

$$([a]_{p,q})_k = \frac{\Gamma_{p,q}(a+k)}{\Gamma_{p,q}(a)}, \quad k = 1,2,\ldots. \tag{7.36}$$

Theorem 241. *Let a be a complex number. Then the (p,q)-analogue of the Legendre duplication formula is given by*

$$\Gamma_{p^2,q^2}\left(\frac{1}{2}\right)\Gamma_{p,q}(2a) = (p^{\frac{1}{2}} + q^{-\frac{1}{2}})^{2a-1}\,\Gamma_{p^2,q^2}(a)\,\Gamma_{p^2,q^2}\left(a+\frac{1}{2}\right). \tag{7.37}$$

Proof. Using the definition of the (p,q)-gamma function (7.22), we have

$$\begin{aligned}
\Gamma_{p^2,q^2}(a) &= q^{\frac{a(1-a)}{2}}(q^{-1}-p^1)^{1-a}(pq;pq)_\infty((pq)^a;pq)_\infty^{-1},\\
\Gamma_{p^2,q^2}\left(a+\frac{1}{2}\right) &= q^{\frac{(2a+1)(1-2a)}{8}}(q^{-1}-p^1)^{1-(a+\frac{1}{2})}(pq;pq)_\infty((pq)^{\frac{1}{2}+a};pq)_\infty^{-1},\\
\Gamma_{p^2,q^2}\left(\frac{1}{2}\right) &= q^{\frac{1}{8}}(q^{-1}-p^1)^{-\frac{1}{2}}(pq;pq)_\infty((pq)^{\frac{1}{2}};pq)_\infty^{-1}.
\end{aligned}$$

With the help of these relations, we get

$$\begin{aligned}
\frac{\Gamma_{p^2,q^2}(a)\,\Gamma_{p^2,q^2}(a+\frac{1}{2})}{\Gamma_{p^2,q^2}(\frac{1}{2})} &= (q^{-\frac{1}{2}}+p^{\frac{1}{2}})^{1-2a}q^{\frac{2a(1-2a)}{4}}(q^{-\frac{1}{2}}-p^{\frac{1}{2}})^{1-2a}((pq)^{\frac{1}{2}};(pq)^{\frac{1}{2}})_\infty\\
&\quad\times((pq)^{\frac{1}{2}(2a)};(pq)^{\frac{1}{2}})_\infty^{-1}.
\end{aligned} \tag{7.38}$$

Using equation (7.22) in (7.38), we get

$$\Gamma_{p^2,q^2}\left(\frac{1}{2}\right)\Gamma_{p,q}(2a) = (p^{\frac{1}{2}} + q^{-\frac{1}{2}})^{2a-1}\,\Gamma_{p^2,q^2}(a)\,\Gamma_{p^2,q^2}\left(a+\frac{1}{2}\right). \tag{7.39}$$

It completes the proof. □

7.1.2 The (p, q)-beta function

In this section, we define the (p, q)-beta function in terms of (p, q)-gamma functions. We also give the (p, q)-integral representation of the (p, q)-beta function and certain recurrence relations satisfied by the (p, q)-beta function.

Theorem 242. *Let a and b be complex numbers. Then, for $|pq| < 1$, the (p, q)-integral*

$$\int_0^1 u^{a-1}(p^{\frac{1}{2}}\, u; (pq)^{\frac{1}{2}})_\infty (q^{-\frac{1}{2}} u\, (pq)^{\frac{1}{2}b}; (pq)^{\frac{1}{2}})_\infty^{-1}\, d_{p,q}u, \tag{7.40}$$

converges absolutely.

Definition 243. Let a and b be complex numbers such that $\Re(a), \Re(b) > 0$. Then the (p, q)-beta function is defined by

$$\beta_{p,q}(a, b) = \left(\int_0^1 u^{a-1}(p^{\frac{1}{2}}\, u; (pq)^{\frac{1}{2}})_\infty (q^{-\frac{1}{2}} u\, (pq)^{\frac{1}{2}b}; (pq)^{\frac{1}{2}})_\infty^{-1}\, d_{p,q}u\right) \times q^{-\frac{1}{2}a(1-b)}. \tag{7.41}$$

Theorem 244. *Let a, b and $a + b$ be complex numbers such that $\Re(a), \Re(b), \Re(a + b) > 0$. Then the (p, q)-beta function can be written in terms of (p, q)-gamma function as*

$$\beta_{p,q}(a, b) = \Gamma_{p,q}(a)\, \Gamma_{p,q}(b)\, \Gamma_{p,q}^{-1}(a + b). \tag{7.42}$$

Proof. The definition of (p, q)-integral for $pq < 1$ yields

$$\begin{aligned}
&\int_0^1 u^{a-1}(p^{\frac{1}{2}}\, u; (pq)^{\frac{1}{2}})_\infty (q^{-\frac{1}{2}} u\, (pq)^{\frac{1}{2}b}; (pq)^{\frac{1}{2}})_\infty^{-1}\, d_{p,q}u \\
&\quad = (q^{-\frac{1}{2}} - p^{\frac{1}{2}}) \sum_{k=0}^{\infty} p^{\frac{1}{2}ka} q^{\frac{1}{2}(k+1)a} ((pq)^{\frac{1}{2}(k+1)}; (pq)^{\frac{1}{2}})_\infty ((pq)^{\frac{1}{2}(b+k)}; (pq)^{\frac{1}{2}})_\infty^{-1} \\
&\quad = (q^{-\frac{1}{2}} - p^{\frac{1}{2}}) q^{\frac{1}{2}a} ((pq)^{\frac{1}{2}}; (pq)^{\frac{1}{2}})_\infty ((pq)^{\frac{1}{2}b}; (pq)^{\frac{1}{2}})_\infty^{-1} \sum_{k=0}^{\infty} \frac{((pq)^{\frac{1}{2}b}, (pq)^{\frac{1}{2}})_k}{((pq)^{\frac{1}{2}}; (pq)^{\frac{1}{2}})_k} ((pq)^{\frac{1}{2}a})^k.
\end{aligned} \tag{7.43}$$

Putting

$$\sum_{k=0}^{\infty} \frac{(b; q)_k}{(q; q)_k}\, a^k = \frac{(ab; q)_\infty}{(a; q)_\infty}, \tag{7.44}$$

in (7.43), we get

$$\begin{aligned}
\beta_{p,q}(a, b) &= (q^{-\frac{1}{2}} - p^{\frac{1}{2}}) q^{\frac{1}{2}a} ((pq)^{\frac{1}{2}}; (pq)^{\frac{1}{2}})_\infty ((pq)^{\frac{1}{2}a}; (pq)^{\frac{1}{2}})_\infty^{-1} ((pq)^{\frac{1}{2}b}; (pq)^{\frac{1}{2}})_\infty^{-1} \\
&\quad \times ((pq)^{\frac{1}{2}(a+b)}; (pq)^{\frac{1}{2}})_\infty\, q^{-\frac{1}{2}a(1-b)}.
\end{aligned} \tag{7.45}$$

Applying the definition of the (p,q)-gamma function given in (7.30) together with (7.41), we get the identity

$$\beta_{p,q}(a,b) = \Gamma_{p,q}(a)\,\Gamma_{p,q}(b)\,\Gamma_{p,q}^{-1}(a+b). \tag{7.46}$$

This completes the proof. □

We now give two recurrence relations satisfied by the (p,q)-beta function.

Theorem 245. *For complex numbers a and b, the (p,q)-beta function satisfies the following recurrence relations:*

$$\beta_{p,q}(a+k,b) = ([a]_{p,q})_k([a+b]_{p,q})_k^{-1}\beta_{p,q}(a,b); \tag{7.47}$$

$$\beta_{p,q}(a+k,b+k) = ([a]_{p,q})_k([b]_{p,q})_k([a+b]_{p,q})_{2k}^{-1}\beta_{p,q}(a,b), \tag{7.48}$$

where k is a nonnegative integer.

Proof. From equation (7.41), we have

$$\begin{aligned}\beta_{p,q}(a+k,b) &= \Gamma_{p,q}(a+k)\,\Gamma_{p,q}(b)\,\Gamma_{p,q}^{-1}(a+b+k)\\ &= \Gamma_{p,q}(a+k)\,\Gamma_{p,q}(a)\,\Gamma_{p,q}^{-1}(a)\,\Gamma_{p,q}(b)\,\Gamma_{p,q}^{-1}(a+b+k)\\ &\quad\times\Gamma_{p,q}(a+b)\,\Gamma_{p,q}^{-1}(a+b)\\ &= ([a]_{p,q})_k\,\Gamma_{p,q}(a)\,\Gamma_{p,q}(b)\,\Gamma_{p,q}^{-1}(a+b)\,([a+b]_{p,q})_k^{-1}\\ &= ([a]_{p,q})_k([a+b]_{p,q})_k^{-1}\beta_{p,q}(a,b).\end{aligned} \tag{7.49}$$

The recurrence relation (7.48) can be proved similarly. □

7.1.3 The (p,q)-Gauss hypergeometric function

The (p,q)-analogue of the generalized hypergeometric function for complex numbers $u_1,\dots,u_r,v_1,\dots,v_s,z$ is given by

$$\begin{aligned}&{}_r\psi_s(u_1,\dots,u_r;v_1,\dots,v_s;p,q;z)\\ &\quad=\sum_{k=0}^{\infty}\frac{([u_1]_{p,q})_k\cdots([u_r]_{p,q})_k}{([v_1]_{p,q})_k\cdots([v_s]_{p,q})_k}\left[\frac{(-1)^k q^{-\frac{1}{2}\binom{k}{2}}}{(p^{\frac{1}{2}}-q^{-\frac{1}{2}})^k}\right]^{1+s-r}\frac{z^k}{[k]_{p,q}!}.\end{aligned} \tag{7.50}$$

The particular cases of the (p,q) generalized hypergeometric function give the (p,q)-analogues of various classical hypergeometric functions. In particular, when $r=1=s$, we get the (p,q)-analogue of the confluent hypergeometric function given by

$${}_1\psi_1(u_1;v_1;p,q;z) = \sum_{k=0}^{\infty}\frac{([u_1]_{p,q})_k}{([v_1]_{p,q})_k}\left[\frac{(-1)^k q^{-\frac{1}{2}\binom{k}{2}}}{(p^{\frac{1}{2}}-q^{-\frac{1}{2}})^k}\right]\frac{z^k}{[k]_{p,q}!}. \tag{7.51}$$

Further, when $r = 2$, $s = 1$, we get the (p,q)-analogue of the Gauss hypergeometric function as

$$ {}_2\psi_1(u_1, u_2; v_1; p, q; z) = \sum_{k=0}^{\infty} \frac{([u_1]_{p,q})_k([u_2]_{p,q})_k}{([v_1]_{p,q})_k} \frac{z^k}{[k]_{p,q}!}. \tag{7.52} $$

Here, in this section, we study the various properties of the (p,q)-Gauss hypergeometric function in detail, namely regions of convergence, integral representations and the difference equation obeyed by this function.

Let a, b and c be complex numbers such that $(pq)^{-\frac{1}{2}k} \neq \mathbb{R}((pq)^{\frac{1}{2}c})$. Then the (p,q)-Gauss hypergeometric function is defined by

$$ {}_2\psi_1(a, b; c; p, q; z) = \sum_{k=0}^{\infty} ([a]_{p,q})_k \, ([b]_{p,q})_k \, ([c]_{p,q})_k^{-1} \frac{z^k}{[k]_{p,q}!}, \tag{7.53} $$

and converges absolutely for $|z| < 1$. In the following theorem, we give the convergence region of the function (7.53) for $|z| = 1$.

Theorem 246. *Let a, b and c be complex numbers such that $(pq)^{-\frac{1}{2}k} \neq \mathbb{R}((pq)^{\frac{1}{2}c})$. Then the (p,q)-Gauss function defined in the equation (7.53) converges absolutely for $|z| = 1$ and*

$$ \mathbb{R}(a) + \mathbb{R}(b) < 1 + \mathbb{R}(c). \tag{7.54} $$

Proof. Let the general term of the series (7.53) be $\mathcal{A}_k(z)$. Then, using the definition of the (p,q)-shifted factorial function and q-shifted factorial function, we have

$$ \begin{aligned} |\mathcal{A}_k(z)| \leq{} & \left| \frac{((pq)^{\frac{1}{2}a}; (pq)^{\frac{1}{2}})_k}{(1-(pq)^{\frac{1}{2}})^k} q^{-\frac{1}{2}ka-\frac{1}{4}(k(k-3))} \right| \left| \frac{((pq)^{\frac{1}{2}b}; (pq)^{\frac{1}{2}})_k}{(1-(pq)^{\frac{1}{2}})^k} q^{-\frac{1}{2}kb-\frac{1}{4}(k(k-3))} \right| \\ & \times \left| ((pq)^{\frac{1}{2}c}; (pq)^{\frac{1}{2}})_k^{-1} (1-(pq)^{\frac{1}{2}})^k q^{\frac{1}{2}kc+\frac{1}{4}(k(k-3))} \right| \frac{q^{\frac{1}{4}(k(k-1))}(1-(pq)^{\frac{1}{2}})^k}{((pq)^{\frac{1}{2}}; (pq)^{\frac{1}{2}})_k} |z|^k \\ \leq{} & \left| \frac{q^{-\frac{1}{2}ka}((pq)^{\frac{1}{2}a}; (pq)^{\frac{1}{2}})_k (q^{-\frac{1}{2}} - p^{\frac{1}{2}})^{1-a} (q^{-\frac{1}{2}} - p^{\frac{1}{2}})^{-1+a} q^{\frac{a(1-a)}{4}} q^{-\frac{a(1-a)}{4}}}{((pq)^{\frac{1}{2}}; (pq)^{\frac{1}{2}})_k} \right| \\ & \times \left| \frac{q^{-\frac{1}{2}kb}((pq)^{\frac{1}{2}b}; (pq)^{\frac{1}{2}})_k (q^{-\frac{1}{2}} - p^{\frac{1}{2}})^{1-b} (q^{-\frac{1}{2}} - p^{\frac{1}{2}})^{-1+b} q^{\frac{b(1-b)}{4}} q^{-\frac{b(1-b)}{4}}}{((pq)^{\frac{1}{2}}; (pq)^{\frac{1}{2}})_k} \right| \\ & \times \left| q^{\frac{1}{2}kc}((pq)^{\frac{1}{2}c}; (pq)^{\frac{1}{2}})_k^{-1} (q^{-\frac{1}{2}} - p^{\frac{1}{2}})^{1-c} (q^{-\frac{1}{2}} - p^{\frac{1}{2}})^{-1+c} q^{\frac{c(1-c)}{4}} q^{-\frac{c(1-c)}{4}} \right| \\ & \times ((pq)^{\frac{1}{2}}; (pq)^{\frac{1}{2}})_k \, q^{\frac{1}{2}k} \, |z|^k \\ \leq{} & N \left| q^{\frac{1}{2}(a+b-c)} \right| \left| (q^{-\frac{1}{2}} - p^{\frac{1}{2}})^{-a-b+c+1} \right| \left| q^{\frac{1}{4}\{a(1-a)+b(1-b)-c(1-c)\}} \right|, \end{aligned} \tag{7.55} $$

where

$$ N \geq \left| \Gamma_{p,q}^{-1}(a) \, \Gamma_{p,q}^{-1}(b) \, \Gamma_{p,q}(c) \right|. \tag{7.56} $$

At $|z| = 1$ and for $\mathbb{R}(a) + \mathbb{R}(b) < 1 + \mathbb{R}(c)$, $|\mathcal{A}_k(z)| \to 0$ as $k \to \infty$. Hence (p,q)-Gauss hypergeometric function converges absolutely at $|z| = 1$. □

Now, we introduce a (p,q)-difference equation satisfied by the (p,q)-Gauss hypergeometric function ${}_2\psi_1(a,b;c;p,q;z)$.

Theorem 247. *Let a, b and c be complex numbers such that $(pq)^{-\frac{1}{2}k} \neq \mathbb{R}((pq)^{\frac{1}{2}c})$. Then the (p,q)-Gauss hypergeometric function satisfies the following (p,q)-difference equation:*

$$\left[D_{p,q}\left[(\mathcal{D})_{c-1}\right]_{p,q} - \left[(\mathcal{D})_a\right]_{p,q}\left[(\mathcal{D})_b\right]_{p,q}\right] {}_2\psi_1(a,b;c;p,q;z) = 0, \tag{7.57}$$

where $[(\mathcal{D})_a]_{p,q}$ is the (p,q)-difference operator, defined by

$$\left[(\mathcal{D})_a\right]_{p,q} f(z) = \frac{f(p^{\frac{1}{2}}z)\,p^{\frac{1}{2}a} - f(q^{-\frac{1}{2}}z)\,q^{-\frac{1}{2}a}}{(p^{\frac{1}{2}} - q^{-\frac{1}{2}})}. \tag{7.58}$$

Proof. Using the definition of (p,q)-difference operators (7.16) and (7.58), we have

$$\begin{aligned}
&D_{p,q}\left[(\mathcal{D})_{c-1}\right]_{p,q}\left({}_2\psi_1(a,b;c;p,q;z)\right)\\
&= D_{p,q}\left[(\mathcal{D})_{c-1}\right]_{p,q}\left(\sum_{k=0}^{\infty}([a]_{p,q})_k([b]_{p,q})_k([c]_{p,q})_k^{-1}\frac{z^k}{[k]_{p,q}!}\right)\\
&= \sum_{k=1}^{\infty}([a]_{p,q})_k([b]_{p,q})_k([c]_{p,q})_{k-1}^{-1}\frac{z^{k-1}}{[k-1]_{p,q}!}\\
&= \sum_{k=0}^{\infty}([a]_{p,q})_{k+1}([b]_{p,q})_{k+1}([c]_{p,q})_k^{-1}\frac{z^k}{[k]_{p,q}!}\\
&= \sum_{k=0}^{\infty}\frac{p^{\frac{1}{2}(a+k)} - q^{-\frac{1}{2}(a+k)}}{p^{\frac{1}{2}} - q^{-\frac{1}{2}}}\,\frac{p^{\frac{1}{2}(b+k)} - q^{-\frac{1}{2}(b+k)}}{p^{\frac{1}{2}} - q^{-\frac{1}{2}}}([a]_{p,q})_k([b]_{p,q})_k([c]_{p,q})_k^{-1}\frac{z^k}{[k]_{p,q}!}\\
&= \left[(\mathcal{D})_a\right]_{p,q}\left[(\mathcal{D})_b\right]_{p,q}\left({}_2\psi_1(a,b;c;p,q,z)\right).
\end{aligned} \tag{7.59}$$

It completes the proof. □

Next, we give the (p,q)-integral representation of the (p,q)-Gauss hypergeometric function using the integral of the (p,q)-beta function.

Theorem 248. *Let a, b, c be complex numbers such that $(pq)^{-\frac{1}{2}k} \neq \mathbb{R}((pq)^{\frac{1}{2}c})$. Then the (p,q)-Gauss hypergeometric function defined in (7.53) can be put in the integral form as*

$$\begin{aligned}
&{}_2\psi_1(a,b;c;p,q;z)\\
&= \Gamma_{p,q}\begin{pmatrix} c \\ b, c-b \end{pmatrix}\int_0^1 \left((pq)^{\frac{1}{2}a}uzq^{-\frac{1}{2}(a+b-c)};(pq)^{\frac{1}{2}}\right)_\infty\left(uzq^{-\frac{1}{2}(a+b-c)};(pq)^{\frac{1}{2}}\right)_\infty^{-1}\\
&\quad\times u^{b-1}\left(q^{-\frac{1}{2}}(pq)^{\frac{1}{2}(c-1)}u;(pq)^{\frac{1}{2}}\right)_\infty^{-1}\left(p^{\frac{1}{2}}u;(pq)^{\frac{1}{2}}\right)_\infty d_{p,q}u.
\end{aligned} \tag{7.60}$$

Proof. In view of the equations (7.36), (7.41), (7.42), we have

$$\begin{aligned}([b]_{p,q})_k\,([c]_{p,q})_k^{-1} &= \Gamma_{p,q}\binom{c}{b,c-b}\,\Gamma_{p,q}(b+k)\Gamma_{p,q}(c-b)\Gamma_{p,q}^{-1}(c+k)\\ &= \Gamma_{p,q}\binom{c}{b,c-b}\,\beta_{p,q}(b+k,c-b)\\ &= \int_0^1 u^{b+k-1}\,(q^{-\frac{1}{2}}(pq)^{\frac{1}{2}(c-b)}u;(pq)^{\frac{1}{2}})_\infty^{-1}\,(p^{\frac{1}{2}}u;(pq)^{\frac{1}{2}})_\infty\,d_{p,q}u\\ &\quad\times\Gamma_{p,q}\binom{c}{b,c-b}\,q^{-\frac{1}{2}(b+k)(1-c+b)}.\end{aligned} \tag{7.61}$$

Using (7.61) into (7.53), we get

$$\begin{aligned}&{}_2\psi_1(a,b;c;p,q;z)\\ &= \Gamma_{p,q}\binom{c}{b,c-b}\,q^{-\frac{1}{2}b(1-c+b)}\sum_{k=0}^{\infty}([a]_{p,q})_k\,\frac{(uzq^{-\frac{1}{2}(1-c+b)})^k}{[k]_{p,q}!}\\ &\times\int_0^1 u^{b-1}\,(q^{-\frac{1}{2}}(pq)^{\frac{1}{2}(c-b)}u;(pq)^{\frac{1}{2}})_\infty^{-1}\,(p^{\frac{1}{2}}u;(pq)^{\frac{1}{2}})_\infty\,d_{p,q}u.\end{aligned} \tag{7.62}$$

With the help of the identity,

$$\sum_{k=0}^{\infty}([a]_{p,q})_k\,\frac{(uzq^{-\frac{1}{2}(1-c+b)})^k}{[k]_{p,q}!} = ((pq)^{\frac{1}{2}a}uzq^{-\frac{1}{2}(a+b-c)};(pq)^{\frac{1}{2}})_\infty(uzq^{-\frac{1}{2}(a+b-c)};(pq)^{\frac{1}{2}})_\infty^{-1},$$

we get

$$\begin{aligned}&{}_2\psi_1(a,b;c;p,q;z)\\ &= \Gamma_{p,q}\binom{c}{b,c-b}\int_0^1((pq)^{\frac{1}{2}a}uzq^{-\frac{1}{2}(a+b-c)};(pq)^{\frac{1}{2}})_\infty(uzq^{-\frac{1}{2}(a+b-c)};(pq)^{\frac{1}{2}})_\infty^{-1}\\ &\times u^{b-1}\,(q^{-\frac{1}{2}}(pq)^{\frac{1}{2}(c-1)}u;(pq)^{\frac{1}{2}})_\infty^{-1}\,(p^{\frac{1}{2}}u;(pq)^{\frac{1}{2}})_\infty\,d_{p,q}u.\end{aligned}$$

It completes the proof. □

7.2 Incomplete Gauss hypergeometric function

The incomplete gamma functions $\gamma(u,x)$ and $\Gamma(u,x)$ are defined by

$$\gamma(u,x) = \int_0^x u^{u-1}\,e^{-u}\,du, \quad \Re(u) > 0; x \geq 0 \tag{7.63}$$

and

$$\Gamma(u,x) = \int_x^{\infty} u^{u-1} e^{-u}\, du, \quad \Re(u) > 0; x \geq 0, \text{when } x = 0, \tag{7.64}$$

respectively, satisfy the following decomposition formula:

$$\gamma(u,x) + \Gamma(u,x) = \Gamma(u), \quad \Re(u) > 0. \tag{7.65}$$

The incomplete gamma functions $\gamma(u,x)$ and $\Gamma(u,x)$ are related to error function $\operatorname{erf}(x)$ and the complimentary error function $\operatorname{erfc}(x)$ as follows:

$$\gamma\left(\frac{1}{2}, u\right) = \sqrt{\pi}\operatorname{erf}(\sqrt{u}) \quad \text{and} \quad \Gamma\left(\frac{1}{2}, u\right) = \sqrt{\pi}\operatorname{erfc}(\sqrt{u}),$$

where $\operatorname{erf}(u)$ is given by

$$\operatorname{erf}(u) = \frac{2}{\sqrt{\pi}} \int_0^u e^{-u^2}\, du, \tag{7.66}$$

and $\operatorname{erfc}(u) = 1 - \operatorname{erf}(u)$.

7.2.1 The incomplete Pochhammer symbols

In terms of the incomplete gamma functions $\gamma(u,x)$ and $\Gamma(u,x)$ given in (7.63) and (7.64), the incomplete Pochhammer symbols are defined as follows:

$$(u;x)_k = \frac{\gamma(u+k,x)}{\Gamma(u)}, \quad u,k \in \mathbb{C}; x \geq 0 \tag{7.67}$$

and

$$[u;x]_k = \frac{\Gamma(u+k,x)}{\Gamma(u)}, \quad u,k \in \mathbb{C}; x \geq 0. \tag{7.68}$$

In view of (7.65), these incomplete Pochhammer symbols $(u;x)_k$ and $[u;x]_k$ satisfy the following decomposition relation:

$$(u;x)_k + [u;x]_k = (u)_k, \quad u,k \in \mathbb{C}; x \geq 0, \tag{7.69}$$

where $(u)_k$ is the Pochhammer symbol given in Chapter 2. For the sake of completeness, we state the following theorems without proof.

Theorem 249. *Each of the following identities hold true:*

$$(u;x)_k = \frac{(-1)^k\, x^{u+k}}{\Gamma(u)} \frac{d^k}{dx^k}\{x^{-u}\, \gamma(u,x)\}, \quad k \in \mathbb{N}_0; \tag{7.70}$$

$$(u;x)_{-k} = \frac{(-1)^k\, e^{-x}}{(1-u)_k\, \Gamma(u)} \frac{d^k}{dx^k}\{e^{x}\, \gamma(u,x)\}, \quad k \in \mathbb{N}_0; \tag{7.71}$$

$$[u;x]_k = \frac{(-1)^k\, x^{u+k}}{\Gamma(u)} \frac{d^k}{dx^k}\{x^{-u}\, \Gamma(u,x)\}, \quad k \in \mathbb{N}_0; \tag{7.72}$$

$$[u;x]_{-k} = \frac{(-1)^k\, e^{-x}}{(1-u)_k\, \Gamma(u)} \frac{d^k}{dx^k}\{e^{x}\, \Gamma(u,x)\}, \quad k \in \mathbb{N}_0. \tag{7.73}$$

Theorem 250. *Each of the following identities holds true:*

$$[1;x]_k = k!\,\{e^{-x}\, e_k(x)\}, \quad k \in \mathbb{N}_0; \tag{7.74}$$

$$[1;x]_{-k} = x^{1-k}\,\{e^{x}\, E_k(x)\}, \quad k \in \mathbb{N}_0; \tag{7.75}$$

$$(1;x)_k = k!\,\{1 - e^{-x}\, e_k(x)\}, \quad k \in \mathbb{N}_0, \tag{7.76}$$

where

$$e_k(x) = \sum_{n=0}^{k} \frac{x^k}{k!}, \quad k \in \mathbb{N}_0$$

and

$$E_k(x) = \int_1^{\infty} e^{-xu}\, u^{-k}\, du, \quad \Re(x) > 0, k \in \mathbb{N}_0.$$

We now give the definition of an incomplete Gauss hypergeometric function as

$${}_2\gamma_1\left[\begin{matrix}(u;x), v; \\ w;\end{matrix}\; z\right] = \sum_{k=0}^{\infty} \frac{(u,x)_k\, (v)_k}{(w)_k} \frac{z^k}{k!} \tag{7.77}$$

and

$${}_2\Gamma_1\left[\begin{matrix}[u;x], v; \\ w;\end{matrix}\; z\right] = \sum_{k=0}^{\infty} \frac{[u;x]_k\, (v)_k}{(w)_k} \frac{z^k}{k!}. \tag{7.78}$$

An immediate consequence of definitions (7.77) and (7.78) is that we have the following decomposition formula:

$${}_2\gamma_1\left[\begin{matrix}(u;x), v; \\ w;\end{matrix}\; z\right] + {}_2\Gamma_1\left[\begin{matrix}[u;x], v; \\ w;\end{matrix}\; z\right] = {}_2F_1\left[\begin{matrix}u, v; \\ w;\end{matrix}\; z\right], \tag{7.79}$$

for the Gauss hypergeometric function. The relation (7.79) allows to write an incomplete form of the Gauss hypergeometric equation. The proof is simple and straightforward so we omit them.

Theorem 251. *The following differential equation is obeyed by the incomplete Gauss hypergeometric functions ${}_2\gamma_1$ and ${}_2\Gamma_1$:*

$$z(1-z)\frac{d^2\mathfrak{w}}{dz^2}+[w-(u+v+1)\,z]\frac{d\mathfrak{w}}{dz}-u\,v\,\mathfrak{w}=0, \tag{7.80}$$

where

$$\mathfrak{w}=\mathfrak{w}(z)={}_2\gamma_1\left[\begin{matrix}(u;x),v; \\ w;\end{matrix}\ z\right]+{}_2\Gamma_1\left[\begin{matrix}[u;x],v; \\ w;\end{matrix}\ z\right].$$

Theorem 252. *The following integral representations for incomplete Gauss hypergeometric functions hold true:*

$$\begin{aligned}&{}_2\gamma_1\left[\begin{matrix}(u;x),v; \\ w;\end{matrix}\ z\right]\\ &\quad=\frac{1}{\Gamma(u)}\int_0^x \mathfrak{u}^{u-1}\,e^{-\mathfrak{u}}\,{}_1F_1\left[\begin{matrix}v; \\ w;\end{matrix}\ z\,\mathfrak{u}\right]d\mathfrak{u},\quad x\geq 0;\ \Re(u)>0\ \textit{when}\ x=0\end{aligned} \tag{7.81}$$

and

$$\begin{aligned}&{}_2\Gamma_1\left[\begin{matrix}[u;x],v; \\ w;\end{matrix}\ z\right]\\ &\quad=\frac{1}{\Gamma(u)}\int_x^\infty \mathfrak{u}^{u-1}\,e^{-\mathfrak{u}}\,{}_1F_1\left[\begin{matrix}v; \\ w;\end{matrix}\ z\,\mathfrak{u}\right]d\mathfrak{u},\quad x\geq 0;\ \Re(u)>0\ \textit{when}\ x=0.\end{aligned} \tag{7.82}$$

Proof. Using the integral representation of the incomplete gamma function, we get

$${}_2\gamma_1\left[\begin{matrix}(u;x),v; \\ w;\end{matrix}\ z\right]=\frac{1}{\Gamma(u)}\sum_{k=0}^{\infty}\frac{(v)_k}{(w)_k}\frac{z^k}{k!}\int_0^x \mathfrak{u}^{u+k-1}\,e^{-\mathfrak{u}}\,d\mathfrak{u}. \tag{7.83}$$

For region $x\geq 0$; $\Re(u)>0$ when $x=0$, the integral and series involved in the above equation converges absolutely, and hence the interchange of summation and integration is possible. Thus, we write

$${}_2\gamma_1\left[\begin{matrix}(u;x),v; \\ w;\end{matrix}\ z\right]=\frac{1}{\Gamma(u)}\int_0^x \mathfrak{u}^{u-1}\,e^{-\mathfrak{u}}\sum_{k=0}^{\infty}\frac{(v)_k}{(w)_k}\frac{(z\mathfrak{u})^k}{k!}\,d\mathfrak{u}$$

$$= \frac{1}{\Gamma(u)} \int_0^x u^{u-1} e^{-u} {}_1F_1\left[\begin{matrix} v; \\ w; \end{matrix} z u\right] du. \tag{7.84}$$

Similarly, equation (7.82) can be derived. It completes the proof. □

Theorem 253. *For incomplete Gauss hypergeometric functions, we have the following derivative formulas.*

$$\frac{d^k}{dz^k}\left\{{}_2\gamma_1\left[\begin{matrix} (u;x), v; \\ w; \end{matrix} z\right]\right\} = \frac{(u)_k\,(v)_k}{(w)_k}\,{}_2\gamma_1\left[\begin{matrix} (u+k;x), v+k; \\ w+k; \end{matrix} z\right], \quad k \in \mathbb{N}_0 \tag{7.85}$$

and

$$\frac{d^k}{dz^k}\left\{{}_2\Gamma_1\left[\begin{matrix} [u;x], v; \\ w; \end{matrix} z\right]\right\} = \frac{(u)_k\,(v)_k}{(w)_k}\,{}_2\Gamma_1\left[\begin{matrix} [u+k;x], v+k; \\ w+k; \end{matrix} z\right], \quad k \in \mathbb{N}_0. \tag{7.86}$$

Proof. On differentiating the incomplete gamma function ${}_2\gamma_1$ with respect to z, we find that

$$\frac{d}{dz}\left\{{}_2\gamma_1\left[\begin{matrix} (u;x), v; \\ w; \end{matrix} z\right]\right\}$$
$$= \frac{v\,\Gamma(u+1)}{w\,\Gamma(u)}\,\frac{1}{\Gamma(u+1)} \int_0^\infty u^{(u+1)-1} e^{-u} {}_1F_1\left[\begin{matrix} v+1; \\ w+1; \end{matrix} z u\right] du, \tag{7.87}$$

which can be written as follows:

$$\frac{d}{dz}\left\{{}_2\gamma_1\left[\begin{matrix} (u;x), v; \\ w; \end{matrix} z\right]\right\}$$
$$= \frac{u\,v}{w}\,\frac{1}{\Gamma(u+1)} \int_0^\infty u^{(u+1)-1} e^{-u} {}_1F_1\left[\begin{matrix} v+1; \\ w+1; \end{matrix} z u\right] du. \tag{7.88}$$

Now, from (7.81) and (7.88), we find that

$$\frac{d}{dz}\left\{{}_2\gamma_1\left[\begin{matrix} (u;x), v; \\ w; \end{matrix} z\right]\right\} = \frac{u\,v}{w}\,{}_2\gamma_1\left[\begin{matrix} (u+1;x), v+1; \\ w+1; \end{matrix} z\right]. \tag{7.89}$$

By repeating the process, we get

$$\frac{d^k}{dz^k}\left\{{}_2\gamma_1\left[\begin{matrix} (u;x), v; \\ w; \end{matrix} z\right]\right\} = \frac{(u)_k\,(v)_k}{(w)_k}\,{}_2\gamma_1\left[\begin{matrix} (u+k;x), v+k; \\ w+k; \end{matrix} z\right], \quad k \in \mathbb{N}_0. \tag{7.90}$$

Similarly, the second formula can be derived. □

Theorem 254. *The following transformation formula holds true:*

$$ {}_2\gamma_1\left[\begin{matrix}(u;x),v;\\ w;\end{matrix}\,z\right] = (1-z)^{-u}\,{}_2\gamma_1\left[\begin{matrix}(u;x(1-z)),w-v;\\ c;\end{matrix}\,-\frac{z}{1-z}\right]. \tag{7.91}$$

Proof. Using Kummer's first formula,

$$ {}_1F_1\left[\begin{matrix}v;\\ w;\end{matrix}\,z\right] = e^z\,{}_1F_1\left[\begin{matrix}w-v;\\ w;\end{matrix}\,-z\right]. \tag{7.92}$$

In (7.81), we find that

$$ {}_2\gamma_1\left[\begin{matrix}(u;x),v;\\ w;\end{matrix}\,z\right] = \frac{1}{\Gamma(u)}\int_0^x \mathrm{u}^{u-1}\,e^{-(1-z)\,\mathrm{u}}\,{}_1F_1\left[\begin{matrix}w-v;\\ w;\end{matrix}\,-z\,\mathrm{u}\right]d\mathrm{u}. \tag{7.93}$$

The substitution $t=(1-z)\,\mathrm{u}$ in the above equation leads to

$$\begin{aligned} {}_2\gamma_1\left[\begin{matrix}(u;x),v;\\ w;\end{matrix}\,z\right] &= \frac{(1-z)^{-u}}{\Gamma(u)}\int_0^{x(1-z)} t^{u-1}e^{-t}\,{}_1F_1\left[\begin{matrix}w-v;\\ w;\end{matrix}\,-\left(\frac{z}{1-z}\right)t\right]dt \\ &= (1-z)^{-u}\,{}_2\gamma_1\left[\begin{matrix}(u;x(1-z)),w-v;\\ w;\end{matrix}\,-\frac{z}{1-z}\right]. \end{aligned} \tag{7.94}$$

It completes the proof. □

As the incomplete hypergeometric functions ${}_2\gamma_1$ and ${}_2\Gamma_1$ are not symmetric in the numerator parameters u and v for $x\neq 0$, we do not find any possibility of having symmetry in the transformation formulas.

Theorem 255. *The value of the incomplete gamma function, at $z=1$, is given by*

$$ {}_2\Gamma_1\left[\begin{matrix}[u;x],v;\\ w;\end{matrix}\,1\right] = \frac{\Gamma(w)\,\Gamma(w-u-v)}{\Gamma(w-u)\,\Gamma(w-v)} - \frac{x^u}{\Gamma(u+1)}\,{}_2F_2\left[\begin{matrix}w-v,u;\\ w,u+1;\end{matrix}\,-x\right]. \tag{7.95}$$

Proof. Putting $z=1$ in the decomposition formula (7.79), we get

$$ {}_2\Gamma_1\left[\begin{matrix}[u;x],v;\\ w;\end{matrix}\,1\right] = {}_2F_1\left[\begin{matrix}u,v;\\ w;\end{matrix}\,1\right] - {}_2\gamma_1\left[\begin{matrix}(u;x),v;\\ w;\end{matrix}\,1\right]. \tag{7.96}$$

Using (7.81), we have

$$ {}_2\Gamma_1\left[\begin{matrix}[u;x],v;\\ w;\end{matrix}\,1\right] = {}_2F_1\left[\begin{matrix}u,v;\\ w;\end{matrix}\,1\right] - \frac{1}{\Gamma(u)}\int_0^x \mathrm{u}^{u-1}e^{-\mathrm{u}}\,{}_1F_1\left[\begin{matrix}v;\\ w;\end{matrix}\,\mathrm{u}\right]d\mathrm{u}. \tag{7.97}$$

In view of the Kummer's formula ${}_1F_1[{}^{v;}_{w;}\,z] = e^z\,{}_1F_1[{}^{w-v;}_{w;}\,-z]$, the above equation yields

$$ {}_2\Gamma_1\left[\begin{matrix}[u;x],v;\\ w;\end{matrix}1\right] = {}_2F_1\left[\begin{matrix}u,v;\\ w;\end{matrix}1\right] - \frac{1}{\Gamma(u)}\int_0^x u^{u-1}e^{-u}\,{}_1F_1\left[\begin{matrix}w-v;\\ w;\end{matrix}-u\right]du. \tag{7.98} $$

Finally, applying the Gauss summation formula,

$$ {}_2F_1\left[\begin{matrix}u,v;\\ w;\end{matrix}1\right] = \frac{\Gamma(w)\,\Gamma(w-u-v)}{\Gamma(w-u)\,\Gamma(w-v)} $$

and using the integral

$$ \frac{1}{\Gamma(u)}\int_0^x u^{u-1}\,{}_1F_1\left[\begin{matrix}w-v;\\ w;\end{matrix}-u\right]du = \frac{x^u}{\Gamma(u+1)}{}_2F_2\left[\begin{matrix}w-v,u;\\ w,u+1;\end{matrix}-x\right], $$

we have

$$ {}_2\Gamma_1\left[\begin{matrix}[u;x],v;\\ w;\end{matrix}1\right] = \frac{\Gamma(w)\,\Gamma(w-u-v)}{\Gamma(w-u)\,\Gamma(w-v)} - \frac{x^u}{\Gamma(u+1)}{}_2F_2\left[\begin{matrix}w-v,u;\\ w,u+1;\end{matrix}-x\right]. \tag{7.99} $$

It completes the proof. □

Theorem 256. *The following recurrence relation holds true:*

$$ [v-(w-1)]\,{}_2\Gamma_1\left[\begin{matrix}[u;x],v;\\ w;\end{matrix}z\right] = v\,{}_2\Gamma_1\left[\begin{matrix}[u;x],v+1;\\ w;\end{matrix}z\right] - (w-1)\,{}_2\Gamma_1\left[\begin{matrix}[u;x],v;\\ w-1;\end{matrix}z\right]. \tag{7.100} $$

Proof. The following contiguous relation is obeyed by the confluent hypergeometric function

$$ (w-v-1)\,{}_1F_1\left[\begin{matrix}v;\\ w;\end{matrix}z\right] = (w-1)\,{}_1F_1\left[\begin{matrix}v;\\ w-1;\end{matrix}z\right] - v\,{}_1F_1\left[\begin{matrix}v+1;\\ w;\end{matrix}z\right]. \tag{7.101} $$

Using (7.82) and (7.101), we get

$$ \begin{aligned} &(w-v-1)\,{}_2\Gamma_1\left[\begin{matrix}[u;x],v;\\ w;\end{matrix}z\right]\\ &= \frac{1}{\Gamma(u)}\int_x^\infty u^{u-1}e^{-u}\left((w-1)\,{}_1F_1\left[\begin{matrix}v;\\ w-1;\end{matrix}zu\right] - v\,{}_1F_1\left[\begin{matrix}v+1;\\ w;\end{matrix}zu\right]\right)du\\ &= \frac{(w-1)}{\Gamma(u)}\int_x^\infty u^{u-1}e^{-u}\,{}_1F_1\left[\begin{matrix}v;\\ w-1;\end{matrix}zu\right]du - \frac{v}{\Gamma(u)}\int_x^\infty u^{u-1}e^{-u}\,{}_1F_1\left[\begin{matrix}v+1;\\ w;\end{matrix}zu\right]du\\ &= (w-1)\,{}_2\Gamma_1\left[\begin{matrix}[u;x],v;\\ w-1;\end{matrix}z\right] - v\,{}_2\Gamma_1\left[\begin{matrix}[u;x],v+1;\\ w;\end{matrix}z\right]. \end{aligned} \tag{7.102} $$

A simple transposition of terms gives the required result. □

Theorem 257. *The following recurrence relation holds true:*

$$\frac{uz}{w}\,{}_2\Gamma_1\left[\begin{matrix}[u+1;x],v+1;\\ w+1;\end{matrix}z\right]={}_2\Gamma_1\left[\begin{matrix}[u;x],v+1;\\ w;\end{matrix}z\right]-{}_2\Gamma_1\left[\begin{matrix}[u;x],v;\\ w;\end{matrix}z\right].\tag{7.103}$$

Proof. We note that

$$uz\,{}_2\Gamma_1\left[\begin{matrix}[u+1;x],v+1;\\ w+1;\end{matrix}z\right]=\frac{u}{\Gamma(u+1)}\int_x^\infty u^{u-1}e^{-u}\left(zu\,{}_1F_1\left[\begin{matrix}v;\\ w-1;\end{matrix}zu\right]\right)du.\tag{7.104}$$

Now, by using the following contiguous relation:

$$w\,{}_1F_1\left[\begin{matrix}v;\\ w;\end{matrix}z\right]-w\,{}_1F_1\left[\begin{matrix}v-1;\\ w;\end{matrix}z\right]=z\,{}_1F_1\left[\begin{matrix}v;\\ w+1;\end{matrix}z\right],$$

in the above equation, we have

$$\begin{aligned}
&uz\,{}_2\Gamma_1\left[\begin{matrix}[u+1;x],v+1;\\ w+1;\end{matrix}z\right]\\
&=\frac{w}{\Gamma(u)}\int_x^\infty u^{u-1}e^{-u}\left({}_1F_1\left[\begin{matrix}v+1;\\ w;\end{matrix}zu\right]-{}_1F_1\left[\begin{matrix}v;\\ w;\end{matrix}zu\right]\right)du\\
&=\frac{w}{\Gamma(u)}\int_x^\infty u^{u-1}e^{-u}\,{}_1F_1\left[\begin{matrix}v+1;\\ w;\end{matrix}zu\right]du-\frac{w}{\Gamma(u)}\int_x^\infty u^{u-1}e^{-u}\,{}_1F_1\left[\begin{matrix}v;\\ w;\end{matrix}zu\right]du\\
&=w\,{}_2\Gamma_1\left[\begin{matrix}[u;x],v+1;\\ w;\end{matrix}z\right]-w\,{}_2\Gamma_1\left[\begin{matrix}[u;x],v;\\ w;\end{matrix}z\right].
\end{aligned}\tag{7.105}$$

On dividing the above equation by w gives the required relation (7.103). □

Theorem 258. *The following relationship holds true with the incomplete gamma function $\gamma(u,x)$ defined by* (7.63)*:*

$$\begin{aligned}
&{}_2\Gamma_1\left[\begin{matrix}[u;x],v;\\ v+1;\end{matrix}-z\right]\\
&=\frac{vz^{-v}}{\Gamma(u)}\int_x^\infty u^{(u-v)-1}e^{-u}\,\gamma(v,zu)\,du,\quad x\geq 0;\Re(u)>0\text{ when }x=0.
\end{aligned}\tag{7.106}$$

Proof. Putting $w=v+1$ in (7.82) and replacing z by $-z$, we get

$${}_2\Gamma_1\left[\begin{matrix}[u;x],v;\\ v+1;\end{matrix}-z\right]=\frac{1}{\Gamma(u)}\int_x^\infty u^{u-1}e^{-u}\,{}_1F_1\left[\begin{matrix}v;\\ v+1;\end{matrix}-zu\right]du.\tag{7.107}$$

Using the well known result,

$$ {}_1F_1\left[\begin{matrix} v; \\ v+1; \end{matrix} -zu\right] = v\, z^{-v}\, \gamma(v,z), $$

gives

$$ {}_2\Gamma_1\left[\begin{matrix} [u;x], v; \\ v+1; \end{matrix} -z\right] = \frac{v\, z^{-v}}{\Gamma(u)} \int_x^{\infty} u^{(u-v)-1}\, e^{-u}\, \gamma(v, zu)\, du. \tag{7.108} $$

This completes the proof. □

Corollary 259. *The following relationship holds true with the incomplete gamma function $\gamma(u;x)$ defined by*

$$ {}_2F_1\left[\begin{matrix} u, v; \\ v+1; \end{matrix} -z\right] = \frac{v\, z^{-v}}{\Gamma(u)} \int_0^{\infty} u^{(u-v)-1}\, e^{-u}\, \gamma(v, zu)\, du. \tag{7.109} $$

Theorem 260. *The following relationship holds true with the incomplete gamma function $\Gamma(u,x)$ defined by*

$$ {}_2\Gamma_1\left[\begin{matrix} [u;x], v; \\ v; \end{matrix} z\right] = {}_1\Gamma_0\left[\begin{matrix} [u;x]; \\ -; \end{matrix} z\right] = \frac{(1-z)^{-u}}{\Gamma(u)}\, \Gamma(u, x(1-z)), \quad |z|<1; x \ge 0. \tag{7.110} $$

Proof. Putting $w = v$ in (7.82), we immediately obtain the following simplified form:

$$ {}_2\Gamma_1\left[\begin{matrix} [u;x], v; \\ v; \end{matrix} z\right] = \frac{1}{\Gamma(u)} \int_x^{\infty} u^{u-1}\, e^{-(1-z)u}\, du. $$

Now, by setting

$$ u = \frac{t}{1-z} \quad \text{and} \quad du = \frac{dt}{1-z}. $$

We get

$$ {}_2\Gamma_1\left[\begin{matrix} [u;x], v; \\ v; \end{matrix} z\right] = \frac{(1-z)^{-u}}{\Gamma(u)} \int_{x(1-z)}^{\infty} t^{u-1}\, e^{-t}\, dt = \frac{(1-z)^{-u}}{\Gamma(u)}\, \Gamma(u, x(1-z)); $$

hence, the proof. □

Theorem 261. *The following integral representation holds true:*

$$ {}_2\Gamma_1\left[\begin{matrix} [u;x], v; \\ w; \end{matrix} z\right] $$
$$ = \frac{\Gamma(w)}{\Gamma(v)\,\Gamma(w-v)} \int_0^1 u^{v-1}\, (1-u)^{w-v-1}\, {}_1\Gamma_0\left[\begin{matrix} [u;x]; \\ -; \end{matrix} zu\right] du, \quad \Re(w) > \Re(v) > 0. \tag{7.111} $$

Proof. Using the integral relation

$$\frac{(v)_k}{(w)_k} = \frac{\Gamma(v+k)\,\Gamma(w)}{\Gamma(w+k)\,\Gamma(v)} = \frac{\Gamma(w)}{\Gamma(v)\,\Gamma(w-v)}\,\beta(v+k,w-v)$$
$$= \frac{\Gamma(w)}{\Gamma(v)\,\Gamma(w-v)} \int_0^1 u^{v+k-1}(1-u)^{w-v-1}\,du. \tag{7.112}$$

We have

$${}_2\Gamma_1\left[\begin{matrix}[u;x],v;\\ w;\end{matrix}\,z\right] = \frac{\Gamma(w)}{\Gamma(v)\,\Gamma(w-v)} \int_0^1 \sum_{k=0}^{\infty} [u;x]_k \frac{(zu)^k}{k!}\, u^{v-1}(1-u)^{w-v-1}\,du$$
$$= \frac{\Gamma(w)}{\Gamma(v)\,\Gamma(w-v)} \int_0^1 u^{v-1}(1-u)^{w-v-1}\,{}_1\Gamma_0\left[\begin{matrix}[u;x];\\ -;\end{matrix}\,zu\right] du. \tag{7.113}$$

It completes the proof. □

7.2.2 Generalized incomplete hypergeometric functions

With the following notation:

$$(A_p)_k = (u_1)_k\,(u_2)_k\cdots(u_p)_k;$$
$$(A_p;x)_k = (u_1;x)_k\,(u_2)_k\cdots(u_p)_k;$$
$$[A_p;x]_k = [u_1;x]_k\,(u_2)_k\cdots(u_p)_k$$

and

$$(A_p)_k = (A_p;x)_k + [A_p;x]_k.$$

We have a pair of incomplete generalized hypergeometric functions, defined by

$${}_p\gamma_q\left[\begin{matrix}(A_p;x);\\ B_q;\end{matrix}\,z\right] = {}_p\gamma_q\left[\begin{matrix}(u_1;x),u_2,\dots,u_p;\\ v_1,\dots,v_q;\end{matrix}\,z\right]$$
$$= \sum_{k=0}^{\infty} \frac{(A_p;x)_k}{(B_q)_k}\frac{z^k}{k!} = \sum_{k=0}^{\infty} \frac{(u_1;x)_k\,(u_2)_k\cdots(u_p)_k}{(v_1)_k\cdots(v_q)_k}\frac{z^k}{k!} \tag{7.114}$$

and

$${}_p\Gamma_q\left[\begin{matrix}[A_p;x];\\ B_q;\end{matrix}\,z\right] = {}_p\gamma_q\left[\begin{matrix}[u_1;x],u_2,\dots,u_p;\\ v_1,\dots,v_q;\end{matrix}\,z\right]$$
$$= \sum_{k=0}^{\infty} \frac{[A_p;x]_k}{(B_q)_k}\frac{z^k}{k!} = \sum_{k=0}^{\infty} \frac{[u_1;x]_k\,(u_2)_k\cdots(u_p)_k}{(v_1)_k\cdots(v_q)_k}\frac{z^k}{k!}, \tag{7.115}$$

provided that the definition of each infinite series is absolutely convergent. Indeed, in their special case when $x = 0$ both ${}_p\gamma_q$ and ${}_p\Gamma_q$ reduces to the generalized hypergeometric function ${}_pF_q$.

As an immediate consequence of the definitions (7.114) and (7.115), we have the following decomposition formula:

$$ {}_p\gamma_q\left[\begin{matrix}(A_p;x); \\ B_q;\end{matrix} z\right] + {}_p\Gamma_q\left[\begin{matrix}[A_p;x]; \\ B_q;\end{matrix} z\right] = {}_pF_q\left[\begin{matrix}A_p; \\ B_q;\end{matrix} z\right]. \tag{7.116}$$

We now list several properties of generalized incomplete hypergeometric functions without their proofs.

Theorem 262. *The following integral representations hold true:*

$$ {}_p\gamma_q\left[\begin{matrix}(u_1;x), u_2, \dots, u_p; \\ v_1, \dots, v_q;\end{matrix} z\right] = \frac{1}{\Gamma(u_1)} \int_0^x u^{u_1-1} e^{-u} {}_{p-1}F_q\left[\begin{matrix}u_2, \dots, u_p; \\ v_1, \dots, v_q;\end{matrix} zu\right] du \tag{7.117}$$

and

$$ {}_p\Gamma_q\left[\begin{matrix}[u_1;x], u_2, \dots, u_p; \\ v_1, \dots, v_q;\end{matrix} z\right] = \frac{1}{\Gamma(u_1)} \int_x^\infty u^{u_1-1} e^{-u} {}_{p-1}F_q\left[\begin{matrix}u_2, \dots, u_p; \\ v_1, \dots, v_q;\end{matrix} zu\right] du. \tag{7.118}$$

Theorem 263. *The following differential formulas hold true:*

$$\begin{aligned} &\frac{d^k}{dz^k}\left\{{}_p\gamma_q\left[\begin{matrix}(u_1;x), u_2, \dots, u_p; \\ v_1, \dots, v_q;\end{matrix} z\right]\right\} \\ &\quad = \frac{(u_1)_k \cdots (u_p)_k}{(v_1)_k \cdots (v_q)_k} {}_p\gamma_q\left[\begin{matrix}(u_1+k;x), u_2+k, \dots, u_p+k; \\ v_1+k, \dots, v_q+k;\end{matrix} z\right] \end{aligned} \tag{7.119}$$

and

$$\begin{aligned} &\frac{d^k}{dz^k}\left\{{}_p\Gamma_q\left[\begin{matrix}[u_1;x], u_2, \dots, u_p; \\ v_1, \dots, v_q;\end{matrix} z\right]\right\} \\ &\quad = \frac{(u_1)_k \cdots (u_p)_k}{(v_1)_k \cdots (v_q)_k} {}_p\Gamma_q\left[\begin{matrix}[u_1+k;x], u_2+k, \dots, u_p+k; \\ v_1+k, \dots, v_q+k;\end{matrix} z\right]. \end{aligned} \tag{7.120}$$

Theorem 264. *The following integral representations hold true:*

$$\begin{aligned} &{}_p\gamma_q\left[\begin{matrix}(u_1;x), u_2, \dots, u_{p-1}, v; \\ v_1, \dots, v_{q-1}, w;\end{matrix} z\right] \\ &\quad = \frac{\Gamma(w)}{\Gamma(v)\,\Gamma(w-v)} \int_0^1 u^{v-1} (1-u)^{w-v-1} \end{aligned}$$

$$\times {}_{p-1}\gamma_{q-1}\left[\begin{matrix}(u_1;x), u_2, \ldots, u_{p-1};\\ v_1, \ldots, v_{q-1};\end{matrix} z\,\mathrm{u}\right] du, \quad \mathbb{R}(w) > \mathbb{R}(v) > 0 \tag{7.121}$$

and

$$\begin{aligned}&{}_p\Gamma_q\left[\begin{matrix}[u_1;x], u_2, \ldots, u_{p-1}, v;\\ v_1, \ldots, v_{q-1}, w;\end{matrix} z\right]\\ &= \frac{\Gamma(w)}{\Gamma(v)\,\Gamma(w-v)}\int_0^1 \mathrm{u}^{v-1}\,(1-\mathrm{u})^{w-v-1}\\ &\times {}_{p-1}\Gamma_{q-1}\left[\begin{matrix}(u_1;x), u_2, \ldots, u_{p-1};\\ v_1, \ldots, v_{q-1};\end{matrix} z\,\mathrm{u}\right] du, \quad \mathbb{R}(w) > \mathbb{R}(v) > 0.\end{aligned} \tag{7.122}$$

The particular cases of the above three theorems produce the results for the generalized hypergeometric function as presented in the following corollary.

Corollary 265. *The following results hold true for* pF_q*:*

$${}_pF_q\left[\begin{matrix}u_1, \ldots, u_p;\\ v_1, \ldots, v_q;\end{matrix} z\right] = \frac{1}{\Gamma(u_1)}\int_0^\infty \mathrm{u}^{u_1-1}\,e^{-\mathrm{u}}\,{}_{p-1}F_q\left[\begin{matrix}u_2, \ldots, u_p;\\ v_1, \ldots, v_q;\end{matrix} z\mathrm{u}\right] du;$$

$$\begin{aligned}&\frac{d^k}{dz^k}\left\{{}_pF_q\left[\begin{matrix}u_1, \ldots, u_p;\\ v_1, \ldots, v_q;\end{matrix} z\right]\right\}\\ &= \frac{(u_1)_k\cdots(u_p)_k}{(v_1)_k\cdots(v_q)_k}\,{}_pF_q\left[\begin{matrix}u_1+k, \ldots, u_p+k;\\ v_1+k, \ldots, v_q+k;\end{matrix} z\right];\end{aligned}$$

$$\begin{aligned}&{}_pF_q\left[\begin{matrix}u_1, \ldots, u_{p-1}, v;\\ v_1, \ldots, v_{q-1}, w;\end{matrix} z\right]\\ &= \frac{\Gamma(w)}{\Gamma(v)\,\Gamma(w-v)}\int_0^1 \mathrm{u}^{v-1}\,(1-\mathrm{u})^{w-v-1}\\ &\times {}_{p-1}F_{q-1}\left[\begin{matrix}u_1, u_2, \ldots, u_{p-1};\\ v_1, \ldots, v_{q-1};\end{matrix} z\,\mathrm{u}\right] du, \quad \mathbb{R}(w) > \mathbb{R}(v) > 0.\end{aligned}$$

7.3 Discrete Gauss hypergeometric function

The discrete generalized hypergeometric series is defined by

$${}_p\mathcal{F}_q(u_1, \ldots, u_p; v_1, \ldots, v_q; u, n, z) = \sum_{k=0}^{\infty}\frac{(u_1)_k\cdots(u_p)_k}{(v_1)_k\cdots(v_q)_k}\,\frac{z^k\,(-1)^{nk}\,(-u)_{nk}}{k!}, \tag{7.123}$$

where $u_i, 1 \le i \le p$ and $v_j, 1 \le j \le q$ are complex numbers such that $v_j + l \ne 0$ for all positive integers l. If $n = 0$ in (7.123), we get

$$
\begin{aligned}
{}_p\mathcal{F}_q(u_1,\ldots,u_p;v_1,\ldots,v_q;u,0,z) &= \sum_{k=0}^{\infty}\frac{(u_1)_k\cdots(u_p)_k}{(v_1)_k\cdots(v_q)_k}\frac{z^k}{k!}\\
&= {}_pF_q(u_1,\ldots,u_p;v_1,\ldots,v_q;z).
\end{aligned}
\tag{7.124}
$$

If $n = 1$, then we have

$$
\begin{aligned}
{}_p\mathcal{F}_q(u_1,\ldots,u_p;v_1,\ldots,v_q;u,1,z) &= \sum_{k=0}^{\infty}\frac{(u_1)_k\cdots(u_p)_k}{(v_1)_k\cdots(v_q)_k}\frac{(-u)_k\,(-1)^k\,z^k}{k!}\\
&= {}_{p+1}F_q(u_1,\ldots,u_p,-u;v_1,\ldots,v_q;-z).
\end{aligned}
\tag{7.125}
$$

To get the relationship between the discrete and classical generalized hypergeometric function, first we see the following lemma.

Lemma 266. *For a natural number n and nonnegative integer k, the following identity holds:*

$$
(-u)_{nk} = n^{nk}\prod_{m=0}^{n-1}\left(\frac{-u+m}{n}\right)_k. \tag{7.126}
$$

Proof. Using the definition of a shifted factorial, we have

$$
\begin{aligned}
(-u)_{nk} &= (-u)\,(-u+1)\cdots(-u+n)(-u+n+1)\cdots(-u+nk-1)\\
&= \left[\prod_{m=0}^{k-1}(-u+m\,n)\right]\left[\prod_{m=0}^{k-1}(-u+m\,n+1)\right]\cdots\left[\prod_{m=0}^{k-1}(-u+m\,n+(n-1))\right]\\
&= n^{nk}\left[\frac{1}{n^k}\prod_{m=0}^{k-1}(-u+m\,n)\right]\left[\frac{1}{n^k}\prod_{m=0}^{k-1}(-u+m\,n+1)\right]\cdots\\
&\quad\times\left[\frac{1}{n^k}\prod_{m=0}^{k-1}(-u+m\,n+(n-1))\right]\\
&= n^{nk}\left[\prod_{m=0}^{k-1}\left(\frac{-u}{n}+m\right)\right]\left[\prod_{m=0}^{k-1}\left(\frac{-u+1}{n}+m\right)\right]\cdots\\
&\quad\times\left[\prod_{m=0}^{k-1}\left(\frac{-u+(n-1)}{n}+m\right)\right]\\
&= n^{nk}\left(\frac{-u}{n}\right)_k\left(\frac{-u+1}{n}\right)_k\cdots\left(\frac{-u+(n-1)}{n}\right)_k\\
&= n^{nk}\prod_{m=0}^{n-1}\left(\frac{-u+m}{n}\right)_k.
\end{aligned}
\tag{7.127}
$$

It completes the proof. □

With the help of the above lemma, the discrete generalized hypergeometric function can be defined in terms of the classical one as

$$
\begin{aligned}
&{}_p\mathcal{F}_q(u_1,\ldots,u_p;v_1,\ldots,v_q;u,n,z)\\
&=\sum_{k=0}^{\infty}\frac{(u_1)_k\cdots(u_p)_k}{(v_1)_k\cdots(v_q)_k}\frac{z^k\,(-1)^{nk}\,(-u)_{nk}}{k!}\\
&=\sum_{k=0}^{\infty}\frac{(u_1)_k\cdots(u_p)_k\,(\frac{-u}{n})_k\cdots(\frac{-u+(n-1)}{n})_k}{(v_1)_k\cdots(v_q)_k}\frac{z^k\,(-n)^{nk}}{k!}\\
&={}_{p+n}F_q\left(u_1,\ldots,u_p,\frac{-u}{n},\ldots,\frac{-u+(n-1)}{n};v_1,\ldots,v_q;(-n)^n\,z\right).
\end{aligned}
\tag{7.128}
$$

Next, we discuss the convergence of the discrete generalized hypergeometric function ${}_p\mathcal{F}_q$ in the following theorem.

Theorem 267. *The discrete generalized hypergeometric series defined in* (7.123) *converges for:*

1. $p+n<q+1$,
2. $p+n=q+1$ *and* $|z|\,n^n<1$, *or*
3. $p+n=q+1$, $|z|\,n^n=1$, *and*

$$\Re\left(\sum_{j=1}^{q}v_j-\sum_{i=1}^{p}u_i-\frac{1}{n}\sum_{m=0}^{n-1}-u+m>0\right).$$

Also, it diverges for

4. $p+n>q+1$, *or*
5. $p+n=q+1$ *and* $|z|\,n^n>1$.

Proof. When u is a natural number, $(-u)_{nk}$ terminates by definition so we assume $u\in\mathbb{C}\setminus\mathbb{N}$. We compute first

$$\frac{(-u)_{n(k+1)}}{(-u)_{nk}}=(-u+nk)\,(-u+nk+1)\cdots(-u+nk+n-1).$$

Now, assume the general term of (7.123) is A_k. By the ratio test, we have

$$
\begin{aligned}
\left|\frac{A_{k+1}}{A_k}\right|&=\left|z\frac{(u_1+k)\cdots(u_p+k)\,(nk-u)\cdots(nk-1+n-u)}{(v_1+k)\cdots(v_q+k)\,(k+1)}\right|\\
&=\left|z\frac{n^n\,k^{n+p}+\cdots}{k^{q+1}+\cdots}\right|.
\end{aligned}
\tag{7.129}
$$

If we take the limit as $k\to\infty$, then the limit is zero when $p+n<q+1$. This yields the convergence of ${}_p\mathcal{F}_q$. The limit is infinity when $p+n>q+1$, yielding the divergence. Similarly, the other criteria mentioned in the theorem can be verified. □

Theorem 268. *Let* Δ *be the forwarding operator such that* $\Delta f(u)=f(u+1)-f(u)$. *Then, the following difference formula holds true:*

$$\Delta^r \left[{}_p\mathcal{F}_q \left(\begin{matrix} u_1, \ldots, u_p \\ v_1, \ldots, v_q \end{matrix}; u, 1, z \right) \right] = \frac{\prod_{i=1}^{p} (u_i)_r}{\prod_{j=1}^{q} (v_j)_r} z^r \, {}_p\mathcal{F}_q \left(\begin{matrix} u_1 + r, \ldots, u_p + r \\ v_1 + r, \ldots, v_q + r \end{matrix}; u, 1, z \right). \tag{7.130}$$

Proof.

$$\begin{aligned} \Delta^r \left[{}_p\mathcal{F}_q \left(\begin{matrix} u_1, \ldots, u_p \\ v_1, \ldots, v_q \end{matrix}; u, 1, z \right) \right] &= \Delta^r \left[\sum_{k=0}^{\infty} \frac{(u_1)_k \cdots (u_p)_k}{(v_1)_k \cdots (v_q)_k} \frac{z^k \, (-1)^k \, (-u)_k}{k!} \right] \\ &= \sum_{k=r}^{\infty} \frac{(u_1)_k \cdots (u_p)_k}{(v_1)_k \cdots (v_q)_k} \frac{z^k \, (-1)^{k-r} \, (-u)_{k-r}}{(k-r)!} \\ &= \sum_{k=0}^{\infty} \frac{(u_1)_{k+r} \cdots (u_p)_{k+r}}{(v_1)_{k+r} \cdots (v_q)_{k+r}} \frac{z^{k+r} \, (-1)^k \, (-u)_k}{k!} \\ &= \frac{\prod_{i=1}^{p} (u_i)_r}{\prod_{j=1}^{q} (v_j)_r} z^r \, {}_p\mathcal{F}_q \left(\begin{matrix} u_1 + r, \ldots, u_p + r \\ v_1 + r, \ldots, v_q + r \end{matrix}; u, 1, z \right). \end{aligned} \tag{7.131}$$

It completes the proof. □

Corollary 269. *We have*

$$\Delta \left[{}_p\mathcal{F}_q \left(\begin{matrix} u_1, \ldots, u_p \\ v_1, \ldots, v_q \end{matrix}; u, 1, z \right) \right] = \frac{u_1 \cdots u_p}{v_1 \cdots v_q} z \, {}_p\mathcal{F}_q \left(\begin{matrix} u_1 + 1, \ldots, u_p + 1 \\ v_1 + 1, \ldots, v_q + 1 \end{matrix}; u, 1, z \right). \tag{7.132}$$

Defining the shift operator ρ by $(\rho f)(u) = f(u-1)$ and define the Θ by $\Theta = u \rho \Delta$, which is a discrete analogue of $\theta = z \frac{\partial}{\partial z}$. We have the following lemma.

Lemma 270.

$$\Theta(-1)^k \, (-u)_k = k \, (-1)^k \, (-u)_k, \quad k \in \mathbb{N}.$$

Proof.

$$\begin{aligned} \Theta(-1)^k \, (-u)_k &= u \, \rho \, \Delta[(-1)^k \, (-u)_k] \\ &= u \, \rho[k \, (-1)^{k-1} \, (-u)_{k-1}] \\ &= u \, [k \, (-1)^{k-1} \, (-u+1)_{k-1}] \\ &= k \, (-1)^k \, (-u)_k. \end{aligned} \tag{7.133}$$

It completes the proof. □

Theorem 271. *The discrete generalized hypergeometric function ${}_p\mathcal{F}_q(z)$ defined in* (7.123) *satisfies the following difference equation:*

$$\left[\Theta\prod_{j=1}^{q}\left(\frac{1}{n}\Theta+v_j-1\right)-nz(-1)^n(-u)_n\rho^n\prod_{i=1}^{p}\left(\frac{1}{n}\Theta+u_i\right)\right]{}_p\mathcal{F}_q(z)=0. \tag{7.134}$$

Proof. Start with the computation

$$\begin{aligned}
&nz(-1)^n(-u)_n\rho^n\prod_{i=1}^{p}\left(\frac{1}{n}\Theta+u_i\right){}_p\mathcal{F}_q(z)\\
&=nz(-1)^n(-u)_n\rho^n\sum_{k=0}^{\infty}\frac{(u_1)_k\cdots(u_p)_k}{(v_1)_k\cdots(v_q)_k}\frac{z^k(-1)^{nk}}{k!}\prod_{i=1}^{p}\left(\frac{1}{n}\Theta+u_i\right)(-u)_{nk}\\
&=nz(-1)^n(-u)_n\rho^n\sum_{k=0}^{\infty}\frac{(u_1)_k\cdots(u_p)_k}{(v_1)_k\cdots(v_q)_k}\frac{z^k(-1)^{nk}(-u)_{nk}}{k!}\prod_{i=1}^{p}(k+u_i)\\
&=nz(-1)^n\sum_{k=0}^{\infty}\frac{(u_1)_k\cdots(u_p)_k}{(v_1)_k\cdots(v_q)_k}\frac{z^k(-1)^{nk}(-u)_{nk+n}}{k!}\prod_{i=1}^{p}(k+u_i).
\end{aligned}\tag{7.135}$$

Now, we compute

$$\begin{aligned}
&\Theta\prod_{j=1}^{q}\left(\frac{1}{n}\Theta+v_j-1\right){}_p\mathcal{F}_q(z)\\
&=\Theta\sum_{k=0}^{\infty}(u_1)_k\cdots(u_p)_k\left(\frac{1}{n}\Theta+v_j-1\right)\frac{1}{(v_1)_k\cdots(v_q)_k}\frac{z^k(-1)^{nk}(-u)_{nk}}{k!}\\
&=\Theta\sum_{k=0}^{\infty}\frac{(u_1)_k\cdots(u_p)_k}{(v_1)_{k-1}\cdots(v_q)_{k-1}}\frac{z^k(-1)^{nk}(-u)_{nk}}{k!}\\
&=nu\rho\sum_{k=0}^{\infty}\frac{(u_1)_k\cdots(u_p)_k}{(v_1)_{k-1}\cdots(v_q)_{k-1}}\frac{z^k(-1)^{nk-1}(-u)_{nk-1}}{(k-1)!}\\
&=nu\rho\sum_{k=0}^{\infty}\frac{(u_1)_{k+1}\cdots(u_p)_{k+1}}{(v_1)_k\cdots(v_q)_k}\frac{z^{k+1}(-1)^{nk+n-1}(-u)_{nk+n-1}}{k!}\\
&=nz(-u)\rho\sum_{k=0}^{\infty}\frac{(u_1)_k\cdots(u_p)_k}{(v_1)_k\cdots(v_q)_k}\frac{z^k(-1)^{nk+n}(-u)_{nk+n-1}}{k!}\prod_{i=1}^{p}(k+u_i)\\
&=nz(-u)\sum_{k=0}^{\infty}\frac{(u_1)_k\cdots(u_p)_k}{(v_1)_k\cdots(v_q)_k}\frac{z^k(-1)^{nk+n}(-u+1)_{nk+n-1}}{k!}\prod_{i=1}^{p}(k+u_i)\\
&=nz(-1)^n\sum_{k=0}^{\infty}\frac{(u_1)_k\cdots(u_p)_k}{(v_1)_k\cdots(v_q)_k}\frac{z^k(-1)^{nk}(-u)_{nk+n}}{k!}\prod_{i=1}^{p}(k+u_i).
\end{aligned}\tag{7.136}$$

The equations (7.135) and (7.136) can be easily put in the form of equation (7.134). It completes the proof. □

Corollary 272. *If $n = 1$, then the following difference equation holds:*

$$\left[\Theta \prod_{j=1}^{q}(\Theta + v_j - 1) - z\,u\,\rho \prod_{i=1}^{p}(\Theta + u_i)\right] {}_p\mathcal{F}_q\left(\begin{matrix}u_1,\dots,u_p\\ v_1,\dots,v_q\end{matrix}; u, 1, z\right) = 0. \tag{7.137}$$

We use the following abbreviations:

$$\begin{aligned}
{}_p\mathcal{F}_q(z) &= {}_p\mathcal{F}_q(u_1,\dots,u_p; v_1,\dots,v_q; u, n, z)\\
&= \sum_{k=0}^{\infty} \frac{(u_1)_k \cdots (u_p)_k}{(v_1)_k \cdots (v_q)_k} \frac{z^k\,(-1)^{nk}\,(-u)_{nk}}{k!},\\
{}_p\mathcal{F}_q(u_i\pm) &= {}_p\mathcal{F}_q(u_1,\dots,u_{i-1}, u_i \pm 1, u_{i+1},\dots,u_p; v_1,\dots,v_q; u, n, z)\\
&= \sum_{k=0}^{\infty} \frac{(u_1)_k \cdots (u_{i-1})_k\,(u_i \pm 1)_k\,(u_{i+1})_k \cdots (u_p)_k}{(v_1)_k \cdots (v_q)_k} \frac{z^k\,(-1)^{nk}\,(-u)_{nk}}{k!},\\
{}_p\mathcal{F}_q(v_j\pm) &= {}_p\mathcal{F}_q(u_1,\dots,u_p; v_1,\dots,v_{j-1}, v_j \pm 1, v_{j+1},\dots,v_q; u, n, z)\\
&= \sum_{k=0}^{\infty} \frac{(u_1)_k \cdots (u_p)_k}{(v_1)_k \cdots (v_{j-1})_k\,(v_j \pm 1)_k\,(v_{j+1})_k \cdots (v_q)_k} \frac{z^k\,(-1)^{nk}\,(-u)_{nk}}{k!},
\end{aligned}$$

to obtain the contiguous relations obeyed by the discrete generalized hypergeometric function listed in the following lemma.

Lemma 273. *For the discrete generalized hypergeometric function, we have*

$$\left(\frac{1}{n}\,\Theta + u_i\right) {}_p\mathcal{F}_q(z) = u_i\,F(u_i+), \quad 1 \le i \le p; \tag{7.138}$$

$$\left(\frac{1}{n}\,\Theta + v_j - 1\right) {}_p\mathcal{F}_q(z) = (v_j - 1)\,F(v_j-), \quad 1 \le j \le q; \tag{7.139}$$

$$(u_1 - u_i){}_p\mathcal{F}_q(z) = u_1\,{}_p\mathcal{F}_q(u_1+) - u_i{}_p\mathcal{F}_q(u_i+), \quad 2 \le i \le p; \tag{7.140}$$

$$(u_1 - v_j + 1){}_p\mathcal{F}_q(z) = u_1\,{}_p\mathcal{F}_q(u_1+) - (v_j - 1)\,{}_p\mathcal{F}_q(v_j-); \tag{7.141}$$

$$\frac{1}{n}\,\Theta\,{}_p\mathcal{F}_q(z) = z\,(-1)^n\,(-u)_n\,\rho^n \sum_{j=1}^{q} U_j\,{}_p\mathcal{F}_q(v_j+),$$
$$p < q, v_j's \text{ are pairwise different;} \tag{7.142}$$

$$u_1\,{}_p\mathcal{F}_q(z) = u_1\,{}_p\mathcal{F}_q(u_1+) - z\,(-1)^n\,(-u)_n\,\rho^n \sum_{j=1}^{q} U_j\,{}_p\mathcal{F}_q(v_j+),$$
$$p < q, v_j's \text{ are pairwise different;} \tag{7.143}$$

$$\begin{aligned}
u_1\,{}_p\mathcal{F}_q(z) + z\,(-1)^n\,(-u)_n\,\rho^n &= u_1\,{}_p\mathcal{F}_q(u_1+)\\
&\quad - z\,(-1)^n\,(-u)_n\,\rho^n \sum_{j=1}^{q} U_j\,{}_p\mathcal{F}_q(v_j+), \quad p = q;
\end{aligned} \tag{7.144}$$

$$
\begin{aligned}
{}_p\mathcal{F}_q(z) &= F(u_i-) + (-1)^n\,(-u)_n\,\rho^n \sum_{j=1}^{q} W_{j,i}\,{}_p\mathcal{F}_q(v_j+),\\
&\quad p \le q, 1 \le i \le p.
\end{aligned} \tag{7.145}
$$

The proofs of these relations are same as for the generalized hypergeometric function, and hence omitted here.

The particular cases of the discrete generalized hypergeometric function lead a number of elementary discrete special functions. In particular, for $p = 2$ and $q = 1$ it gives the discrete version of the Gauss hypergeometric function as

$$
{}_2\mathcal{F}_1(z) = {}_2\mathcal{F}_1(u_1, u_2; v_1; u, n, z) = \sum_{k=0}^{\infty} \frac{(u_1)_k\,(u_2)_k}{(v_1)_k}\,\frac{z^k\,(-1)^{nk}\,(-u)_{nk}}{k!}. \tag{7.146}
$$

For $p = 1$ and $q = 1$, it gives the discrete version of the confluent hypergeometric function as

$$
{}_1\mathcal{F}_1(z) = {}_1\mathcal{F}_1(u_1; v_1; u, n, z) = \sum_{k=0}^{\infty} \frac{(u_1)_k}{(v_1)_k}\,\frac{z^k\,(-1)^{nk}\,(-u)_{nk}}{k!}. \tag{7.147}
$$

Various properties of the discrete hypergeometric Gauss and confluent functions can be derived in the same way as for the discrete generalized hypergeometric function. We list few of them here.

Theorem 274. *The following difference formula holds true:*

$$
\Delta^r \left[{}_2\mathcal{F}_1\left(\begin{matrix} u_1, u_2 \\ v_1 \end{matrix}; u, 1, z \right) \right] = \frac{(u_1)_r\,(u_2)_r}{(v_1)_r}\, z^r\, {}_2\mathcal{F}_1\left(\begin{matrix} u_1 + r, u_2 + r \\ v_1 + r \end{matrix}; u, 1, z \right). \tag{7.148}
$$

Theorem 275. *The discrete Gauss hypergeometric function ${}_2\mathcal{F}_1(z)$ satisfies the following difference equation:*

$$
\left[\Theta\left(\frac{1}{n}\Theta + v_1 - 1\right) - n\,z\,(-1)^n\,(-u)_n\,\rho^n \left(\frac{1}{n}\Theta + u_1\right)\left(\frac{1}{n}\Theta + u_2\right) \right] {}_2\mathcal{F}_1(z) = 0. \tag{7.149}
$$

The proof of the above two theorems is similar to the proof for the generalized discrete hypergeometric function, and hence we are not producing it. Now, we give the integral representation of the discrete Gauss hypergeometric function.

Theorem 276. *Let u_1, u_2, v_1 and u be complex numbers. Then for $|z| < 1$ and $n \in \mathbb{N}$, the discrete Gauss hypergeometric function can be represented in the following integral form:*

$$
{}_2\mathcal{F}_1(u_1, u_2; v_1; u, n, z) = \Gamma\left(\begin{matrix} v_1 \\ u_1, v_1 - u_1 \end{matrix} \right) \int_0^1 t^{u_1 - 1}(1-t)^{v_1 - u_1 - 1}\, {}_1\mathcal{F}_0(u_2; -; u, n, tz)\,dt. \tag{7.150}
$$

Proof. Using

$$\frac{(u_1)_m}{(v_1)_m} = \Gamma\left(\begin{matrix} v_1 \\ u_1, v_1 - u_1 \end{matrix}\right) \int_0^1 t^{u_1+m-1} (1-t)^{v_1-u_1-1}\, dt, \tag{7.151}$$

one can write

$$\begin{aligned}
&{}_2\mathcal{F}_1(u_1, u_2; v_1; u, n, z) \\
&= \Gamma\left(\begin{matrix} v_1 \\ u_1, v_1 - u_1 \end{matrix}\right) \sum_{m\geq 0} \int_0^1 t^{u_1+m-1} (1-t)^{v_1-u_1-1} \\
&\quad \times \frac{(u_2)_m (-1)^{mn} (-u)_{mn}}{m!} z^m\, dt \\
&= \Gamma\left(\begin{matrix} v_1 \\ u_1, v_1 - u_1 \end{matrix}\right) \int_0^1 t^{u_1-1} (1-t)^{v_1-u_1-1} \sum_{m\geq 0} \frac{(u_2)_m (-1)^{mn} (-u)_{mn}}{m!} (zt)^m\, dt \\
&= \Gamma\left(\begin{matrix} v_1 \\ u_1, v_1 - u_1 \end{matrix}\right) \int_0^1 t^{u_1-1}(1-t)^{v_1-u_1-1}\, {}_1\mathcal{F}_0(u_2; -; u, n, tz)\, dt.
\end{aligned} \tag{7.152}$$

It completes the proof. □

Finally, we list the differential recursion formulae satisfied by ${}_2\mathcal{F}_1(u_1, u_2; v_1; u, n, z)$. Using the list of simple differential relations as follows:

$$u_1\, {}_2\mathcal{F}_1(u_1 + 1) = (u_1 + \theta)\, {}_2\mathcal{F}_1; \tag{7.153}$$

$$(u_1 + \theta - 1)\, {}_2\mathcal{F}_1(u_1 - 1) = (u_1 - 1)\, {}_2\mathcal{F}_1; \tag{7.154}$$

$$u_2\, {}_2\mathcal{F}_1(u_2 + 1) = (u_2 + \theta)\, {}_2\mathcal{F}_1; \tag{7.155}$$

$$(u_2 + \theta - 1)\, {}_2\mathcal{F}_1(u_2 - 1) = (u_2 - 1)\, {}_2\mathcal{F}_1; \tag{7.156}$$

$$(v_1 - 1)\, {}_2\mathcal{F}_1(v_1 - 1) = (v_1 + \theta - 1)\, {}_2\mathcal{F}_1; \tag{7.157}$$

$$(v_1 + \theta)\, {}_2\mathcal{F}_1(v_1 + 1) = v_1\, {}_2\mathcal{F}_1. \tag{7.158}$$

We get first-order or second-order differential recursion relations as

$$u_1 (u_1 - 1)\, {}_2\mathcal{F}_1(u_1 + 1) - (u_1 + \theta)(u_1 + \theta - 1) {}_2\mathcal{F}_1(u_1 - 1) = 0; \tag{7.159}$$

$$u_1 (u_2 - 1)\, {}_2\mathcal{F}_1(u_1 + 1) - (u_1 + \theta)(u_2 + \theta - 1) {}_2\mathcal{F}_1(u_2 - 1) = 0; \tag{7.160}$$

$$u_1 v_1\, {}_2\mathcal{F}_1(u_1 + 1) - (u_1 + \theta)(v_1 + \theta) {}_2\mathcal{F}_1(v_1 + 1) = 0; \tag{7.161}$$

$$u_1 (u_2 + \theta)\, {}_2\mathcal{F}_1(u_1 + 1) - u_2 (u_1 + \theta)\, {}_2\mathcal{F}_1(u_2 + 1) = 0; \tag{7.162}$$

$$u_1 (v_1 + \theta - 1)\, {}_2\mathcal{F}_1(u_1 + 1) - (v_1 - 1)(u_1 + \theta)\, {}_2\mathcal{F}_1(v_1 - 1) = 0; \tag{7.163}$$

$$(u_1 + \theta - 1)(u_2 + \theta)\, {}_2\mathcal{F}_1(u_1 - 1) - u_2 (u_1 - 1)\, {}_2\mathcal{F}_1(u_2 + 1) = 0; \tag{7.164}$$

$$(u_1 + \theta - 1)(v_1 + \theta - 1)\, {}_2\mathcal{F}_1(u_1 - 1) - (v_1 - 1)(u_1 - 1)\, {}_2\mathcal{F}_1(v_1 - 1) = 0; \tag{7.165}$$

$$(u_2-1)(u_1+\theta-1)\,{}_2\mathcal{F}_1(u_1-1)-(u_1-1)(u_2+\theta-1){}_2\mathcal{F}_1(u_2-1)=0; \tag{7.166}$$
$$v_1(u_1+\theta-1)\,{}_2\mathcal{F}_1(u_1-1)-(u_1-1)(v_1+\theta){}_2\mathcal{F}_1(v_1+1)=0; \tag{7.167}$$
$$u_2(u_2-1)\,{}_2\mathcal{F}_1(u_2+1)-(u_2+\theta)(u_2+\theta-1){}_2\mathcal{F}_1(u_2-1)=0; \tag{7.168}$$
$$u_2(v_1+\theta-1)\,{}_2\mathcal{F}_1(u_2+1)-(v_1-1)(u_2+\theta){}_2\mathcal{F}_1(v_1-1)=0; \tag{7.169}$$
$$u_2\,v_1\,{}_2\mathcal{F}_1(u_2+1)-(v_1+\theta)(u_2+\theta){}_2\mathcal{F}_1(v_1+1)=0; \tag{7.170}$$
$$(u_2+\theta-1)(v_1+\theta-1)\,{}_2\mathcal{F}_1(u_2-1)-(v_1-1)(u_2-1){}_2\mathcal{F}_1(v_1-1)=0; \tag{7.171}$$
$$v_1(u_2+\theta-1)\,{}_2\mathcal{F}_1(u_2-1)-(u_2-1)(v_1+\theta)\,{}_2\mathcal{F}_1(v_1+1)=0; \tag{7.172}$$
$$v_1(v_1-1)\,{}_2\mathcal{F}_1(v_1-1)-(v_1+\theta-1)(v_1+\theta)\,{}_2\mathcal{F}_1(v_1+1)=0. \tag{7.173}$$

Similarly, using the difference-differential relations,

$$u_1\,{}_2\mathcal{F}_1(u_1+1)=\left(u_1+\frac{1}{n}\Theta_u\right){}_2\mathcal{F}_1; \tag{7.174}$$
$$\left(u_1+\frac{1}{n}\Theta_u-1\right){}_2\mathcal{F}_1(u_1-1)=(u_1-1)\,{}_2\mathcal{F}_1; \tag{7.175}$$
$$u_2\,{}_2\mathcal{F}_1(u_2+1)=\left(u_2+\frac{1}{n}\Theta_u\right){}_2\mathcal{F}_1; \tag{7.176}$$
$$\left(u_2+\frac{1}{n}\Theta_u-1\right){}_2\mathcal{F}_1(u_2-1)=(u_2-1)\,{}_2\mathcal{F}_1; \tag{7.177}$$
$$(v_1-1)\,{}_2\mathcal{F}_1(v_1-1)=\left(v_1+\frac{1}{n}\Theta_u-1\right){}_2\mathcal{F}_1; \tag{7.178}$$
$$\left(v_1+\frac{1}{n}\Theta_u\right){}_2\mathcal{F}_1(v_1+1)=v_1\,{}_2\mathcal{F}_1, \tag{7.179}$$

we get the difference differential recursion relations as

$$u_1(u_1-1)\,{}_2\mathcal{F}_1(u_1+1)-\left(u_1+\frac{1}{n}\Theta_u\right)\left(u_1+\frac{1}{n}\Theta_u-1\right){}_2\mathcal{F}_1(u_1-1)=0; \tag{7.180}$$
$$u_1(u_2-1)\,{}_2\mathcal{F}_1(u_1+1)-\left(u_1+\frac{1}{n}\Theta_u\right)\left(u_2+\frac{1}{n}\Theta_u-1\right){}_2\mathcal{F}_1(u_2-1)=0; \tag{7.181}$$
$$u_1\,v_1\,{}_2\mathcal{F}_1(u_1+1)-\left(u_1+\frac{1}{n}\Theta_u\right)\left(v_1+\frac{1}{n}\Theta_u\right){}_2\mathcal{F}_1(v_1+1)=0; \tag{7.182}$$
$$u_1\left(u_2+\frac{1}{n}\Theta_u\right){}_2\mathcal{F}_1(u_1+1)-u_2\left(u_1+\frac{1}{n}\Theta_u\right){}_2\mathcal{F}_1(u_2+1)=0; \tag{7.183}$$
$$u_1\left(v_1+\frac{1}{n}\Theta_u-1\right){}_2\mathcal{F}_1(u_1+1)-(v_1-1)\left(u_1+\frac{1}{n}\Theta_u\right){}_2\mathcal{F}_1(v_1-1)=0; \tag{7.184}$$
$$\left(u_1+\frac{1}{n}\Theta_u-1\right)\left(u_2+\frac{1}{n}\Theta_u\right){}_2\mathcal{F}_1(u_1-1)-u_2(u_1-1)\,{}_2\mathcal{F}_1(u_2+1)=0; \tag{7.185}$$
$$\left(u_1+\frac{1}{n}\Theta_u-1\right)\left(v_1+\frac{1}{n}\Theta_u-1\right){}_2\mathcal{F}_1(u_1-1)-(v_1-1)(u_1-1)\,{}_2\mathcal{F}_1(v_1-1)=0; \tag{7.186}$$

$$(u_2-1)\left(u_1+\frac{1}{n}\Theta_u-1\right){}_2\mathcal{F}_1(u_1-1)-(u_1-1)\left(u_2+\frac{1}{n}\Theta_u-1\right){}_2\mathcal{F}_1(u_2-1)=0; \quad (7.187)$$

$$v_1\left(u_1+\frac{1}{n}\Theta_u-1\right){}_2\mathcal{F}_1(u_1-1)-(u_1-1)\left(v_1+\frac{1}{n}\Theta_u\right){}_2\mathcal{F}_1(v_1+1)=0; \quad (7.188)$$

$$u_2\left(v_1+\frac{1}{n}\Theta_u-1\right){}_2\mathcal{F}_1(u_2+1)-(v_1-1)\left(u_2+\frac{1}{n}\Theta_u\right){}_2\mathcal{F}_1(v_1-1)=0; \quad (7.189)$$

$$u_2\,v_1\,{}_2\mathcal{F}_1(u_2+1)-\left(v_1+\frac{1}{n}\Theta_u\right)\left(u_2+\frac{1}{n}\Theta_u\right){}_2\mathcal{F}_1(v_1+1)=0; \quad (7.190)$$

$$\left(u_2+\frac{1}{n}\Theta_u-1\right)\left(v_1+\frac{1}{n}\Theta_u-1\right){}_2\mathcal{F}_1(u_2-1)-(v_1-1)\,(u_2-1){}_2\mathcal{F}_1(v_1-1)=0; \quad (7.191)$$

$$v_1\left(u_2+\frac{1}{n}\Theta_u-1\right){}_2\mathcal{F}_1(u_2-1)-(u_2-1)\left(v_1+\frac{1}{n}\Theta_u\right){}_2\mathcal{F}_1(v_1+1)=0; \quad (7.192)$$

$$v_1\,(v_1-1)\,{}_2\mathcal{F}_1(v_1-1)-\left(v_1+\frac{1}{n}\Theta_u-1\right)\left(v_1+\frac{1}{n}\Theta_u\right){}_2\mathcal{F}_1(v_1+1)=0. \quad (7.193)$$

Exercises

1. Prove that the (p,q)-difference operator follows the following product and quotient rule of differentiation:

$$D_{p,q}\left[g(u)\,h(u)\right]=g(p^{\frac{1}{2}}\,u)\,D_{p,q}\,h(u)+h(q^{-\frac{1}{2}}\,u)\,D_{p,q}\,g(u);$$
$$D_{p,q}\left[g(u)\,h(u)\right]=h(p^{\frac{1}{2}}\,u)\,D_{p,q}\,g(u)+g(q^{-\frac{1}{2}}\,u)\,D_{p,q}\,h(u);$$
$$D_{p,q}\left\{\frac{g(u)}{h(u)}\right\}=\frac{h(q^{-\frac{1}{2}}\,u)\,D_{p,q}\,g(u)-g(q^{-\frac{1}{2}}\,u)\,D_{p,q}\,h(u)}{h(p^{\frac{1}{2}}u)\,h(q^{-\frac{1}{2}}u)};$$
$$D_{p,q}\left\{\frac{g(u)}{h(u)}\right\}=\frac{h(p^{\frac{1}{2}}\,u)\,D_{p,q}\,g(u)-g(p^{\frac{1}{2}}\,u)\,D_{p,q}\,h(u)}{h(p^{\frac{1}{2}}u)\,h(q^{-\frac{1}{2}}u)}.$$

2. For a complex number u, obtain the (p,q)-analogue of the Gauss multiplication formula given by

$$\Gamma_{p,q}(ku)\prod_{n=1}^{k-1}\Gamma_{p^k,q^k}\left(\frac{n}{k}\right)=\left(p^{\frac{1}{2}(k-1)}+p^{\frac{1}{2}(k-2)}q^{-\frac{1}{2}}+\cdots+q^{-\frac{1}{2}(k-1)}\right)^{2u-1}\times\prod_{n=0}^{k-1}\Gamma_{p^k,q^k}\left(u+\frac{n}{k}\right). \quad (7.194)$$

3. Using the (p,q)-difference operator, prove that the (p,q)-confluent hypergeometric function satisfies the following (p,q)-difference equation:

$$\left[D_{p,q}\left[(\mathcal{D})_{v_1-1}\right]_{p,q}\right]{}_1\phi_1(z)=\left[\frac{-1}{(p^{\frac{1}{2}}-q^{-\frac{1}{2}})}\right]\left[(\mathcal{D})_{u_1}\right]_{p,q}\,{}_r\phi_s(q^{-\frac{1}{2}}z).$$

Further, it can be verified that the generalized (p,q)-hypergeometric function ${}_r\phi_s(z) = {}_r\psi_s(u_1,\dots,u_r;v_1,\dots,v_s;p,q;z)$ obeys the equation:

$$\left[D_{p,q}\left[(\mathcal{D})_{v_1-1}\right]_{p,q}\cdots\left[(\mathcal{D})_{v_s-1}\right]_{p,q}\right]{}_r\phi_s(z)$$
$$=\left[\frac{-1}{(p^{\frac{1}{2}}-q^{-\frac{1}{2}})}\right]^{1+s-r}\left[(\mathcal{D})_{u_1}\right]_{p,q}\cdots\left[(\mathcal{D})_{u_r}\right]_{p,q}\,{}_r\phi_s(q^{-\frac{1}{2}(1+s-r)}z).$$

4. Verify the following difference formulas obeyed by the (p,q)-confluent and Gauss hypergeometric functions:

$$D_{p,q}\,{}_1\psi_1(u_1;v_1;p,q;z)=\left[\frac{-1}{(p^{\frac{1}{2}}-q^{-\frac{1}{2}})}\right]\frac{([u_1]_{p,q})}{([v_1]_{p,q})}\,{}_1\psi_1(u_1+1;v_1+1;p,q;q^{-\frac{1}{2}}z);$$
$$D_{p,q}\,{}_2\psi_1(u_1,u_2;v_1;p,q;z)=\frac{([u_1]_{p,q})\,([u_2]_{p,q})}{([v_1]_{p,q})}\,{}_2\psi_1(u_1+1,u_2+1;v_1+1;p,q;z).$$

In general,

$$(D_{p,q})^n\,{}_1\psi_1(u_1;v_1;p,q;z)=\left[\frac{-1}{(p^{\frac{1}{2}}-q^{-\frac{1}{2}})}\right]^n\frac{([u_1]_{p,q})_n}{([v_1]_{p,q})_n}\,{}_1\psi_1(u_1+n;v_1+n;p,q;q^{-\frac{n}{2}}z);$$
$$(D_{p,q})^n\,{}_2\psi_1(u_1,u_2;v_1;p,q;z)=\frac{([u_1]_{p,q})_n\,([u_2]_{p,q})_n}{([v_1]_{p,q})_n}\,{}_2\psi_1(u_1+n,u_2+n;v_1+n;p,q;z).$$

5. Verify the following transformation formula:

$${}_2\Gamma_1\left[\begin{matrix}[u;x],v;\\ w;\end{matrix}\;z\right]=(1-z)^{-u}\,{}_2\Gamma_1\left[\begin{matrix}[u;x(1-z)],w-v;\\ w;\end{matrix}\;-\frac{z}{1-z}\right].$$

6. Obtain the following recurrence relation obeyed by the incomplete hypergeometric function given by

$$[v-(w-1)]\,{}_2\gamma_1\left[\begin{matrix}(u;x),v;\\ w;\end{matrix}\;z\right]$$
$$=v\,{}_2\gamma_1\left[\begin{matrix}(u;x),v+1;\\ w;\end{matrix}\;z\right]-(w-1)\,{}_2\gamma_1\left[\begin{matrix}(u;x),v;\\ w-1;\end{matrix}\;z\right];$$
$$\frac{uz}{w}\,{}_2\gamma_1\left[\begin{matrix}(u+1;x),v+1;\\ w+1;\end{matrix}\;z\right]={}_2\gamma_1\left[\begin{matrix}(u;x),v+1;\\ w;\end{matrix}\;z\right]-{}_2\gamma_1\left[\begin{matrix}(u;x),v;\\ w;\end{matrix}\;z\right].$$

7. Prove that the discrete confluent hypergeometric function ${}_1\mathcal{F}_1(z)$ satisfies the following difference equation:

$$\left[\Theta\left(\frac{1}{n}\Theta+v_1-1\right)-nz\,(-1)^n\,(-u)_n\,\rho^n\left(\frac{1}{n}\Theta+u_1\right)\right]{}_1\mathcal{F}_1(z)=0.$$

8. Verify the following difference and differential formulas:

$$\Delta\left[{}_1\mathcal{F}_1\left({u_1 \atop v_1}; u, 1, z\right)\right] = \frac{(u_1)_1}{(v_1)_1}\, z\, {}_1\mathcal{F}_1\left({u_1+1 \atop v_1+1}; u, 1, z\right);$$

$$\theta\left[{}_2\mathcal{F}_1\left({u_1, u_2 \atop v_1}; u, n, z\right)\right] = \frac{(-1)^n\,(u_1)\,(u_2)}{(v_1)}\, z\, {}_2\mathcal{F}_1\left({u_1+1, u_2+1 \atop v_1+1}; u, n, z\right);$$

$$\theta\left[{}_1\mathcal{F}_1\left({u_1 \atop v_1}; u, n, z\right)\right] = \frac{(-1)^n\,(u_1)}{(v_1)}\, z\, {}_1\mathcal{F}_1\left({u_1+1 \atop v_1+1}; u, n, z\right).$$

In general,

$$\Delta^r\left[{}_1\mathcal{F}_1\left({u_1 \atop v_1}; u, 1, z\right)\right] = \frac{(u_1)_r}{(v_1)_r}\, z^r\, {}_1\mathcal{F}_1\left({u_1+r \atop v_1+r}; u, 1, z\right);$$

$$\theta^r\left[{}_2\mathcal{F}_1\left({u_1, u_2 \atop v_1}; u, n, z\right)\right] = \frac{(-1)^{rn}\,(u_1)_r\,(u_2)_r}{(v_1)_r}\, z^r\, {}_2\mathcal{F}_1\left({u_1+r, u_2+r \atop v_1+r}; u, n, z\right);$$

$$\theta^r\left[{}_1\mathcal{F}_1\left({u_1 \atop v_1}; u, n, z\right)\right] = \frac{(-1)^{rn}\,(u_1)_r}{(v_1)_r}\, z\, {}_1\mathcal{F}_1\left({u_1+r \atop v_1+r}; u, n, z\right).$$

9. Obtain the following integral forms for the discrete Gauss hypergeometric function:

$$\begin{aligned} {}_2\mathcal{F}_1(u_1, u_2; v_1; u, n, z) &= \frac{1}{\Gamma(u_1)} \int_0^\infty e^{-t}\, t^{u_1-1}\, {}_1\mathcal{F}_1(u_2; v_1; u, n, tz)dt; \\ & \frac{1}{\Gamma(u_2)} \int_0^\infty e^{-t}\, t^{u_2-1}\, {}_1\mathcal{F}_1(u_1; v_1; u, n, tz)dt; \\ & \frac{1}{\Gamma(-u)} \int_0^\infty e^{-t}\, t^{-u-1}\, {}_2F_1(u_1, u_2; v_1; (-t)^n\, z)dt. \end{aligned}$$

10. Verify the following difference and differential recurrence relations for the discrete confluent hypergeometric function:

$$\begin{aligned} u_1\,(u_1-1)\,{}_1\mathcal{F}_1(u_1+1) - (u_1+\theta)\,(u_1+\theta-1)\,{}_1\mathcal{F}_1(u_1-1) &= 0; \\ u_1\, v_1\, {}_1\mathcal{F}_1(u_1+1) - (u_1+\theta)\,(v_1+\theta)\,{}_1\mathcal{F}_1(v_1+1) &= 0; \\ u_1\,(v_1+\theta-1)\,{}_1\mathcal{F}_1(u_1+1) - (v_1-1)\,(u_1+\theta)\,{}_1\mathcal{F}_1(v_1-1) &= 0; \\ (u_1+\theta-1)\,(v_1+\theta-1)\,{}_1\mathcal{F}_1(u_1-1) - (v_1-1)\,(u_1-1)\,{}_1\mathcal{F}_1(v_1-1) &= 0; \\ v_1\,(u_1+\theta-1)\,{}_1\mathcal{F}_1(u_1-1) - (u_1-1)\,(v_1+\theta)\,{}_1\mathcal{F}_1(v_1+1) &= 0; \\ v_1\,(v_1-1)\,{}_1\mathcal{F}_1(v_1-1) - (v_1+\theta-1)(v_1+\theta)\,{}_1\mathcal{F}_1(v_1+1) &= 0 \end{aligned}$$

and

$$u_1\,(u_1-1)\,{}_1\mathcal{F}_1(u_1+1)-\left(u_1+\frac{1}{n}\Theta_u\right)\left(u_1+\frac{1}{n}\Theta_u-1\right){}_1\mathcal{F}_1(u_1-1)=0;$$

$$u_1\,v_1\,{}_1\mathcal{F}_1(u_1+1)-\left(u_1+\frac{1}{n}\Theta_u\right)\left(v_1+\frac{1}{n}\Theta_u\right){}_1\mathcal{F}_1(v_1+1)=0;$$

$$u_1\left(v_1+\frac{1}{n}\Theta_u-1\right){}_1\mathcal{F}_1(u_1+1)-(v_1-1)\left(u_1+\frac{1}{n}\Theta_u\right){}_1\mathcal{F}_1(v_1-1)=0;$$

$$\left(u_1+\frac{1}{n}\Theta_u-1\right)\left(v_1+\frac{1}{n}\Theta_u-1\right){}_1\mathcal{F}_1(u_1-1)-(v_1-1)\,(u_1-1)\,{}_1\mathcal{F}_1(v_1-1)=0;$$

$$v_1\left(u_1+\frac{1}{n}\Theta_u-1\right){}_1\mathcal{F}_1(u_1-1)-(u_1-1)\left(v_1+\frac{1}{n}\Theta_u\right){}_1\mathcal{F}_1(v_1+1)=0;$$

$$v_1\,(v_1-1)\,{}_1\mathcal{F}_1(v_1-1)-\left(v_1+\frac{1}{n}\Theta_u-1\right)\left(v_1+\frac{1}{n}\Theta_u\right){}_1\mathcal{F}_1(v_1+1)=0.$$

11. Derive the following transformation formulas for the discrete Gauss hypergeometric function:

$$(1-t)^{-u_1}\,{}_2\mathcal{F}_1\left(u_1,u_2;v_1;u,n,\frac{z}{1-t}\right)=\sum_{r=0}^{\infty}\frac{(u_1)_r}{r!}\,z^r\,{}_2\mathcal{F}_1(u_1+r,u_2;v_1;u,n,z);$$

$$(1-t)^{-u_2}\,{}_2\mathcal{F}_1\left(u_1,u_2;v_1;u,n,\frac{z}{1-t}\right)=\sum_{r=0}^{\infty}\frac{(u_2)_r}{r!}\,z^r\,{}_2\mathcal{F}_1(u_1,u_2+r;v_1;u,n,z).$$

A Sets, relations and functions

In this first Appendix, we explore the concepts of sets, relations and functions briefly. These are fundamental concepts used throughout the book.

It will not be an exaggeration if we say that a set is the root of the branch of mathematics called algebra. It was introduced by a German mathematician, George Cantor, while he was working on a trigonometric series. A set is defined as the collection of objects or numbers having some specific meaning. A set is always denoted by the capital letter of English alphabets. Any belonging to a set is called the element of the set and all the elements of a set are always put in the bracket $\{\}$. Set builder and roaster forms are two common ways of representing a set. In the set builder form, elements are represented by a common property. For example, the set of vowels will be represented as

$$A = \{x \mid x \text{ is a vowel}\}.$$

In roaster form, we put each element separated by commas. The above example can be written in roster form as

$$A = \{a, e, i, o, u\}.$$

If an element x belongs to a set A, we represent it mathematically by $x \in A$. Whenever a set B is contained in a set C, we write $B \subseteq C$ or B is the subset of C. In the case an element does not belong to a set, we use the symbol $\notin$ to represent it mathematically. Similarly, the symbol $\not\subset$ or $\not\subseteq$ is used for 'not a subset'. For examples, the elements of set $B = \{1, 2, 3\}$ are contained in set $C = \{1, 2, 3, 4, 5\}$. We can say that $B \subset C$ or $C \supset B$. The number of elements contained in set A represent the cardinality of A and it is denoted by $n(A)$. On the basis of cardinality and nature of a set, we have the following type of sets:

1. *Empty set:* A set without any elements is called the empty set. It is denoted by either ϕ or $\{\cdot\}$.
2. *Singleton set:* A set containing only one element is known as a singleton set. For example,

$$A = \{x \mid x \text{ is an even prime number}\}.$$

3. *Finite set:* If a set has a finite number of elements, it is called a finite set. For example,

$$A = \{a, e, i, o, u\};$$
$$B = \{x \mid x \text{ is an even natural number and } x \leq 15\}.$$

4. *Infinite set:* If a set has an infinite number of elements, it is called an infinite set. For example,

$$A = \{2, 4, 6, \dots\};$$
$$B = \{x \mid x \text{ is an odd natural number}\}.$$

https://doi.org/10.1515/9783111324586-008

5. *Equivalent sets:* If two or more sets have the same number of elements, they will be called equivalent sets. For example, the sets $A = \{a, e, i, o, u\}$ and $B = \{1, 2, 3, 4, 5\}$ are equivalent sets because they have an equal number of elements.
6. *Equal sets:* Two sets have exactly the same elements excluding the repetition are called equal sets. For example, the set $A = \{a, e, i, o, u\}$ and $B = \{a, a, i, u, o, e, e\}$ are equal sets.
7. *Universal set:* A set in some context contains all of the sets called the universal set. For sets $A = \{1, 2, 3\}$, $B = \{3, 4, 5, 6, 7\}$ and $C = \{1, 2, 3, 4, 5\}$, the set $\mathbb{U} = \{1, 2, 3, 4, 5, 6, 7, 8\}$ will be considered as the universal set.
8. *Power set:* A collection of all subsets of a given set is called the power set. For a given set A, the power set is denoted by $P(A)$. Consider the set $A = \{a, b, c\}$, then its subsets are

$$\phi, \{a\}, \{b\}, \{c\}, \{a, b\}, \{b, c\}, \{a, c\}, \{a, b, c\}.$$

Putting these subsets together in a brace bracket, it will be the power set of the set A. Thus, we have

$$P(A) = \{\phi, \{a\}, \{b\}, \{c\}, \{a, b\}, \{b, c\}, \{a, c\}, \{a, b, c\}\}.$$

Similarly, for set $A = \{1, 2, 3, 4\}$ the power set is given by

$$\begin{aligned} P(A) = \{&\phi, \{1\}, \{2\}, \{3\}, \{4\}, \{1, 2\}, \{1, 3\}, \{1, 4\}, \{2, 3\}, \{2, 4\}, \{3, 4\}, \{1, 2, 3\}, \\ &\{1, 2, 4\}, \{1, 3, 4\}, \{2, 3, 4\}, \{1, 2, 3, 4\}\}. \end{aligned}$$

In the first example, we have seen that the power set has eight elements when A has three elements, while in second example it has sixteen elements when the set has four elements. With these two, we can conclude that the cardinality of power set $P(A)$ is 2^n, where n is the cardinality of set A. Among all the subsets of A, the empty set and the set itself are called improper sets, while others are known as proper subsets.

Here is a list of some standard examples of sets.
- *The set of natural numbers:* $\mathbb{N} = \{1, 2, 3, 4, \dots\}$.
- *The set of integers:* $\mathbb{Z}$ or $\mathbb{I} = \{\dots, -2, -1, 0, 1, 2, \dots\}$.
- *The set of rational numbers:* $\mathbb{Q} = \{\frac{x}{y} \mid x, y \in \mathbb{Z} \text{ and } y \neq 0\}$.
- *The set of real numbers:* It is denoted by $\mathbb{R}$ and has a collection of numbers, which can be presented on the number line.
- *The set of complex numbers:* $\mathbb{C} = \{z \mid z = u + \iota v, u, v \in \mathbb{R}\}$, where ι is called as iota and defined by $\iota^2 = -1$ or $\iota = \sqrt{-1}$.

As in the addition and subtraction of numbers, two sets can be combined together using the following set operations:

1. *Union of sets:* The union of two sets P and Q is denoted by $P \cup Q$ and defined by

$$P \cup Q = \{x \mid x \in P \text{ or } x \in Q\}.$$

It follows from definition that $P \subseteq P \cup Q$, $Q \subseteq P \cup Q$. Further, if P and Q are two sets such that $P \subset Q$, then $P \cup Q = Q$.
Note that if $P_1, P_2, \ldots, P_k$ is a family of sets, then their union is obtained by $\bigcup_{i=1}^{k} P_i$.

2. *Intersection of sets:* The intersection of two sets P and Q is denoted by $P \cap Q$ and defined as

$$P \cap Q = \{x \mid x \in P \text{ and } x \in Q\}.$$

As a consequence of the definition, $P \cap Q \subseteq P$, $P \cap Q \subseteq Q$. Further, if P and Q are two sets such that $P \subset Q$, then $P \cap Q = P$. Two sets P and Q are disjoint if and only if $P \cap Q = \phi$.
Note that if $P_1, P_2, \ldots, P_k$ is a family of sets, then their intersection is obtained by $\bigcap_{i=1}^{k} P_i$.

3. *Compliment of a set:* The compliment of a set P is represented by P' or P^C and defined by

$$P' = \{x \mid x \in \mathbb{U} \text{ and } x \notin A\},$$

where $\mathbb{U}$ denotes the universal set.

4. *The difference of sets:* The difference of a set Q from P is defined by

$$P - Q = \{x \mid x \in P \text{ but } x \notin Q\}.$$

5. *Symmetric difference:* The symmetric difference of two sets P and Q is given by

$$P \Delta Q = (P - Q) \cup (Q - P).$$

The multiplication of two sets P and Q is called the cross-product denoted by $P \times Q$ and defined as

$$P \times Q = \{(x, y) \mid x \in P \text{ and } y \in Q\}.$$

The elements of $P \times Q$ are known as ordered pairs because of their fixed order. Two ordered pairs are equal in the case that their corresponding entries are equal. Similarly,

$$Q \times P = \{(y, x) \mid x \in P \text{ and } y \in Q\}.$$

Example 277. Let $\mathbb{U} = \{a, b, c, d, e, f, g, h, i\}$ be the universal set and $P = \{a, b, c, d\}$, $Q = \{b, d, f, h\}$, $R = \{c, d, e, f\}$. Then find:

(i) P' (ii) Q' (iii) $(P \cap R)'$

(iv) $(P \cup Q)'$ (v) $(P')'$ (vi) $(Q - R)'$.

Solution. The set P' can be obtained by deleting the element of P from the universal set $\mathbb{U}$, i. e.,

$$P' = \mathbb{U} - P = \{e, f, g, h, i\}.$$

Similarly,

$$Q' = \mathbb{U} - Q = \{a, c, e, g, i\}.$$

For (iii), we first obtain

$$P \cap R = \{x \mid x \in P \text{ and } x \in R\} = \{c, d\}.$$

Then

$$(P \cap R)' = \mathbb{U} - (P \cap R) = \{a, b, e, f, g, h, i\}.$$

Similarly,

$$(P \cup Q)' = \mathbb{U} - (P \cup Q) = \{e, g, i\}.$$

By the property $(P')' = P$, we have

$$(P')' = P = \{a, b, c, d\}.$$

The set $Q - R$ can be obtained from Q by deleting the elements of Q, which are already in R, and thus we get

$$(Q - R)' = \mathbb{U} - (Q - R) = \{a, c, d, e, f, g, i\}.$$

Now, we see some laws based on the operations of the sets:

1. *Idempotent law:* For any set P,

   (i) $P \cap P = P$ (ii) $P \cup P = P$.

2. *Identity law:* For any set P and universal set $\mathbb{U}$,

   (i) $P \cup \phi = P$ (ii) $P \cap \mathbb{U} = P$,

   i. e., ϕ and $\mathbb{U}$ are identity elements for union and intersection, respectively.

3. *Commutative law:* For any two sets P and Q,

$$\text{(i) } P \cap Q = Q \cap P \qquad \text{(ii) } P \cup Q = Q \cup P.$$

That means the union and intersection are commutative.

4. *Associative law:* The operations union and intersection obeys the associative law, i. e., for any three sets P, Q, R,

$$\text{(i) } P \cap (Q \cap R) = (P \cap Q) \cap R \qquad \text{(ii) } P \cup (Q \cup R) = (P \cup Q) \cup R.$$

5. *Distributive laws:* The union is distributive over the intersection and in same way the intersection is distributive over the union. For any three sets P, Q, R:

$$\text{(i) } P \cup (Q \cap R) = (P \cup Q) \cap (P \cup R)$$
$$\text{(ii) } P \cap (Q \cup R) = (P \cap Q) \cup (P \cap R).$$

6. *De-Morgan's laws:* For any two sets P and Q,

$$\text{(i) } (P \cap Q)' = Q' \cup P' \qquad \text{(ii) } (P \cup Q)' = Q' \cap P'.$$

Next, we list results based on the cardinality of sets. Let P, Q, R be finite sets and $\mathbb{U}$ be the universal finite set. Then the following results hold true:

1. $n(P \cup Q) = n(P) + n(Q) - n(P \cap Q)$;
2. $n(P \cup Q) = n(P) + n(Q) \iff P$ and Q are disjoint;
3. $n(P - Q) = n(P) - n(P \cap Q)$;
4. $n(P \Delta Q) = n(P) + n(Q) - 2n(P \cap Q)$;
5. $n(P \cup Q \cup R) = n(P) + n(Q) + n(R) - n(P \cap Q) - n(Q \cap R) - n(P \cap R) + n(P \cap Q \cap R)$;
6. $n(P' \cup Q') = n(P \cap Q)' = n(\mathbb{U}) - n(P \cap Q)$;
7. $n(P' \cap Q') = n(P \cup Q)' = n(\mathbb{U}) - n(P \cup Q)$;
8. The number of elements in exactly two of the sets P, Q, R is equal to

$$n(P \cap Q) + n(Q \cap R) + n(P \cap R) - 3n(P \cap Q \cap R);$$

9. The number of elements in exactly one of the sets P, Q, R is equal to

$$n(P) + n(Q) + n(R) - 2n(P \cap Q) - 2n(Q \cap R) - 2n(P \cap R) + 3n(P \cap Q \cap R).$$

Relations

A relation R from set P to Q is a set of ordered pairs taken from $P \times Q$ and follows a special statement of property. Thus,

$$R \subseteq P \times Q.$$

An element $(u, v) \in R$ if and only if u is related to v, and in short we write uRv. If a relation R is defined from set P to P, we simply say R is defined on P or R is a binary relation. If R is a relation on P, then

$$\text{Domain of } R = \{x \mid (x, y) \in R\},$$

and

$$\text{Range of } R = \{y \mid (x, y) \in R\}.$$

On the basis of nature and number of elements, a relation can be categorized in the following types:

1. *Empty relation:* A relation without any element is called an empty relation. If R is a relation on set $P = \{1, 3, 5\}$, defined as $R = \{(u, v) \mid u - v \text{ is an odd number}\}$. Then R is an empty relation.
2. *Universal relation:* A relation R defined on set P is called universal if $R = P \times P$. For example, if R is a relation defined on set $P = \{2, 4, 6\}$ by $R = \{(u, v) \mid u - v \text{ is an even integer}\}$, then R will be an universal relation.
3. *Identity relation:* A relation on a set P, defined by $R = \{(u, u) \mid \forall u \in P\}$ is called an identity relation.
4. *Inverse relation:* If R is a relation from set P to Q, then R^{-1} is a relation from Q to P and defined by

$$R^{-1} = \{(v, u) \mid \forall (u, v) \in R\}.$$

The domain and range of inverse relation R^{-1} is interchanged with the domain and range of R, i. e.,

$$\text{Domain of } R^{-1} = \text{Range of } R;$$

and

$$\text{Range of } R^{-1} = \text{Domain of } R.$$

5. *Reflexive relation:* A relation R on P is called reflexive if $(a, a) \in R, \forall a \in P$. For example, if $P = \{1, 2, 3\}$ is a set and R is a relation on P, then R is reflexive if $(1, 1), (2, 2), (3, 3) \in R$.
6. *Symmetric relation:* A relation R is symmetric if $\forall (a, b) \in R \implies (b, a) \in R$. As an example, we can see the relation R on $P = \{1, 2, 3\}$ given by

$$R = \{(1, 1), (1, 2), (2, 1), (2, 2)\}$$

is a symmetric relation because for every $(a, b) \in R$ there is a $(b, a) \in R$.

7. *Asymmetric relation:* A relation R defined on P is called symmetric if

$$(a, b) \in R \text{ and } (b, a) \in R, \text{ then } a = b.$$

8. *Transitive relation:* A relation R is called transitive if

$$(a, b) \in R \text{ and } (b, c) \in R \implies (a, c) \in R.$$

9. *Equivalence relation:* A relation which is reflexive, symmetric as well as transitive called an equivalence relation.

Example 278. Let R_1, R_2, R_3 be relations defined on the set $P = \{1, 2, 3\}$ as follows:

$$R_1 = \{(1,1), (1,2), (1,3), (2,2), (2,3), (3,1), (3,2), (3,3)\};$$
$$R_2 = \{(1,1)\};$$
$$R_3 = \{(2,1), (1,3)\}.$$

Find whether the relations R_1, R_2, R_3 on P is (i) reflexive; (ii) symmetric; and (iii) transitive.

Solution. The relation R_1 defined on P is

$$R_1 = \{(1,1), (1,2), (1,3), (2,2), (2,3), (3,1), (3,2), (3,3)\}.$$

It can be seen clearly that $\forall a \in P, a R_1 a \implies R_1$ is reflexive. There is an ordered pair $(1, 2) \in R_1$ but $(2, 1) \notin R_1$, so R_1 is not symmetric. Similarly, we can see $(2, 3)$ and $(3, 1) \in R_1$ but $(2, 1) \notin R_1$. So, R_1 is not transitive.

The relation $R_2 = \{(1, 1)\}$ satisfies all the axioms of reflexive, symmetric and transitive, so it is reflexive, symmetric and transitive. Similarly, we can check the relation $R_3 = \{(2, 1), (1, 3)\}$ is neither reflexive nor symmetric nor transitive.

Functions

A function f is a relation from a set P to a set Q such that if f has unique image for all $u \in P$ in Q. The function is represented by

$$f : P \to Q,$$

and image of $u \in P$ with respect to function f will be denoted by $f(u) = v \in Q$. As we have seen in a relation that the first value of each ordered pair $(u, v) \in R$ is the domain while second is the range. In function, a domain is a set of all $u \in P$, which has a possible output $v \in Q$, thus Domain $(f) \subseteq P$. If f is a function from a set P to a set Q, then the range of f is the set of elements from Q, which has a preimage in set P. Thus, the range of f is a subset of codomain Q.

Some standard real functions

The specific behavior and output of a given function allow us to categorize functions in several ways:

1. *Constant function:* When outputs of a function remain the same for every input, it is called as constant function. In other words, a function $f : A \to B$ defined by

$$f(x) = k \text{ (a scalar value)}, \quad \forall x \in P,$$

and is considered as constant. For example, we have the following.

Example 279. If $f : \mathbb{R} \to \mathbb{R}$ defined by $f(x) = 3$, $\forall x \in \mathbb{R}$, then f is a constant function.

Example 280. For a set $P = \{\frac{\pi}{2}, \frac{3\pi}{2}, \frac{5\pi}{2}, \dots\}$, the function defined by $f(x) = \cos x$, $\forall x \in P$ is a constant function.

2. *Identity function:* If the output of a function is the same as a given input, then it is called an identity function. In symbols, a function $f : P \to P$ defined by

$$f(x) = x, \quad \forall x \in P,$$

and is an identity function.

3. *One-One function (Injective):* Let $f : P \to Q$ be a function and all distinct elements of P has distinct images in Q. Then it is called a one-one or injective function. Mathematically, $f : P \to Q$ will be a one-one function if

$$f(x) = f(y) \implies x = y, \quad \forall x, y \in P.$$

In other words, $f : P \to Q$ will be a one-one function if

$$x \neq y \implies f(x) \neq f(y), \quad \forall x, y \in P.$$

Example 281. If $P = \{a, b, c\}$ and $Q = \{1, 2, 3\}$, then a function f defined by $f(a) = 1$, $f(b) = 2$ and $f(c) = 3$ is called a one-one function because all the distinct elements from P has the distinct images in Q.

4. *Many-one function:* If two or more elements from a set P has same image in Q with respect to the function $f : P \to Q$, then f is called many-one function. Thus, we can say $f : P \to Q$ is many-one if

$$x \neq y \implies f(x) = f(y), \quad \text{for some } x, y \in P.$$

Example 282. If $P = \{a, b, c, d\}$ and $Q = \{1, 2, 3\}$, then a function f defined by $f(a) = 1$, $f(b) = 2, f(c) = 2$ and $f(d) = 3$ is called a many-one function.

5. *Onto function:* If $f : P \to Q$ is a function and each element from Q has a pre-image in P, then f is called an onto function. In other words, if set Q is equal to $f(x), \forall x \in P$, then it is called an onto function.

   Example 283. If $P = \{-3, -2, -1, 1, 2, 3\}$, $Q = \{1, 4, 9\}$ and $f : P \to Q$ is a function defined by $f(x) = x^2$, then f will be an onto mapping.

6. *Into function:* If $f : P \to Q$ is a function and there exists at least one element in Q such that it has not a preimage in P, then f is called an into function. Mathematically, it has same meaning as range of $f \subset Q$.
7. *Polynomial function:* A function from $\mathbb{R} \to \mathbb{R}$ defined by

$$f(x) = a_0 x^n + a_1 x^{n-1} + \cdots + a_n, \quad a_0 \neq 0,$$

   is called a polynomial function. Here, the index n is called the degree of $f(x)$ and $a_0, a_1, \ldots, a_n$ are constants. For example, $f(x) = 3x^2 + 2x + 5$, $f(x) = x^3 + 3x + 2$. On the basis of degrees, the polynomial functions can be further divided into linear functions if degree is one; quadratic if degree is two; cubic when the degree is three and quartic if the degree is four.
8. *Modulus function:* A modulus function $f : \mathbb{R} \to \mathbb{R}^+$ is defined by

$$f(x) = |x| = \begin{cases} x, & \text{if } x \geq 0, \\ -x, & \text{if } x < 0. \end{cases}$$

   We list some useful properties of modulus function:
   (i) For any real number u, $\sqrt{u^2} = |u|$.
   (ii) If u and v are positive real numbers, then

$$\begin{aligned} x^2 \leq u^2 &\iff |x| \leq u \iff -u \leq x \leq u; \\ x^2 \geq u^2 &\iff |x| \geq u \iff -u \geq x \text{ or } x \geq u; \\ u^2 \leq x^2 \leq v^2 &\iff u \leq x \leq v \iff x \in [-v, -u] \cup [u, v]; \\ u^2 < x^2 < v^2 &\iff u < x < v \iff x \in (-v, -u) \cup (u, v). \end{aligned}$$

   (iii) For real numbers u and v, we have

$$\begin{aligned} |u + v| = |u| + |v| &\iff (u \geq 0 \text{ and } v \geq 0) \text{ or } (u \leq 0 \text{ and } v \leq 0); \\ |u - v| = |u| - |v| &\iff (u \geq 0, v \geq 0 \text{ and } |u| \geq |v|) \text{ or} \\ &\qquad (u \leq 0, v \leq 0 \text{ and } |u| \geq |v|); \\ |u \pm v| &\leq |u| + |v|; \\ |u \pm v| &\geq \big||u| - |v|\big|. \end{aligned}$$

9. *Signum function:* A function $f : \mathbb{R} \to \{-1, 1\}$ defined by

$$f(x) = \frac{|x|}{x} = \begin{cases} 1, & \text{if } x \geq 0, \\ -1, & \text{if } x < 0 \end{cases}$$

is called signum function.

10. *Greatest integer function:* The greatest integer function is also known as the floor function denoted by the $[u]$ or $\lfloor u \rfloor$ for real number u. It represents the greatest integer less than or equal to u. For example,

$$\lfloor 2.3 \rfloor = 2, \quad \lfloor 8 \rfloor = 8, \quad \lfloor -5.43 \rfloor = -6.$$

If k is an integer and u is a real number between k and $k+1$, then the greatest integer function obeys the following properties:

(i) $\lfloor -k \rfloor = -\lfloor k \rfloor$;

(ii) $\lfloor u + k \rfloor = \lfloor u \rfloor + k$;

(iii) $\lfloor -u \rfloor = -\lfloor u \rfloor - 1$;

(iv) $\lfloor u \rfloor + \lfloor -u \rfloor = \begin{cases} -1, & \text{if } u \notin \mathbb{Z}, \\ 0, & \text{if } u \in \mathbb{Z}; \end{cases}$

(v) $\lfloor u \rfloor - \lfloor -u \rfloor = \begin{cases} 2\lfloor u \rfloor + 1, & \text{if } u \notin \mathbb{Z}, \\ 2\lfloor u \rfloor, & \text{if } u \in \mathbb{Z}; \end{cases}$

(vi) $\lfloor u \rfloor \geq k \implies u \geq k$;

(vii) $\lfloor u \rfloor \leq k \implies u < k + 1$;

(viii) $\lfloor k\,u \rfloor = \lfloor u \rfloor + \lfloor u + \frac{1}{k} \rfloor + \cdots + \lfloor u + \frac{k-1}{k} \rfloor, k \in \mathbb{N}$.

11. *Smallest integer function:* Also known as a ceiling function denoted by $\lceil u \rceil$, for some real u defined as the smallest integer greater than or equal to u. For example,

$$\lceil 4.753 \rceil = 5, \quad \lceil -13.246 \rceil = -13, \quad \lceil 10 \rceil = 10.$$

For an integer k and real number u, we have the following properties satisfied by the smallest integer function:

(i) $\lceil -k \rceil = -\lceil k \rceil$;

(ii) $\lceil -u \rceil = -\lceil u \rceil + 1, u \in \mathbb{R} - \mathbb{Z}$;

(iii) $\lceil u + k \rceil = \lceil u \rceil + k, u \in \mathbb{R} - \mathbb{Z}$;

(iv) $\lceil u \rceil + \lceil -u \rceil = \begin{cases} 1, & \text{if } u \notin \mathbb{Z}, \\ 0, & \text{if } u \in \mathbb{Z}; \end{cases}$

(v) $\lceil u \rceil - \lceil -u \rceil = \begin{cases} 2\lceil u \rceil - 1, & \text{if } u \notin \mathbb{Z}, \\ 2\lceil u \rceil, & \text{if } u \in \mathbb{Z}. \end{cases}$

12. *Fractional part function:* The function $f : \mathbb{R} \to \mathbb{R}$ defined by

$$f(u) = \{u\} = u - \lfloor u \rfloor, \quad u \in \mathbb{R},$$

is known as the fractional part function. For example,

$$\{3.45\} = 0.45, \quad \{4\} = 0, \quad \{-5\} = 0, \quad \{-6.32\} = 0.68.$$

Operations on real functions

We mainly discuss the addition, subtraction, multiplication, division and composition of real functions.

1. *Addition/Subtraction:* Let f and g be two functions defined on domain D_1 and D_2, respectively. Then

$$(f \pm g)(u) = f(u) \pm g(u), \quad \forall u \in D_1 \cap D_2.$$

2. *Product:* For two functions f and g defined on domain D_1 and D_2, respectively, the product or pointwise multiplication is defined on $D_1 \cap D_2$ by

$$(f \cdot g)(u) = f(u) \cdot g(u), \quad \forall u \in D_1 \cap D_2.$$

3. *Division:* Let f be a function defined on domain D_1. Then the reciprocal of f is given by

$$\left(\frac{1}{f}\right)(u) = \frac{1}{f(u)}, \quad \forall u \in D_1 - \{u \mid f(u) = 0\}.$$

Further, for a function g defined on D_2, we have

$$\left(\frac{g}{f}\right)(u) = \frac{g(u)}{f(u)}, \quad \forall u \in D_1 \cap D_2 - \{u \mid f(u) = 0\}.$$

4. *Composition of two functions:* Let f and g be two functions defined on domains D_1 and D_2, respectively. Then composition of f and g is denoted by gof and fog, defined as

$$(gof)(u) = g(f(u)), \quad \forall u \in D_1 \text{ and } f(u) \in D_2;$$
$$(fog)(u) = f(g(u)), \quad \forall u \in D_2 \text{ and } g(u) \in D_1.$$

For any two functions f and g, there is a possibility that either gof or fog or both exist. Also, there is a possibility that both gof and fog exists but are unequal, i. e., in general,

$$gof \neq fog.$$

Inverse of a function

Let $f : X \to Y$ be a bijective (one-one and onto) function. Then a function $g : Y \to X$ is called inverse of f if g associates each element of Y to a unique element in X. The inverse of f is denoted by f^{-1}. Thus,

$$f(u) = v \iff f^{-1}(v) = u.$$

Following properties can be verified easily:

1. The inverse of a bijective function is unique and bijective.
2. If $f : X \to Y$ is a bijective function and $g : Y \to X$ is the inverse of f, then

   $$fog = I_Y \quad \text{and} \quad gof = I_X,$$

   where I_X and I_Y are identity functions on X and Y, respectively. In the case $X = Y$, then $fog = gof = I_X$.
3. If $f : X \to Y$ and $g : Y \to Z$ are two bijective maps, then $gof : X \to Z$ is bijective and

   $$(gof)^{-1} = f^{-1}og^{-1}.$$

4. If $f : X \to Y$ and $g : X \to Z$ be two functions such that

   $$fog = I_Y \quad \text{and} \quad gof = I_X,$$

   then f and g are bijections and $g = f^{-1}$.
5. If f is an invertible function, then $(f^{-1})^{-1} = f$.

Solved examples

Example 284. Show that the function $f : \mathbb{R} \to \mathbb{R}$, given by $f(u) = u^2 + 1$ is not invertible.

Solution. One can easily show that

$$f(-5) = f(5) = 26, \quad \text{but } -5 \neq 5.$$

Thus, f is not one-one, and hence not invertible.

Example 285. If function $f : [1, \infty) \to [1, \infty)$ defined by $f(u) = 3^{(u-1)}$ is invertible, then find f^{-1}.

Solution. We know that

$$(fof^{-1})(u) = u$$
$$\Rightarrow f(f^{-1}(u)) = u$$

$$\Longrightarrow 3^{f^{-1}(u)-1} = u$$
$$\Longrightarrow f^{-1}(u) - 1 = \log_3(u), \quad f^{-1}(u) \geq 1$$
$$\Longrightarrow f^{-1}(u) = 1 + \log_3(u).$$

Example 286. If $f(u) = \frac{4u+3}{6u-4}$, $u \neq \frac{2}{3}$, then show that $f(f(u)) = u$ for all $u \neq \frac{2}{3}$. Also, find f^{-1}.

Solution. Using the definition,

$$\begin{aligned}(fof)(u) &= f(f(u)) \\ &= f\left(\frac{4u+3}{6u-4}\right) \\ &= \frac{4 \times \frac{4u+3}{6u-4} + 3}{6 \times \frac{4u+3}{6u-4} - 4} \\ &= \frac{34u}{34} = u.\end{aligned}$$

It completes the proof of first part. For second, consider $f(u) = v \implies u = f^{-1}(u)$. This gives

$$\begin{aligned}&\frac{4u+3}{6u-4} = v \\ \Longrightarrow\ & 4u + 3 = 6uv - 4v \\ \Longrightarrow\ & u = f^{-1}(v) = \frac{4v+3}{6v-4}.\end{aligned}$$

Thus, $f^{-1}(u) = \frac{4u+3}{6u-4}$.

Example 287. Let $f(u) = \lfloor u \rfloor$ and $g(u) = |u|$. Find

(i) $(gof)(-8)$ (ii) $(fog)(3.45)$ (iii) $(f + 2g)(-3)$ (iv) $(gof)\left(\frac{3}{2}\right) - (fog)\left(\frac{3}{2}\right)$.

*Solution.*The first, second and third can be computed as

$$\begin{aligned}(gof)(-8) &= g(f(-8)) = g(\lfloor u \rfloor) = g(-8) = |-8| = 8, \\ (fog)(3.45) &= f(g(3.45)) = f(|3.45|) = f(3.45) = \lfloor 3.45 \rfloor = 3, \\ (f + 2g)(-3) &= f(-3) + 2g(-3) = -3 + 2 \times 3 = 3.\end{aligned}$$

The fourth can be computed as

$$
\begin{aligned}
(gof)\left(\frac{3}{2}\right) - (fog)\left(\frac{3}{2}\right) \\
&= g\left(f\left(\frac{3}{2}\right)\right) - f\left(g\left(\frac{3}{2}\right)\right) \\
&= g\left(\left\lfloor\frac{3}{2}\right\rfloor\right) - f\left(\left|\frac{3}{2}\right|\right) \\
&= g(1) - f\left(\frac{3}{2}\right) = 1 - 1 = 0.
\end{aligned}
$$

Complex numbers

A set of complex numbers is the collection of numbers of the form $z = u + \iota v$. In a complex number $z = u + \iota v$, u and v are known as real and imaginary parts of z denoted by $\Re(z)$ and $\Im(z)$ respectively.

The algebra of complex numbers:

1. *Addition\Subtraction:* If $z_1 = u_1 + \iota v_1, z_2 = u_2 + \iota v_2$ are complex numbers, then

$$z_1 + z_2 = z_2 + z_1 = (u_1 + u_2) + \iota(v_1 + v_2), \tag{A.1}$$

$$z_1 - z_2 = (u_1 - u_2) + \iota(v_1 - v_2). \tag{A.2}$$

2. *Multiplication:* For complex numbers $z_1 = u_1 + \iota v_1, z_2 = u_2 + \iota v_2$, the product is given by

$$z_1 z_2 = z_2 z_1 = (u_1 u_2 - v_1 v_2) + \iota(u_1 v_2 + u_2 v_1).$$

3. *Division:* If $z = u + \iota v$, then

$$\frac{1}{z} = \frac{1}{u + \iota v} = \frac{1}{u + \iota v} \times \frac{u - \iota v}{u - \iota v} = u' + \iota v',$$

where $u' = \frac{u}{u^2+v^2}, v' = \frac{-v}{u^2+v^2}$.

4. *Distributive:* The multiplication in complex numbers is distributive over addition. Let z, z_1 and z_2 be complex numbers. Then

$$z\,(z_1 + z_2) = z\,z_1 = z\,z_2 \quad \text{(left distribution)},$$

$$(z_1 + z_2)\,z = z_1\,z = z_2\,z \quad \text{(right distribution)}.$$

The conjugate of a complex number is represented by $\bar{z}$ and can be obtained by replacing ι with $-\iota$. Thus, for a complex number $z = u + \iota v$, the conjugate $\bar{z}$ is $u - \iota v$.

The absolute value of a complex number $z = u + \iota v$ is denoted by $|z|$ and defined as

$$|z| = \sqrt{u^2 + v^2}.$$

An immediate relation between absolute value and conjugate of a complex number can be obtained as

$$|z|^2 = z\,\bar{z} \quad \text{or} \quad |z| = \sqrt{z\,\bar{z}}.$$

The argument of a complex number $z = u + \iota v$ is defined by

$$\arg(z) = \tan^{-1}\left(\frac{v}{u}\right).$$

We now list some relations in terms of absolute value and arguments of a complex number:

1. $\overline{z_1 \pm z_2} = \overline{z_1} \pm \overline{z_2}$;
2. $\overline{z_1 \cdot z_2} = \overline{z_1} \cdot \overline{z_2}$;
3. $\overline{(\frac{z_1}{z_2})} = \frac{\overline{z_1}}{\overline{z_2}}$, $z_2 \neq 0$;
4. $|z_1\, z_2| = |z_1|\, |z_2|$;
5. $|\frac{z_1}{z_2}| = \frac{|z_1|}{|z_2|}$, $z_2 \neq 0$;
6. $|z_1 \pm z_2|^2 = |z_1|^2 + |z_2|^2 \pm 2\,\Re(z_1\,\overline{z_2})$;
7. $\arg(z_1\, z_2) = \arg(z_1) + \arg(z_2)$;
8. $\arg(\frac{z_1}{z_2}) = \arg(z_1) - \arg(z_2)$.

B Matrices

Matrix

A matrix is basically defined as a rectangular arrangements of numbers, functions or any other objects, we put this arrangements in brackets like

$$\begin{bmatrix} m_{11} & m_{12} & \dots & m_{1n} \\ m_{21} & m_{22} & \dots & m_{2n} \\ \vdots & \vdots & \vdots & \vdots \\ m_{p1} & m_{p2} & \dots & m_{pn} \end{bmatrix} \quad \text{or} \quad \begin{pmatrix} m_{11} & m_{12} & \dots & m_{1n} \\ m_{21} & m_{22} & \dots & m_{2n} \\ \vdots & \vdots & \vdots & \vdots \\ m_{p1} & m_{p2} & \dots & m_{pn} \end{pmatrix}. \tag{B.1}$$

The horizontal lines in the presentation are called rows while the vertical are columns and the order of matrix is given by the number of rows times the number of columns. In the above presentation the order is $p \times n$; The inputs $m_{11}, m_{12}, \dots$ are called the entries of matrix.

Special matrices

Definition 288 (Row and column matrices). A matrix of order $1 \times n$ is called a row matrix while a matrix of order $m \times 1$ called a column matrix, e. g.,

$$X = [a_{11}, \quad a_{12}, \dots, \quad a_{1n}]_{1\times n} \quad \text{and} \quad Y = \begin{bmatrix} b_{11} \\ b_{21} \\ \vdots \\ b_{m1} \end{bmatrix}_{m\times 1} \tag{B.2}$$

are row and column matrices, respectively.

Definition 289 (Null matrix). A matrix is called null matrix, if its all entries are zero. For example, the matrices

$$\begin{bmatrix} 0 & 0 \\ 0 & 0 \end{bmatrix}, \quad \begin{bmatrix} 0 & 0 & 0 \\ 0 & 0 & 0 \end{bmatrix} \quad \text{and} \quad \begin{bmatrix} 0 & 0 & 0 \\ 0 & 0 & 0 \\ 0 & 0 & 0 \end{bmatrix} \tag{B.3}$$

all are null matrices.

Definition 290 (Square matrix). A matrix with equal number of rows and columns is called square matrix. In other words, when $m = n$, the matrix P will be square matrix of order n or n-square matrix. For example,

https://doi.org/10.1515/9783111324586-009

$$[a]_{1\times1}, \quad \begin{bmatrix} a & b \\ c & d \end{bmatrix}_{2\times2}, \quad \begin{bmatrix} 1 & 0 & 2 \\ 2 & -1 & 0 \\ -2 & 0 & 5 \end{bmatrix}, \quad \text{etc.} \tag{B.4}$$

Definition 291 (Diagonal matrix). A square matrix whose all nonzero entries present in the diagonal only is called a diagonal matrix, i. e.,

$$P = [a_{ij}], \quad 1 \le i, j \le n \text{ and } a_{ij} = 0, \text{for } i \ne j. \tag{B.5}$$

For example,

$$\begin{bmatrix} 3 & 0 \\ 0 & 4 \end{bmatrix}, \quad \begin{bmatrix} 1 & 0 & 0 \\ 0 & -1 & 0 \\ 0 & 0 & 4 \end{bmatrix}, \quad \begin{bmatrix} 3 & 0 & 0 & 0 \\ 0 & -4 & 0 & 0 \\ 0 & 0 & 7 & 0 \\ 0 & 0 & 0 & 11 \end{bmatrix}, \quad \text{etc.} \tag{B.6}$$

Definition 292 (Triangular matrix). A square matrix whose all entries either below or above the main diagonal are zero is called triangular matrix. The matrices

$$\begin{bmatrix} 1 & 0 & 0 \\ 2 & -1 & 0 \\ 3 & -2 & 4 \end{bmatrix} \quad \text{and} \quad \begin{bmatrix} 1 & -2 & -1 \\ 0 & -1 & 3 \\ 0 & 0 & 4 \end{bmatrix} \tag{B.7}$$

are upper and lower triangular matrices, respectively.

Definition 293 (Scalar matrix). A diagonal matrix is called a scalar matrix, if all the diagonal entries are same, i. e., for a nonzero scalar c and a square matrix $P = [a_{ij}]$, where

$$a_{ij} = \begin{cases} 0, & \text{if } i \ne j, \\ c, & \text{if } i = j. \end{cases} \tag{B.8}$$

For example,

$$\begin{bmatrix} 3 & 0 \\ 0 & 3 \end{bmatrix}, \quad \begin{bmatrix} -1 & 0 & 0 \\ 0 & -1 & 0 \\ 0 & 0 & -1 \end{bmatrix}, \quad \begin{bmatrix} 7 & 0 & 0 & 0 \\ 0 & 7 & 0 & 0 \\ 0 & 0 & 7 & 0 \\ 0 & 0 & 0 & 7 \end{bmatrix}, \quad \text{etc.} \tag{B.9}$$

Definition 294 (Unit or identity matrix). A scalar matrix is called an identity matrix if the scalar value is 1. For example,

$$\begin{bmatrix} 1 & 0 \\ 0 & 1 \end{bmatrix}, \quad \begin{bmatrix} 1 & 0 & 0 \\ 0 & 1 & 0 \\ 0 & 0 & 1 \end{bmatrix}, \quad \begin{bmatrix} 1 & 0 & 0 & 0 \\ 0 & 1 & 0 & 0 \\ 0 & 0 & 1 & 0 \\ 0 & 0 & 0 & 1 \end{bmatrix}, \quad \text{etc.} \tag{B.10}$$

Definition 295 (Submatrix). A matrix obtained by deleting some row or column or both from a given matrix is called a submatrix of the original matrix. For example, the matrices

$$\begin{bmatrix} 3 & 1 \\ 2 & -3 \end{bmatrix}, \quad \begin{bmatrix} 1 & 3 & 1 \\ 0 & 2 & -3 \end{bmatrix}, \quad \begin{bmatrix} 3 & 1 \\ 2 & -3 \\ 0 & 5 \end{bmatrix}, \tag{B.11}$$

are the submatrices of the matrix

$$P = \begin{bmatrix} 1 & 3 & 1 & 0 \\ 0 & 2 & -3 & 0 \\ 0 & 0 & 5 & 0 \\ 4 & 0 & 0 & 11 \end{bmatrix}$$

because each matrix in (B.11) can be obtained from P by deleting some rows and columns.

Definition 296 (Equal matrices). Two matrices P and Q are equal if and only if
- The order of P and Q is same;
- Corresponding entries of each matrix are equal.

For example, the matrices $P = \left[\begin{smallmatrix} x & y \\ z & w \end{smallmatrix}\right]$ and $Q = \left[\begin{smallmatrix} 1 & 3 \\ 4 & 11 \end{smallmatrix}\right]$ are equal if and only if $x = 1, y = 3$, $z = 4$ and $w = 11$.

Definition 297 (Transpose of a matrix). A matrix obtained by interchanging the rows and columns is called the transpose of the original matrix. The transpose of a matrix P is denoted by P' or P^T. For example, if $P = \left[\begin{smallmatrix} a & b & c \\ d & e & f \\ g & h & i \end{smallmatrix}\right]$, then $P' = \left[\begin{smallmatrix} a & d & g \\ b & e & h \\ c & f & i \end{smallmatrix}\right]$.

Note that if P is a matrix of order $m \times n$, then P' is the matrix of order $n \times m$.

Definition 298 (Properties of transpose). The transpose of a matrix satisfies the following properties:
- For any scalar k, $(kP)' = kP'$;
- For matrices P and Q, $(P + Q)' = P' + Q'$;
- For matrices P and Q, $(PQ)' = Q'P'$.

Definition 299 (Symmetric and skew-symmetric matrices). A square matrix P is called symmetric if $P' = P$ and skew-symmetric if $P' = -P$. In other words, a matrix $P = [a_{ij}]$ is symmetric or skew-symmetric accordingly,

$$a_{ij} = \begin{cases} a_{ji}, \\ \\ -a_{ji}, \end{cases} \quad \text{or} \tag{B.12}$$

respectively for all i and j.

It can be concluded that, for skew symmetric matrix $a_{ii} = -a_{ii}$, this implies $a_{ii} = 0$ for each i. Hence, in skew symmetric matrix the diagonal entries are zero. For example, the matrices

$$P = \begin{bmatrix} a & b & c \\ b & e & f \\ c & f & i \end{bmatrix} \quad \text{and} \quad Q = \begin{bmatrix} 0 & b & c \\ -b & 0 & f \\ -c & -f & 0 \end{bmatrix} \tag{B.13}$$

are symmetric and skew-symmetric, respectively.

Theorem 300. *Every square matrix can be written uniquely as a sum of symmetric and skew-symmetric matrices.*

Proof. Let A be a square matrix. Then we can write A as

$$A = \left(\frac{A + A'}{2}\right) + \left(\frac{A - A'}{2}\right) = P + Q. \tag{B.14}$$

Now, using the properties of transpose of a matrix, one can easily show that $P' = P$ and $Q' = -Q$. To prove the uniqueness, lets assume another expression $A = P_1 + Q_1$ such that P_1 is symmetric and Q_1 is skew-symmetric. This implies

$$A' = P_1' + Q_1' = P_1 - Q_1 \tag{B.15}$$

and hence

$$P_1 = \frac{A + A'}{2} = P, \quad Q_1 = \frac{A - A'}{2} = Q. \tag{B.16}$$

It completes the proof. □

Algebra of matrices

Definition 301 (Scalar multiplication in matrices). Let P be a matrix and k be any scalar. Then the scalar multiplication kP is defined by

$$kP = [k\, a_{ij}]_{m \times n}, \quad 1 \le i \le m, 1 \le j \le n. \tag{B.17}$$

That means, in scalar multiplication, each entry is multiplied by the scalar. For example, if

$$P = \begin{bmatrix} 4 & 2 & 1 \\ 2 & 2 & -3 \\ 1 & 0 & 5 \end{bmatrix}, \quad \text{then} \quad 3P = \begin{bmatrix} 12 & 6 & 3 \\ 6 & 6 & -9 \\ 3 & 0 & 15 \end{bmatrix}.$$

Definition 302 (Addition and subtraction). Let P and Q be two matrices of same order. Then their sum and difference is defined by

$$P \pm Q = [a_{ij} \pm b_{ij}]_{m\times n}, \quad 1 \le i \le m, 1 \le j \le n. \tag{B.18}$$

For example, if

$$P = \begin{bmatrix} -1 & 2 & 1 \\ 2 & -1 & -3 \\ 1 & 0 & 5 \end{bmatrix} \quad \text{and} \quad Q = \begin{bmatrix} 2 & 6 & 3 \\ 6 & 6 & -9 \\ 3 & 0 & 5 \end{bmatrix},$$

then

$$P + Q = \begin{bmatrix} 1 & 8 & 4 \\ 8 & 5 & -12 \\ 4 & 0 & 10 \end{bmatrix} \quad \text{and} \quad P - Q = \begin{bmatrix} -3 & -4 & -2 \\ -4 & -7 & 6 \\ -2 & 0 & 0 \end{bmatrix}.$$

Definition 303 (Matrix multiplication). Let P and Q be two matrices of order $m \times n$ and $n \times p$, respectively (the number of columns in P is equal to the number of rows in Q). Then PQ is the product matrix of order $m \times p$. Let

$$P = \begin{bmatrix} a_{11} & a_{12}\dots & a_{1n} \\ a_{21} & a_{22}\dots & a_{2n} \\ \vdots & \vdots & \vdots \\ a_{m1} & a_{m2}\dots & a_{mn} \end{bmatrix}_{m\times n} \quad \text{and} \quad Q = \begin{bmatrix} b_{11} & b_{12}\dots & b_{1p} \\ b_{21} & b_{22}\dots & b_{2p} \\ \vdots & \vdots & \vdots \\ b_{n1} & b_{n2}\dots & b_{np} \end{bmatrix}_{n\times p}. \tag{B.19}$$

Then

$$PQ = \begin{bmatrix} a_{11}\cdot b_{11} + a_{12}\cdot b_{21} + \dots + a_{1n}\cdot b_{n1} & \dots & a_{11}\cdot b_{1p} + a_{12}\cdot b_{2p} + \dots + a_{1n}\cdot b_{np} \\ a_{21}\cdot b_{11} + a_{22}\cdot b_{21} + \dots + a_{2n}\cdot b_{n1} & \dots & a_{21}\cdot b_{1p} + a_{22}\cdot b_{2p} + \dots + a_{2n}\cdot b_{np} \\ \vdots & \vdots & \vdots \\ a_{m1}\cdot b_{11} + a_{m2}\cdot b_{21} + \dots + a_{mn}\cdot b_{n1} & \dots & a_{m1}\cdot b_{1p} + a_{m2}\cdot b_{2p} + \dots + a_{mn}\cdot b_{np} \end{bmatrix}_{m\times p}. \tag{B.20}$$

Example 304. Let $P = \left[\begin{smallmatrix} 1 & 2 \\ 3 & 0 \end{smallmatrix}\right]$ and $Q = \left[\begin{smallmatrix} 0 & -2 \\ 4 & 2 \end{smallmatrix}\right]$. Then

$$PQ = \begin{bmatrix} 1 & 2 \\ 3 & 0 \end{bmatrix} \begin{bmatrix} 0 & -2 \\ 4 & 2 \end{bmatrix} = \begin{bmatrix} 1\cdot 0 + 2\cdot 4 & 1\cdot(-2) + 2\cdot 2 \\ 3\cdot 0 + 0\cdot 4 & 3\cdot(-2) + 0\cdot 2 \end{bmatrix} = \begin{bmatrix} 8 & 2 \\ 0 & -6 \end{bmatrix}. \tag{B.21}$$

Example 305. If

$$P = \begin{bmatrix} 4 & -1 & 0 \\ 1 & 2 & -1 \\ 3 & 0 & -2 \end{bmatrix} \quad \text{and} \quad Q = \begin{bmatrix} 1 & 3 & 0 \\ 0 & -2 & 1 \\ 4 & 0 & 1 \end{bmatrix},$$

then

$$PQ = \begin{bmatrix} 4 & -1 & 0 \\ 1 & 2 & -1 \\ 3 & 0 & -2 \end{bmatrix} \begin{bmatrix} 1 & 3 & 0 \\ 0 & -2 & 1 \\ 4 & 0 & 1 \end{bmatrix}$$

$$= \begin{bmatrix} 4\cdot 1 + (-1)\cdot 0 + 0\cdot 4 & 4\cdot 3 + (-1)\cdot(-2) + 0\cdot 0 & 4\cdot 0 + (-1)\cdot 1 + 0\cdot 1 \\ 1\cdot 1 + 2\cdot 0 + (-1)\cdot 4 & 1\cdot 3 + 2\cdot(-2) + (-1)\cdot 0 & 1\cdot 0 + 2\cdot 1 + (-1)\cdot 1 \\ 3\cdot 1 + 0\cdot 0 + (-2)\cdot 4 & 3\cdot 3 + 0\cdot(-2) + (-2)\cdot 0 & 3\cdot 0 + 0\cdot 1 + (-2)\cdot 1 \end{bmatrix}$$

$$= \begin{bmatrix} 4 & 14 & -1 \\ -3 & -1 & 1 \\ -5 & 9 & -2 \end{bmatrix}. \tag{B.22}$$

Properties of matrix multiplication

- In general, for any two matrices P and Q, the matrix multiplication is not commutative, *i. e.*, $PQ \neq QP$;
- Matrix multiplication is associative, *i. e.*, $P(QR) = (PQ)R$;
- Matrix multiplication is distributive over addition from left as well as from right, *i. e.*, $P(Q + R) = PQ + PR$ and $(Q + R)P = QP + RP$;
- $P^2 = P \cdot P$, $P^3 = P \cdot P \cdot P$, in general $P^p = P \cdot P \cdots P$ (p-times).

Elementary operations

The following operations on a matrix are called an elementary operations:
1. Interchange of two rows (columns);
2. Multiplication of a row (column) by a nonzero scalar k;
3. Replacement of a row (column) with the sum of that row (column) and k-times of another row (column).

We can also consider elementary operation as a function (e) from a matrix P of order $m \times n$ to a matrix Q of order $m \times n$, *i. e.*,

$$e : [P]_{m\times n} \to [Q]_{m\times n}$$

defined by $e(P) = Q$. The matrix Q, obtained from a matrix P by one or more elementary operations, is said to be equivalent to P, written $P \equiv Q$.

We restrict ourself on elementary row operations. The following symbols will be used to denote elementary row operations:
1. R_{ij}, for interchange of ith row and j^{th} row;
2. kR_i, for multiplication of ith row by scalar k;
3. $R_i + cR_j$, for sum of ith row and c-times the jth row, where c is a nonzero scalar.

Similarly, one can write these operations for column by replacing R with C. Points to be remembered:

1. Any elementary row operation on the product of two matrices P and Q is equal to the product of elementary row transform of P and Q, i. e., $R(PQ) = R(P)Q$;
2. An elementary column transformation on the product of two matrices is equal to the product of P and elementary column transform of Q that is $C(PQ) = PC(Q)$.

Row reduced echelon matrices

A matrix R of order $m \times n$ is called a row-reduced echelon matrix if:

1. R is row reduced;
2. Every row of R, which has all its entries, 0 occurs below every row, which has a nonzero entry;
3. If rows $1, 2, \ldots, r$ are the nonzero rows of R and if the leading nonzero entry is occurs in column k_i, $i = 1, 2, \ldots, r$, then $k_1 < k_2 < \cdots < k_r$.

Example 306. The identity matrices,

$$\begin{bmatrix} 1 & 0 \\ 0 & 1 \end{bmatrix}, \quad \begin{bmatrix} 1 & 0 & 0 \\ 0 & 1 & 0 \\ 0 & 0 & 1 \end{bmatrix}, \quad \begin{bmatrix} 1 & 0 & 0 & 0 \\ 0 & 1 & 0 & 0 \\ 0 & 0 & 1 & 0 \\ 0 & 0 & 0 & 1 \end{bmatrix}, \tag{B.23}$$

and the zero matrices of any order are trivial row-reduced echelon matrix. Other nontrivial examples of row reduced echelon matrices are

$$\begin{bmatrix} 0 & 1 & 2 & 3 & 0 \\ 0 & 0 & 0 & 4 & 6 \\ 0 & 0 & 0 & 0 & 0 \end{bmatrix}, \quad \begin{bmatrix} 1 & 2 & -1 & 0 \\ 0 & -1 & 0 & 2 \\ 0 & 0 & 0 & 1 \end{bmatrix}. \tag{B.24}$$

Theorem 307. *Every $m \times n$ matrix P is row equivalent to a row-reduced echelon matrix.*

Definition 308 (Elementary matrix). A matrix P is said to be an elementary matrix if it can be obtained from an identity matrix by means of a single elementary row operation.

Example 309. The matrices

$$\begin{bmatrix} 1 & 0 & 0 \\ 0 & 2 & 0 \\ 0 & 0 & 1 \end{bmatrix} \quad \text{and} \quad \begin{bmatrix} 1 & 0 \\ 3 & 1 \end{bmatrix}$$

are elementary matrices because they can be obtained by applying a single elementary row transformation on 2×2 and 3×3 identity matrices, respectively.

Theorem 310. *Let e be an elementary row operation and let E be the $m \times m$ elementary matrix such that $E = e(I)$. Then for every $m \times n$ matrix P,*

$$e(P) = E\,P.$$

Corollary 311. *Let P and Q be the $m \times n$ matrices over the field F. Then Q is row equivalent to P if and only if $Q = B^{-1} P A$, where A and B are some invertible matrices of order $n \times n$ and $m \times m$ respectively.*

Determinants

We define determinant as a scalar function, which takes each square matrix to a scalar. Let P be a n-square matrix. Then the determinant of P, denoted by $\det(P)$ or $|P|$ or

$$\begin{vmatrix} a_{11} & a_{12} & \cdots & a_{1n} \\ a_{21} & a_{22} & \cdots & a_{2n} \\ \vdots & \vdots & \cdots & \vdots \\ a_{n1} & a_{n2} & \cdots & a_{nn} \end{vmatrix}.$$

Lets start with the finding of determinants of order 1 and 2. The determinant of a matrix of order 1 is the scalar entry itself, i. e., $\det([a]) = a$. The determinant of a 2-square matrix P is defined as

$$|P| = \begin{vmatrix} a & b \\ c & d \end{vmatrix} = ad - bc. \tag{B.25}$$

Now, the determinant of a matrix P of order 3 is given by

$$\begin{aligned} |P| &= \begin{vmatrix} a_{11} & a_{12} & a_{13} \\ a_{21} & a_{22} & a_{23} \\ a_{31} & a_{32} & a_{33} \end{vmatrix} \\ &= a_{11} \begin{vmatrix} a_{22} & a_{23} \\ a_{32} & a_{33} \end{vmatrix} - a_{12} \begin{vmatrix} a_{21} & a_{23} \\ a_{31} & a_{33} \end{vmatrix} + a_{13} \begin{vmatrix} a_{21} & a_{22} \\ a_{31} & a_{32} \end{vmatrix} \\ &= a_{11}a_{22}a_{33} + a_{12}a_{23}a_{31} + a_{13}a_{21}a_{32} - a_{13}a_{22}a_{31} - a_{12}a_{21}a_{33} - a_{11}a_{23}a_{32}. \end{aligned} \tag{B.26}$$

Similarly, the determinant of order 4 can be defined

$$\begin{aligned} |P| &= \begin{vmatrix} a_{11} & a_{12} & a_{13} & a_{14} \\ a_{21} & a_{22} & a_{23} & a_{24} \\ a_{31} & a_{32} & a_{33} & a_{34} \\ a_{41} & a_{42} & a_{43} & a_{44} \end{vmatrix} \\ &= a_{11} \begin{vmatrix} a_{22} & a_{23} & a_{24} \\ a_{32} & a_{33} & a_{34} \\ a_{42} & a_{43} & a_{44} \end{vmatrix} - a_{12} \begin{vmatrix} a_{21} & a_{23} & a_{24} \\ a_{31} & a_{33} & a_{34} \\ a_{41} & a_{43} & a_{44} \end{vmatrix} + a_{13} \begin{vmatrix} a_{21} & a_{22} & a_{24} \\ a_{31} & a_{32} & a_{34} \\ a_{41} & a_{42} & a_{44} \end{vmatrix} \\ &\quad - a_{14} \begin{vmatrix} a_{21} & a_{22} & a_{23} \\ a_{31} & a_{32} & a_{33} \\ a_{41} & a_{42} & a_{43} \end{vmatrix}. \end{aligned} \tag{B.27}$$

Thus, we can observe that the determinant of n-square matrix is the linear combination of determinant of order $(n-1)$ with respect to any row or column.

Properties of determinant

The following properties of determinant can be seen immediately:

1. Suppose Q is equivalent to P (that means Q is obtained from P by an elementary row (column) operation). Then
   (a) $\det(Q) = -\det(P)$, if Q is obtained by interchange of two rows (columns).
   (b) $\det(Q) = k\det(P)$, if a row (column) of P is multiplied by a scalar k.
   (c) $\det(Q) = \det(P)$, if a multiple of a row (column) of P were added to another row (column) of P.
2. The determinant of product of two matrices is the product of their determinants, i. e.,

$$\det(P\,Q) = \det(P)\,\det(Q),$$

which can be generalized for any number of matrices as

$$\det(P_1\,P_2\cdots P_s) = \det(P_1)\,\det(P_2)\cdots\det(P_s).$$

3. The determinant of a triangular/diagonal matrix is equal to the product of the diagonal entries.
4. If any two rows (columns) of a matrix P are identical, then $|P| = 0$.
5. If any row (column) of a matrix P is zero, then $|P| = 0$.
6. The transpose of a matrix does not affect the value of determinant, i. e., $\det(P) = \det(P')$.

Example 312. Evaluate the following determinants:

$$\text{(i)}\ \begin{vmatrix} 0 & 2 \\ 1 & 3 \end{vmatrix} \qquad \text{(ii)}\ \begin{vmatrix} -1 & -2 & -3 \\ 2 & 4 & 6 \\ 3 & 0 & 5 \end{vmatrix} \qquad \text{(iii)}\ \begin{vmatrix} 1 & 2 & 3 \\ 2 & 0 & -1 \\ 1 & 2 & 0 \end{vmatrix}.$$

Solution. The determinant of 2×2 matrix $\left[\begin{smallmatrix} a & b \\ c & d \end{smallmatrix}\right]$ is given by $a\,d - b\,c$. Using this, we get

$$\begin{vmatrix} 0 & 2 \\ 1 & 3 \end{vmatrix} = -2.$$

For (ii), we notice that

$$\begin{vmatrix} -1 & -2 & -3 \\ 2 & 4 & 6 \\ 3 & 0 & 5 \end{vmatrix} = -2 \begin{vmatrix} -1 & -2 & -3 \\ -1 & -2 & -3 \\ 3 & 0 & 5 \end{vmatrix} = -2 \times 0 = 0,$$

since two rows are identical. For (iii), we expand along the first row and get

$$\begin{vmatrix} 1 & 2 & 3 \\ 2 & 0 & -1 \\ 1 & 2 & 0 \end{vmatrix} = 1 \cdot \begin{vmatrix} 0 & -1 \\ 2 & 0 \end{vmatrix} - 2 \cdot \begin{vmatrix} 2 & -1 \\ 1 & 0 \end{vmatrix} + 3 \cdot \begin{vmatrix} 2 & 0 \\ 1 & 2 \end{vmatrix}$$
$$= 2 - 2 \times 1 + 3 \times 4 = 12.$$

Minors and cofactors

The determinants of $(n-1)$-square submatrices of a n-square matrix are called minors. The minor corresponding to an element a_{ij} of matrix P, denoted by M_{ij}, is the determinant of a submatrix obtained by deleting the ith row and j^{th} column. The cofactorcorresponding to an element a_{ij} is denoted by P_{ij} and defined by

$$P_{ij} = (-1)^{i+j} M_{ij}.$$

Adjoint of a square matrix

The adjoint of a square matrix P is the transposed of the matrix of the cofactors of P, i. e.,

$$\operatorname{adj}(P) = \begin{bmatrix} P_{11} & P_{12} & P_{13} \\ P_{21} & P_{22} & P_{23} \\ P_{31} & P_{32} & P_{33} \end{bmatrix}' = \begin{bmatrix} P_{11} & P_{21} & P_{31} \\ P_{12} & P_{22} & P_{32} \\ P_{13} & P_{23} & P_{33} \end{bmatrix},$$

where $P_{ij}'s$ are the cofactors of corresponding a_{ij} of P.

Example 313. If $P = \left[\begin{smallmatrix} 2 & -5 \\ 3 & 4 \end{smallmatrix}\right]$, then evaluate the minors $M_{11}, M_{12}, M_{21}, M_{22}$. Also, obtain the cofactors and adj(P).

Solution. The minors are given by

$$\begin{aligned} M_{11} &= \text{minor of } a_{11} = 4, \\ M_{12} &= \text{minor of } a_{12} = 3, \\ M_{21} &= \text{minor of } a_{21} = -5, \\ M_{22} &= \text{minor of } a_{22} = 2. \end{aligned}$$

The cofactors P_{ij} are obtained by $P_{ij} = (-1)^{i+j} M_{ij}$. Hence,

$$P_{11} = (-1)^{1+1}\, M_{11} = 4, \quad P_{12} = (-1)^{1+2}\, M_{12} = -3,$$
$$P_{21} = (-1)^{2+1}\, M_{21} = 5, \quad P_{22} = (-1)^{2+2}\, M_{22} = 2.$$

The adjoint of the matrix P is given by

$$\operatorname{adj}(P) = \begin{bmatrix} P_{11} & P_{21} \\ P_{12} & P_{22} \end{bmatrix} = \begin{bmatrix} 4 & 5 \\ -3 & 2 \end{bmatrix}.$$

Example 314. Find the minors and cofactors of the matrix

$$P = \begin{bmatrix} 1 & 2 & 3 \\ -1 & 0 & 2 \\ 2 & 1 & 3 \end{bmatrix}.$$

Solution. If M_{ij} and P_{ij}, respectively, are minors and cofactors, then

$$M_{11} = \begin{vmatrix} 0 & 2 \\ 1 & 3 \end{vmatrix} = -2 \implies P_{11} = (-1)^{1+1} M_{11} = -2;$$

$$M_{12} = \begin{vmatrix} -1 & 2 \\ 2 & 3 \end{vmatrix} = -7 \implies P_{12} = (-1)^{1+2} M_{12} = 7;$$

$$M_{13} = \begin{vmatrix} -1 & 0 \\ 2 & 1 \end{vmatrix} = -1 \implies P_{13} = (-1)^{1+3} M_{13} = -1;$$

$$M_{21} = \begin{vmatrix} 2 & 3 \\ 1 & 3 \end{vmatrix} = 3 \implies P_{21} = (-1)^{2+1} M_{21} = -3;$$

$$M_{22} = \begin{vmatrix} 1 & 3 \\ 2 & 3 \end{vmatrix} = -3 \implies P_{22} = (-1)^{2+2} M_{22} = -3;$$

$$M_{23} = \begin{vmatrix} 1 & 2 \\ 2 & 1 \end{vmatrix} = -3 \implies P_{23} = (-1)^{2+3} M_{23} = 3;$$

$$M_{31} = \begin{vmatrix} 2 & 3 \\ 0 & 2 \end{vmatrix} = 4 \implies P_{31} = (-1)^{3+1} M_{31} = 4;$$

$$M_{32} = \begin{vmatrix} 1 & 3 \\ -1 & 2 \end{vmatrix} = 5 \implies P_{32} = (-1)^{3+2} M_{32} = -5;$$

$$M_{33} = \begin{vmatrix} 1 & 2 \\ -1 & 0 \end{vmatrix} = 2 \implies P_{33} = (-1)^{3+3} M_{33} = 2.$$

The adjoint is now given by

$$\operatorname{adj}(P) = \begin{bmatrix} P_{11} & P_{21} & P_{31} \\ P_{12} & P_{22} & P_{32} \\ P_{13} & P_{23} & P_{33} \end{bmatrix} = \begin{bmatrix} -2 & 7 & -1 \\ -3 & -3 & 3 \\ 4 & -5 & 2 \end{bmatrix}.$$

Invertible matrices

Left inverse of a matrix

Let P be an $m \times n$ matrix. A matrix Q of order $n \times m$ is called the left inverse of P if $QP = [I]_{n\times n}$.

Right inverse of a matrix

Let P be an $m \times n$ matrix. A matrix Q of order $n \times m$ is called the right inverse of P if $PQ = [I]_{m\times m}$.

A square matrix P is called invertible if there exists a matrix Q, which is left inverse as well as the right inverse of P, i. e., $QP = PQ = I$. The matrix Q is called as the inverse of P written as $Q = P^{-1}$.

Theorem 315. *Let P and Q be two square matrices. Then:*
1. *If P is invertible, so is P^{-1} and $(P^{-1})^{-1} = P$.*
2. *If both P and Q are invertible, so is PQ and $(PQ)^{-1} = Q^{-1}P^{-1}$.*

The following result can be produced immediately.

Corollary 316. *A product of invertible matrices is invertible.*

Theorem 317. *If P is a square matrix, then the following are equivalent*
(i) *P is invertible;*
(ii) *P is row equivalent to identity matrix;*
(iii) *P is a product of elementary matrices.*

Corollary 318. *If P is an invertible matrix and if a sequence of elementary row operations reduces P to the identity matrix, then the same sequence of operations when applied to I yields P^{-1}.*

Some other useful results on the inverse and adjoint of a matrix are as follows. If P, Q, R are square matrices of the same order and P is nonsingular, then:
1. $PQ = PR \implies Q = R$;
2. $QP = RP \implies Q = R$;
3. $(P')^{-1} = (P^{-1})'$;
4. $|P^{-1}| = |P|^{-1}$ or $\frac{1}{|P|}$;
5. $|\operatorname{adj}(P)| = |P|^{p-1}$, p is the order of P;
6. $\operatorname{adj} P' = (\operatorname{adj} P)'$;
7. $\operatorname{adj}(\operatorname{adj} P) = |P|^{p-2} P$;
8. $|\operatorname{adj}(\operatorname{adj} P)| = |P|^{(p-2)^2}$.

Example 319. Find the inverse of the matrix $P = \begin{bmatrix} 2 & -5 \\ 3 & 4 \end{bmatrix}$.

One of the ways to get the inverse is by using the adjoint. The adj(P) is given in Example 313:

$$\operatorname{adj}(P) = \begin{bmatrix} 4 & 5 \\ -3 & 2 \end{bmatrix}$$

and the determinant is 23. Then the inverse is given by

$$P^{-1} = \frac{1}{23}\begin{bmatrix} 4 & 5 \\ -3 & 2 \end{bmatrix}.$$

Example 320. By using elementary row transformations, find the inverse of the matrix $P = \begin{bmatrix} 1 & 4 \\ 2 & 7 \end{bmatrix}$.

Solution. We know that

$$P = IP$$

$$\text{or} \quad \begin{bmatrix} 1 & 4 \\ 2 & 7 \end{bmatrix} = \begin{bmatrix} 1 & 0 \\ 0 & 1 \end{bmatrix} P,$$

$$\Longrightarrow \begin{bmatrix} 1 & 4 \\ 0 & -1 \end{bmatrix} = \begin{bmatrix} 1 & 0 \\ -2 & 1 \end{bmatrix} P, \quad \text{by } R_2 \to R_2 + (-2)R_1;$$

$$\Longrightarrow \begin{bmatrix} 1 & 4 \\ 0 & 1 \end{bmatrix} = \begin{bmatrix} 1 & 0 \\ 2 & -1 \end{bmatrix} P, \quad \text{by } R_2 \to (-1)R_2;$$

$$\Longrightarrow \begin{bmatrix} 1 & 0 \\ 0 & 1 \end{bmatrix} = \begin{bmatrix} -7 & 4 \\ -2 & 1 \end{bmatrix} P, \quad \text{by } R_1 \to R_1 + (-4)R_2.$$

Thus,

$$P^{-1} = \begin{bmatrix} -7 & 4 \\ -2 & 1 \end{bmatrix}.$$

Example 321. Find the inverse of the matrix

$$P = \begin{bmatrix} 1 & 1 & -1 \\ -1 & 3 & 0 \\ 0 & -2 & 1 \end{bmatrix}.$$

Solution. We know

$$P = IP$$

$$\text{or} \quad \begin{bmatrix} 1 & 1 & -1 \\ -1 & 3 & 0 \\ 0 & -2 & 1 \end{bmatrix} = \begin{bmatrix} 1 & 0 & 0 \\ 0 & 1 & 0 \\ 0 & 0 & 1 \end{bmatrix} P,$$

$$\Rightarrow \begin{bmatrix} 1 & 1 & -1 \\ 0 & 4 & -1 \\ 0 & -2 & 1 \end{bmatrix} = \begin{bmatrix} 1 & 0 & 0 \\ 1 & 1 & 0 \\ 0 & 0 & 1 \end{bmatrix} P, \quad \text{by } R_2 \to R_2 + R_1;$$

$$\Rightarrow \begin{bmatrix} 1 & 1 & -1 \\ 0 & 4 & -1 \\ 0 & 0 & 1 \end{bmatrix} = \begin{bmatrix} 1 & 0 & 0 \\ 1 & 1 & 0 \\ 1 & 1 & 2 \end{bmatrix} P, \quad \text{by } R_3 \to 2R_3 + R_2;$$

$$\Rightarrow \begin{bmatrix} 1 & 1 & -1 \\ 0 & 1 & \frac{-1}{4} \\ 0 & 0 & 1 \end{bmatrix} = \begin{bmatrix} 1 & 0 & 0 \\ \frac{1}{4} & \frac{1}{4} & 0 \\ 1 & 1 & 2 \end{bmatrix} P, \quad \text{by } R_2 \to \frac{1}{4}R_2;$$

$$\Rightarrow \begin{bmatrix} 1 & 1 & 0 \\ 0 & 1 & 0 \\ 0 & 0 & 1 \end{bmatrix} = \begin{bmatrix} 2 & 1 & 2 \\ \frac{1}{2} & \frac{1}{2} & \frac{1}{2} \\ 1 & 1 & 2 \end{bmatrix} P, \quad \text{by } R_1 \to R_1 + R_3, R_2 \to R_2 + \frac{1}{4}R_3;$$

$$\Rightarrow \begin{bmatrix} 1 & 0 & 0 \\ 0 & 1 & 0 \\ 0 & 0 & 1 \end{bmatrix} = \begin{bmatrix} \frac{3}{2} & \frac{1}{2} & \frac{3}{2} \\ \frac{1}{2} & \frac{1}{2} & \frac{1}{2} \\ 1 & 1 & 2 \end{bmatrix} P, \quad \text{by } R_1 \to R_1 + (-1)R_2.$$

Thus,

$$P^{-1} = \begin{bmatrix} \frac{3}{2} & \frac{1}{2} & \frac{3}{2} \\ \frac{1}{2} & \frac{1}{2} & \frac{1}{2} \\ 1 & 1 & 2 \end{bmatrix} = \frac{1}{2} \begin{bmatrix} 3 & 1 & 3 \\ 1 & 1 & 1 \\ 2 & 2 & 4 \end{bmatrix}.$$

Rank of a matrix

A matrix P is said to be of rank r, written as $\rho(P) = r$, if:

1. It has at least one nonzero minor of order of r.
2. All the minors of order $(r + 1)$ or higher than r are zero.

From the definition of the rank of a matrix P, we conclude the following:

1. If P is a null matrix, then $\rho(P) = 0$.
2. If P is not a null matrix, then $\rho(P) \geq 1$.
3. If P is a nonsingular $p \times p$ matrix, then $\rho(P) = p$.
4. If P is an $p \times n$ matrix, then $\rho(P) \leq \min(p, n)$.
5. If all minors of order r are equal to zero, then $\rho(P) < r$.

Methods of finding rank: We discuss three methods for finding the rank of a matrix:

1. In the first method, we use the definition of the rank of a matrix P.
2. The second method is the *Normal form Method.* In this method, we convert a given matrix P into the following forms, called normal or canonical form,

$$\begin{bmatrix} I_r & \vdots & 0 \\ \cdots & \cdots & \cdots \\ 0 & \vdots & 0 \end{bmatrix}, \quad \begin{bmatrix} I_r & \vdots & 0 \end{bmatrix}, \quad \begin{bmatrix} I_r \\ \cdots \\ 0 \end{bmatrix}, \quad [I_r]$$

using the series of elementary (row or column or both) transformations. One can follow the following steps to get the normal forms:

(i) Interchange rows (columns) to obtain a nonzero element in the first row and first column of given matrix.

(ii) Make this nonzero element as 1.

(iii) Obtain zeros in the remainder of the I-row and I-column.

(iv) Continue the process down the main diagonal either until it reaches the end of the diagonal or all the remaining elements of the matrix are zero. The order of the identity matrix in the normal form is the rank of the matrix.

3. *Echelon form method.* A matrix is reduced in the Echelon form as:
   (i) The first nonzero element in a row should be unity.
   (ii) All the nonzeros, if any, precede the zero rows.

Then the rank of the matrix is equal to the number of nonzero diagonal elements or non-zero rows.

Linearly independence and dependence vectors

A set of k vectors $\{v_1, v_2, \ldots, v_k\}$ are linearly independent if and only if there exists a set of k-scalars $c_1, c_2, \ldots, c_k$ such that

$$c_1 v_1 + c_2 v_2 + \cdots + c_k v_k = 0.$$

This implies $c_1 = c_2 = \cdots = c_k = 0$. If at least one of the scalar c_i is nonzero, then the set of vectors is linearly dependent.

Example 322. The set of vectors $\{(1,2), (0,-3)\}$ is linearly independent because

$$c_1 (1,2) + c_2 (0,-3) = 0$$

if and only if $c_1 = c_2 = 0$.

Some facts about linearly independent and dependent vectors are:

1. If a set of vectors is linearly dependent, then at least one member of the set can be expressed as a linear combination of the remaining vectors.
2. If a set of vectors contains a zero vector, then it is linearly dependent.
3. The p vectors $X_1, X_2, \ldots, X_p$ of dimension n are linearly dependent if the rank of the matrix $[X_1, X_2, \ldots, X_p]$ with the given vectors as columns is less than p.

4. Any nonempty subset of a linearly independent set is linearly independent.
5. Any super set of a linearly dependent set is linearly dependent.

System of linear equations

An equation in unknowns $x_1, x_2, \dots, x_n$, which can be written in the form,

$$a_1x_1 + a_2x_2 + \cdots + a_nx_n = b, \tag{B.28}$$

is called a linear equation in n-variables. Here, $a_1, a_2, \dots, a_n$ are constant coefficients of the variables $x_1, x_2, \dots, x_n$ and b is the constant term.

A set of linear equations in the same unknowns is called a system of linear equations. Let us consider a system of m-equations in n-unknowns $x_1, x_2, \dots, x_n$,

$$\begin{aligned} a_{11}x_1 + a_{12}x_2 + \cdots + a_{1n}x_n &= b_1, \\ a_{21}x_1 + a_{22}x_2 + \cdots + a_{2n}x_n &= b_2, \\ \vdots \qquad \vdots \qquad\quad \vdots& \\ a_{m1}x_1 + a_{m2}x_2 + \cdots + a_{mn}x_n &= b_m, \end{aligned} \tag{B.29}$$

where a_{ij} and b_i are constants.

The system (B.29) is called homogeneous or nonhomogeneous accordingly $b_i = 0$ for each i or at least one b_i is nonzero.

A set of values of unknowns, which satisfy the linear equation is called the *solution* of the linear equation. Similarly, if a set of values of unknowns satisfy every linear equation in the system, then it is called the solution of the system of linear equations.

Example 323. $x = -6, y = 5, z = -1$ is the solution of the system of linear equations,

$$\begin{aligned} x + 2y + 3z &= 1, \\ x + 3y + 6z &= 3, \\ 2x + 6y + 13z &= 5, \end{aligned} \tag{B.30}$$

because this set of values of unknowns satisfy each linear equation in the system.

On the basis of existence of a solution, the system of linear equations can be categorized as follows.

Inconsistent/consistent

If a system of linear equations has at least one solution, then it is called consistent, and otherwise inconsistent. The consistent system can have either one (unique) solution or an infinite number of solutions. The system (B.29) can also be written as

$$PX = Q, \tag{B.31}$$

where

$$P = \begin{bmatrix} a_{11} & a_{12} & \cdots & a_{1n} \\ a_{21} & a_{22} & \cdots & a_{2n} \\ \vdots & \vdots & \cdots & \vdots \\ a_{m1} & a_{m2} & \cdots & a_{mn} \end{bmatrix}$$

is the coefficient matrix and

$$Q = \begin{bmatrix} b_1 \\ b_2 \\ \vdots \\ b_n \end{bmatrix}$$

is the column matrix of constant terms. Now, we construct the augmented matrix M as

$$M = \begin{bmatrix} a_{11} & a_{12} & \cdots & a_{1n} & \vdots & b_1 \\ a_{21} & a_{22} & \cdots & a_{2n} & \vdots & b_2 \\ \vdots & \vdots & \cdots & \vdots & \vdots & \vdots \\ a_{m1} & a_{m2} & \cdots & a_{mn} & \vdots & b_n \end{bmatrix}.$$

Theorem 324. *The following can be calculated for a nonhomogeneous system:*
(i) *If* rank(M) = rank(P) = *number of unknowns, then the system is consistent and has a unique solution.*
(ii) *If* rank(M) = rank(P) < *number of unknowns, then the system is still consistent and has an infinite number of solutions.*
(iii) *If* rank(M) $\neq$ rank(P), *then the system is inconsistent,* i. e., *it has no solution.*

We can write a theorem similar to Theorem 324 for a homogeneous system.

Theorem 325. *A homogeneous system*

$$\begin{aligned} a_{11}x_1 + a_{12}x_2 + \cdots + a_{1n}x_n &= 0, \\ a_{21}x_1 + a_{22}x_2 + \cdots + a_{2n}x_n &= 0, \\ \vdots \qquad \vdots \qquad & \vdots \\ a_{m1}x_1 + a_{m2}x_2 + \cdots + a_{mn}x_n &= 0 \end{aligned} \tag{B.32}$$

or

$$PX = 0$$

is always consistent. It has either a unique solution or an infinite number of solutions. We conclude this from:
1. *If* $\operatorname{rank}(P) = n$ *(number of unknowns), then the system has unique solution.*
2. *If* $\operatorname{rank}(P) < n$*, then the system has an infinite number of solutions.*

For a square system, i. e., number of equations = number of unknowns, the consistency can be determined using the determinant. If the system is nonhomogeneous, i. e., $Q \neq 0$ in $PX = Q$, then the system has a unique solution when $\det(P) \neq 0$ and $P^{-1}Q$ is the unique solution. Similarly, in the homogeneous system, $PX = 0$ has a unique solution if $\det(P) \neq 0$ and infinite number of solutions if $\det(P) = 0$.

Complex matrices

If at least one element of a matrix is a complex number, then the matrix is called a complex matrix. The matrix obtained by replacing the elements of a complex matrix P by the corresponding conjugate complex numbers is called the conjugate of the matrix P and is denoted by $\bar{P}$. The conjugate of the transpose of P, i. e., $\overline{(P')}$ and the transposed conjugate of P, i. e., $(\overline{P})'$ are equal and denoted by P^*.

Hermitian and skew-Hermitian matrix

A square matrix $P = [a_{ij}]$ is said to be Hermitian if $P^* = P$ or $a_{ij} = \overline{a_{ji}}$ and skew-Hermitian if $P^* = -P$ or $a_{ij} = -\overline{a_{ji}}$. Here are some facts about Hermitian and skew-Hermitian matrices:
1. The diagonal elements of a Hermitian matrix are all real, while the other element is the conjugate complex of the element in the transposed position.
2. The diagonal elements of a skew-Hermitian matrix are zero or purely imaginary numbers, while the other element is the negative of the conjugate complex of the element in the transposed position.

Let P and Q be complex square matrices and λ be any complex number. Then the following results hold:

(i) $\overline{(\overline{P})} = P$ (ii) $\overline{P+Q} = \overline{P} + \overline{Q}$ (iii) $\overline{\lambda P} = \bar{\lambda}\,\overline{P}$

(iv) $\overline{PQ} = \overline{P}\,\overline{Q}$ (v) $(P^*)^* = P$ (vi) $(P+Q)^* = P^* + Q^*$

(vii) $(\lambda P)^* = \bar{\lambda} P^*$ (viii) $(PQ)^* = Q^* P^*$.

Here are a collection of matrices that represent a unique space, namely:

1. $M_{p\times n}$; the set of all matrices of order $p \times n$.
2. $\mathbb{R}^{p\times p}$ = the set of all p-square matrices with real entries.
3. $\mathbb{C}^{p\times p}$ = the set of all p-square matrices with complex entries.
4. $\mathrm{GL}(p, \mathbb{R})$ = the set of all invertible matrices with real entries.
5. $\mathrm{GL}(p, \mathbb{C})$ = the set of all invertible matrices with complex entries.
6. $\mathrm{SL}(p, \mathbb{R})$ = the set of all real matrices with det = 1.
7. $\mathrm{SL}(p, \mathbb{C})$ = The set of all complex matrices with det = 1.

Eigenvalues and eigenvectors

Let P be a square matrix and X be a nonzero vector. Then X is called an eigenvector or characteristic vector if it satisfies the relation:

$$PX = \lambda X,$$

where λ is a scalar value. This relation can be rewritten as

$$PX - \lambda X = 0 \implies (P - \lambda I)X = 0,$$

which is a homogeneous system of linear equations and it will have a solution when $\det(P - \lambda I) = 0$. The expression $\det(P - \lambda I)$ is called the characteristic polynomial and equation $\det(P - \lambda I) = 0$ is called a characteristic equation. The solutions of a characteristic equation are known as characteristic roots or latent roots or eigenvalues and for each eigenvalue there is an eigenvector, which can be obtained by solving $(P - \lambda X) = 0$.

Example 326. Let $P = \left[\begin{smallmatrix} 1 & 3 \\ 4 & 5 \end{smallmatrix}\right]$. Then find the characteristic values and characteristic vectors.

Solution. The characteristic equation is given by

$$\begin{aligned} \det(P - \lambda I) &= 0 \\ &\text{or} \\ \det\begin{pmatrix} 1-\lambda & 3 \\ 4 & 5-\lambda \end{pmatrix} &= 0 \\ &\text{or} \\ (1-\lambda)(5-\lambda) - 12 &= 0 \\ \implies \lambda^2 - 6\lambda - 7 &= 0, \end{aligned}$$

which on solving gives $\lambda_1 = 7, \lambda_2 = -1$.

Let us consider $X_1 = \left[\begin{smallmatrix} x \\ y \end{smallmatrix}\right]$ as the characteristic vector corresponding to eigenvalue $\lambda_1 = 7$, and thus we have

$$(P - 7I)X_1 = 0$$

$$\Rightarrow \left\{\begin{bmatrix} 1 & 3 \\ 4 & 5 \end{bmatrix} - \begin{bmatrix} 7 & 0 \\ 0 & 7 \end{bmatrix}\right\} \begin{bmatrix} x \\ y \end{bmatrix} = \begin{bmatrix} 0 \\ 0 \end{bmatrix}$$

or

$$\begin{bmatrix} -6 & 3 \\ 4 & -2 \end{bmatrix} \begin{bmatrix} x \\ y \end{bmatrix} = \begin{bmatrix} 0 \\ 0 \end{bmatrix}.$$

On solving this homogeneous system, we get one of the nonzero solutions as

$$X_1 = \begin{bmatrix} x \\ y \end{bmatrix} = \begin{bmatrix} 1 \\ 2 \end{bmatrix}.$$

Similarly corresponding to the characteristic value $\lambda_2 = -1$, there exists a characteristic vector $X_2 = \left[\begin{smallmatrix} x \\ y \end{smallmatrix}\right]$ such that

$$\left\{\begin{bmatrix} 1 & 3 \\ 4 & 5 \end{bmatrix} - \begin{bmatrix} -1 & 0 \\ 0 & -1 \end{bmatrix}\right\} \begin{bmatrix} x \\ y \end{bmatrix} = \begin{bmatrix} 0 \\ 0 \end{bmatrix}$$

or

$$\begin{bmatrix} 2 & 3 \\ 4 & 6 \end{bmatrix} \begin{bmatrix} x \\ y \end{bmatrix} = \begin{bmatrix} 0 \\ 0 \end{bmatrix}.$$

The solution of the above system is given by $X_2 = \left[\begin{smallmatrix} -3 \\ 2 \end{smallmatrix}\right]$.

Example 327. Find the eigenvalues and eigenvectors of the matrix $P = \left[\begin{smallmatrix} 1 & 2 \\ 2 & 4 \end{smallmatrix}\right]$.

Solution. The characteristic equation corresponding to matrix P is given by

$$\det(P - \lambda I) = 0.$$

On simplification, we get two distinct eigenvalues as $\lambda_1 = 0$, $\lambda_2 = 5$. Corresponding to eigenvalue $\lambda_1 = 0$, consider $X_1 = \left[\begin{smallmatrix} x \\ y \end{smallmatrix}\right]$ as the eigenvector, and hence we have

$$\begin{bmatrix} 1 & 2 \\ 2 & 4 \end{bmatrix} \begin{bmatrix} x \\ y \end{bmatrix} = \begin{bmatrix} 0 \\ 0 \end{bmatrix},$$

solving the system yields $X_1 = \left[\begin{smallmatrix} -2 \\ 1 \end{smallmatrix}\right]$. Similarly, the eigenvalue $\lambda_2 = 5$ produces another eigenvector $X_2 = \left[\begin{smallmatrix} x \\ y \end{smallmatrix}\right]$ given by

$$\left\{\begin{bmatrix} 1 & 2 \\ 2 & 4 \end{bmatrix} - \begin{bmatrix} 5 & 0 \\ 0 & 5 \end{bmatrix}\right\} \begin{bmatrix} x \\ y \end{bmatrix} = \begin{bmatrix} 0 \\ 0 \end{bmatrix}$$

or

$$\begin{bmatrix} -4 & 2 \\ 2 & -1 \end{bmatrix} \begin{bmatrix} x \\ y \end{bmatrix} = \begin{bmatrix} 0 \\ 0 \end{bmatrix}.$$

Which after simplification gives $X_2 = \left[\begin{smallmatrix} 1 \\ 2 \end{smallmatrix}\right]$ as the eigenvector corresponding to the eigenvalue $\lambda_2 = 5$.

Besides the straightforward calculation, we have simple formulas to get the characteristic polynomial as well as the equation. Let $P = \left[\begin{smallmatrix} c_{11} & c_{12} \\ c_{21} & c_{22} \end{smallmatrix}\right]$ be a square matrix of order 2. Then the formula for a characteristic polynomial is given by

$$\Delta(\lambda) = \lambda^2 - (c_{11} + c_{22})\,\lambda + \det(P)$$

or

$$\Delta(\lambda) = \lambda^2 - \operatorname{tr}(P)\,\lambda + \det(P).$$

For a matrix of order 3, say $P = \begin{bmatrix} c_{11} & c_{12} & c_{13} \\ c_{21} & c_{22} & c_{23} \\ c_{31} & c_{32} & c_{33} \end{bmatrix}$, the characteristic polynomial is given by

$$\Delta(\lambda) = \lambda^3 - (c_{11} + c_{22} + c_{33})\,\lambda^2 + (P_{11} + P_{22} + P_{33}) - \det(P)$$

or

$$\Delta(\lambda) = \lambda^3 - \operatorname{tr}(P)\,\lambda^2 + (P_{11} + P_{22} + P_{33}) - \det(P),$$

where P_{11}, P_{22}, P_{33} are cofactors corresponding to the diagonal elements c_{11}, c_{22}, c_{33}.

In general, for a p-square matrix the characteristic polynomial is given by

$$\Delta(\lambda) = \lambda^p - S_1\,\lambda^{p-1} + S_2\,\lambda^{p-2} + \cdots + (-1)^p \det(P),$$

where S_p is the principal minor of order $k, 1 \le k \le p$.

Now, we list some properties of eigenvalues and eigenvectors:

1. If a p-square matrix P has eigenvalues $\lambda_1, \lambda_2, \ldots, \lambda_p$, then P^r has eigenvalues $\lambda_1^r, \lambda_2^r, \ldots, \lambda_p^r$. If P is invertible, then P^{-1} has eigenvalues $\lambda_1^{-1}, \lambda_2^{-1}, \ldots, \lambda_p^{-1}$; the matrix $P + rI$ has eigenvalues $\lambda_1 + r, \lambda_2 + r, \ldots, \lambda_p + r$.
2. The trace of P or tr (P) = sum of diagonal entries = $\lambda_1 + \lambda_2 + \cdots + \lambda_p$.
3. $\det(P) = \lambda_1\,\lambda_2 \cdots \lambda_p$.

Diagonalization

A p-square matrix P is said to be diagonalizable if there exists a matrix Q such that

$$R = Q^{-1}\,P\,Q$$

is a diagonal matrix. There are several ways to check the diagonalizability of a matrix. One of the efficient ways is by using the characteristic vectors stated in the following theorem.

Theorem 328. *A p-square matrix P is said to be diagonalizable or similar to a diagonal matrix if and only if P has p-linearly independent characteristic vectors. In such case, the diagonal element of* $R = Q^{-1} P Q$ *are the eigenvalues and Q is the matrix whose columns are characteristic vectors of P.*

Example 329. Check whether $P = \left[\begin{smallmatrix} 3 & 1 \\ 2 & 2 \end{smallmatrix}\right]$ the matrix is diagonalizable or not. If yes, find the diagonal matrix similar to P.

Solution. Solving the characteristic equation

$$|P - \lambda I| = 0$$

gives the characteristic values $\lambda_1 = 1, \lambda_2 = 4$ and corresponding characteristic vectors are

$$X_1 = \begin{bmatrix} 1 \\ -2 \end{bmatrix}, \quad X_2 = \begin{bmatrix} 1 \\ 1 \end{bmatrix}.$$

The two characteristic vectors X_1, X_2 are linearly independent, and hence the matrix P is diagonalizable. Now, the matrix

$$Q = \begin{bmatrix} 1 & 1 \\ -2 & 1 \end{bmatrix} \quad \text{and} \quad Q^{-1} = \begin{bmatrix} \frac{1}{3} & \frac{-1}{3} \\ \frac{2}{3} & \frac{1}{3} \end{bmatrix}$$

are obtained so

$$R = Q^{-1} P Q = \begin{bmatrix} \frac{1}{3} & \frac{-1}{3} \\ \frac{2}{3} & \frac{1}{3} \end{bmatrix} \begin{bmatrix} 3 & 1 \\ 2 & 2 \end{bmatrix} \begin{bmatrix} 1 & 1 \\ -2 & 1 \end{bmatrix} = \begin{bmatrix} 1 & 0 \\ 0 & 4 \end{bmatrix}$$

is a diagonal matrix similar to P.

This factorization is helpful in the calculation of various matrix based computations. The matrix P can be written in the form

$$P = Q R Q^{-1}.$$

Now, one can compute

$$P^3 = Q R^3 Q^{-1} = Q \begin{bmatrix} 1 & 0 \\ 0 & 64 \end{bmatrix} Q^{-1}.$$

Similarly, for any polynomial $f(x) = x^3 - 3x + 2$, the matrix function $f(P)$ can be evaluated with the simple matrix multiplication

$$f(P) = Q f(R) Q^{-1} = Q \begin{bmatrix} 0 & 0 \\ 0 & 54 \end{bmatrix} Q^{-1},$$

and so on.

We now list some facts related to characteristic values, characteristic vectors and diagonalizability.

1. The following statements are equivalent:
   (a) A scalar λ is a characteristic value of the matrix P;
   (b) The matrix $Q = P - \lambda I$ is singular;
   (c) The scalar λ is a root of the characteristic polynomial of the matrix P.
2. If $X_1, X_2, \ldots, X_p$ are nonzero eigenvectors corresponding to the distinct eigenvalues $\lambda_1, \lambda_2, \ldots, \lambda_p$, then $X_1, X_2, \ldots, X_p$ are linearly independent and matrix P is diagonalizable.
3. If the characteristic polynomial of the matrix P can be factorized into the linear factor, i. e.,

$$\Delta(u) = (u - a_{11})(u - a_{22}) \cdots (u - a_{pp}),$$

then P is similar to a diagonal matrix R given by

$$R = \begin{bmatrix} a_{11} & \cdots & 0 \\ \vdots & \ddots & \vdots \\ 0 & \cdots & a_{pp} \end{bmatrix}.$$

C Groups, rings, fields and vector spaces

A special operation which takes each element of $A \times A$ to an element of A is called the binary operation. In such case, the set A is called a closed set. For example, the set of natural numbers is closed under addition but it is not closed with respect to subtraction because there exists two elements a and b in $\mathbb{N}$ such that $a - b \notin \mathbb{N}$. Similarly, it can be checked that the set of natural numbers is closed with respect to multiplication but not with division.

The next elementary property to be discussed is associative. A binary operation $\star$ is called associative if any three elements a, b, c of a set preserves the order of composition, i. e.,

$$a \star (b \star c) = (a \star b) \star c. \tag{C.1}$$

For example, the set of natural numbers, integers, rational numbers, real numbers and complex numbers with the binary operations of addition and multiplication preserve the order, and hence are associative.

Further, if there is an element e in a set A such that

$$a \star e = e \star a = a, \quad \forall a \in A, \tag{C.2}$$

then e is called the identity element. Also, if there exists an element $b \in A$ such that

$$a \star b = b \star a = e, \tag{C.3}$$

then b is called the inverse of element a.

These elementary properties lead to define an enriched concept of mathematics known as a group.

Group

A set G together with a binary operation $\star$ is called a group if:
1. "$\star$" is associative;
2. There exists an identity element;
3. The inverse of each element in G exists.

The group G becomes more specific when the commutative property ($a \star b = b \star a$, for every $a, b \in G$) holds. Such special groups are called the Abelian group. We enlighten this concept with a rich list of examples.

Under addition, the set of integers, rational numbers, real numbers, complex numbers and set of rectangular matrices forms a group. In the same way, the rational num-

https://doi.org/10.1515/9783111324586-010

bers, real numbers, complex numbers, set of invertible matrices and Gaussian integers form a group under multiplication. Besides, we discuss other examples in detail.

Example 330. The infinite set $G = \{u + v\sqrt{2} \mid u, v \in \mathbb{Q}\}$ is a group with respect to the addition.

Solution. To prove G is a group, we need to verify the following axioms:

1. *Closure:* Let $a = u_1 + v_1\sqrt{2}$, $b = u_2 + v_2\sqrt{2}$, $u_1, u_2, v_1, v_2 \in \mathbb{Q}$ be any two elements of G. Then

$$a + b = (u_1 + u_2) + (v_1 + v_2)\sqrt{2} \in G,$$

   because the sum of two rational numbers is again a rational number. Thus, for every $a, b \in G \implies a + b \in G$, so G is closed.
2. *Associative:* For any three elements a, b and c from G, it can be verified that

$$a + (b + c) = (a + b) + c.$$

3. *Existence of identity:* There exists an element $e = 0 = 0 + 0 \cdot \sqrt{2} \in G$ such that

$$e + a = a + e, \quad \forall a \in G.$$

4. *Existence of inverse:* For an element $a = u_1 + v_1\sqrt{2} \in G$, there exists an element $b = -a = -u_1 - v_1\sqrt{2} \in G$ which satisfies

$$a + b = b + a = e.$$

Hence, G is a group with respect to addition.

Example 331. The infinite set $\mathbb{Q} - \{1\}$ of all rational numbers forms a group with respect to the binary operation defined by

$$a * b = a + b - ab.$$

Solution. To show G is a group, we verify the following:

1. *Closure:* If $a, b \in \mathbb{Q} - \{1\}$, then $a * b \in \mathbb{Q} - \{1\}$. Note that $a * b$ is equal to 1 if and only if $a = 1$ and $b = 1$.
2. *Associative:* For any three elements a, b and c in $\mathbb{Q} - \{1\}$, we have

$$\begin{aligned}(a * b) * c &= (a + b - ab) * c \\ &= a + b - ab + c - (a + b - ab)c \\ &= a + b + c - ab - ac - bc + abc \\ &= a + (b + c - bc) - a(b + c - bc) \\ &= a * (b * c).\end{aligned}$$

   Thus, $*$ is associative.

3. *Existence of identity:* Let $e \in \mathbb{Q} - \{1\}$ be the identity element. Then

$$e * a = a * e = a, \quad \forall a \in \mathbb{Q} - \{1\}.$$

Thus,

$$\begin{aligned} e * a = e + a - ea &= a \\ &\Longrightarrow e(1-a) = 0 \\ &\Longrightarrow e = 0 \text{ or } 1 - a = 0 \end{aligned}$$

but $a \neq 1$, hence $e = 0$ is the identity element.

4. *Existence of inverse:* Let $a, b \in \mathbb{Q} - \{1\}$ be the inverse of each other. Then

$$a * b = b * a = 0.$$

This gives

$$b = \frac{a}{a-1} \text{ or } a = \frac{b}{b-1} \in \mathbb{Q} - \{1\}, \quad \text{since } a, b \neq 1. \tag{C.4}$$

Thus, the set of rational numbers except 1 is a group with respect to the binary operation $*$.

Example 332. The set of matrices

$$P_\alpha = \begin{bmatrix} \cos\alpha & -\sin\alpha \\ \sin\alpha & \cos\alpha \end{bmatrix}, \quad \alpha \in \mathbb{R},$$

is a group with respect to the matrix addition and multiplication.

Solution. Let us verify the group's axioms:

1. *Closure:* For real numbers α, β, we have

$$P_\alpha = \begin{bmatrix} \cos\alpha & -\sin\alpha \\ \sin\alpha & \cos\alpha \end{bmatrix}, \quad P_\beta = \begin{bmatrix} \cos\beta & -\sin\beta \\ \sin\beta & \cos\beta \end{bmatrix}. \tag{C.5}$$

Then

$$\begin{aligned} P_\alpha P_\beta &= \begin{bmatrix} \cos\alpha & -\sin\alpha \\ \sin\alpha & \cos\alpha \end{bmatrix} \begin{bmatrix} \cos\beta & -\sin\beta \\ \sin\beta & \cos\beta \end{bmatrix} \\ &= \begin{bmatrix} \cos(\alpha+\beta) & -\sin(\alpha+\beta) \\ \sin(\alpha+\beta) & \cos(\alpha+\beta) \end{bmatrix} \\ &= P_{\alpha+\beta}. \end{aligned}$$

Hence, the set is closed with respect to the usual matrix multiplication.

2. *Associative:* We already know the matrix multiplication is associative. So, associative law holds.
3. *Existence of identity:* The identity matrix of order 2×2 is obtained by

$$P_0 = \begin{bmatrix} \cos 0 & -\sin 0 \\ \sin 0 & \cos 0 \end{bmatrix} = \begin{bmatrix} 1 & 0 \\ 0 & 1 \end{bmatrix}$$

and satisfies

$$P_0 P_\alpha = P_\alpha P_0 = P_\alpha.$$

4. *Existence of inverse:* The determinant of the matrix P_α for any real α is given by $|P_\alpha| = \cos^2 \alpha + \sin^2 \alpha = 1 \neq 0$. Hence, the inverse of the matrix P_α exists and given by

$$P_\alpha^{-1} = \begin{bmatrix} \cos(-\alpha) & -\sin(-\alpha) \\ \sin(-\alpha) & \cos(-\alpha) \end{bmatrix}.$$

5. *Commutative law:* One can also verify that

$$P_\alpha P_\beta = P_{\alpha+\beta} = P_{\beta+\alpha} = P_\beta P_\alpha.$$

Thus, the given set of matrices is an Abelian group.

Example 333 (Permutation group or symmetric group). If S is a finite set with k elements, then the set S_k of all one-one onto functions from S to S forms a group with respect to the composition of functions.

Solution. First, we will see the structure and presentation of the elements of permutation set S, then we verify the axioms of group. Let $S = \{u_1, \ldots, u_k\}$. Then an element f of S_k is presented in the following form:

$$f - \begin{pmatrix} u_1 & \cdots & u_k \\ f(u_1) & \cdots & f(u_k) \end{pmatrix}$$

The rearrangements of the elements $\{u_1, \ldots, u_k\}$ and their unique images $f(u_1), f(u_k)$ give distinct elements of S_k and so it is called a set of permutations. There are $k!$ such elements in S_k. Let f and g be any two elements of a permutation set given by

$$f = \begin{pmatrix} u_1 & \cdots & u_k \\ v_1 & \cdots & v_k \end{pmatrix}, \quad g = \begin{pmatrix} v_1 & \cdots & v_k \\ w_1 & \cdots & w_k \end{pmatrix}.$$

Then the composition or product of two permutations will be given by

$$gf = gof = \begin{pmatrix} v_1 & \cdots & v_k \\ w_1 & \cdots & w_k \end{pmatrix} \begin{pmatrix} u_1 & \cdots & u_k \\ v_1 & \cdots & v_k \end{pmatrix} = \begin{pmatrix} u_1 & \cdot & u_k \\ w_1 & \cdots & w_k \end{pmatrix}.$$

Now, one can verify the axioms of group using this composition as a binary operation in the following ways:

1. *Closure:* We have shown above that for any two permutations f and g in S_k, the permutation $gof \in S_k$. Hence, the set of permutations is closed with respect to the composition.
2. *Associative:* The composition of function is associative so it is obvious.
3. *Existence of identity:* The permutation

$$I = \begin{pmatrix} u_1 & \cdots & u_k \\ u_1 & \cdots & u_k \end{pmatrix} \tag{C.6}$$

acts as an identity element of S_k because

$$foI = Iof = f, \quad \forall f \in S_k.$$

4. *Existence of inverse:* Let

$$f = \begin{pmatrix} u_1 & \cdots & u_k \\ v_1 & \cdots & v_k \end{pmatrix}$$

be an element of S_k. Then the permutation

$$g = \begin{pmatrix} v_1 & \cdots & v_k \\ u_1 & \cdots & u_k \end{pmatrix} \in S_k$$

will be the inverse of f such that $fog = gof$.

Since the composition of functions is noncommutative in general, so the set S_k is a non-Abelian finite group.

We now list a few properties of a group:

1. The identity element of a group is unique.
2. The inverse of each element in a group is unique. Also, if u^{-1} is the inverse of $u \in G$, then $(u^{-1})^{-1} = u$.
3. The inverse of the product is the product of inverse in reverse order, i. e.,

$$(u\,v)^{-1} = v^{-1}u^{-1}, \quad u, v \in G.$$

4. Both left and right cancellation laws hold in a group, i. e.,

$$u\,v = u\,w \implies v = w, \quad \text{left cancellation law,}$$
$$v\,u = w\,u \implies v = w, \quad \text{right cancellation law.}$$

5. For any two elements $u, v \in G$, the equations $u\,x = v$ and $y\,u = v$ have unique solutions in G.

The concept of a group is being structured with one binary operation. The structure of a group can be generalized to next milestone called ring when another binary operation is included.

Ring

A structure which contains a set, say R, and two binary operations, namely $\oplus$, $\otimes$ becomes a ring or called a ring if it satisfies the following axioms:

1. $(R, \oplus)$ must be Abelian, i. e.,
   (i) The binary operation $\oplus$ is associative

   $$(x \oplus y) \oplus z = x \oplus (y \oplus z), \quad \forall x, y, z \in R.$$

   (ii) There exists an identity element $e_{\oplus}$ such that

   $$x \oplus e_{\oplus} = e_{\oplus} \oplus x = x.$$

   (iii) For each element $x \in R$, there is an element $y \in R$ such that

   $$x \oplus y = y \oplus x = e_{\oplus}.$$

   Then we say x and y are inverse of each other.
   (iv) The elements of R must obey the commutative law with respect to the operation $\oplus$, which means

   $$x \oplus y = y \oplus x.$$

2. The set R with second binary operation $\otimes$ satisfies the following two properties:
   (i) The operation $\otimes$ is associative

   $$(x \otimes y) \otimes z = x \otimes (y \otimes z).$$

   (ii) An identity element $e_{\otimes}$ is there such that

   $$x \otimes e_{\otimes} = e_{\otimes} \otimes x = x.$$

3. The operation $\otimes$ is distributive over $\oplus$, which means

   $$x \otimes (y \oplus z) = (x \otimes y) \oplus (x \otimes z);$$
   $$(y \oplus z) \otimes x = y \otimes x \oplus z \otimes x.$$

Most of the time binary operations are taken as usual addition and multiplication. In view of that, we can define the ring in a simple way as:

1. $(R, +)$ is an Abelian group.
2. With respect to multiplication, R is associative and must have multiplicative identity.
3. The multiplication will be distributive over addition.

Some authors consider a set without multiplicative identity as a ring. When it has identity, it is called a ring with unity. If we recall from the history, we have different notation for each type. The term "rng" was used for ring without unity and "ring" is used for a ring with unity. Now, look at this list of examples on a ring.

Example 334 (The ring of integers). The infinite set of integers forms a ring with respect to the usual addition "+" and multiplication "×" of integers. Since integers commute with each other and multiplicative identity $1 \in \mathbb{Z}$, so $(\mathbb{Z}, +, \times)$ is called a commutative ring with unity.

Example 335 (The Gaussian ring). The infinite set

$$\mathbb{Z}(\iota) = \{u + \iota v \mid u, v \in \mathbb{Z}\},$$

forms a ring with usual complex addition and multiplication as described in Appendix A. Since the elements of $\mathbb{Z}(\iota)$ commute and $1 = 1+\iota \cdot 0 \in \mathbb{Z}(\iota)$. So, $(\mathbb{Z}(\iota), +, \times)$ is a commutative ring with unity.

Example 336 (The ring $\mathbb{Z}_k$). The finite set

$$\mathbb{Z}_k = \{0, 1, \ldots, k-1\},$$

with respect to the addition modulo k "$+_k$" and multiplication modulo "$\times_k$" forms a commutative ring with unity $1 \in \mathbb{Z}_k$.

Example 337 (The set of polynomials). The set of polynomials

$$p(z) = u_0 + u_1 z + \cdots + u_k z^k, \quad k \in \mathbb{N}_0,$$

where coefficients $u_0, u_1, \ldots, u_k \in \mathbb{R}$ or $\mathbb{C}$, form a commutative ring with unity with respect to the addition of polynomial

$$p(z) + q(z) = (u_0 + v_0) + (u_1 + v_1) z + \cdots + (u_k + v_k) z^k,$$

and multiplication

$$p(z) \cdot q(z) = \sum_{i,j=0}^{k} u_i v_j z^{i+j},$$

where

$$q(z) = v_0 + v_1 z + \cdots + v_k z^k, \quad k \in \mathbb{N}_0.$$

Example 338 (The ring of square matrices of order p). The set of all p-square real or complex matrices forms a ring under usual matrix addition and multiplication. The matrix multiplication in general does not commute, and hence it is a noncommutative ring. Identity matrix $[I]_{p\times p}$ is there so it becomes a noncommutative ring with unity.

Example 339 (The ring of even integers). As we know, an infinite set of integers forms a commutative ring with unity with usual addition and multiplication. Similarly, the set of even integers (integers divisible by 2) forms a commutative ring without unity with respect to usual addition and multiplication.

As we have seen, a ring is Abelian with respect to operation $\oplus$ or simply say addition but with respect to multiplication only associative and the existence of identity holds. When additional properties with respect to multiplication are included, we get different structure. As such, if multiplication is commutative, then the ring is called a commutative ring. A commutative ring without zero divisor ($a \neq 0$, $b \neq 0$ do not implies $a\,b \neq 0$) becomes an integral domain. Further, a commutative ring where multiplicative inverse of each non-zero elements exist is called a field. We discuss the elementary definitions and the properties of a field in the succeeding section.

Field

The field is the most generalized form of a set. For the concept of a field in mathematics, the credit goes to mathematicians, Neils Henrik Abel and Évariste Galois, who used their work on a solution of equations with a rational coefficient of five or more degrees. The term "field" was first used by Ellakim Hastings Moore in 1893. Nowadays, field is studied under the field theory. The theory is enriched with beautiful topics, namely fields, extension of fields, finite fields and perfect fields to name a few.

A set F with two binary operations "+" and "·" is called a field if

1. $(F, +)$ is Abelian, i. e.,
   (i) The operation "+" is associative;
   (ii) Additive identity be there;
   (iii) Additive inverse of each element should exist;
   (iv) Addition must be commutative.
2. $(\mathbb{F}^*, \cdot)$ is a group, i. e.,
   (i) The operation "·" is associative;
   (ii) The identity element with respect to "·" must be there;
   (iii) Nonzero elements are invertible.
3. The operation "·" is distributive with respect to + that means

$$x \cdot (y + z) = x \cdot y + x \cdot z;$$
$$(y + z) \cdot x = y \cdot x + z \cdot x.$$

These are some fruitful examples on a field.

Example 340 (The finite set $(\mathbb{Z}_4, +_4, \times_4)$ is not a field). We have already shown that $(\mathbb{Z}_k, +_k, \times_k)$ forms a ring with respect to the addition and multiplication modulo k. So, $(\mathbb{Z}_4, +_4, \times_4)$ is a ring. Also, it has an unity element with respect to multiplication but $\mathbb{Z}_4^* = \{1, 2, 3\}$ is not a group with respect to the multiplication because $2 \times_4 2 = 0 \notin \mathbb{Z}_4$. Hence, $(\mathbb{Z}_4, +_4, \times_4)$ is not a field.

Example 341 (The infinite set $(\mathbb{Z}, +, \times)$). The set of integers with respect to addition forms an Abelian group. Also, has unity but except 1 and −1 no other nonzero elements in $\mathbb{Z}$ had a multiplicative inverse, and hence $(\mathbb{Z}, +, \times)$ is not a field.

Example 342 (The finite set $\mathbb{Z}_p$, where p is a prime). For a prime number $p \geq 2$, the set

$$\mathbb{Z}_p = \{0, 1, 2, \ldots, p - 1\}$$

endowed with addition and multiplication modulo p is a field with p elements. We denote this field by either $(\mathbb{Z}_p, +, \times)$ or $\mathbb{Z}_p$ or $\mathbb{Z} \setminus p\,\mathbb{Z}$ or $\mathbb{F}_p$ and called as a field of residue classes of integer modulo p.

Example 343 (The field of rational numbers). The infinite set of rational numbers $\mathbb{Q}$ with respect to usual addition form an Abelian group and $\mathbb{Q} \setminus \{0\}$ forms a group with multiplication. Distribution law can be verified easily. Also, the elements of $\mathbb{Q}$ commutes with respect to multiplication. Hence, the set of rational numbers is a commutative field with respect to usual addition and multiplication.

Example 344 (The field of real or complex numbers). The set of real numbers $\mathbb{R}$ is a commutative field with respect to usual addition and multiplication. Also, following the addition and multiplication of complex numbers, one can easily verify $\mathbb{C}$ is an Abelian group with addition and $\mathbb{C} - \{0\}$ is a group with multiplication. Hence, $\mathbb{C}$ is a commutative field.

Example 345 (The infinite set $\mathbb{Q}\sqrt{p}$, p is a prime). The infinite set $\mathbb{Q}\sqrt{p}$ is given by

$$\mathbb{Q}\sqrt{p} = \{u + v\sqrt{p} \mid u, v \in \mathbb{Q}\}, \quad p \geq 2 \text{ is a prime.}$$

Let $a = u_1 + v_1\sqrt{p}$, $b = u_2 + v_2\sqrt{p}$ be two elements of $\mathbb{Q}\sqrt{p}$. Then addition and multiplication are defined by

$$a + b = (u_1 + u_2) + (v_1 + v_2)\sqrt{p};$$
$$a \cdot b = (u_1 u_2 + p\, v_1 v_2) + (u_1 v_2 + u_2 v_1)\sqrt{p}.$$

Now, using the above defined addition and multiplication, one can verify that $\mathbb{Q}\sqrt{p}$ is an Abelian group with respect to addition and $[\mathbb{Q}\sqrt{p}]^*$ is a group with respect to multiplication defined above. The multiplication inverse of an element is given by the fact that

$$a\,a^{-1} = a^{-1}\,a = 1,$$

which gives $a^{-1} = u_1' + v_1'\sqrt{p}$, where

$$u_1' = \frac{u_1}{u_1^2 - p\,v_1^2}, \qquad v_1' = \frac{-v_1}{u_1^2 - p\,v_1^2}.$$

The distribution law can be verified easily. Hence, $\mathbb{Q}\sqrt{p}$ forms a commutative field with respect to usual addition and multiplication.

Example 346 (The field of quaternions). The set $\mathbb{H}$ of Hamiltonian quaternion consists the element of the form

$$q = t + u\,i + v\,j + w\,k, \quad t, u, v, w \in \mathbb{R}$$

and i, j, k are given by

$$\begin{aligned} &i \times i = 1 = j \times j = k \times k, \\ &i \times j = -j \times i = k, \quad j \times k = -k \times j = i, \quad k \times i = -i \times k = j. \end{aligned}$$

The addition and multiplication of two elements $q_1, q_2 \in \mathbb{H}$ are given by

$$\begin{aligned} q_1 + q_2 &= (t_1 + u_1\,i + v_1\,j + w_1\,k) + (t_2 + u_2\,i + v_2\,j + w_2\,k) \\ &= (t_1 + t_2) + (u_1 + u_2)\,i + (v_1 + v_2)\,j + (w_1 + w_2)\,k, \\ q_1 \cdot q_2 &= t_1\,t_2 - u_1\,u_2 - v_1\,v_2 - w_1\,w_2 \\ &\quad + (t_2\,u_1 + t_1\,u_2 + v_1\,w_2 - v_2\,w_1)\,i \\ &\quad + (t_2\,v_1 + t_1\,v_2 + u_2\,w_1 - w_2\,u_1)\,j \\ &\quad + (t_2\,w_1 + t_1\,w_2 + u_1\,v_2 - v_1\,u_2)\,k. \end{aligned}$$

Using addition and multiplication, one can verify that $(\mathbb{H}, +)$ is an Abelian group with $0 = 0 + 0\,i + 0\,j + 0\,k \in \mathbb{H}$ and $(\mathbb{H}^*, \times)$ is a noncommutative group with unity $1 = 1 + 0\,i + 0\,j + 0\,k \in \mathbb{H}$. The distribution law can be checked easily, and hence $(\mathbb{H}^*, +, \times)$ is a field.

Next, we list some properties held in a field:

1. The additive and multiplicative identity in a field is unique.
2. The additive inverse of each element and multiplicative inverse of a nonzero element is unique in a field.

3. In a field, the product of two nonzero elements is different from zero, i. e.,

$$a \neq 0, b \neq 0 \implies a \times b \neq 0.$$

4. A field is commutative if the multiplicative operation is commutative.

Finite field

A field with a finite number of elements will be known as a finite field. A finite field is also known as a Galois field. In this section, we explore basic phenomena of a finite field.

In the previous section, based on the field, we illustrated various examples, and if you examine closely, you see the field with a finite number of elements is commutative. We have the following proposition.

Proposition 347. *A finite field is necessarily commutative.*

Now, we can define a finite field or a Galois field in the following manner.

Definition 348. A finite field or Galois field $\mathbb{F}$ with additive identity 0 and multiplicative identity 1 satisfies the following axioms:
1. $\mathbb{F}$ is Abelian with respect to addition;
2. $\mathbb{F}^*$ is Abelian with respect to multiplication;
3. The multiplication is distributive with respect to addition.

A finite field with q elements is usually denoted by $\mathrm{GF}(q)$ or $\mathbb{F}_q$. The following proposition describes the possible cardinality of a finite field.

Proposition 349. *The cardinality q of a finite field necessarily has the form $q = p^k$, for some prime p and positive integer k. Also, for every prime p and positive integer, there exists a field containing p^k elements.*

Consequently, one can say there exists no finite field with a number of elements different from a prime power.

A finite field with p^k elements will be denoted by $\mathrm{GF}(p^k)$ if $k > 2$ and $\mathbb{F}_p$ if $k = 1$. Further properties of finite fields:
1. All finite fields with same cardinality are isomorphic;
2. A finite field with p elements is isomorphic to $\mathbb{Z} \setminus p\,\mathbb{Z}$;

Example 350 (A finite field of prime cardinality ($\mathbb{F}_2, \mathbb{F}_3, \mathbb{F}_5$)). We already proved that $\mathbb{Z}_p$ is a field and so are $\mathbb{F}_2$, $\mathbb{F}_3$, $\mathbb{F}_5$. $\mathbb{F}_2$ is the smallest finite field while $\mathbb{F}_3$ is the smallest finite field with an odd number of elements.

Example 351 (A field of cardinality of p^2). The set with elements of the form

$$x = u + v\sqrt{2}, \quad u, v \in \mathbb{F}_p$$

is a finite field with addition and multiplication modulo p.

Vector spaces

A set V of vectors with respect to the vector addition "+" and scalar multiplication becomes a vector space over a field $\mathbb{F}$, if it satisfies the following axioms

1. With respect to vector addition, the set V is Abelian:
   (i) The set V is closed with respect to vector addition, i. e., for any $x, y \in V$, the vector sum $x + y \in V$.
   (ii) The vector addition is associative, i. e.,

$$(x + y) + z = x + (y + z), \quad \forall x, y, z \in V.$$

   (iii) Existence of an identity, an element $e \in V$ is called the identity element if

$$a + e = e + a = a, \quad \forall a \in V.$$

   (iv) Existence of inverse, for each element $a \in V$ there exists an element $b \in V$ such that

$$a + b = b + a = e,$$

   then b is the inverse of a and vice versa.
   (v) The elements of V must follow the commutative law, i. e.,

$$a + b = b + a, \quad \forall a, b \in V.$$

2. With respect to scalar multiplication:
   (i) The set V is closed with respect to scalar multiplication. This means for every $\alpha \in \mathbb{F}$ and $v \in V$, the scalar multiplication of $\alpha v \in V$.
   (ii) The scalar multiplication obeys the distributive law as

$$a \cdot (\alpha + \beta) = a \cdot \alpha + a \cdot \beta, \quad \forall a \in V, \alpha, \beta \in \mathbb{F};$$
$$\alpha \cdot (a + b) = \alpha \cdot a + \alpha \cdot b, \quad \forall a, b \in V, \alpha \in \mathbb{F}.$$

   (iii) For all vectors $v \in V$, there is a unity such that $1 \cdot v = v \cdot 1 = v$.

We denote a vector space over field $\mathbb{F}$ by $V(\mathbb{F})$. Let us see the following list of examples of vector spaces.

Example 352. Let $\mathbb{F}$ be any field and $\mathbb{F}^p$ denotes the set of all p-tuples $(u_1, \dots, u_p)$ of elements in $\mathbb{F}$. Then $\mathbb{F}^p$ forms a vector space over $\mathbb{F}$ with respect to the following operations:

1. *Vector addition:*

$$(u_1, \dots, u_p) + (v_1, \dots, v_p) = (u_1 + v_1, \dots, u_p + v_p);$$

2. *Scalar multiplication:*

$$\lambda (u_1, \dots, u_p) = (\lambda u_1, \dots, \lambda u_p).$$

Example 353. Let $P(u)$ be the space of all polynomials of the form

$$p(u) = a_0 + a_1 u + a_2 u^2 + \cdots + a_p u^p, \quad p = 1, 2, \dots,$$

where the coefficients are from a field $\mathbb{F}$. Then $P(u)$ forms a vector space over $\mathbb{F}$ with respect to the following operations:

1. *Addition:*

$$\begin{aligned} p(u) + q(u) &= (a_0 + a_1 u + a_2 u^2 + \cdots + a_p u^p) + (b_0 + b_1 u + b_2 u^2 + \cdots + b_p u^p) \\ &= (a_0 + b_0) + (a_1 + b_1) u + (a_2 + b_2) u^2 + \cdots + (a_p + b_p) u^p; \end{aligned}$$

2. *Scalar multiplication:*

$$\lambda p(u) = \lambda a_0 + \lambda a_1 u + \lambda a_2 u^2 + \cdots + \lambda a_p u^p.$$

Example 354. The set of all matrices of order $p \times n$ with entries from a field $\mathbb{F}$ forms a vector space with respect to usual addition of matrices and scalar multiplication.

Example 355. Let f be a function defined from a set U to a field $\mathbb{F}$ and $F(u)$ denotes the collection of such function. Then $F(u)$ forms a vector space over $\mathbb{F}$ with respect to the following operations:

1. *Vector addition:*

$$(f + g)(u) = f(u) + g(u), \quad u \in U;$$

2. *Scalar multiplication:*

$$(\lambda f)(u) = \lambda f(u), \quad \forall u \in U.$$

Bibliography

[1] H. Abd-Elmageed, M. Abdalla, M. Abul-Ez, N. Saad, *Some results on the first Appell matrix function*. Linear Multilinear Algebra 68 (2020), no. 2, 278–292.

[2] M. Abdalla, *Special matrix functions: characteristics, achievements and future directions*. Linear Multilinear Algebra 68 (2020), no. 1, 1–28.

[3] M. Abramowitz, I. A. Stegun, *Handbook of mathematical functions with formulas, graphs, and mathematical tables*. For sale by the superintendent of documents, National Bureau of Standards applied mathematics series, 55. U. S. Government Printing Office, Washington, DC, 1964. xiv+1046 pp.

[4] A. Altin, B. Cekim, R. Sahin, *On the matrix versions of Appell hypergeometric functions*. Quaest. Math. 37 (2014), no. 1, 31–38.

[5] G. E. Andrews, R. Askey, R. Roy, *Special functions*. Encyclopedia of mathematics and its applications, 71. Cambridge University Press, Cambridge, 1999.

[6] P. Appell, J. Kampé de Fériet, *Fonctions hypergéometriques et hypersphériques. Polynômes d'Hermite*. Gauthier-Villars, Paris, 1926.

[7] R. Askey, *Orthogonal polynomials and special functions*. Society for Industrial and Applied Mathematics, Philadelphia, PA, 1975. vii+110 pp.

[8] W. N. Bailey, *Generalized hypergeometric series*. Cambridge University Press, Cambridge, 1935.

[9] A. Baker, *Matrix groups. An introduction to Lie group theory*. Springer undergraduate mathematics series. Springer-Verlag London, Ltd., London, 2002. xii+330 pp.

[10] R. Bellman, *Introduction to matrix analysis*. McGraw-Hill, New York, 1969.

[11] M. Bohner, T. Cuchta, *The Bessel difference equation*. Proc. Am. Math. Soc. 145 (2017), no. 4, 1567–1580.

[12] M. Bohner, T. Cuchta, *The generalized hypergeometric difference equation*. Demonstr. Math. 51 (2018), no. 1, 62–75.

[13] C. Brezinski, *Other manifestations of the Schur complement*. Linear Algebra Appl. 111 (1988), 231–247.

[14] R. Bru, L. Elsner, M. Neumann, *Convergence of infinite products of matrices and inner-outer iteration schemes*. Electron. Trans. Numer. Anal. 2 (1994), 183–193.

[15] I. M. Burban, A. U. Klimyk, *(p,q)-Differentiation, (p,q)-integration and (p,q)-hypergeometric functions related to quantum groups*. Integral Transforms Spec. Funct. 2 (1994), 15–36.

[16] J. L. Burchnall, T. W. Chaundy, *Expansions of Appell's double hypergeometric functions*. Q. J. Math., Oxf. Ser. 11, (1940). 249–270.

[17] R. W. Butler, A. T. A. Wood, *Laplace approximations for hypergeometric functions with matrix argument*. Ann. Stat. 30 (2002), no. 4, 1155–1177.

[18] R. W. Butler, A. T. A. Wood, *Laplace approximation for Bessel functions of matrix argument*. J. Comput. Appl. Math. 155 (2003), no. 2, 359–382.

[19] R. W. Butler, A. T. A. Wood, *Laplace approximations to hypergeometric functions of two matrix arguments*. J. Multivar. Anal. 94 (2005), no. 1, 1–18.

[20] R. Chakrabarti, R. Jagannathan, *A (p,q)-oscillator realization of two-parameter quantum algebras*. J. Phys. A 24 (1991), no. 13, L711–L718.

[21] M. A. Chaudhry, A. Qadir, *Incomplete exponential and hypergeometric functions with applications to the non central χ^2-distribution*. Commun. Stat., Theory Methods 34 (2005), no. 3, 525–535.

[22] M. A. Chaudhry, A. Qadir, M. Rafique, S. M. Zubair, *Extension of Euler's beta function*. J. Comput. Appl. Math. 78 (1997), 19–32.

[23] M. A. Chaudhry, A. Qadir, H. M. Srivastava, R. B. Paris *Extended hypergeometric and confluent hypergeometric functions*. Appl. Math. Comput. 159 (2004), no. 2, 589–602.

[24] M. A. Chaudhry, N. M. Temme, E. J. M. Veling, *Asymptotic and closed form of a generalized incomplete gamma function*. J. Comput. Appl. Math. 67 (1996), 371–379.

[25] M. A. Chaudhry, S. M. Zubair, *Generalized incomplete gamma functions with applications*. J. Comput. Appl. Math. 55 (1994), 99–124.

https://doi.org/10.1515/9783111324586-011

[26] M. A. Chaudhry, S. M. Zubair, *On the decomposition of generalized incomplete gamma functions with applications of Fourier transforms*. J. Comput. Appl. Math. 59 (1995), no. 3, 253–284.

[27] M. A. Chaudhry, S. M. Zubair, *On a class of incomplete functions with applications*. CRC Press (Chapman and Hall), Boca Raton, FL, 2002.

[28] Y. Chikuse, *Generalized Hermite and Laguerre polynomials in multiple symmetric matrix arguments and their applications*. Linear Algebra Appl. 176 (1992), 261–287.

[29] Y. Chikuse, *Properties of Hermite and Laguerre polynomials in matrix argument and their applications*. Linear Algebra Appl. 176 (1992), 237–260.

[30] Y. Chikuse, *Hermite and Laguerre polynomials with complex matrix arguments*. Linear Algebra Appl. 388 (2004), 91–105.

[31] J. Choi, A. K. Rathie, R. K. Parmar, *Extension of extended beta, hypergeometric and confluent hypergeometric functions*. Honam Math. J. 36 (2014), no. 2, 357–385.

[32] A. G. Constantine, *Some non-central distribution problems in multivariate analysis*. Ann. Math. Stat. 34 (1963), 1270–1285.

[33] A. G. Constantine, R. J. Muirhead, *Partial differential equations for hypergeometric functions of two argument matrices*. J. Multivar. Anal. 2 (1972), 332–338.

[34] J. C. Cortés, L. Jódar, F. J. Solís, R. Ku-Carrillo, *Infinite matrix products and the representation of the matrix gamma function*. Abstr. Appl. Anal. 2015 (2015), Art. ID 564287, 8 pp.

[35] T. Cuchta, D. Grow, N. Wintz, *Divergence criteria for matrix generalized hypergeometric series*. Proc. Am. Math. Soc. 150 (2022), no. 3, 1235–1240.

[36] T. Cuchta, D. Grow, N. Wintz, *Discrete matrix hypergeometric functions*. J. Math. Anal. Appl. 518 (2023), no. 2, Paper No. 126716, 14 pp.

[37] T. Cuchta, R. Luketic, *Discrete hypergeometric Legendre polynomials*. Mathematics 9 (2021), 2546.

[38] W. J. Culver, *On the existence and uniqueness of the real logarithm of a matrix*. Proc. Am. Math. Soc. 17 (1966), 1146–1151.

[39] I. Daubechies, J. C. Lagarias, *Sets of matrices all infinite products of which converge*. Linear Algebra Appl. 161 (1992), 227–263.

[40] A. W. Davis, *Invariant polynomials with two matrix arguments extending the zonal polynomials: applications to multivariate distribution theory*. Ann. Inst. Stat. Math. 31 (1979), no. 3, 465–485.

[41] A. W. Davis, *Invariant polynomials with two matrix arguments, extending the zonal polynomials*. Multivariate analysis, V (Proc. Fifth Internat. Sympos., Univ. Pittsburgh, Pittsburgh, Pa., 1978), pp. 287–299. North-Holland, Amsterdam–New York, 1980.

[42] P. J. Davis, *Leonhard Euler's integral: a historical profile of the gamma function*. Am. Math. Mon. 66 (1959), 849–869.

[43] E. Defez, L. Jodar, *Chebyshev matrix polynomials and second order matrix differential equations*. Util. Math. 61 (2002), 107–123.

[44] N. Dunford, J. Schwartz, *Linear operators, Part-I*. Addison-Wesley, New York, 1957.

[45] A. J. Duran, *Markov's theorem for orthogonal matrix polynomials*. Can. J. Math. 48 (1996), no. 6, 1180–1195.

[46] A. J. Duran, P. Lopez-Rodriguez, *Orthogonal matrix polynomials: zeros and Blumenthal's theorem*. J. Approx. Theory 84 (1996), no. 1, 96–118.

[47] A. J. Durán, W. Van Assche, *Orthogonal matrix polynomials and higher-order recurrence relations*. Linear Algebra Appl. 219 (1995), 261–280.

[48] R. Dwivedi, V. Sahai, *On the hypergeometric matrix functions of two variables*. Linear Multilinear Algebra 66 (2018), no. 9, 1819–1837.

[49] R. Dwivedi, V. Sahai, *On the hypergeometric matrix functions of several variables*, J. Math. Phys. 59 (2018), no. 2, 023505, 15 pp.

[50] R. Dwivedi, V. Sahai, *On the basic hypergeometric matrix functions of two variables*. Linear Multilinear Algebra 67 (2019), no. 1, 1–19.

[51] R. Dwivedi, V. Sahai, *On certain properties and expansions of zeta matrix function, digamma matrix function and polygamma matrix function*. Quaest. Math. 43 (2020), no. 1, 97–105.

[52] R. Dwivedi, V. Sahai, *A note on the Appell matrix functions*. Quaest. Math. 43 (2020), no. 3, 321–334.

[53] A. Erdélyi, W. Magnus, F. Oberhettinger, F. G. Tricomi, *Higher transcendental functions*, Vol. I. McGraw-Hill, New York, London, 1953.

[54] R. J. Evans, *Identities for products of Gauss sums over finite fields*. Enseign. Math. (2) 27 (1981), no. 3–4, 197–209 (1982).

[55] R. J. Evans, *Character sum analogues of constant term identities for root systems*. Isr. J. Math. 46 (1983), no. 3, 189–196.

[56] R. J. Evans, *Hermite character sums*. Pac. J. Math. 122 (1986), no. 2, 357–390.

[57] H. Exton, *On certain confluent hypergeometric functions of three variables*. Ganita 21 (1970), no. 2, 79–92.

[58] H. Exton, *Multiple hypergeometric functions and applications*. Wiley, New York, 1976.

[59] H. Exton, *Hypergeometric functions of three variables*. J. Indian Acad. Math. 4 (1982) 113–119.

[60] H. Exton, *q-Hypergeometric functions and applications*. Ellis Horwood, 1983.

[61] G. B. Folland, *Fourier analysis and its applications*. Wadsworth and Brooks, Pacific Grove, CA, 1992.

[62] G. Gasper, M. Rahman, *Basic hypergeometric series*, Second edition. Encyclopedia of mathematics and its applications, 96. Cambridge University Press, Cambridge, 2004.

[63] C. F. Gauss, *Disquisitiones generales circa seriem infinitam* $1 + \frac{\alpha \cdot \beta}{1 \cdot \gamma} x + \cdots$, 2. Commentationes Societatis Regiae Scientarum Gottingensis Recentiores (in Latin), Göttingen, 1813.

[64] G. H. Ghany, *q-derivative of basic hypergeometric series with respect to parameters*. Int. J. Math. Anal. (Ruse) 3 (2009) 1617–1632.

[65] G H. Golub, C. F. Van Loan, *Matrix Computations*. Johns Hopkins Univ. Press, Baltimore, MD, 1989.

[66] J. Greene, *Hypergeometric functions over finite fields*. Trans. Am. Math. Soc. 301 (1987), no. 1, 77–101.

[67] J. Greene, D. Stanton, *A character sum evaluation and Gaussian hypergeometric series*. J. Number Theory 23 (1986), no. 1, 136–148.

[68] J. R. Greene, *Character sum analogues for hypergeometric and generalized hypergeometric functions over finite fields*, Thesis (Ph. D.), University of Minnesota. ProQuest LLC, Ann Arbor, MI, 1984. 133 pp.

[69] K. I.Gross, D. St. P. Richards, *Special functions of matrix argument. I. Algebraic induction, zonal polynomials, and hypergeometric functions*. Trans. Am. Math. Soc. 301 (1987), no. 2, 781–811.

[70] A. K. Gupta, D. K. Nagar, *Matrix-variate beta distribution*. Int. J. Math. Math. Sci. 24 (2000), no. 7, 449–459.

[71] A. K. Gupta, D. K. Nagar, *Matrix variate distributions*. Chapman & Hall/CRC monographs and surveys in pure and applied mathematics, 104. Chapman & Hall/CRC, Boca Raton, FL, 2000. x+367 pp.

[72] A. K. Gupta, D. K. Nagar, *Matrix variate generalization of a bivariate beta type* 1 *distribution*. J. Stat. Manag. Syst. 12 (2009), no. 5, 873–885.

[73] A. K. Gupta, D. K. Nagar, *Properties of matrix variate beta type* 3 *distribution*. Int. J. Math. Math. Sci. 2009 (2009), Art. ID 308518, 18 pp.

[74] A. K. Gupta, D. K. Nagar, *Some bimatrix beta distributions*. Commun. Stat., Theory Methods 41 (2012), no. 5, 869–879.

[75] A. K. Gupta, D. K. Nagar, *Matrix-variate Gauss hypergeometric distribution*. J. Aust. Math. Soc. 92 (2012), no. 3, 335–355.

[76] E. Heine, *Untersuchungen über die Reihe* $1 + \frac{(1-q^\alpha)(1-q^\beta)}{(1-q)(1-q^\gamma)} x + \frac{(1-q^\alpha)(1-q^{\alpha+1})(1-q^\beta)(1-q^{\beta+1})}{(1-q)(1-q^2)(1-q^\gamma)(1-q^{\gamma+1})} x^2 + \cdots$. J. Reine Angew. Math. 34 (1847), 285–328.

[77] C. S. Herz, *Bessel functions of matrix argument*. Ann. Math. 61 (1955), 474–523.

[78] I. Higueras, B. Garcia-Celayeta, *Logarithmic norms for matrix pencils*. SIAM J. Matrix Anal. Appl. 20 (1999), 646–666.

[79] O. Holtz, *On convergence of infinite matrix products*. Electron. J. Linear Algebra 7 (2000), 178–181.

[80] R. A. Horn, C. R. Johnson, *Matrix analysis*. Cambridge University Press, Cambridge, 1985. xiii+561 pp.

[81] G. D. Hu, M. Liu, *The weighted logarithmic matrix norm and bounds of the matrix exponential*. Linear Algebra Appl. 390 (2004), 145–154.

[82] P. Humbert, *The confluent hypergeometric functions of two variables*. Proc. R. Soc. Edinb. 41 (1920–21), 73–96.

[83] K. F. Ireland, M. I. Rosen, *A classical introduction to modern number theory. Revised edition of Elements of number theory*. Graduate texts in mathematics, 84. Springer-Verlag, New York–Berlin, 1982. xiii+341 pp.

[84] F. H. Jackson, *On basic double hypergeometric functions*. Q. J. Math., Oxf. Ser. 13, (1942). 69–82.

[85] F. H. Jackson, *Basic double hyper-geometric functions. II*. Q. J. Math., Oxf. Ser. 15, (1944). 49–61.

[86] R. Jagannathan, *(P, Q)-special functions*. Special functions and differential equations (Madras, 1997), pp. 158–164. Allied Publ., New Delhi, 1998.

[87] A. T. James, *Zonal polynomials of the real positive definite symmetric matrices*. Ann. Math. (2) 74 (1961), 456–469.

[88] A. T. James, *Distributions of matrix variates and latent roots derived from normal samples*. Ann. Math. Stat. 35 (1964), 475–501.

[89] A. T. James, *Special functions of matrix and single argument in statistics. Theory and application of special functions*. Proc. Advanced Sem., Math. Res. Center, Univ. Wisconsin, Madison, Wis., 1975, pp. 497–520. Math. Res. Center, Univ. Wisconsin, Publ., 35. Academic Press, New York, 1975.

[90] L. Jódar, *Explicit solutions for second order operator differential equations with two boundary value conditions*. Linear Algebra Appl. 103 (1988), 73–86.

[91] L. Jódar, R. Company, *Hermite matrix polynomials and second order matrix differential equations*. Approx. Theory Appl. (N. S.) 12 (1996), no. 2, 20–30.

[92] L. Jódar, R. Company, E. Navarro, *Laguerre matrix polynomials and systems of second-order differential equations*. Appl. Numer. Math. 15 (1994), no. 1, 53–63.

[93] L. Jódar, R. Company, E. Ponsoda *Orthogonal matrix polynomials and systems of second order differential equations*. Differ. Equ. Dyn. Syst. 3 (1995), no. 3, 269–288.

[94] L. Jódar, J. C. Cortés, *Some properties of gamma and beta matrix functions*. Appl. Math. Lett. 11 (1998), no. 1, 89–93.

[95] L. Jódar, J. C. Cortés, *On the hypergeometric matrix function*. Proceedings of the VIIIth symposium on orthogonal polynomials and their applications (Seville, 1997). J. Comput. Appl. Math. 99 (1998), no. 1–2, 205–217.

[96] L. Jódar, J. C. Cortés, *Closed form general solution of the hypergeometric matrix differential equation*. Mathematical and computer modelling in engineering sciences. Math. Comput. Model. 32 (2000), no. 9, 1017–1028.

[97] L. Jódar, E. Defez, *A connection between Laguerre's and Hermite's matrix polynomials*. Appl. Math. Lett. 11 (1998), no. 1, 13–17.

[98] L. Jódar, J. Sastre, *On Laguerre matrix polynomials*. Util. Math. 53 (1998), 37–48.

[99] V. Kac, P. Cheung, *Quantum calculus*. Springer, Berlin (2001).

[100] M. A. Khan, *Discrete hypergeometric functions and their properties*. Demonstr. Math. 30, (1997), no. 3, 537–544.

[101] C. G. Khatri, *On Certain distribution problems based on positive definite quadratic functions in normal vectors*. Ann. Math. Stat. 37 (1966), 468–479.

[102] P. Koev, A. Edelman, *The efficient evaluation of the hypergeometric function of a matrix argument*. Math. Comput. 75 (2006), no. 254, 833–846.

[103] C. Krattenthaler, H. M. Srivastava, *Summations for basic hypergeometric series involving a q-analogue of the digamma function*. Comput. Math. Appl. 32 (1996), no. 3, 73–91.

[104] E. E. Kummer, *Über die hypergeometrische Reihe*. J. Reine Angew. Math. 15 (1836), 39–83 (German).

[105] E. E. Kummer, *Über die hypergeometrische Reihe. (Fortsetzung)*. J. Reine Angew. Math. 15 (1836), 127–172 (German).

[106] G. Lauricella, *Sulle funzioni ipergeometriche a piú variabili*. Rend. Circ. Mat. Palermo 7 (1893), 111–158.

[107] N. N. Lebedev, *Special functions and their applications*. Revised edition, translated from the Russian and edited by Richard A. Silverman. Unabridged and corrected republication. Dover Publications, Inc., New York, 1972. xii+308 pp.

[108] D. M. Lee, A. K. Rathie, R. K. Parmar, Y. S. Kim, *Generalization of extended beta function, hypergeometric and confluent hypergeometric functions*. Honam Math. J. 33 (2011), 187–206.

[109] R. Lidl, H. Niederreiter, *Finite fields*. With a foreword by P. M. Cohn. Encyclopedia of mathematics and its applications, 20. Addison-Wesley Publishing Company, Advanced Book Program, Reading, MA, 1983. xx+755 pp.

[110] Y. L. Luke, *The special functions and their approximations*, Vol. II. Mathematics in science and engineering, 53. Academic Press, New York–London, 1969. xx+485 pp.

[111] A. M. Mathai, *Some results on functions of matrix argument*. Math. Nachr. 84 (1978), 171–177.

[112] A. M. Mathai, *Appell's and Humbert's functions of matrix arguments*. Linear Algebra Appl. 183 (1993), 201–221.

[113] A. M. Mathai, *Lauricella functions of real symmetric positive definite matrices*. Indian J. Pure Appl. Math. 24 (1993), no. 9, 513–531.

[114] A. M. Mathai, *Jacobians of matrix transformations and functions of matrix argument*. World Scientific Publishing Co., Inc., River Edge, NJ, 1997.

[115] A. M. Mathai, G. Pederzoli, *Some transformations for functions of matrix arguments*. Indian J. Pure Appl. Math. 27 (1996), no. 3, 277–284.

[116] A. M. Mathai, G. Pederzoli, *A Whittaker function of matrix argument*. Linear Algebra Appl. 269 (1998), 91–103.

[117] R. J. Muirhead, *Expressions for some hypergeometric functions of matrix argument with applications*. J. Multivar. Anal. 5 (1975), no. 3, 283–293.

[118] R. J. Muirhead, *Aspects of multivariate statistical theory*. Wiley series in probability and mathematical statistics. John Wiley & Sons, Inc., New York, 1982. xix+673 pp.

[119] S. Nadarajah, *A note on: "Incomplete exponential and hypergeometric functions with applications to the non central χ^2-distribution [Comm. Statist. Theory Methods 34 (2005), no. 3, 523--535; MR2181199] by M. A. Chaudhry and A. Qadir*. Arab. J. Sci. Eng., Sect. A Sci. 32 (2007), no. 2, 223–224.

[120] D. K. Nagar, S. A. Gómez-Noguera, A. Gómez, *Integrals of functions of matrix arguments and their applications*. Int. J. Math. Anal. 11 (2017), no. 10, 479–492.

[121] D. K. Nagar, A. K. Gupta, *Matrix-variate Kummer-beta distribution*. J. Aust. Math. Soc. 73 (2002), no. 1, 11–25.

[122] D. K. Nagar, S. Nadarajah, *Appell's hypergeometric functions of matrix arguments*. Integral Transforms Spec. Funct. 28 (2017), no. 2, 91–112.

[123] P. Njionou Sadjang, *On two (p,q)-analogues of the Laplace transform*. J. Differ. Equ. Appl. 23 (2017), no. 9, 1562–1583.

[124] P. Njionou Sadjang, *On the fundamental theorem of (p,q)-calculus and some (p,q)-Taylor formulas*. Results Math. 73 (2018), no. 1, Paper No. 39, 21 pp.

[125] P. Njionou Sadjang, *On the (p,q)-Gamma and the (p,q)-Beta functions*, https://arxiv.org/pdf/1506.07394.

[126] E. Özergin, M. A. Özarslan, A. Altin, *Extension of gamma, beta and hypergeometric functions*. J. Comput. Appl. Math. 235 (2011), no. 16, 4601–4610.

[127] L. S. Pontrjagin, *Topological groups*. Princeton Univ. Press, Princeton, New Jersey, 1958.

[128] E. D. Rainville, *Special functions*. Chelsea, New York, 1960.

[129] D. Richards, Q. Zheng, *A reflection formula for the Gaussian hypergeometric function of matrix argument*, 2023, arXiv:2002.05248.

[130] V. Sahai, A. Verma, *nth-order q-derivatives of multivariable q-hypergeometric series with respect to parameters*. Asian-Eur. J. Math. 7 (2014), no. 2, 1450019, 29 pp.

[131] V. Sahai, S. Yadav, *On models of certain p, q-algebra representations: the quantum Euclidean algebra $\mathcal{E}_{p,q}(2)$*. J. Math. Anal. Appl. 308 (2008), 1043–1053.

[132] A. Salem, *On a q-gamma and a q-beta matrix functions*. Linear Multilinear Algebra 60 (2012), no. 6, 683–696.

[133] A. Salem, *Some properties and expansions associated with the q-digamma function*. Quaest. Math. 36 (2013), no. 1, 67–77.

[134] A. Salem, *The basic Gauss hypergeometric matrix function and its matrix q-difference equation*. Linear Multilinear Algebra 62 (2014), no. 3, 347–361.

[135] S. Saran, *Hypergeometric functions of three variables*. Ganita 5 (1954), 71–91.

[136] J. Sastre, E. Defez, *On the asymptotics of Laguerre matrix polynomials for large x and n*. Appl. Math. Lett. 19 (2006), no. 8, 721–727.

[137] J. Sastre, L. Jódar, *Asymptotics of the modified Bessel and the incomplete gamma matrix functions*. Appl. Math. Lett. 16 (2003), no. 6, 815–820.

[138] R. K. Saxena, P. L. Sethi, O. P. Gupta, *Appell functions of matrix arguments*. Indian J. Pure Appl. Math. 28 (1997), no. 3, 371–380.

[139] J. B. Seaborn, *Hypergeometric functions and their applications*. Springer, New York, 1991.

[140] L. J. Slater, *Generalized hypergeometric functions*. Cambridge University Press, Cambridge, London and New York, 1966.

[141] F. Soleyman, P. Njionou Sadjang, M. Masjed-Jamei, I. Area, *(p, q)-Sturm–Liouville problems and their orthogonal solutions*. Math. Methods Appl. Sci. 41 (2018), no. 18, 8997–9009.

[142] H. M. Srivastava, *Hypergeometric functions of three variables*. Ganita 15 (1964), 97–108.

[143] H. M. Srivastava, *Some integrals representing triple hypergeometric functions*. Rend. Circ. Mat. Palermo 16 (1967), no. 2, 99–115.

[144] H. M. Srivastava, M. A. Chaudhry, R. P. Agarwal, *The incomplete Pochhammer symbols and their applications to hypergeometric and related functions*. Integral Transforms Spec. Funct. 23 (2012), no. 9, 659–683.

[145] H. M. Srivastava, J. Choi, *Series associated with the zeta and related functions*. Kluwer Academic Publishers, Dordrecht, 2001.

[146] H. M. Srivastava, J. Choi, *Zeta and q-Zeta functions and associated series and integrals*. Elsevier, Inc., Amsterdam, 2012.

[147] H. M. Srivastava, P. W. Karlsson, *Multiple Gaussian hypergeometric series*. Ellis Horwood Limited, Chichester, 1985.

[148] H. M. Srivastava, H. L. Manocha, *A treatise on generating functions*. John Wiley and Sons, 1984.

[149] R. Srivastava, *Some properties of a family of incomplete hypergeometric functions*. Russ. J. Math. Phys. 20 (2013), no. 1, 121–128.

[150] R. Srivastava, N. E. Cho, *Generating functions for a certain class of incomplete hypergeometric polynomials*. Appl. Math. Comput. 219 (2012), no. 6, 3219–3225.

[151] A. Takemura, *Zonal polynomials*, Institute of Mathematical Statistics lecture notes—monograph series, 4. Institute of Mathematical Statistics, Hayward, CA, 1984. ii+104 pp.

[152] N. M. Temme, *Special functions. An introduction to the classical functions of mathematical physics*. A Wiley-Interscience Publication. John Wiley & Sons, Inc., New York, 1996. xiv+374 pp.

[153] A. J. Tirao, *The matrix-valued hypergeometric equation*. Proc. Natl. Acad. Sci. USA 100 (2003), no. 6, 8138–8141.

[154] L. N. Trefethen, M. Embree, *Spectra and pseudospectra: the behaviour of nonnormal matrices and operators*. Princeton University Press, Princeton, NJ, 2005.

[155] W. F. Trench, *Invertibly convergent infinite products of matrices, with applications to difference equations*. Comput. Math. Appl. 30 (1995), no. 11, 39–46.

[156] C. Van Loan, *The sensitivity of the matrix exponential*. SIAM J. Numer. Anal. 14 (1977), no. 6, 971–981.

Index

https://doi.org/10.1515/9783111324586-012

THE WRITE WAY

Dein Ghostwriter als Guide durch den Buch-Dschungel

Lisa Keskin

There's no wrong way ... there's only the Write Way!

ost

Umschlaggestaltung und Innenlayout: Lisa Keskin

Lektorat: Conny Strumberger-Sellner

Endlektorat und Projektmanagement: Monika Lexa

Druck und Vertrieb im Auftrag der Autorin:
Buchschmiede von Dataform Media GmbH, Wien

ISBN Softcover: 978-3-99139-100-5

ISBN Hardcover: 978-3-99139-099-2

ISBN E-Book: 978-3-99165-832-0

PRINTED IN
AUSTRIA

This is for you

Wenn du in dir den Drang zu schreiben spürst, dann tu es. Auch, wenn du irgendwann einmal gehört hast, dass du es nicht kannst.

Vielleicht schaffst du es ganz allein – du wärst nicht die Erste. Es gibt viele, die diesen Weg bereits hinter sich haben.

Vielleicht brauchst du auch ab einem gewissen Punkt Unterstützung – da befindest du dich ebenfalls in bester Gesellschaft.

Und vielleicht schlummert etwas in dir, das es wert ist, erweckt zu werden. Ganz ehrlich? Ich bin sogar ziemlich sicher, dass das so ist!

Also trau dich, schreib, fang an, tu es!

Meine Tipps:

Wirf niemals, niemals etwas Geschriebenes weg!
Und wirf vor allem niemals, niemals das Handtuch, nur weil du glaubst, es nicht zu können!

Ich wünsch dir viel Freude beim Lesen und noch mehr Freude beim Umsetzen.

INHALT

01
Idee
02
Erst-
gespräch
Ziel-
Workshop
12
Druck
Veröffent-
lichung
11
Upload
Verlag
Freigabe
Grafik
10
End-
Lektorat
09
Cover
Layout
Grafik
07

EINLEITUNG

Angebot Konzept
Auftrags-erteilung
Lese-probe
04
Feedback
05
Feedback-schleifen
Über-arbeitung Manuskript
06
Lektorat Korrektorat
08

Einleitung

Warum dieses Buch?

„Waaaas? Die hat das Buch gar nicht selbst geschrieben? Ist das nicht Betrug?“

„Wenn ich meine Idee einem Ghostwriter gebe, wie kann ich sicher sein, dass sie nicht gestohlen wird?“

„Ghostwriting ist, wenn sich jemand seine Diplomarbeit von jemand anderem schreiben lässt!“

„Ghostwriten lassen? Das ist doch hochgradig unmoralisch!“

Diese und viele ähnliche Aussagen habe ich im Laufe der vergangenen Jahre gehört. Doch das ist nur die eine Seite der Medaille. Auf der anderen ist es noch immer für viele Menschen spannend, wenn ich mich als Ghostwriter zu erkennen gebe, mit der Erläuterung: „Ich schreibe Bücher für andere Menschen, speziell für Unternehmerinnen und Selbstständige!“

Meist prasseln da jede Menge Fragen auf mich ein. *„Wie kann man für jemand anderen schreiben, geht das überhaupt?“ „Stört es dich nicht, wenn dein Name nicht auf dem Buchdeckel steht?“ „Hast du schon viele Bestseller geschrieben?“*

Mit anderen Worten: Das Interesse ist groß, aber viele Menschen verlieren sich im Dschungel der Möglichkeiten und Ideen. Daher habe ich mich entschieden, ein Buch darüber zu schreiben, das hoffentlich die dringlichsten Fragen beantwortet – und dir so nebenbei die Gelegenheit gibt, die Hintergründe des Ghostwritings ein bisschen kennenzulernen. Außerdem habe ich noch einen anderen Grund, dieses Buch zu schreiben, der mir fast ein bisschen peinlich ist.

Ich predige meinen Kundinnen immer, sie mögen doch endlich ihr Expertenbuch schreiben. Einfach aus dem Grund, damit sie auf dem Markt besser wahrgenommen und noch mehr als Expertin anerkannt werden.

Ich selbst habe viele Anläufe genommen, mein eigenes Buch zu schreiben, aber es ist mir so ergangen wie dem Schuster mit den kaputten Leisten: Alle anderen Projekte hatten Vorrang.

Und so habe ich es geschafft, in nicht einmal zehn Jahren weit über 100 Bücher zu schreiben, zu betreuen und in die Buchhandlungen zu bringen – nur mein eigenes ist immer auf der Strecke geblieben.

Wie du siehst, hat sich das hiermit geändert. Und weißt du was? Es ist deswegen kein einziges meiner zahlreichen Kundenprojekte liegengeblieben oder schlechter behandelt worden!

Das "Genderg'satzl"

Ach so, fast hätte ich es vergessen. Aus Gründen der Lesbarkeit ist dieses Buch durchgehend in der weiblichen Form geschrieben. Personen aller anderen Geschlechter mögen sich in aller Liebe und Wertschätzung mit angesprochen fühlen.

Es gibt eine Ausnahme: Das Wort „**Ghostwriter**" gendere ich nicht durchgehend. Es kommt aus dem Englischen, und ich finde, dass es in diesem Fall nicht zwingend in der weiblichen Form geschrieben werden sollte.

Das heißt: Ich verwende die weibliche Form nur dann, wenn das Wort mit einem weiblichen Artikel (oder mit mir persönlich) verbunden ist.

Wie du siehst, falle ich schon mal gerne aus dem Rahmen.

Über die Autorin

Lisa Keskin

Beruf: Ghostwriterin, BuchMacherin und Buchmentorin

Persönliches:

- ★ Hörbuch-Addict
- ★ Espressotrinkerin
- ★ Fauler Technikfreak
- ★ Lehrgangsleiterin der Ghostwriting Academy in Wien
  Ich habe gemeinsam mit Monika Lexa in den vergangenen Jahren über 30 Ghostwriter ausgebildet, von denen viele heute eng mit uns zusammenarbeiten.

Mein Motto lautet „MITeinander statt GEGENeinander".

Meine Special Effects:

Als klassischer „Bunter Vogel" habe ich die letzten 35 Jahre damit verbracht, neue Sachen zu lernen. Von Naturheilkunde über Tätowieren bis hin zu Reinkarnationstherapie.

In meinem heutigen Beruf kann ich allerdings meine abgeschlossene Ausbildung im Bereich Werbegrafik und Design am besten anwenden: Liebend gerne entwerfe ich Layouts und Covers und mache Bücher, wie auch das dir hier vorliegende, druckfertig.

Und wenn es sein muss – aber nur im höchsten Notfall! – kann ich meinen Kundinnen ihr Buchcover auf den Bauch tätowieren. ;-)

Was gibt es sonst über mich zu sagen?

Nachdem zu meinen diversen Ausbildungen auch die zur diplomierten Mental- und Resilienztrainerin gehört – das ist als Ghostwriter nie verkehrt! –, habe ich mich intensiv mit den Werten auseinandergesetzt, die mich ausmachen und die ich auch in Kundenbeziehungen lebe.

★ Ehrlichkeit

In meiner Arbeit lege ich großen Wert auf einen humorvollen, sanften, wertschätzenden und vor allem ehrlichen Umgang miteinander. Gegenseitiges Vertrauen und im Zweifelsfall immer vom Besten auszugehen, gehören für mich zum Business ebenso wie zum Privaten. Ich möchte glauben, dass die Welt im Großen und Ganzen gut ist – auch, wenn manchmal etwas Unangenehmes passiert.

★ Klarheit

Ich tue mein Bestes, Klarheit und Transparenz zu vermitteln und meine Kundinnen – soweit diese das möchten – aktiv in die Entstehung des Buches miteinzubeziehen. Umgekehrt brauche ich diese Klarheit, um gut arbeiten zu können. Ich kann nicht schreiben, wenn in meinem Kopf die Nebelsuppe der Fragezeichen herumwabert.

★ Respekt

ist etwas, das ich gebe und einfordere. Respekt vor meiner Zeit, den immer weniger werdenden Tagen und Stunden, die ich noch auf dieser Welt habe. Respekt vor meiner Einstellung und Meinung.

Umgekehrt respektiere ich auch die Meinung meiner Kundinnen, nehme nicht mehr ihrer Zeit in Anspruch als unbedingt notwendig und gebe mir Mühe, meist freundlich und immer wertschätzend zu sein.

★ Humor!

Vorletzter Punkt – aber beileibe nicht der unwichtigste! Ich lache gerne, ich liebe Menschen mit Humor, und ich brauche den Spaßfaktor. Auch in meiner Arbeit.

Denn man liest es heraus, wenn ein Buch mit Liebe und Lachen entstanden ist. Ebenso wie es zwischen den Zeilen spürbar ist, wenn es, wie wir in der Steiermark sagen, ein „G'wirks“ war, es fertigzustellen. Und deswegen gibt es meinen

★ Soul Tribe

Als meinen „Soul Tribe“ bezeichne ich jene Lebewesen, die die Welt ein bisschen anders sehen und wahrnehmen, und jede Person so sein lassen, wie sie ist.

Die mit der fünfspurigen Autobahn im Kopf, die manchmal vor Ideen bersten, Visionen haben und Großes erreichen wollen, auf der anderen Seite jedoch auch einfach glücklich sein können, wenn sie mit einer Tasse Kaffee in der Sonne sitzen.

Und jene, die nicht immer klaglos „funktionieren“, weil manchmal eben zu viele Eindrücke auf einmal auf sie einprasseln – zu viele Menschen oder zu viel Lärm.

So, nun aber genug von mir – hier kommt meine Co-Autorin Sascha!

Über die Co-Autorin

Sascha Beinhart

Sascha ist keine wirkliche Person (sorry, Sascha!). Sie ist eine Mischung aus vielen meiner Kundinnen, Interessentinnen und Teilnehmerinnen unserer Lehrgänge.

Sie ist wissbegierig, ein bisschen misstrauisch und möchte immer alles ganz genau wissen. Sascha ist ein wenig wie du und ich. Sie stellt all die Fragen, die vielleicht genau dir auf der Zunge brennen. Daher habe ich sie gebeten, mich bei diesem Buch zu unterstützen.

So haben wir eine gute Chance, dass sich beim Lesen all deine Fragezeichen in Rufzeichen verwandeln. Und sollten am Ende doch Fragezeichen übrigbleiben, dann hast du immer noch die Möglichkeit, mich persönlich zu kontaktieren!

Ich habe Sascha eine weibliche Identität gegeben. Sie könnte jedoch genauso gut männlich oder non-binär oder divers sein.

Mir ist es egal – Hauptsache, ich kann mit liebenswerten, freundlichen und offenen Lebewesen zusammenarbeiten!

Kleine Gebrauchsanweisung für dieses Buch

Sascha und ich werden dich durch diese Seiten begleiten. Sie hat eine Buchidee, die umgesetzt werden soll. Aber – ist Ghostwriting das Richtige für sie? Was kostet das? Gibt es Fallstricke oder gängige Irrtümer, vor denen sie sich hüten sollte? Und ist es am Ende überhaupt Saschas Buch, wenn sie nicht (alles) selbst schreibt?

> *Ich habe Teile des Buches als Gespräch zwischen Sascha und mir geschrieben. Sie stellt viele der Fragen, die mir meine Kundinnen immer wieder stellen. Wenn Sascha dran ist, erkennst du das daran, dass der Text leicht nach innen versetzt und etwas heller geschrieben ist so wie das hier.*

Zunächst gehen wir den Entstehungszyklus eines Buches durch – von der Idee bis zum fertigen Buch in deiner Hand. Danach werfen wir einen kurzen Blick darauf, was du mit deinem Buch anstellen kannst, und worauf du noch achten solltest. Und am Ende findest du eine kurze Zusammenfassung mit den häufigsten Fragen und Antworten.

Wenn du also nur schnell ein paar Lösungen suchst, dann kannst du ins Inhaltsverzeichnis schauen und dir die entsprechende Stelle heraussuchen.

Es wäre allerdings schade, wenn du das Buch auf diese Weise „lesen" würdest. Nicht nur, weil ich mir dann umsonst heftigst mein hübsches Köpfchen zerbrochen hätte, um dir einen möglichst klaren Überblick über den Ablauf deines Buchprojektes zu vermitteln, sondern auch, weil du sicherlich einiges verpasst, das du nicht auf dem Radar gehabt hast, und das dir Zeit, Geld und Nerven sparen kann.

Ich habe meine ganze Erfahrung als Ghostwriterin und Autorin in diese Seiten gestopft. Naja – nicht alle Erfahrungen! Einerseits stelle ich meine Kundinnen nicht bloß, wenn sie mal Mist bauen, andererseits kann ich ja nicht mein ganzes Pulver verschießen.

Außerdem kommen – wie du dir sicher vorstellen kannst – täglich neue Erfahrungen dazu.

Zwischendurch findest du einige Grafiken und Bilder. Manche haben einen tieferen Sinn, andere sind einfach dazu da, um die Seiten ein wenig aufzulockern und dir das Lesen schmackhafter zu machen.

Wenn du öfter umblättern kannst, hast du mehr Erfolgserlebnisse, in deinem Gehirn wird eine Dopaminausschüttung veranlasst, und du fühlst dich besser. Gern geschehen!

Wir verlegen selbst!

Ein wichtiger Punkt, den ich gleich jetzt am Anfang ansprechen möchte: Ich gehe mit dem Großteil meiner Kundinnen den Weg des On-Demand-Verlags. Das bedeutet, wir nutzen zwar die Infrastruktur eines Verlages, sind aber selbst für Cover, Innengestaltung und Lektorat zuständig

Dadurch haben wir die notwendige Freiheit, zeitlich und thematisch unabhängig zu veröffentlichen und Dinge zu wagen, für die ein traditioneller Verlag weder Ressourcen noch Interesse aufbringt.

Das bedeutet aber nicht, dass dieses Buch nur dann für dich interessant ist, wenn du den gleichen Weg gehen willst. Es bedeutet lediglich, dass du Sequenzen in diesem Buch finden wirst, in denen von Layout, Covergestaltung und Lektorat die Rede ist. Falls du vorhast, auf Verlagssuche zu gehen, überspring diese paar Seiten einfach.

Wer braucht denn heute noch Ghostwriter? Wir haben doch die KI!

> *„Ich muss dich das jetzt fragen. Nachdem ChatGPT und die anderen Programme immer besser werden – wozu brauch ich dann noch einen Ghostwriter? Kann ich nicht einfach meine Idee dort eingeben, und dann kommt ein Buch raus? Das machen andere doch auch! Kostet nichts und funktioniert sicher – oder?“*

„Haha, ja, das kannst du tun. Dann hast du ein … Buch. Oder zumindest viele Buchstaben, die teilweise sicherlich auch Sinn ergeben. Nur wird es nichts mit dir zu tun haben, dich nicht repräsentieren und nicht wirklich gut lesbar sein.

Du möchtest ein Buch, das wirklich emotional, stilistisch und sprachlich auf dich zugeschnitten ist? Dann kommst du an einem echten Menschen nicht vorbei.

Warum ich mir da so sicher bin? Ich habe den Versuch mit einer KI auch mit diesem Buch gemacht – einfach, weil ich neugierig war.

Dabei habe ich sehr schnell festgestellt, dass die Vorschläge eh okay sind, aber mit mir als Person, meinem Stil, meinem Wesen und vor allem meiner Einzigartigkeit nichts zu tun haben.

Doch genau das ist es, was dein Buch so wichtig macht: nicht, dass es geschrieben wird, sondern, dass es dich, deinen einzigartigen Standpunkt, deine Sichtweise auf ein Thema widerspiegelt.

Nur so kann es dich dabei unterstützen, dich zu zeigen. Entweder deine Expertise hervorzuheben, die dich von anderen abhebt. Oder eben deine persönliche Reise (in einer Biografie zum Beispiel) abzubilden.

Wenn du vorhast, einen Roman zu schreiben, gilt das umso mehr: Du möchtest dich ausdrücken, Bilder in den Köpfen deiner Leserinnen erzeugen, sie in deine Welt entführen?

Das geht nur, wenn du vor allem auch Emotionen deiner Leserinnen ansprichst. Nicht auf die plumpe Art, sondern sehr subtil, in feinen Zwischentönen.

Auch dein Wortwitz geht in der Arbeit mit der KI verloren. Weil sie eben – wie der Name schon sagt – künstlich ist. Sie kann nichts Neues erschaffen, nur schon Gelerntes wiederkäuen.

Etwas Einzigartiges kannst du auf diese Weise nicht in die Welt bringen! Das bedeutet nicht, dass du dir von der KI keine Ideen holen sollst, aber einen Menschen ersetzt sie nicht.“

> *„Oh! Ja, da hast du recht! So habe ich das noch nie betrachtet! Aber ... es werden derzeit sehr viele Bücher mit KI geschrieben und veröffentlicht. Gerade auf Amazon sieht man das.“*

„Ja, das ist korrekt. Deswegen ist es mittlerweile auf Amazon und anderen Plattformen, die etwas auf sich halten, obligatorisch, anzugeben, ob ein Buch mit KI geschrieben wurde. Wenn diese Bücher gleich gut wären wie selbst geschriebene, wäre das überflüssig, oder?“

> *„Du meinst, die KI wird euch Ghostwriter oder uns Autorinnen nicht so schnell ersetzen?“*

„Ich sage nicht, dass sich die Technik nicht weiterentwickelt. Aber das heißt nicht, dass sie uns in annähernd absehbarer Zeit ersetzen wird.

Eine Freundin von mir ist als Übersetzerin seit vielen Jahren mit dem Thema ‚die KI wird uns ersetzen' konfrontiert.

Sie selbst meint, dass das Übersetzen gewisser Passagen mit KI zwar recht gut funktioniert, aber das Programm in jenem Augenblick, in dem es um mehr als einen reinen Sachtext geht, auch heute noch restlos überfordert ist.

Und noch etwas kommt dazu: Was glaubst du denn, woher die KI ihre Texte nimmt, die sie dir ausspuckt? Die KI lernt ... von den Usern. Das bedeutet, wenn du nicht aufpasst und vergisst, die richtigen Häkchen zu setzen, findest du auch deine Texte irgendwann in einem anderen Buch wieder.

Daher kann ich dir von einem Buch, das rein mit KI geschrieben wurde, nur dringend abraten. Wenn du es trotzdem probieren willst, feel free! Ich bin auch später noch für dich da, wenn du es ausprobiert hast!

Ach ja – wenn du mehr darüber wissen möchtest, wann du KI einsetzen kannst und wann lieber nicht, dann rate ich dir, das Buch ‚Künstliche Intelligenz im Autorenstübchen'[1] zu lesen."

[1] Gabriele Herbst, Künstliche Intelligenz im Autorenstübchen – Von der ersten Idee bis zum Klappentext
Buchschmiede 2024, ISBN: 978-3-99165-561-9

01
Idee
02
Erst-
gespräch
Ziel-
Workshop
12
Druck
Veröffent-
lichung
11
Upload
Verlag
07
Freigabe
Grafik
10
End-
Lektorat
09
Cover
Layout
Grafik

DER ABLAUF EINES BUCH-PROJEKTS

Der Ablauf eines Buchprojekts

Die Idee – und viele schräge Gedanken

> *„So. Ich hab da eine Idee. Ich möchte endlich mein eigenes Buch schreiben oder schreiben lassen. Immerhin habe ich eine ziemliche Expertise auf meinem Gebiet. Schad' nur, dass das noch nicht so viele Leute wissen.*
>
> *Und wenn ich ein Buch habe, dann kann ich mich eh erst einmal gemütlich zurücklehnen und mich über die Tantiemen freuen."*

„Ups, Sascha! Ich fürchte, wir müssen reden! Das mit den Tantiemen ist so eine Sache ... man kann von den Tantiemen eines Buches nicht leben, sagt man. Und weißt du was? Das ist leider kein Gerücht, sondern die bittere Wahrheit. Außer, du bist Sebastian Fitzek oder jemand anderer, dessen Bücher in mehrere Sprachen übersetzt und idealerweise auch verfilmt wurden.

Solltest du also vorhaben, ein Buch zu schreiben und von dessen Verkauf zu leben, rate ich dir dringendst: Lass es. Bitte!"

> *„Aha. Wenn ich nichts damit verdiene, warum sollte ich dann Zeit und Geld investieren, um an ein eigenes Buch zu kommen? Ist doch Schwachsinn!"*

„Tja. Ganz unrecht hast du da nicht. Es ist viel Arbeit, ein Buch zu schreiben, und kurzfristig wird es dich eher ärmer als reicher machen. Du solltest es also eher als Investition betrachten – eine, die sich hundertmal lohnen kann.

Doch es ist wie bei einer Diät: Schnelle Erfolge mit einem Buch sind eine Illusion. Und wer immer sie dir verspricht, ist ein Scharlatan!

Schwachsinn ist das Schreiben eines Buches jedoch absolut nicht. Wenn du das Buch richtig nutzt, kannst du vor allem beruflich in vielerlei Hinsicht davon profitieren.

Und auch, wenn du es ‚nur' für dich selbst schreibst, wirst du viel Nutzen davon haben. Im zweiteren Fall wird der Nutzen allerdings eher ideell sein als materiell."

„Na gut, ich hab's verstanden! Ich werde nicht reich mit meinem Buch. Aber ... dann will ich wenigstens einen Bestseller schreiben – das wirst du als Ghostwriter wohl hinbekommen, oder?"

„Die Frage lautet: Was erwartest du dir von einem Bestseller?"

Was ist ein Bestseller?

Ab ca. 25.000 Stück verkaufter Exemplare gilt ein Buch heutzutage als Bestseller. Früher waren es 100.000 Stück, die verkauft werden mussten. Quelle: https://www.deutschlandfunkkultur.de/wie-charts-die-buchbranche-steuern-100.html

Bei gewissen Bestsellerlisten, wie zum Beispiel beim Spiegel, spielt zudem der Zeitraum, in dem die Bücher verkauft wurden, eine erhebliche Rolle. Das ist jedoch eine Wissenschaft für sich, auf die ich hier nicht eingehen möchte.

„Aber es gibt enorm viele Bestseller-Autorinnen, ich sehe das immer auf LinkedIn – jede dritte Person hat gefühlt ein ‚Bestseller-Autorin' unter ihrem Namen stehen. Wie haben die das denn alle geschafft?"

„Okay. Ich sehe, ich muss genauer werden. Es gibt diese offizielle Definition von Bestseller, die du im Kasten oben findest, aber auch jede Menge andere Versionen, die wesentlich mehr Menschen glücklich – wenn auch nicht wohlhabend – machen. Zum Beispiel Amazon."

„Aha, wie ist das bei Amazon?“

„Hier sieht es etwas anders aus. Extrem anders sogar. Bei Amazon gibt es einen sogenannten ‚Sales Rank'. Rang Nummer eins ist der erstrebenswerteste Wert.

Die Nummer-eins-Bücher werden zusätzlich mit einem eigenen orangefarbenen Bestseller-Badge ausgewiesen.

Die Rangfolge wird einerseits für die Unterkategorie erhoben, in der das Buch der Autorin eingegeben wurde, andererseits aber auch in der dazugehörigen Hauptkategorie. Das heißt, mit ein bisschen Taktik ist es super easy, einen Amazon-Bestseller zu haben.

Für Qualität oder gar Renommee ist jedoch nicht automatisch gesorgt. Leider werden die Bücher von Amazon nicht auf ihre Qualität überprüft – was man gut an den teilweise von Fehlern strotzenden E-Books erkennen kann, die man zum Beispiel auf Kindle Unlimited findet.

Aber egal, ich schweife ab!

Worauf ich eigentlich hinauswollte: Ein Amazon-Bestseller ist zwar eh nett, aber weder lukrativ noch ein großes Renommee. Das heißt allerdings nicht, dass man sich nicht darüber freuen und ihn auch nutzen kann! Und es ist etwas, das sich gut unter dem eigenen Profilbild auf LinkedIn und Co. macht.“

„Aber wenn ich einen Bestseller bei einem Verlag lande, dann lohnt sich das sicher auch finanziell für mich, oder?“

„Nochmals nein. Leider. Je nachdem, welchen Vertrag du mit deinem Verlag hast, bekommst du pro Buch einen sehr geringen Prozentsatz des Verkaufs- oder Einkaufspreises.

Der liegt bei Neuautorinnen im Schnitt bei circa einem Euro und ein paar ‚Zerquetschten', wie wir in Wien sagen. Also recht niedrig.

Solltest du also aus welchen Gründen auch immer eine sehr große Menge an Büchern verkaufen, dann würde sich das schon lohnen, und du bekämst im Laufe der Jahre immerhin eine Summe von ein paar tausend Euro ausbezahlt. Das ist aber noch immer weniger, als du investierst.

Und du musst davon ausgehen, dass ein Verlag zwar Buchgestaltung und Lektorat übernimmt, aber nicht großartig für dich ins Marketing investiert.

Ich weiß, man hört immer was von Lesereisen, die Verlage organisieren. Tun sie auch – aber eben nur für bereits arrivierte Autorinnen!"

„Also macht mich nicht einmal ein Bestseller reich?"

„Sorry, wenn ich dich jetzt auf den Teppich zurückhole, aber das ist leider sehr unwahrscheinlich. Im Durchschnitt werden von Business- und Sachbüchern maximal ein paar hundert Exemplare pro Buch und Jahr verkauft.

Das reicht also nicht, um sich den Traum von der Schreibklause in der Karibik oder in Frankreich finanzieren zu können. Und auch nicht annähernd, um den Ghostwriter deines Vertrauens zu finanzieren."[2]

„Und wie verdiene ich dann Geld mit meinem Buch?"

„Wie genau es funktioniert, kann ich dir jetzt noch nicht sagen – das ist wirklich von Fall zu Fall unterschiedlich. Es gibt zwar diese ‚Allheil-Antwort-Gurus', die eine Lösung für alle Menschen dieser Welt anbieten.

[2] Anmerkung: Sollte ein Ghostwriter so etwas behaupten, dann sollte er nicht (länger) der deines Vertrauens sein. Doch wenn du dein Projekt gut aufziehst, dann wirst du davon auf alle Fälle profitieren – auch finanziell! Und eines spricht definitiv dafür: Du kannst zwar nicht mit deinem Buch, sehr wohl aber durch dein Buch Geld verdienen. Gutes Geld!

Aber seien wir uns ehrlich: Das ist, wie wenn du zum Friseur gehst, der dir einen Kochtopf überstülpt und einfach der Kante entlang deine neue Frisur bastelt. Geht. Ist aber scheiße.

Ich zeige dir im letzten Kapitel ein paar Möglichkeiten, wie sich ein Buch für dich rechnen kann. Wenn nicht du dein Buch verkaufst, sondern dein Buch dich, dann hast du vieles richtig gemacht. Und dann kann es richtig abgehen!

Das setzt allerdings gute Vorarbeit voraus: Hirnschmalz, Strategie, Ehrlichkeit, Einfühlen in deine Zielgruppe. Und noch vieles mehr. Doch dazu später.“

„Okay, bevor wir jetzt weitermachen, noch eine Frage: Was kostet ein Buch?“

„Haha, das ist eine meiner Lieblingsfragen – nicht! Warum? Weil die Preise für ein Buch in etwa so unterschiedlich sind wie die Preise für ein Auto.

Möchtest du einen gebrauchten Minivan mit 500.000 Kilometern?

Oder einen nagelneuen Mercedes?

Umgelegt auf ein Buch: Brauchst du eine Überarbeitung von 20 schon vorgeschriebenen Seiten, oder möchtest du ein Komplettpaket für ein 280-Seiten-Buch inklusive Layout, Cover und Veröffentlichung? Das ist schon ein Unterschied!

Generell kannst du damit rechnen, dass du für ein Buch bei einem Profi eine fünfstellige Summe bezahlst. Wie hoch diese Summe wirklich ist, hängt natürlich nicht nur vom Arbeitsaufwand ab, sondern auch davon, mit wem du zusammenarbeitest.

Ich gebe zu, ich bin preislich gesehen eher Mercedes- als Trabi-Klasse. Das gilt allerdings auch für mein Angebot.

Wäre es für dich okay, wenn wir diese Frage noch ein bisschen nach hinten verschieben? Du wirst bald sehen, warum!“

„Ja, passt! Ich kann es vermutlich abwarten! Schieß los, wie geht es weiter?“

„Jetzt schauen wir uns gemeinsam an, warum Menschen überhaupt Bücher schreiben möchten. Denn das Schreibmotiv ist mit ausschlaggebend für Erfolg oder Misserfolg eines Buches."

Wozu brauchst du ein eigenes Buch?

„Okay … und du meinst, ich sollte trotzdem ein Buch schreiben (lassen)? Auch, wenn es vermutlich teuer ist? Ich meine: Mercedes? Hallo? Also, was zum Geier kann mir ein eigenes Buch bringen, das diesen Aufwand wert ist?“

„Oh, dafür gibt es eine ganze Menge Gründe! Einerseits höchst menschliche, andererseits wirtschaftliche. Das heißt, zuallererst solltest du dir überlegen, warum DU ein Buch haben möchtest.

Ich habe auf den nächsten Seiten für dich einen Überblick, warum andere Menschen ein Buch schreiben möchten. Die Auflistung ist natürlich nicht vollständig, und viele Gründe überschneiden sich. Aber so bekommst du einmal eine Idee. Was dein eigenes Motiv ist – und es dürfen durchaus auch mehrere sein! –, kannst nur du wissen. Oder wir können es gemeinsam herausfinden.“

❖ „Ich habe Angst, dass ich es vergesse, wenn ich es nicht niederschreibe."

Gerade, wenn du etwas Außergewöhnliches erlebt hast, möchtest du es wahrscheinlich aufschreiben, bevor es in Vergessenheit gerät. Es kann jedoch auch sein, dass du einfach einen Teil deines Lebens niederschreiben möchtest, damit du dich später daran erinnern kannst.

❖ „Ich möchte, dass meine Enkelinnen oder Kinder sich später an mich erinnern können."

Wie schön ist das! Ja, ich verstehe dich sehr gut! Ich wünschte, ich hätte meine Großeltern noch interviewen können, um ihre ganze Geschichte zu kennen. Es kommt übrigens auch oft vor, dass Kinder ihren Eltern oder Großeltern einen Ghostwriter schenken, der ihre Erinnerungen niederschreibt.

❖ „Ich möchte meine Liebesgeschichte aufschreiben!"

Du hast sie erlebt, die eine große Liebe, die alles überstrahlt? Vielleicht bist du noch mittendrin und auf Wolke sieben?

Oder das Schicksal hat euch getrennt, und du möchtest diese Liebe für die Nachwelt konservieren? Deinen Emotionen eine Stimme geben? Auch das ist ein großartiges Motiv, ein Buch zu schreiben.

❖ „Ich möchte wahrgenommen werden."

I feel you! Wer möchte das nicht? Gerade in einer Zeit, in der viel mehr Menschen als je zuvor es wagen, sich zu zeigen, wie sie wirklich sind – sei es in ihrer Neurodiversität, in ihrer Genderdefinition oder einfach in dem, was sie ausmacht –, ist es nicht nur für dich selbst wichtig, gesehen zu werden.

Es kann auch für andere eine Inspiration sein, wenn sie dich in all deinen Farben leuchten sehen.

Als Mentaltrainerin habe ich gelernt, gut zuzuhören, hinzuspüren und zwischen den Zeilen zu lesen. Dennoch achte ich darauf, niemals mehr als eine Kundin gleichzeitig zu haben, mit der ich eine sehr persönliche Geschichte schreibe. Es ist eine sehr herausfordernde Tätigkeit, die viel Einfühlungsvermögen benötigt, und daher auch energetisch enorm intensiv.

❖ *„Ich möchte meine Eitelkeit befriedigen."*

Passt – ich bin dabei! Schreiben und malen wir deine Geschichte in den strahlendsten Tönen! Es ist dein Leben, du hältst den Stift in der Hand, und du erzählst deine Story so, wie du es brauchst.

Ach ja ... solltest du zu weit übers Ziel hinausschießen, hole ich dich mit zarter Hand zurück, okay? Schließlich wollen wir, dass deine Leserinnen dich lieben!

❖ *„Ich möchte berühmt werden."*

Hm. Ja. Tja. Ein Buch kann ein guter Einstieg sein, aber das allein reicht bei weitem nicht aus, um berühmt zu werden. 2022 gab es z. B. in Deutschland ca. 65.000 Neuerscheinungen, in der Schweiz knapp 13.000.[3]

In Österreich sind es jedes Jahr ca. 90.000 Neuerscheinungen, davon ca. 9.000 über österreichische Verlage.[4] Um da herauszustechen, braucht man schon ein bisschen Kreativität und Initiative.

Wenn es wirklich dein einziges Anliegen ist, berühmt zu werden, dann bist du – fürchte ich – bei mir nicht gut aufgehoben.

---

[3] Quelle: https://de.statista.com
[4] Quelle: https://buchhandel.at/buchhandlung

❖ *„Ich möchte Menschen berühren."*

Das ist ein wunderschönes Motiv, ein Buch zu schreiben! Du solltest dir allerdings dabei im Klaren sein, dass du wahrscheinlich Geld investieren wirst, das du über den Verkauf nicht zurückbekommst.

Wenn du es dir finanziell erlauben kannst, dein Buch als Hobby zu schreiben, dann hau rein! Ich freu mich schon darauf, es zu lesen (oder mit dir zu schreiben)!

❖ *„Ich möchte informieren, entweder über Missstände oder über etwas ganz Großartiges, das alle wissen / kennen sollten."*

Dieses Motiv kann in zwei Richtungen gehen. Wenn du etwas ganz Tolles mit der Welt teilen willst, kann ich nur sagen: Juhu, ich bin dabei!

Wenn du allerdings nur deinem Unmut über Missstände Luft machen oder gar jemanden anpatzen willst, kann ich dich nicht unterstützen. Ich persönlich schreibe keine „Aufdeckerbücher", weil mir die Energie, die dahintersteht, nicht guttut.

Das heißt aber nicht, dass es nicht für eine meiner Kolleginnen das Richtige sein kann. Ich kann dir also möglicherweise jemand anderen empfehlen, der dich dabei unterstützt.

❖ *„Ich möchte andere gesund, fit oder glücklich machen."*

Klingt doch gut, oder? Ist es auch! Und Bücher, die einen konkreten Nutzen für die Leserinnen haben, sind beliebt. Wenn du ein solches Buch schreiben möchtest, ohne im entsprechenden Bereich zu arbeiten, dann ist es wahrscheinlich am besten, du schreibst einen Ratgeber oder Ratgeberroman (siehe Infobox).

❖ ***„Ich möchte für das, was ich erlitten habe, bemitleidet werden. Die Welt soll meinen Schmerz sehen!“***

Sei mir nicht böse, aber wenn es dir wirklich darum geht, zu zeigen, wie arm du bist und wie sehr dir das Leben mitgespielt hat, werde ich dein Buch nicht schreiben oder begleiten.

Einerseits fürchte ich, dass sich das Interesse der Leserschaft an einem solchen Buch sehr in Grenzen halten wird, sofern du nicht mega berühmt bist. Andererseits behalte ich lieber die Sonne in meinem Herzen und meine Unschuld.

Das bedeutet nicht, dass ich nicht über krasse Lebensgeschichten schreibe. Viele Menschen haben ein hartes Leben hinter sich oder befinden sich gerade mittendrin und sind sehr reflektiert.

Und mit diesen Menschen, die meist eine ganze Menge Optimismus und Resilienz mitbringen, arbeite ich sehr gerne, weil ich von ihnen sehr viel lernen kann.

Was ist ein Ratgeberroman?

Bei einem Ratgeberroman oder Sachbuchroman geht es darum, dein Thema in eine Geschichte zu verpacken.

Du erzählst – wie in einem klassischen Roman – eine Story, mit einer Protagonistin, mit Hochs und Tiefs und mit Learnings. So unterhältst du deine Leserschaft, während du unterschwellig und so nebenher deine Botschaft unterbringst.

Zudem kannst du in beiden Genres auch Infoboxen einbauen, Grafiken oder andere Goodies, die deine Botschaft unterstützen.

Wirtschaftliche Motive, ein Buch zu schreiben

❖ ***„Ich möchte meinen Mitbewerberinnen einen Schritt voraus sein bzw. aufholen."***

Auch hier ist Expertise gefragt. Oft haben diese Kundinnen jedoch schon einen genauen Plan bzw. ein „Vorbild".

Achtung: Was ich sicherlich nicht mache, ist etwas von anderen 1:1 abkupfern, auch wenn das noch so verlockend erscheinen mag! Man darf schon mal anderen über die Schulter schauen, etwas übernehmen und dann ein eigenes Ding draus machen.

Aber „genauso ein Buch wie XY, nur mit meinem Namen drauf", das spielt's nicht! Nicht nur, weil es schwer illegal wäre, sondern auch, weil es dir nicht guttut! Du hast einen anderen Stil, einen anderen Zugang, und du bist jemand anderes! Und wenn du noch nicht genau weißt, wer du bist, dann sollten wir das herausfinden, bevor wir zu schreiben beginnen. Sonst wirfst du viel Geld raus für etwas, das dir später nichts bringt.

Eine meiner Kundinnen hatte die Herausforderung, dass ihr ein Mitbewerber nicht nur dicht auf den Fersen klebte, sondern auch immer wieder versuchte, bei ihr abzuschauen.

Mit dem Buch schaffte sie den entscheidenden Schritt nach vorne, und ihr Mitbewerber hatte die Nase auf jeden Fall um ein gutes Jahr hinten.

In solchen Fällen kann man übrigens zu 100 Prozent auf meine Loyalität zählen: Wenn zwei in Konkurrenz zueinander stehen, schreibe ich nur mit einer von beiden ein Buch!

❖ *„Ich möchte mich in einem (neuen) Markt etablieren und positionieren."*

Dies ist ein Wunsch, der oft an mich herangetragen wird. Betreffende Kundinnen freuen sich, wenn sie mit jemandem über ihre Ideen sprechen und diese reflektieren können.

Nun muss ich sagen: Ich bin keine Unternehmensberaterin. Und ich berate auch nicht. Ich habe allerdings über dreißig Jahre Erfahrung als Unternehmerin in unterschiedlichen Sparten und stelle mich gerne als „Sparring-Partnerin" zur Verfügung.

Nachdem ich immer alles etwas anders mache und gerne Möglichkeiten auslote, sehe ich manchmal Optionen, die außerhalb der Box und gerade deswegen erfolgversprechend sind.

Jedoch schlage ich im Zweifelsfall immer vor, eine Steuerberaterin, Anwältin oder eine Unternehmensberaterin zu befragen, bevor wir ein Projekt wirklich angehen.

Ein Trainer aus Deutschland hat sich mit seinem Buch ein Alter Ego aufgebaut, das als Sprecher auf die Bühne geht.

Dieses Buch, zusammen mit einer Kostprobe seines Auftritts, schickte er an diverse Großfirmen – seine Erfolgsquote lag bei 75 Prozent, das ist um ein Vielfaches höher als in der normalen Akquise!

❖ „Ich möchte meinen Expertenstatus erhöhen!"

Das ist eine tolle Idee! Achte einmal bewusst auf deine eigene Reaktion, wenn du feststellst, dass jemand, den du kennst oder, sagen wir, im LinkedIn-Netzwerk hast, ein Buch geschrieben hat. Wie geht es dir damit? Was denkst du über diese Person?

Möglicherweise kommen dir Gedanken wie „Na, die Meier muss ja ganz schön was draufhaben, dass die ein Buch zu ihrem Thema schreibt!"

Du hast schon eine tolle Expertise auf deinem Gebiet. Das wissen allerdings noch nicht genug Menschen. Wenn du dein Buch veröffentlichst und dies ausreichend in deinem Netzwerk und darüber hinaus bekanntmachst, dann wächst automatisch auch deine Autorität auf deinem Gebiet.

Ein Kunde von mir war schon lange in seiner Branche als Mentaltrainer tätig. Seine Angebote wurden jedoch von größeren Unternehmen meist übersehen.

Das lag zum Teil daran, dass er sich seiner menschlichen und fachlichen Special Effects (in diesem Fall enorm viel Humor und einer Spezialisierung auf neurodiverse Menschen) noch nicht ausreichend bewusst war.

Auf der anderen Seite hatte er aus genau diesem Grund auch sein Angebotsprofil noch nicht so zugespitzt, dass er sich von anderen abheben konnte. Durch sein Erstlingswerk – ein Sachbuch, das exakt auf die Zielgruppe Human Resources in Unternehmen zugeschnitten ist – konnte er einen Fuß in die Tür bekommen und ist mittlerweile auf Monate hinaus ausgebucht.

❖ ***„Ich möchte das Buch super verkaufen, damit ich nie wieder arbeiten muss."***

Haha, das hatten wir schon! Kein Kommentar! Lies bitte weiter vorne nochmal nach, falls du es vergessen hast!

❖ ***„Ich möchte das Buch nutzen, um meine Produkte oder Dienstleistungen besser zu verkaufen."***

Wunderbar! Dafür gibt es Applaus! Das kann super funktionieren, wenn wir vorher gemeinsam überlegen, wohin die Reise gehen soll. Und wer weiß, welche neue Optionen und Ideen sich in diesem Prozess noch zeigen!

Darüber hinaus kannst du deinen zukünftigen Kundinnen zahlreiche Informationen über dich geben:

Warum schreibst du dieses Buch? Was ist dein Zugang zu deinem Thema? Wie bist du überhaupt darauf gekommen, genau in diesem Bereich arbeiten zu wollen? Gibt es Gründe dafür? Auch wenn du Quereinsteigerin bist, kann es für deine Leserinnen spannend sein, mehr über dich zu erfahren.

Und vor allem hast du einen großen Vorteil: Du kannst die Geschichte aus deiner Perspektive erzählen!

Eine Trainerin, mit der ich ein Buch geschrieben habe, hatte vor unserer Zusammenarbeit mit der Herausforderung zu kämpfen, dass ihre Workshops nicht allzu gut besucht waren. Mit dem Buch als Zugpferd konnte sie die doppelte Menge Workshops voll bekommen – zu einem wesentlich höheren Preis! Der Unterschied: Sie legte für jede Anmeldung ein signiertes Buch drauf.

❖ „Ich möchte mein Buch als Vision Board nutzen!“

„Waaaas?“, denkst du nun vielleicht. „Wie? Warum?“ oder „Was ist ein Vision Board überhaupt genau?“

Ein klassisches Vision Board ist ein Tool aus dem Mentaltraining. Dazu nimmst du ein Stück Papier, eine Leinwand oder eine leere Wand und klebst, schreibst oder malst all das drauf, was du in einem bestimmten Bereich oder generell erreichen möchtest. Und vor allem, wie es sich anspürt, wenn du dein Ziel erreicht hast.

Du kannst auch mit einem Buch Visionsarbeit betreiben. Als Mentaltrainerin kann ich dich dabei gut unterstützen.

Und vor allem darauf achten, dass wir auf dem Weg bleiben und nicht zu sehr abschweifen. Dieses Buch kann zum Beispiel ein Roman sein über eine Person, die es geschafft hat (was immer du schaffen möchtest), oder ein Sachbuch.

> Ich hatte die Ehre, mit einem jungen Mann an seiner Positionierung als Coach zu arbeiten. Er hatte große Pläne, und gemeinsam haben wir die richtigen Worte gefunden, um ihm selbst zu zeigen, was er schon erreicht hatte, und wo es hingehen soll.

Achtung: Es ist nicht klug, einen Ratgeber darüber zu schreiben, wie man ein Ziel erreicht, das du selbst noch nicht erreicht hast!

Man sagt zwar, dass man durch Lehren am besten lernt, aber diese Coaches und Beraterinnen, die selbst nur mit Wasser kochen, anderen aber Espresso oder Daiquiri versprechen, sind unglaubwürdig und – zumindest für mich – mittlerweile unerträglich.

❖ *„Ich möchte meine Zielgruppe besser erreichen und die richtigen Kundinnen finden!“*

Das kannst du – und es ist ein guter Zugang! Mit der Art, wie du schreibst, welchen Stil du bevorzugst, und welche Worte du benutzt, kannst du deine Zielgruppe genau ansprechen.

Wir alle sind einzigartig. Doch das sieht man oft nicht auf den ersten Blick. Daher ist es durchaus sinnvoll, wenn du deine Geschichte erzählst. Und dadurch an Menschen kommst, die perfekt zu dir passen. Solche, mit denen das Arbeiten Freude macht. Solche, die dir mehr Energie geben als nehmen. Und die dich als Mensch schätzen, einfach, weil du bist, wie du bist, ohne Verstellen, Verstecken, Verbiegen und Verbeulen.

Eine ältere Dame, deren Buchprojekt ich begleiten durfte, hatte die Herausforderung, dass sie in der Öffentlichkeit ganz anders wahrgenommen wurde, als sie ist.

Während sie auf den ersten Blick sehr ernst und seriös wirkt, schlummert in ihr ein Vulkan der Ideen, eine ganze Menge Esprit und sehr viel Lachen.

Durch das Buch und eine entsprechende grafische Gestaltung blieben keine Zweifel an ihrem Wesen – und die entsprechenden Kundinnen fanden sich bald ein. Auf diese Weise durfte sie ihre eher langweiligen Kundinnen zurücklassen und sich mit neuer Freude auf ihre zukünftige Zielgruppe einlassen.

❖ *„Ich möchte aus einem ganz anderen Grund ein Buch schreiben!“*

Gut erkannt: Diese Liste ist nicht vollständig. Jeder Fall ist einzigartig, so wie auch jede Ghostwriter-Kundinnen-Beziehung einzigartig ist. Was möglicherweise daran liegt, dass jeder Mensch anders tickt, und wir alle nur so nehmen können (oder sollten), wie sie nun mal eben sind.

Wozu ein Ghostwriter?

Der Unterschied zwischen Inhalt und Form

„Okay! Soweit ist alles klar. Aber ... ich überlege immer noch, ob es wirklich eine gute Idee ist, dich mein Buch schreiben zu lassen. Damit meine ich nicht dich als Person, sondern einen Ghostwriter generell. Ist das nicht so, als würde ich mich mit fremden Federn schmücken?“

„Ich verstehe schon. Und ja, solche Einwände oder Vorbehalte höre ich öfter. Zunächst einmal keine Sorge: Auch wenn ich schreibe, kommen die Inhalte ja trotzdem von dir!

Da sind wir wieder bei meiner Frage von vorhin: Welche Gründe hast du dafür, ein Buch zu schreiben? In deinem Fall kann ich mir vorstellen, was die Hintergründe sind – du hast ja auch schon so etwas angedeutet.

Also lass mich raten: Du hast etwas zu sagen und möchtest Informationen unter die Leute bringen. Du hast wahrscheinlich viele Ausbildungen gemacht, Erfahrungen gesammelt, arbeitest erfolgreich mit deiner eigenen Methode oder hast schon unzählige Menschen mit deinen Vorträgen oder Trainings begeistert. Du hast bereits viele begleitet, vielen geholfen.

Du hast damit deinen Teil der Leistung erbracht!

Es ist nicht zwingend dein Job, das alles in perfekte Worte zu fassen. Das kannst du getrost anderen überlassen, die so etwas beruflich machen. Mir oder meinen Kolleginnen zum Beispiel.“

Ein Buch zu schreiben ist eine große Aufgabe. Man muss strukturieren, Inhalte aufbereiten und in einen sinnvollen Zusammenhang bringen, einen roten Faden durch das ganze Buch halten, alles leicht lesbar und vielleicht auch noch unterhaltsam niederschreiben.

Dazu kommen möglicherweise noch die Gestaltung des Layouts und des Covers sowie die ganzen Entscheidungen, wie das Buch in die Buchhandlung kommen soll.

Wenn du da kein Hintergrundwissen hast, kann dich das schnell überfordern. Ein Ghostwriter kann dir vieles abnehmen, dich unterstützen und im Idealfall nicht nur schreiben, sondern auch das gesamte Projektmanagement übernehmen. Ich gehe davon aus, dass du als Fulltime-Unternehmerin vermutlich keine oder nicht ausreichend Zeit dafür hast.

Wie sag ich's meinen Freundinnen?

„Du sagst also, es ist legitim, sich ein Buch ghostwriten zu lassen? Ich weiß nämlich nicht, wie ich das nach außen kommunizieren soll.

Dann habe ich dieses tolle Buch, aber ich habe es nicht selbst geschrieben. Wie soll ich meinen Bekannten und Geschäftspartnerinnen gegenüber damit umgehen?

Ich will ja auch nicht lügen und sagen, dass ich alles ganz allein gemacht habe!“

„Das ist ganz einfach. Es ist dein Buch, du hast alle Inhalte dazu geliefert, und möglicherweise machst du sogar mehr, schreibst vielleicht ein paar Passagen oder gibst mir Skripten von deinen Workshops.

Du kannst also mit Fug und Recht behaupten, dass es deins ist, dass du es gemacht hast. Wenn du dir bei der Umsetzung helfen lässt, dann ist das keineswegs ehrenrührig.

★ *Ziehst du jedes Jahr deine Winterreifen selbst auf?*

★ *Hast du dein Logo selbst entworfen und ins Reine gezeichnet?*

★ *Und kochst du jede Mahlzeit selbst?*

Wenn du auf alle diese Fragen mit ‚Ja!' antwortest, dann hast du eventuell ein Problem. Wahrscheinlich ein Zeitproblem, weil du deine wertvollen Ressourcen auf Dinge verschwendest, die nicht Teil deiner Kernkompetenz sind.

Oder du hast – bei allem Respekt – zu wenige Aufträge und daher viel Freizeit.

Aber ganz im Ernst: Hast du ein Problem damit zuzugeben, dass du bei der Erstellung deiner Webseite professionelle Hilfe in Anspruch genommen hast?

Hast du Bedenken, dass deine Freundinnen dich für völlig inkompetent halten, wenn du deine Küche nicht komplett allein gefliest und eingerichtet hast?

Oder genierst du dich, wenn du dir eine tolle Torte für ein Fest backen und verzieren lässt? Wahrscheinlich nicht.

Dann sollte es dich ebenso wenig kratzen, die Dienste eines Ghostwriters in Anspruch zu nehmen.

Keine Sorge, von mir erfährt es ohnehin niemand. Aber wenn du in die Verlegenheit kommst, eine Antwort auf die Frage ‚Selbst geschrieben?'zu geben, dann denke an meine Worte.

Und bedenke eines: Man muss erst mal Inhalte und Ideen für ein Buch haben – das Schreiben ist lediglich das Handwerk, das deine Ideen und den Inhalt deines Kopfes zu Papier bringt!

Nachdem du nicht jedes Handwerk selbst ausüben kannst (da hättest du ja viel zu tun!), wendest du dich hierfür einfach an den Ghostwriter deines Vertrauens. Alles klar?"

„Gut, damit kann ich leben! Danke!"

Wenn du das möchtest, schweige ich wie ein Grab über unsere Zusammenarbeit!

Über Geheimhaltung und Urheberrechte

Disclaimer: Ich bin keine Anwältin und verfüge über keine rechtliche Ausbildung. Ich gebe hier nach bestem Wissen und Gewissen wieder, was mir von Anwältinnen gesagt wurde. Wenn du es niet- und nagelfest brauchst, dann kann ich dir gerne einen Anwalt für Urheberrecht empfehlen.

> *„Weil du es vorhin angesprochen hast ... ich habe da zwei Fragen bezüglich Geheimhaltung.*
>
> *Wenn ich dich richtig verstanden habe, kannst du mir schriftlich und verbindlich zusagen, dass niemand erfährt, inwieweit du mich bei meinem Buchprojekt unterstützt hast. Und du würdest mir sogar die Urheberrechte abtreten?“*

„Ja und ja. Bevor wir beginnen, frage ich dich, ob du eine Geheimhaltungsvereinbarung haben möchtest. In diesem Fall schweige ich wie ein Grab und wenn wir uns irgendwo treffen sollten, tue ich so, als würde ich dich nicht kennen – außer, du kommst auf mich zu.

Und was die Urheberrechte betrifft: Die sind normalerweise nicht übertragbar. Das bedeutet, wenn jemand etwas schafft – auch im Auftrag einer anderen Person –, dann kann man zwar die Nutzungsrechte abtreten, nicht jedoch das Urheberrecht[5], weil dies untrennbar mit der Urheberin verbunden ist.

Ghostwriting bildet hier jedoch die Ausnahme: Ich kann und werde dir das Urheberrecht abtreten.

---

[5] https://www.anwalt.de/rechtstipps/the-ghost-behind-spare-ghostwriter-vertraege-207723.html

Geheimhaltung ist möglich bezüglich

- ★ Status als Ghostwriter (Erwähnung im Buch als "mit Unterstützung von ..." oder Ähnlichem)
- ★ jeglicher Beteiligung am Projekt
  - ◆ keine Erwähnung im Buch oder auf dem Cover (Co-Autorin, Unterstützung, Danksagung, ...)
  - ◆ keine Nennung auf meiner Webseite, in den Social Media, etc.
  - ◆ Schweigepflicht gegenüber Dritten (Ausnahmen siehe unten)
- ★ Inhalte, die beispielsweise nicht vor der Veröffentlichung publik werden sollen
- ★ Inhalte, die du mir im Vertrauen erzählst, die aber nicht im Buch landen sollen, behalte ich auf jeden Fall für mich – ob mit oder ohne Vereinbarung.

Ausnahmen von der Geheimhaltung

- ★ notwendige Infos für Kooperationspartnerinnen, die direkt an der Entstehung des Buches beteiligt sind (diese unterschreiben natürlich ebenfalls eine Geheimhaltungsvereinbarung)
- ★ Daten für Rechnungen, Steuer, Buchhaltung, etc.
- ★ Informationen, die bereits öffentlich sind
- ★ Informationen, die ohne Verschulden des Ghostwriters bekannt werden
- ★ Cover, Grafiken, o.ä. – Lektorin, Grafikerin, Fotografin und Übersetzerin haben immer ein Recht auf Nennung.

Damit ich dir das Urheberrecht abtreten kann, brauchen wir nicht einmal einen Vertrag, weil es für mich Ehrensache ist. Ich kann dir allerdings auch das schriftlich geben, wenn du dann besser schlafen kannst!"

> *"Danke! Dann komme ich nun zu meiner zweiten Frage: Wie kann ich sicher sein, dass du nicht meine Infos und Ideen klaust, um ein eigenes Buch daraus zu machen?"*

„Sorry, dass ich gerade so herzlich lachen muss! Weißt du, einerseits ist es so, dass diese Story so eng mit dir verbunden ist, dass ich damit gar nichts anfangen könnte.

Andererseits habe ich selbst so viele Ideen für Bücher, die dann auch wirklich zu meiner eigenen Expertise passen, dass ich den Rest meines Lebens nur an meinem eigenen Zeug schreiben könnte.

Und drittens: Wenn ich nachweislich deine Idee oder deinen Plot stehle, dann darfst du mich mit Fug und Recht verklagen. Glaub mir: So etwas lohnt sich nie!"

Eckdaten unserer Zusammenarbeit

„Sollten wir zusammenarbeiten, worauf muss ich sonst noch achten? Hast du gewisse Vorgaben oder Anforderungen?“

„Danke dir herzlich für diese großartige Frage! Ja, es gibt ein paar Punkte, die mir in einer Zusammenarbeit wichtig sind. Reden wir zunächst übers Technische."

★ Schreibprogramm

Ich arbeite in WORD, und zwar in der aktuellsten Version. Nicht Pages, nicht Skribener, nicht Papyrus. Auch nicht gemeinsam im Google Doc.

Letzteres habe ich einmal mit einem Kunden getestet und dabei festgestellt, dass es zwar schon cool ist, wenn der andere in Echtzeit meine Änderungen sehen kann, dass dem Programm aber viele Features fehlen, die ich für schnelles und effizientes Arbeiten brauche. WORD ist ein Programm, das eingespielt ist. Und das spart Unmengen von Arbeit. Wenn ich das Dokument für jedes Feedback von dir in ein anderes Dateiformat konvertieren muss, kostet das nicht nur Zeit und Nerven, es ist vor allem eine zusätzliche Fehlerquelle.

Außerdem kommt jedes Manuskript in WORD nach Abschluss zur Lektorin. Daher bin ich in diesem Punkt unnachgiebig – auch in deinem Sinne!

★ Zeitdruck – nein danke!

Außerdem lege ich bei einer möglichen Zusammenarbeit sehr großen Wert auf folgende Tatsache: Ich brauche Zeit, um gute Arbeit zu leisten!

Schreiben ist zwar ein Handwerk, dennoch geht jedem Buch idealerweise ein kreativer Prozess voraus.

Das bedeutet: Es kann gerade in der Anfangsphase sein, dass du ein, zwei Monate nichts von mir hörst – und in dieser Zeit auch nichts passiert. Zumindest nichts, das ich dir zeigen könnte.

Ich trage dein Buch in meinem Kopf spazieren, es fährt mit mir auf der Autobahn oder im Zug, ich habe es beim Einkaufen oder am Abend vor dem Fernseher in meinen Gedanken. Ich sammle Ideen, baue und verwerfe Konzepte, achte auf Formulierungen und sprachliche Bilder, die zu deinem Buch passen, und irgendwann ist es dann so weit: Ich schreibe los, und du bekommst die ersten Textproben.

Was ich von dir brauche, ist Geduld und Vertrauen. Daher sollte dir klar sein, dass ein Buch nicht über Nacht entsteht. Auch nicht in zwei, drei Monaten. Ein gutes Buch braucht Zeit. Einige Monate, manchmal ein Jahr oder länger. Wenn du dir darüber im Klaren bist, dann können wir starten!

> *„Ah ja – eine Frage hab ich noch. Ich bin ja nicht gerade eine stromlinienförmige Person, sondern habe so meine Ecken und Kanten. Kannst du das im Buch dann so abmildern, dass man das nicht gleich herausliest?"*

„Ja, kann ich. Werde ich aber nicht! Gerade darum geht es ja: Deine Special Effects ins richtige Licht zu rücken, damit du authentisch in all deiner Pracht strahlst, das ist meine Aufgabe! Du sollst glaubwürdig sein, nicht austauschbar! Und du kannst mit deinem Buch nur dann die richtigen, für dich passenden Kundinnen anziehen, wenn du dich zeigst und nicht versuchst, anders zu sein!"

> *„Uff. Das hatte ich nicht auf dem Radar. Aber … wer soll mich denn so wollen, wie ich bin?"*

„Sascha, du bist liebenswert, speziell, intelligent, hast viel Erfahrung in deinem Metier und bringst durch das, wie du bist, eine völlig neue Qualität in dein Buch. Es wurde schon viel zu diesem Thema geschrieben – aber eben nicht von dir! Du bist es, die dieses Buch und damit auch deine Dienstleistung einzigartig macht!"

> *„Okay … so habe ich das noch nicht gesehen. Danke! Ich hab grad super das Flattern, aber ich vertraue dir und lasse mich darauf ein."*

„Ich danke dir! Und du wirst sehen: Wenn du dich zeigst, wirst du gesehen. Und du bist großartig so wie du bist. Also, gehen wir's an!"

Erstgespräch

„Okay, sagen wir einmal, ich will das Ding mit dir durchziehen. Und – Achtung, Spoileralarm! – ja, ich will! Wie geht es jetzt weiter?“

„Zunächst lernen wir uns ein bisschen kennen – bei einem Online-Treffen via Zoom, am Telefon oder bei einem Kaffee in meinem Büro. Dann passiert das, was wir beide hier seit Beginn dieses Buches machen.

Wir plaudern, du bringst deine Fragen vor, redest darüber, was dir im Kopf herumgeht – Wünsche ebenso wie Befürchtungen.

Dabei können wir ein Gefühl dafür bekommen, ob unsere Zusammenarbeit funktionieren kann. Die passende Chemie ist sehr wichtig, denn wir werden eine Menge Zeit miteinander verbringen.

Wenn wir beide jetzt – wie in unserem Fall – feststellen, dass es mit uns passt, dann geht es weiter.

Wenn eine von uns nicht so happy wäre mit der jeweils anderen, dann gibt es zwei Möglichkeiten: Wenn du möchtest, kann ich dir eine Kollegin empfehlen, mit der du persönlich oder vom Thema her besser harmonierst. Ich habe ein ziemlich gutes Gespür, wer mit wem kann!

Und wenn es gar nicht geht mit uns, und deine Vorstellungen ein bisschen zu weit an der Realität oder am meiner Meinung nach Machbaren vorbeigehen (oder umgekehrt), dann trennen wir uns an dieser Stelle, sagen zum Abschied leise Servus und bleiben Fremde.“

Wenn du dich dafür entscheidest, dein Buch mit mir zu machen, stecken wir die groben Eckdaten ab.

Wir überlegen uns, wie dick das Buch sein soll, auf welche Art du dein Wissen mit mir teilen möchtest, und eventuell weißt du sogar schon, wie du das Buch herausbringen (lassen) willst. Es kann auch sein, dass es andere Dinge gibt, die du mir jetzt unbedingt mitteilen möchtest.

Ich habe für unser erstes Gespräch ungefähr 30 Minuten eingeplant – auf den dazugehörigen Kaffee (oder Tee) bist du eingeladen, und für unser erstes Kennenlernen verrechne ich natürlich nichts.

Was ich beim Erstgespräch von dir wissen möchte

★ Dein Thema (Grundidee)

Nicht jede Person, die zu mir kommt, weiß schon genau, was sie will. Und das ist okay so. Gut wäre es, wenn du eine Idee hättest, wo es hingehen soll.

Also: Möchtest du mit meiner Unterstützung ein Sachbuch schreiben, einen Roman, deine Biografie? Alles weitere können wir später im gemeinsamen Workshop definieren.

★ Was muss gemacht werden, und wie kann ich dich unterstützen?

Brauchst du nur das Manuskript oder ein fertiges Buch?

Hast du schon eine Vorstellung, wie du es veröffentlichen wirst?

Soll ich das komplette Projektmanagement übernehmen? Mich also um alles kümmern, bis das Buch fertig ist?

Und vor allem:

Wie viel hast du schon geschrieben? Gar nichts? Ein paar Seiten? Fast alles?

★ Hast du Unterlagen, an denen ich mich orientieren kann?

Falls du irgendetwas hast (Skripten, PowerPoint, Videos, Diplomarbeit), mit dem ich arbeiten kann, dann hilft das natürlich sehr.

Das wird nicht immer der Fall sein. Wenn wir zum Beispiel deine Lebensgeschichte schreiben, wirst du mir wahrscheinlich das meiste erzählen wollen. Auch das ist fein – ich sollte es nur vorab wissen.

★ *Wann brauchst du das Buch?*

Viele meiner Kundinnen haben eine Vorstellung davon, wann ihr Buch erscheinen soll. Manche möchten es beispielsweise an ihrem Geburtstag, bei einer Firmenfeier, zu Weihnachten oder anlässlich einer Neueröffnung präsentieren.

Wie lange ein Buchprojekt tatsächlich dauert, hängt von vielen Faktoren ab, wie beispielsweise:

- ★ Wie viel ist schon da?
- ★ Wie viel muss gemacht werden?
- ★ Wie dick soll das Buch werden?

Außerdem gibt es hinsichtlich der Entstehungszeit weitere Punkte, die ich im Voraus nicht abschätzen kann: Wenn zum Beispiel von dir kein (zeitnahes) Feedback kommt, kann sich alles verzögern.

Vergiss bitte nicht, dass ein Buch nicht fertig ist, wenn das Manuskript abgeschlossen ist. Da fehlen noch einige Arbeitsschritte, bis du dein fertiges Buch in der Hand hältst – und das kann mehrere Monate dauern!

★ *Geplanter Umfang*

Dieser ist für die geplante Deadline natürlich wichtig. Das anvisierte konkrete Fertigstellungsdatum erarbeiten wir gemeinsam im Zielworkshop. Im ersten Gespräch können wir schon mal herausfinden, ob die Deadline, die dir vorschwebt, für mich umsetzbar ist.

★ *Ist es dir meine Dienstleistung wert?*

Bei diesem ersten Termin bekommst du von mir einen ungefähren Preis. Dieser wird eine von-bis-Zahl sein. Genauer kann ich es jetzt noch nicht sagen, weil wir davor noch vieles abklären müssen.

Es geht mir darum, dir eine ungefähre Vorstellung zu vermitteln, was auf dich zukommt, damit du entscheiden kannst, ob du dir mich leisten kannst und möchtest. Wenn wir uns soweit einig sind, vereinbaren wir einen Termin für einen ungefähr dreistündigen Zielworkshop.

Was wir beim Erstgespräch herausfinden

- ★ Passt es persönlich und finanziell für dich?
- ★ Kommst du mit meiner Arbeitsweise klar?
- ★ Was brauchst du, und wie kann ich dich dabei unterstutzen?
- ★ Passen das Thema und deine Vorstellungen für mich?

> „Sag mal, bekomme ich eigentlich auch einen kostenlosen Probetext von dir? Ich meine, woher soll ich wissen, ob du so schreibst, wie ich will?"

„Liebe Sascha, ich schreibe dein Buch in deinem Stil. Natürlich kann ich dir gerne Leseproben aus schon vorhandenen Texten und Büchern zukommen lassen.

Aber ich kann dir keine Leseprobe aus deinem eigenen Buch anfertigen. Es geht nicht um die halbe Seite oder Seite, die ich dir vorab schicken soll.

Bevor ich beginne, dein Buch zu schreiben, befasse ich mich intensiv mit dir, deinem Thema und deinem Stil.

Das heißt, ich brauche eine ganze Menge Zeit, um mich darauf vorzubereiten. Du verstehst sicherlich, dass ich das nicht kostenlos machen kann!"

> „Ah ja, klar, das verstehe ich voll. Aaaaaber ... was ist, wenn mir das, was du schreibst, dann nicht gefällt?"

„Um das herauszufinden, gibt es zu Beginn unserer Zusammenarbeit ganz viele Feedbackschleifen. Da geht der von mir verfasste Text so lange zwischen uns hin und her, bis er für dich passt.

Wenn wir dieses Fundament gelegt haben, dann geht es zügiger. Aber grad zu Beginn bin ich auf deine Mitarbeit angewiesen.

Sollten wir feststellen, dass ich deinen Stil absolut nicht treffe, dann bin ich übrigens gerne bereit, unseren Vertrag auf dem Kulanzweg zu beenden. Das ist mir noch nie passiert, aber ich schließe nicht aus, dass es eines Tages passieren könnte."

Zielworkshop

„Gut, nun haben wir das geklärt. Ja, ich will mit dir arbeiten. Und ja, ich betrachte dich als meinen kleinen Luxus. Ich möchte ein wirklich gutes Buch haben und nicht allein mittels Versuch und Irrtum herumwurschteln. Die Zeit habe ich nicht – es darf ruhig smooth und bequem gehen! Was ist das jetzt mit diesem Zielworkshop?“

„Das ist ein Workshop, in dessen Verlauf wir alle nötigen Eckdaten festlegen. Wie soll dein Buch gestaltet werden? Welchen Stil soll es haben? Aber auch: Was soll es für dich und für deine zukünftige Leserschaft tun? Welchen kommerziellen und persönlichen Nutzen hat es für dich? Wie willst du von deinem Buch profitieren?

Dieser Workshop ist kostenpflichtig; solltest du dich danach entscheiden, einen anderen Weg einzuschlagen oder dein Buch doch selbst zu schreiben, hast du diese Möglichkeit natürlich auch.

Du bekommst im Anschluss von mir ein Konzept für dein Buch. Dieses beinhaltet ein erstes Inhaltsverzeichnis sowie alle wichtigen Eckpunkte, die wir gemeinsam erarbeitet haben. Welche das genau sind, erfährst du auf den folgenden Seiten. Mit diesem Konzept kannst du auch allein oder mit einem anderen Ghostwriter weiterarbeiten.“

Was erarbeiten wir im Zielworkshop?

Hier klären wir alle Fragen rund um dein Buch. Es geht um die Basics, die wir brauchen, bevor eine effiziente Arbeit möglich ist. Worum soll es in deinem Buch genau gehen?

Welchen Nutzen möchtest du für dich bzw. für deine Leserinnen erreichen?

Kennst du deine Zielgruppe und deine emotionale Tribe (dazu weiter hinten mehr)?

Und wie kannst du mithilfe des Buches dein Business wirklich boosten?

Manche der Fragen sind für dich möglicherweise schon glasklar, mit anderen kannst du vielleicht noch gar nichts anfangen. Wichtig ist, dass wir am Ende des dreistündigen Zielworkshops gemeinsam die wesentlichen Fragen beantwortet haben – oder zumindest wissen, wo wir später noch genauer hinschauen sollten.

★ *Thema*

Nun geht's ans Eingemachte. Wir schauen uns dein Thema im Detail an. Worum geht es konkret? Was ist deine Sicht auf dieses Thema? Hast du eine eigene Methode entwickelt? Wie sieht es mit Workshops aus?

Wenn du deine Biografie schreiben möchtest: Wo liegen deine Schwerpunkte? Willst du dein Leben vom ersten Tag bis heute aufschreiben? (Ich schlage vor: nein!) Auf welche Aspekte legst du Wert? Auch zu anderen Themen habe ich hier sicherlich ein paar Fragen an dich.

★ *Genre*

Was soll es denn werden? Ein Sachbuch? Ein Roman? Ein Sachbuchroman? Ein Krimi? Eine Biografie? Oder etwas völlig anderes?

Wir müssen das Genre nicht genau benennen, es ist jedoch schon ein Unterschied, ob wir eine durchgehende Geschichte erzählen oder unsere Informationen eher sachlicher übermitteln.

Natürlich darf auch in diesem Fall Storytelling nicht fehlen. Fallbeispiele, Geschichten aus deinen Workshops oder kurze Einblicke in dein (Berufs-)Leben müssen unbedingt rein, werden aber völlig anders aufgebaut sein als in einem Buch mit einer durchgängigen Geschichte.

★ *Ziele deines Buches*

Wir haben uns schon eingehend mit unterschiedlichen Motiven befasst, warum jemand ein Buch schreiben möchte. Was ist dein Motiv? Was möchtest du mit dem Buch erreichen?

Wir definieren die Ziele nicht nur für dich, wir klären auch, was die Leserinnen davon haben, dass sie dein Buch kaufen.

★ *Zielgruppe*

Apropos Leserinnen: Für wen willst du eigentlich schreiben? Für Privatpersonen? Unternehmerinnen? Frauen? Männer? Kinder? Alte? Junge? Oder alle? (Also, von „alle“ rate ich dringend ab. Warum, das findest du unter "Zielgruppe: ALLE!".)

Außerdem sprechen wir über deine emotionale Zielgruppe, die meiner Meinung wichtiger ist als alle anderen.

★ *Botschaft bzw. Kernaussage*

Wichtig ist hier auch, die Botschaft bzw. Kernaussage deines Buches herauszuarbeiten. Das hilft, den roten Faden zu bewahren, nicht zu sehr abzuschweifen und auch Unterthemen, die vielleicht dazu passen, aber den Rahmen sprengen würden, auszusortieren.

★ *Wie kommen die Infos von dir zu mir?*

Vielleicht hast du ja schon Skripten, ein begonnenes Manuskript oder diverse Videos? Oder du bietest Workshops an, an denen ich teilnehmen soll? Vielleicht aber brauchen wir einfach einige Interviews, um den gesamten Stoff gemeinsam zu erarbeiten.

Wie auch immer – in irgendeiner Form kommt dein Wissen zu mir.

★ *Die erste Struktur*

Aus diesen Informationen erstellen wir gemeinsam eine Struktur, die all das beinhaltet, was dein Buch am Ende enthalten soll. Wir haben einen Haufen Überschriften, die uns das Skelett für dein Werk bieten. Dein erstes Inhaltsverzeichnis, könnte man sagen.

★ *Stil*

Wie hättest du es denn gerne geschrieben? Salopp oder förmlich? Per du oder per Sie? Humorvoll oder ergreifend? Gegendert? Und wenn ja, wie?

Das alles muss nicht bis ins Letzte geklärt sein, da sich vieles im Schreibprozess noch ändern kann. Es ist jedoch gut, wenn du eine grobe Ahnung hast, wie es laufen soll.

★ *Gibt es Bücher, die dir besonders gut gefallen?*

Was genau gefällt dir in Sachen Stil, Sprache, Inhalt oder Design? Und welche Bücher treffen deinen Geschmack?

Bitte bring sie zu unserem Workshop mit oder schick mir die Links dazu, falls wir uns online sehen!

Nur, um es hier noch einmal zu erwähnen: Es geht nicht ums Abkupfern, sondern darum, ein besseres Gespür zu bekommen, was dir zusagt.

★ Buchformat, Größe & Umfang

Zunächst legen wir fest, welches Format dir vorschwebt. Soll dein Buch als E-Book publiziert werden oder nur als Soft- oder Hardcover? Oder alles auf einmal?

Diese Entscheidung hat Auswirkungen auf Grafik und Layout sowie auf den Prozess des Hochladens und natürlich darauf, wie schnell das Buch fertig ist.

Wichtig ist auch die ungefähre Größe des Buches – also beispielsweise Taschenbuchformat, Minibuch oder Coffee Table Book –, damit ich die ungefähre Seitenanzahl (plus/minus 10 Prozent) im Angebot klar und deutlich angeben kann. Wenn nur steht „150 Seiten“, weißt du nicht, worauf sich das bezieht.

Wie viele Seiten es genau am Ende werden, können wir zu diesem Zeitpunkt beide nicht sagen. Aber wir sollten wissen, ob wir von einem klein-aber-oho-Buch sprechen oder von einem Wälzer. Oder von irgendwas dazwischen.

★ Recherche

Wird es bei deinem Buch notwendig sein, etwas zu recherchieren? Wenn ja – machst du das oder möchtest du es an mich und mein Team auslagern?

★ Art der Veröffentlichung

Hast du schon eine Idee, wie du dein Buch auf den Markt bringen möchtest? Wenn nicht, dann ist jetzt ein guter Zeitpunkt, um darüber zu reden. Weiter hinten, unter „Wege der Veröffentlichung“ findest du ein paar Hard Facts zum Thema, die dir die Entscheidung eventuell erleichtern. Wenn du dir noch absolut unsicher bist, macht es auch nichts; dann werde ich dir einfach ein flexibles Angebot erstellen (auch dazu kommen wir etwas später).

★ *Cover, Layout, Grafiken*

Wenn du über einen traditionellen Verlag gehst, dann musst du dich damit nicht auseinandersetzen. Du bekommst alles vom Verlag gestellt.

Wenn nicht, dann sollten wir reden: Was brauchst du? Welcher Stil passt zu dir? Hast du gewisse Vorstellungen? Was gefällt dir? Und welche Farben spiegeln dich oder dein Unternehmen gut wider? Hast du schon ein Corporate Design? In welchen Farben sind deine Webseite, Visitenkarten und andere Drucksachen gestaltet? Wie sieht dein Logo aus, wie deine Unternehmensschrift?

★ *Deadline*

Die Deadline für Buchprojekte hängt von vielen Faktoren ab.

- ★ Was brauchst du genau?
- ★ Was ist schon da?
- ★ Wie viel Arbeit möchtest du selbst übernehmen – und wie viel Zeit hast du realistisch dafür vorgesehen?
- ★ Wie umfangreich soll es werden?

Wir müssen auch klären, was zur vereinbarten Deadline fertig sein soll: Geht es um das Manuskript, das druckfertige PDF oder das Buch in deiner Hand?

Realistisch gesehen benötigt man ein absolutes Minimum von vier bis sechs Monaten für ein fertiges Manuskript ohne Endabnahme durch dich, ohne Lektorat, etc., wenn schon ausreichend Material vorhanden ist. Und dann dauert es immer noch eine ganze Weile, bis dein Buch fertig ist.

Hier ein paar Richtwerte zur besseren Orientierung.

Achtung! Dieser Zeitplan kann sich je nach Veröffentlichungsoption stark verändern!

Zeitplan Buch	
Lektorat / Korrektorat	2-6 Wochen, je nach Aufwand
Grafik / Layout	1-3 Monate, je nach Aufwand und Feedbackschleifen
Upload beim Verlag	ca. eine Stunde
Feedback vom Verlag (falls im Angebot)	1-10 Tage, je nach Auslastung
Korrekturschleife durch Grafikerin	1-10 Tage, je nach Auslastung
endgültige Freigabe durch dich	Das hangt von dir ab!
Druck	unterschiedlich; üblicherweise 2-6 Wochen
im Handel verfügbar	auf der Verlagswebseite innerhalb kurzer Zeit Listung auf Amazon und Co. kann bis zu zwei Monate dauern

Nicht übersehen: Auch der Druck dauert seine Zeit. Bei einem Taschenbuch circa zwei Wochen, bei Hardcovers circa einen Monat.

Auch die Verteilung im Onlinebuchhandel geht nicht von heute auf morgen, sie dauert zudem bei den einzelnen Anbietern (Amazon, Thalia, Hugendubel etc.) unterschiedlich lange. Gerade bei Amazon kann es unter Umständen Monate dauern, bis dein Buch dort als erhältlich gelistet wird. Und ja, es gibt Workarounds, aber dazu muss ich wissen, worum es dir geht und was genau du wann benötigst.

Kurz gesagt: Nach Abschluss des Manuskripts ist das Buch noch lange nicht fertig! Es müssen mindestens zwei bis drei zusätzliche Monate eingeplant werden, bis das Buch wirklich verfügbar ist!

Das ist vor allem dann wichtig, wenn eine Firmenfeier, Werbekampagne, Lesung und dergleichen geplant sind. Es gibt nichts Unangenehmeres, als wenn du befürchten musst, dass dein heiß ersehntes Buch nicht bis zur Firmenfeier fertig ist, oder du eine Werbekampagne gelauncht und bezahlt hast – und dann das Buch noch Wochen auf sich warten lässt, weil im Vorfeld schlecht geplant wurde.

★ *Marketing*

Dafür ist es noch etwas früh, schon klar. Dennoch hat es durchaus Vorteile, wenn wir jetzt schon über Marketing reden. Was liegt dir? Vorträge, Lesungen, Social Media? YouTube oder Podcasts? Vieles ist möglich, aber was ist sinnvoll? Und brauchst du irgendwo Unterstützung?

Als Expertin kann ich dir erste Ideen zur Vermarktung geben und dir Vorschläge machen, wenn du das wünschst.

Ziele des Zielworkshops

- ★ Alle wichtigen Punkte für dein Buchprojekt sind geklärt.
- ★ Du hast sämtliche Infos, die du benötigst, um das Buchprojekt zu starten (mit mir oder allein).
- ★ Mission Statement
  - Kontrakt mit dir selbst: Während des Zielworkshops schafft du für dich Klarheit und gehst einen Kontrakt mit dir selbst ein: „Ich mache dieses Buch, und so wird es in etwa aussehen!“
  - Zusätzlich gehst du einen Kontrakt mit mir ein. Wir schaffen Klarheit darüber, was geschehen soll.
- ★ Ich habe eine vernünftige Basis, um dir ein von meiner Seite aus verbindliches Angebot stellen zu können.

Mission Statement: Ich vereinbare mit dir, was gemacht wird, bis wann und wie. Es ist wichtig, dass alle Beteiligten wissen, wohin es gehen soll. Auf diese Weise kann das Buchprojekt auch weiterlaufen, wenn du einmal keine Zeit hast oder ich auf Urlaub bin. Natürlich kann sich immer noch etwas ändern, das ist das Leben. Aber zumindest haben wir diese Punkte geklärt und halten sie auch fest.

Ich mache immer Tonaufnahmen – entweder via Zoom oder mit dem Handy. Zusätzlich schreibe ich die wichtigsten Eckpunkte mit. Das Protokoll und das Konzept gehen an dich, und ich speichere die Dateien in meinem Ordner ab, in dem ich alle deine Unterlagen sammle. So können wir im Falle eines Missverständnisses schneller herausfinden, wo wir ansetzen müssen. Sollten wir nicht zusammenkommen, lösche ich selbstverständlich alles, was dich betriffft, wieder.

SIGN HERE!

Angebot und Auftragsbestätigung

> „Wow, da steckt ganz schön viel Arbeit drinnen – und wir haben noch immer keinen einzigen Buchstaben geschrieben. Ich weiß noch nicht mal, was mich das ganze Vergnügen kostet! Zwischen 15.000 und 35.000 Euro für ein Buch in meinem gewünschten Format ist zwar schon aussagekräftig, aber ich brauche es etwas genauer! Wann kannst du mir jetzt endlich sagen, wie viel ich bezahlen muss?"

„Um dir einen ungefähren oder konkreten Preis nennen zu können, ist es für mich wichtig, meinen Arbeitsaufwand einzuschätzen. Das kann ich nun nach unserem Zielworkshop machen. Das heißt, du bekommst ungefähr eine Woche nach unserem Workshop ein konkretes Angebot.

Üblicherweise schlage ich dir einen Pauschalpreis vor, der sich auf eine bestimmte Leistung (zum Beispiel das Schreiben des Buches, die Erstellung eines Layouts oder auch ein All-Inclusive-Angebot) bezieht. So hast du das Budget im Überblick und kannst dich entspannen, was die Finanzen angeht. Denn du weißt im Vorhinein, wie viel dich das Ganze kostet."

Ich schlüssle mein Angebot immer auf. Das heißt, du hast – je nachdem, was du brauchst – die Posten

- ★ Manuskripterstellung
- ★ Lektorat (das ist übrigens immer dabei – selbst, wenn du das Manuskript an einen Verlag gibst!)
- ★ Cover und Layout (inklusive Klappentext, Titelfindung, Grafiken etc.)
- ★ Endlektorat (nochmalige Überprüfung des druckfertigen PDFs durch eine andere Lektorin)
- ★ Upload (inklusive Preisgestaltung, Keyword-Definition, Unterstützung beim Autorenprofil, u.v.m.)
- ★ Marketing
- ★ Geheimhaltung, falls gewünscht
- ★ Sonstiges, das von Person zu Person unterschiedlich ist

Wenn du noch nicht sicher bist, wie du das Buch herausbringen möchtest, ob du das Design von mir und meinem Team oder lieber woanders gestalten lassen möchtest, dann kann ich dir manche Punkte auch optional anbieten, und du entscheidest im Laufe des Projekts, ob du diese optionalen Punkte in Anspruch nehmen möchtest oder nicht.

Zusätzlich findest du auf meinem Angebot die Option zur Teilzahlung. Das heißt, ein Teil der Summe ist sofort fällig, den Restbetrag kannst du in größeren oder kleineren Teilbeträgen bezahlen. Sobald du die Anzahlung überweist, gilt das Angebot als angenommen, und wir können beginnen.

„Aber ... das sind schon viele Optionen! Jetzt weiß ich gar nicht, wie ich damit umgehen soll! Ich bin verwirrt!“

„Kein Problem! Wir können über alles plaudern. Es ist mir wichtig, dass du weißt, was abgeht. Nur wenn du dich sicher fühlst und mir vertraust, können wir wirklich gut zusammenarbeiten.“

Wenn alle Unklarheiten restlos beseitigt sind und du mit den Kosten einverstanden bist, geht es in die nächste Runde: Du bekommst eine Auftragsbestätigung, die alle unsere Vereinbarungen detailliert festhält.

Wenn du eine persönliche Geschichte zu erzählen hast, und es sich bei deinem Buch um ein Herzensprojekt handelt, brauchst du eine andere Art der Begleitung.

Meine liebe Kollegin Conny Strumberger-Sellner ist hier die beste Ansprechperson. Conny, die ich zu meinen engsten Freundinnen und den begabtesten mir bekannten Schreiberinnen zähle, ist die perfekte Partnerin, wenn du jemanden suchst, der dich mit viel Herz und Einfühlungsvermögen dabei unterstützt, deine persönliche Geschichte zu erzählen.

Wieder andere Kolleginnen sind auf wirtschaftliche Themen, Kunst, Medizin, Reiseberichte, Urban Fantasy oder Mode spezialisiert.

Das ist an unserem Beruf das Schöne: Es gibt für jedes Gebiet den richtigen Ghostwriter, und für jeden Ghostwriter die richtigen Themen!

01
Idee
02
Erst-
gespräch
Ziel-
Workshop
07
09
Cover
Layout
Grafik
10
End-
Lektorat
Freigabe
Grafik
11
Upload
Verlag
12
Druck
Veröffent-
lichung

JETZT GEHT'S LOS!
Angebot Konzept
Auftrags-erteilung
Lese-probe 04
Feedback
Erstellung Manuskript 05
Über-arbeitung Manuskript 06
Lektorat Korrektorat 08

Jetzt geht's los!

„Jawoll! Endlich können wir starten! Wie geht's jetzt weiter?"

„Ich habe ja schon einige Unterlagen und Skripten von dir bekommen, und wir haben auch eine grobe Struktur erstellt.

Jetzt bin ich dran. Die nächsten Wochen werde ich damit verbringen, die ersten Inhalte einzufügen, umzuschreiben und zu überarbeiten. Sobald ich etwas vorzuweisen habe, schicke ich es dir und erwarte dein Feedback.

Wenn du möchtest, kannst du dafür wieder in mein Büro kommen. Aber es geht genauso gut, wenn wir telefonieren oder ein Zoom-Gespräch vereinbaren. Oder du schreibst mir deine Anmerkungen einfach ins Dokument und schickst es mir zurück.

***Achtung!** Wenn du das machst, bitte speichere es immer als neue Version und mit deinem Kürzel ab, damit wir keinen Kuddlmuddl zusammenbekommen!*

Sollte ich Fragen dazu haben, melde ich mich, ansonsten schreibe ich einfach weiter, und du bekommst zeitnah das nächste Stück deines Buches. Keine Sorge, es ist nur am Anfang etwas zeitaufwändiger für dich, bis wir uns auf einen Stil eingeschwungen haben. Meist geht das sehr schnell, und manchmal sind es überhaupt nur ein paar Wörter, die ich anpassen muss.

***Was mir wichtig ist:** Dieses Dokument, das ich dir schicke, ist nicht Korrektur gelesen! Ich achte natürlich darauf, dass es nicht vor Fehlern strotzt, aber bitte bedenke, dass es in der Feedbackrunde nicht darum geht, die Kommafehler des Ghostwriters zu finden! Das darf die Lektorin erledigen, sobald wir mit allem fertig sind!"*

„Haha, verstanden! Wenn ich Fehler finde, werde ich sie behalten!“

„Außer, ich habe etwas falsch verstanden. In diesem Fall bitte schrei sofort, damit sich das Missverständnis nicht durchs halbe Buch zieht, bis wir es gefunden und erlegt haben!“

Mit manchen Kundinnen sitze ich regelmäßig zusammen, und wir gehen die Geschichte Schritt für Schritt durch (besonders häufig mache ich das bei Biografien).

Manche schreiben mehr oder weniger große Teile des Manuskriptes selbst. Mit anderen führe ich lange Zoom-Gespräche, und wieder andere geben mir Unterlagen, anhand derer ich nach ein paar Stunden oder einem Tag Briefing selbstständig arbeiten kann.

Wichtig: Die Informationen, die ich von dir bekomme, werden von mir nicht auf ihre Richtigkeit überprüft oder recherchiert – es sei denn, dass du mich in dem einen oder anderen Punkt darum bittest, etwas nachzuschlagen. Darüber haben wir uns im Zielgespräch schon unterhalten.

Ansonsten vertraue ich voll und ganz auf deine Expertise. Das betrifft speziell Sachbücher und Ratgeber.

Du bist die Expertin auf deinem Gebiet – mein Job ist es, dein Wissen in schöne und gut leserliche Worte zu fassen.

Möglichkeiten der Zusammenarbeit mit einem Ghostwriter

> *„Aber sag mal … wie funktioniert denn so eine Zusammenarbeit mit dir oder deinen Kolleginnen überhaupt? Ich glaube, das hätte ich schon früher fragen können, aber es ist mir ja noch gerade rechtzeitig eingefallen, oder?"*

„Haha, ja absolut! Und ich hätte es auch angesprochen. Einfach der Vollständigkeit halber. Wir haben ja unseren Stil der Zusammenarbeit schon gefunden. Aber wer weiß – vielleicht brauchst du beim nächsten Buch etwas ganz anderes! Daher habe ich unterschiedliche Möglichkeiten der Kooperation hier für dich zusammengefasst.

Aber Achtung: Das sind die Optionen, die **ich** *biete – das läuft bei jedem Ghostwriter etwas anders, je nach Typ und Arbeitsweise! Und das ist auch gut so! Das bedeutet: Wenn du dir irgendwann doch eine andere Ghostwriterin suchen solltest, frag bitte unbedingt nach, wie sie es handhabt!"*

Diese Optionen der Zusammenarbeit kann ich dir anbieten:

★ *Ich schreibe für dich*

Wahrscheinlich möchtest du ein Buch, auf dem mit Fug und Recht dein Name steht. Weil es sich so liest, wie du sprichst oder schreibst. Es also deine Worte beinhaltet, deine speziellen Formulierungen, deinen „Sprach-Code".

Es soll dein Buch werden, dir auf den Leib geschneidert sein. Und dabei genau das repräsentieren, was dich und deine Methoden, deine Schwerpunkte und dein Wertesystem ausmacht. Das heißt, ich werde dir bei unserem Zielworkshop „auf den Mund schauen" und mir deine Sprache ausborgen.

Wie schon erwähnt, nehme ich unsere Gespräche auf, weil ich leider eine Schweinsklaue habe und oft meine eigene Schrift nicht lesen kann. (Böse Zungen behaupten, ich hätte, wenn man von meiner Handschrift ausgeht, eigentlich Ärztin werden sollen.) Außerdem bleiben auf diese Weise deine speziellen Ausdrücke und Phrasen nicht nur in meinem Gedächtnis, sondern auch digital konserviert.

„Meine" Bücher tragen übrigens fast nie meinen Namen. Weil es auch nicht die meinen sind. Ich versuche, eine Strecke in den Schuhen anderer zu gehen, ich sehe die Welt durch ihre Brille, und ich übersetze – so gut ich eben kann – den Inhalt ihrer Köpfe, ihrer Seelen und ihrer Herzen für die, die es interessiert.

Ich tauche ganz tief ins Wesen meiner Kundinnen ein – so tief, wie sie es zulassen. Um anschließend in ihren Worten ihre Geschichten zu schreiben, ihre Erkenntnisse weiterzugeben und ihre Lehren zu verbreiten.

Ich kann dein Buch in deinem Namen schreiben und nirgendwo aufscheinen. Ich bin dann also deine Ghost-Writerin.

Alternativ kann ich dich auch offiziell beim Schreiben unterstützen, wir promoten das Buch schon während des Schreibens gemeinsam auf Social Media, dein Buch hat einen Platz auf meiner Webseite, und ich bin bei Bedarf auch bei deiner Buchpräsentation an deiner Seite – im Rampenlicht oder außerhalb davon. Wie vielleicht schon erwähnt, ich bin nicht eitel.

★ *Wir schreiben miteinander*

Ab und zu kommt es vor, dass jemand sagt: „Lisa, schreiben wir das Buch doch gemeinsam. So voll und ganz. Mit unseren beiden Namen auf dem Cover."

Und ja, manchmal kommt es vor, dass ich „Ja" sage. Vor allem dann, wenn ich mich mit einem Thema identifizieren kann, die Person besonders spannend finde, und wir außer einer Sympathie noch einiges andere gemeinsam haben.

In diesem Fall kann ich für kurze Zeit aus meinem Elfenbeinturm treten, mich in der Sonne des Erfolgs sonnen und genießen – nur um mich etwas später wieder an der Stille und Einsamkeit meines Schreibkämmerchens zu laben.

★ *Editoring – mit dem feinen Staubwedel*

Das Manuskript ist schon fertig. Aber es ist nicht gerade berauschend. Alles ist da, alle Schreibtipps wurden berücksichtigt, die du so im Internet gefunden hast.

Und trotzdem ... irgendwie flutscht es noch nicht! Das ist ein Zeichen. Dafür, dass du Editoring benötigst. Dabei wird dein Text nicht grundlegend geändert, sondern nur „mit einem feinen Staubwedel bearbeitet und so zum Glänzen gebracht".

Diese Arbeit geht über ein erweitertes Lektorat hinaus, weil auch ein liebevoller Blick auf die Struktur und den roten Faden geworfen wird. Das ist es, was ein fabelhaftes Buch von einem eh recht guten Buch unterscheidet.

★ *Schreib-Coaching*

Manche Menschen möchten ihr Buch gar nicht schreiben lassen, sondern brauchen nur Unterstützung bei Struktur, Aufbau und Stil. Oder sie möchten ihre Leserinnen besser erreichen und sich optimal verständlich machen – auch auf einem Gebiet, das möglicherweise vielen anderen noch Rätsel aufgibt. Auch das ist möglich.

In einem Schreib-Coaching schauen wir uns an, was du noch brauchst, um dein Buch gut fertigstellen zu können – vom Schreibritual über die Buchstabenblockade bis hin zur gemeinsamen Überarbeitung deines Textes.

Informationstransfer

„Schön und gut – jetzt sollten wir wirklich starten. Aber wie kannst du überhaupt wissen, was ich sagen will? Du wohnst ja nicht in meinem Kopf!“

„Ha, da hast du recht! Wie du dir vorstellen kannst, brauche ich von dir jede Menge Input, um dein Buch schreiben zu können. Es gibt unterschiedliche Methoden, um an dein Wissen zu kommen. Wir haben uns ja schon im Zielworkshop entschieden, wie wir es machen. Aber ich zähle dir gerne auch die anderen Möglichkeiten auf!“

Ab und zu tritt der Glücksfall ein, dass meine Kundinnen schon etwas Fertiges haben. Dann kann ich das wunderbar als Basis verwenden, um an Wissen zu kommen.

Allerdings gehe ich die Inhalte dennoch gerne mit meiner Kundin durch. Denn oft verändert sich der Blickwinkel im Laufe der Zeit, und das wurde in einem speziellen Dokument vielleicht noch nicht berücksichtigt.

❖ *Diplomarbeit oder Ähnliches*

Eine Diplomarbeit hat viele Vorzüge. Sie ist (meist) strukturiert, enthält all dein Fachwissen und ist bereits in ganzen Sätzen zusammengefasst. Sogar ein Literaturverzeichnis gibt es schon!

Natürlich wird sich das Wording extrem ändern, wenn wir deine wissenschaftliche Arbeit in ein Buch verwandeln. Immerhin sollte das Buch ja gerne gelesen werden.

Und die meisten Diplomarbeiten sind in einer akademischen Sprache verfasst, die trocken wie Knäckebrot ist und dir im Hals steckenbleibt.

Es kann auch sein, dass Begrifflichkeiten geklärt werden müssen – manches kann ich googeln, aber sicher nicht alles, weil manche Begriffe in unterschiedlichen Kontexten unterschiedliche Bedeutungen haben. Das heißt, ein bisschen wirst du schon mit mir reden müssen.

❖ *Skripten*

Viele meiner Kundinnen verfügen über Skripten für Lehrgänge, Seminare oder Workshops. Diese Unterlagen sind zwar meist nicht vollständig, reichen aber oft aus, um eine erste Struktur zu erstellen, die von dir im zweiten Schritt mit zusätzlichem Inhalt gefüllt wird.

Das kann mündlich passieren oder schriftlich. Hier spielen wir manchmal Ping-Pong, ab und zu reicht der Inhalt der Skripten aber auch aus, um dein Buch anzufertigen.

❖ *Buch vom Buch*

Manche meiner Kundinnen haben bereits ein Buch, das sie entweder überarbeitet haben möchten oder von dem es eine Fortsetzung, ein Arbeitsbuch oder Ähnliches geben soll. Kalkuliere hierbei bitte ein, dass du mindestens zwei- bis dreimal mit mir zusammensitzen wirst, bevor ich alles habe, um den Auftrag zu deiner Zufriedenheit abschließen zu können.

Wenn dein Buch das erste Mal z. B. in Eigenregie herausgegeben wurde und „nur" lektoriert und neugestaltet werden muss, ist es wesentlich einfacher und auch preisgünstiger für dich.

❖ *Blogartikel*

Wenn du thematisch passende Blogartikel auf deiner Webseite hast, so bilden diese eine gute Basis für ein Buch. Hier kann ich auch deinen Stil gut erkennen.

Mündliches

❖ *Workshops, Vorträge, Seminare etc.*

Wenn mich eine Kundin einlädt, beispielsweise eines ihrer Seminare persönlich zu besuchen, nutze ich diese Gelegenheit fast immer.

Ich stelle diese Zeit normalerweise nicht in Rechnung, weil ich es liebe, neue Sachen zu lernen – ich verlange allerdings Fahrt- und Hotelspesen bzw. schlage diese auf die Pauschale drauf, wenn dir das sympathischer ist.

Wenn möglich, solltest du mir entweder eine Aufzeichnung von der Veranstaltung zur Verfügung stellen, oder du erteilst mir die Erlaubnis, selbst mitzuschneiden. Andernfalls sollten die Unterlagen und Skripten, die du mir gibst, sehr ausführlich sein.

❖ *Videos und Podcasts*

Der große Vorteil daran ist, dass ich deinen Stil hier schon gut heraushören kann. Wie zum Beispiel: Wie trittst du auf? Welche Wörter verwendest du gerne?

Aus einem guten Podcast mit mehreren Folgen bzw. aus einer Videoserie kann ich manchmal schon fast alle Informationen herausholen, die ich für das Buch benötige.

❖ *Interview*

Das ist die aufwändigste Art, um an Informationen zu kommen, aber ist es eine wunderschöne Arbeit, weil ich dich dabei am besten kennenlernen und gleich nachfragen kann, wenn ich etwas nicht verstanden habe.

Außerdem haben wir dabei wahrscheinlich gemeinsam noch ein paar neue Geistesblitze.

Übermittlung der Dateien

Wenn du mir Dateien schickst, gibt es dafür unterschiedliche Wege. Mir ist es wichtig, dass wir uns auf maximal zwei Kommunikationskanäle beschränken, andernfalls verliere ich den Überblick.

- ★ Per E-Mail als Anhang oder via WeTransfer, o. Ä.: Das ist für mich die beste Variante, da ich am schnellsten fündig werde, wenn ich etwas suche.
- ★ Per Sprachnachricht via WhatsApp, Messenger, o. Ä.: Eine kurze Info via Sprachnachricht ist sicher kein Problem. Wenn es aber (zu) lang wird, kann es unübersichtlich werden. Und vor allem: Ich habe bei diesen Nachrichten keinen Betreff, muss also unter Umständen sehr lange suchen, bis ich etwas Bestimmtes finde.
- ★ Cloud: Grundsätzlich spricht nichts dagegen, mit einer Cloud-Lösung zu arbeiten. Allerdings nur, um Dokumente abzuspeichern. Nicht zum gemeinsamen Online-Arbeiten!
- ★ Projektmanagement-Programme, o. Ä.: Ich arbeite mit Asana. Und wenn du möchtest, kann ich dir zu bestimmten Abläufen einen Zugang gewähren, sodass wir darüber kommunizieren können. Das setzt allerdings voraus, dass du mit der Arbeit in diesem Programm vertraut bist. Wenn nicht, dann halten wir uns lieber an andere Wege der Dateiübermittlung!

Über Feedback und Nacharbeiten

„So, ich habe jetzt die ersten Texte von dir bekommen. Sie gefallen mir großteils schon ganz gut. Was mache ich damit?“

„Wie schon vorher besprochen, hast du nun einen Probetext vor dir liegen. Jetzt ist es an dir: Lies dir alles durch. Spür hinein, ob es für dich passt. Und das, was nicht passt, markierst du bitte. Das ist ganz wichtig: Wenn dir etwas nicht gefällt oder du dich nicht damit identifizieren kannst, dann sag es bitte! Wenn du jetzt freundlich sein und mich schonen möchtest, fällt uns das später auf den Kopf. Also raus damit!“

Von einem meiner ersten Kunden habe ich als Feedback prinzipiell ein „Okay“ oder „Passt!“ bekommen. Als das Manuskript fertig war, habe ich dann doch nachgehakt: „Das Buch ist fertig – es ist nur noch nicht dein Buch! Ich brauche deine Meinung!“ Doch überraschenderweise kam auch jetzt nicht viel mehr: „Auf Seite xy steht ein Wort, das würde ich nicht verwenden. Bitte austauschen. Und die Überschrift auf Seite yx gefällt mir nicht. Sonst eh okay!“

Ich habe dieses Projekt abgeschlossen – und für mich entschieden, dass ich auf diese Art nie wieder arbeiten möchte. Weil es zutiefst unbefriedigend ist. Mein Kunde wollte einfach „ein Buch“. Mit seinem Namen drauf. Alles andere war ihm wurscht.

Mir allerdings nicht – ich will glückliche Kundinnen, leuchtende Augen und strahlende Gesichter, wenn das Buch da ist. Und nicht ein: „Ach, jetzt hab ich das auch gemacht.“

Ich weiß, dass wir darüber schon weiter vorne in diesem Buch gesprochen haben. Aber es ist zu wichtig, als dass ich riskieren möchte, dass du es eventuell übersiehst. Daher sage ich es nochmals ganz deutlich: Ich brauch dein Feedback, um dein Buch so gut wie möglich umzusetzen!

Wenn von deiner Seite aus alles passt, schreibe ich weiter. In regelmäßigen Abständen bekommst du von mir das Manuskript zur Kontrolle. Es liegt in deinem Interesse, das bisher Geschriebene so genau wie möglich durchzulesen (speziell, wenn es sich um eine Biografie oder ein fachliches Buch handelt), damit eventuelle Verständigungsfehler keine Chance haben.

Wenn dir etwas nicht gefällt, ändere ich das schon Vorhandene ab, und wir feilen einfach so lange, bis du dich damit wohlfühlst ... oder eigentlich, bis wir beide uns damit wohlfühlen. Wenn wir irgendwo hängenbleiben und auf keinen grünen Zweig kommen, setzen wir uns zusammen. Entweder live oder via Zoom. Manchmal kann man mit geschriebenen Worten nicht genau ausdrücken, was einen stört – im Gespräch geht es viel einfacher.

Ein besonders lieber Kunde war mit einem Kapitel seiner Biografie mega unglücklich. So bekam ich eine Sprachnachricht: „Lisa, ich finde mich in dem Text nicht wieder, das bin nicht ich! Ich weiß auch nicht!"

Nach einer Schrecksekunde fragte ich nach: „Was ist es, was genau stört dich?" und staunte nicht schlecht, als kurz darauf die Antwort kam: „Du schreibst immer ‚Mama', wenn du meine Mutter erwähnst. Ich hab sie ‚Mutti' genannt!

Und im nächsten Absatz schreibst du, ich ‚rutschte' über den Boden. Da würde ich ‚schlitterte' sagen. Oder ‚glitt'. Das sind eher meine Wörter!"

Kurzerhand tauschte ich die beiden Wörter aus und schickte ihm das Kapitel noch einmal.

Als ich seine Reaktion hörte, musste ich grinsen: „Lisa, das hast du großartig gemacht! Das bin so voll ich! So kann es weitergehen!"

Ich bin mir dessen bewusst, dass nicht alle Begriffe zu allen Menschen passen. Deswegen ist es ja so wichtig, dass ich dir beim Sprechen „auf den Mund schaue". Das heißt aber nicht, dass solche Dinge nicht passieren können. Es gehört dazu, dass manche Abschnitte als Ping-Pong zwischen uns hin- und hergehen. Das ist wie beim Polieren: Auch da reicht es nicht, einmal leicht drüberzustreichen. Meist braucht man etwas mehr Einsatz, bis auch die letzten kleinen Schlieren weg sind.

Lektorat und Korrektorat – wozu das denn?

„Wieso soll das Buch jetzt noch zur Lektorin? Kannst du nicht selbst rechtschreiben?“

„Oh ja, das kann ich, aber mein Hauptaugenmerk liegt auf dem Inhalt. Überleg dir einmal, was dein Buch kosten würde, wenn ich nichts delegieren würde! Mein Stundensatz ist um ein Vielfaches höher als der einer Lektorin oder Korrektorin.

Außerdem ist es leider eine Tatsache, dass wir unsere eigenen Fehler nicht entdecken, auch wenn wir noch so oft drüberlesen!

Was glaubst du, wie das WORD-Dokument dieses Buches ausgesehen hat, als ich es von der Lektorin zurückbekommen habe! Dafür habe ich sie ja – und ich schätze es sehr, dass sie mir diesen Schritt der Arbeit abnimmt!“

Wenn wir fertig sind – und damit meine ich, wenn wir beide übereinstimmen, dass in dem Buch alles drinnen ist, was es braucht –, bekomme ich von dir die schriftliche Freigabe für das Manuskript. Danach kommt es ins Lektorat.

Achtung: Von jetzt an wird inhaltlich nichts mehr geändert! Vergewissere dich also bitte unbedingt, ob alles, was jetzt drinnen steht, wirklich rein soll, und ob alles, was rein soll, schon drinnen steht.

Ich werde immer wieder Fragen stellen, schon während der Erstellung des Manuskripts. Wenn aus meiner Sicht etwas bedenklich ist, dann spreche ich es an. Sei es, dass du kurz dein Mütchen kühlen musst, weil deine Schwiegermutter ein alter „Grantscherm" ist oder weil du eventuell ein bisschen zu viel von dir zeigst. Ich werde dir nichts ausreden – außer, es wäre etwas, mit dem du dich strafbar machst –, aber ich werde vieles hinterfragen. Es kann auch sein, dass du erst mal einiges loswerden musst, und wir es auch aufschreiben, es in der letzten Version des Dokuments jedoch gestrichen wird.

Sobald das gesamte Manuskript fertig und von dir abgesegnet ist, lasse ich es von einer Lektorin meines Vertrauens auf allfällige Schreibfehler, grammatikalische Unebenheiten, Logikfehler und den roten Faden checken.

Wenn es zu mir zurückkommt, werde ich zunächst die eindeutigen Fehlerkorrekturen übernehmen und mir die Kommentare ansehen.

An diesem Punkt könnte wieder deine Unterstützung gefragt sein: Manchmal schlägt die Lektorin Änderungen vor, die über ein Korrektorat hinausgehen. Es kann sein, dass wir im Eifer des Gefechtes vergessen haben zu gendern, dass eine Passage zu deinen Ungunsten missverständlich gelesen werden könnte, oder dass wir irgendwo einen Denkfehler drin haben. Manche dieser Entscheidungen kann ich selbst treffen; bei anderen ist es wichtig, dass du deinen Senf dazugibst.

Doch wie auch immer: Auf jeden Fall kommt das fertig lektorierte Dokument als nächstes wieder zu dir, und es ist dein Job, alles noch einmal zu lesen.

Du glaubst es jetzt wahrscheinlich noch nicht, aber du wirst diese Aufgabe möglicherweise gar nicht so gerne erledigen, wie du dir jetzt vorstellst. Immerhin hast du zu diesem Zeitpunkt jede Passage in deinem eigenen Buch zigmal gelesen und kennst alles auswendig. Dennoch wird es dir nicht erspart bleiben, dich noch einmal damit auseinanderzusetzen.

Denn wenn du das Dokument gelesen hast, bekomme ich von dir wieder eine Freigabe, damit ich das Dokument zu Layout und Covergestaltung schicken kann.

Grafik und Layout

„Jetzt will ich aber langsam wissen, wie mein Buch aussehen wird! Wann erfahre ich das?“

„Ja, ich verstehe, dass du gespannt bist! Möglicherweise haben wir zu diesem Zeitpunkt ohnehin schon die ersten Entwürfe für dein Cover.

Gerade, wenn du schon Wünsche geäußert oder andere Bücher als Vorlagen mitgebracht hast, hat sich die Grafikerin wahrscheinlich bereits mit der Gestaltung deines Buches auseinandergesetzt und dir die ersten Vorschläge geschickt.“

Deine CI? Her damit!

Falls du eine bestimmte Farbgebung möchtest oder bevorzugte Schriftarten hast – her damit! Auch den Stil der Grafiken kannst du bestimmen. Ich schicke dir dazu den Link zu einer Webseite, auf der du dich umschauen kannst. Ich kann dich aber auch überraschen, falls dir das lieber ist.

Du hast deine eigene Illustratorin, die deinen Stil kennt und Grafiken oder Doodles für dich macht? Wunderbar! Ich schließe mich mit ihr kurz, damit sie die entsprechenden Druckvorgaben beachtet, und schon kann es losgehen!

Du hast deine eigene Grafikerin? Auch das ist möglich! Hat sie schon Bücher gemacht? Ja? Großartig!

Wenn deine Grafikerin allerdings noch keine Erfahrung mit der Erstellung druckfertiger PDFs eines Buches hat, dann würde ich dir dringend raten, das Endergebnis vor dem Upload von einem Profi checken zu lassen. Es gibt bei Büchern etliche kleine Dinge, auf die man achten sollte, damit das Ergebnis professionell aussieht. Und es wäre doch schade, wenn du so viel Zeit und Geld investierst, nur um dann erst recht am falschen Platz zu sparen.

„Heißt das, dass wir meine Firmenfarben und mein Logo auch in das Buch einbauen können? Ich hätte außerdem gerne eine ähnliche Schrift wie auf meiner Webseite, wenn das geht!"

„Aber klar können wir das! Wir werden zumindest eine ähnliche Schrift finden wie jene, die du auf deinen Unterlagen oder deiner Webseite verwendest!"

Bildrechte und andere Rechte

Ich schicke mal wieder voraus, dass ich keine Anwältin bin. Sollte sich also irgendwo eine kleine Ungereimtheit oder Ungenauigkeit einschleichen, möge man mir verzeihen!

„Sehr fein! Ich habe da auch ein paar Fotos von Kundinnen, die ich ebenfalls gerne im Buch verwenden möchte!"

„Uiuiui, das wird vermutlich leider nicht gehen! Und ich kann dir dazu auch nicht wirklich raten! Denn selbst, wenn deine Kundinnen dir hier und heute schriftlich geben, dass du ihre Bilder verwenden darfst, können sie diese Einwilligung jederzeit zurückziehen.

Und was machst du dann mit deinem Buch – außer einstampfen? Außerdem kannst du selten sicher sein, dass deine Kundinnen wirklich die Veröffentlichungsrechte an den Fotos besitzen. Siehe weiter hinten."

Bilder von Menschen

Wenn du Bilder von Menschen verwenden möchtest, gibt es zwei Optionen:

- ★ Du verwendest Stockfotos[6]. Achte dabei unbedingt darauf, dass du die richtige Lizenz (für Bücher und Medien, Magazine, etc.) erwirbst.
- ★ Du bezahlst Personen dafür, dass sie sich fotografieren lassen. Schließe in diesem Fall unbedingt einen Vertrag mit jeder Person ab, und lass dir auch von jeder eine Rechnung geben, damit du im Falle des Falles später belegen kannst, dass alles rechtens ist. Beachte bitte auch unbedingt, dass Fotografinnen die Rechte an ihren Bildern haben – dazu gleich mehr!

„Aber wie komm ich dann an Bilder? Ah, ich kann ja die nehmen, die mein Fotograf letztes Jahr von mir gemacht hat!"

„Okay. Wir müssen reden."

Bilder von einer Fotografin

Du hast vor kurzem echt tolle Bilder machen lassen, die du jetzt für dein Buch verwenden möchtest? Dann schau einmal ganz genau nach, was auf der Rechnung der Fotografin steht.

Hast du die schriftliche Genehmigung dafür? Normalerweise findest du auf der Rechnung vermerkt, wofür genau du die Lizenz erworben hast.

---

[6] Stockfotos sind professionelle Fotos, die von Fotografinnen aufgenommen und anschließend auf speziellen Webseiten verkauft oder zur Verfügung gestellt werden. Unternehmen und Einzelpersonen können diese Fotos kaufen oder lizenzieren, um sie in ihren Projekten zu verwenden.

Im Zweifelsfall gilt: Lass es dir schriftlich geben, dass du die Bilder im Buch verwenden darfst! Es kann sein, dass das nochmal eine Kleinigkeit kostet, aber das ist ein feuchter Furz gegen das, was dich eine Copyrightverletzung kosten kann.

„Aber ich bin auf den Bildern, und ich hab die Fotos bezahlt! Die gehören mir! Ich habe alle Rechte dafür!"

„Ja – aber nein. Du hast an deinen eigenen Bildern, die die Fotografin gemacht hat, lediglich jenes Nutzungsrecht, für das du bezahlt hast.

Doch du hast niemals das Urheberrecht. Das bleibt untrennbar mit der Fotografin verbunden und kann nicht abgetreten werden. Du erinnerst dich? Das geht nur bei Ghostwritern."

Das bedeutet für dich: Wenn du ein Foto von der Fotografin verwendest, bist du verpflichtet, eine Quellenangabe zu machen; sprich: ihren Namen oder den ihrer Webseite zu nennen. Das ist gesetzlich geregelt. Der Name muss nicht riesig, aber doch leserlich neben oder unter dem Foto ersichtlich sein. Wenn du sehr viele Fotos in einem Buch verwendest, kannst du die Angabe stattdessen auch im Impressum ganz vorne im Buch eintragen.

„Und wie ist das mit Zitaten? Das kann ja kein Problem sein, oder? Wenn etwas schon von Zeitungen abgedruckt wurde und so!"

„Großes Problem. Leider. Zitate, Liedtexte und Auszüge aus Büchern fallen unter Copyright. Aber es gibt Ausnahmen."

Die Verwendung von Zitaten ist nur unter der Voraussetzung zulässig, dass du dieses Zitat „diskutierst" und nicht ohne Kontext verwendest. Von Liedtexten – sofern sie nicht uralt sind und damit aus den Copyrightbedingungen rausfallen – würde ich generell die Finger lassen.

Und auch Auszüge aus Büchern dürfen nicht verwendet werden – außer, du zeichnest sie eigens aus und gibst die Quelle an[7]. Im Zweifelsfall lass ein Zitat oder einen Text lieber weg, als dich angreifbar zu machen.

Und mein Tipp zu Zitaten lautet ohnehin: Formuliere deine eigenen! Es ist immerhin dein Buch und nicht das von Henry Ford, Tony Robbins oder Marc Aurel[8].

„Okay, dann ist alles klar. Ist zwar schade, aber damit kann ich leben. Gehen wir jetzt drucken?"

„Aber ja, jetzt ist es bald so weit. Nur noch ein paar letzte kleine Schritte ..."

Wenn du mit den Entwürfen einverstanden bist, wird dein gesamtes Buch gesetzt. Sobald es fertig ist und alle Grafiken, Fotos, etc. eingefügt wurden, kommt es wieder zu dir – diesmal als PDF.

Du kannst jetzt schon virtuell durch dein Buch blättern und noch Änderungswünsche kundtun: eine Grafik mehr oder weniger? Eine Extraseite mit einem Zitat? Das ist jetzt noch alles möglich.

Sobald du dein Okay gibst, schließt die Grafikerin die Entwürfe ab. Zunächst das Innenleben, also den Kern.

Dann, sobald sie weiß, wie viele Seiten der Kern hat, kann sie die Buchrückenstärke berechnen oder berechnen lassen – je nach Verlag. Und damit kann auch das Cover fertiggestellt werden.

[7] Achtung: Es gibt Autorinnen, die darauf bestehen, dass ihr Buch nicht zitiert werden darf. Recherchiere also im Zweifelsfall lieber einmal öfter als einmal zu wenig.

[8] Apropos Marc Aurel: Seine Zitate dürftest du problemlos verwenden, weil er länger als 70 Jahre tot ist und seine Werke damit den Urheberschutz verloren haben.

Sobald alles abgeschlossen ist, geht dein Buch noch einmal zu dir, und du gibst mal wieder dein Okay. Ich baue hier gerne noch einen letzten Schritt vor dem Abschluss ein: ein Endlektorat. Meiner Meinung nach sollte dieses Lektorat von einer außenstehenden Person vorgenommen werden. Also jemandem, der das Manuskript bisher nicht gelesen hat. Auch wenn es nicht primär um Schreibfehler geht, sondern eher darum, ob mit dem Buchsatz alles stimmt, können auch hier nochmal ein paar letzte Kleinigkeiten gefunden werden.

> *„Aber wie jetzt? Wozu machen wir dann ein Lektorat, wenn nun erst recht noch Fehler im Buch sein können?“*

„Das hat zwei Gründe. Der erste ist rein menschlich. Jede Lektorin, auch die allerbeste, kann etwas übersehen. Weil wir eben alle Menschen sind. Der zweite Grund ist ein optischer. Im fertigen Buchsatz sieht alles anders aus. Möglicherweise stehen zwei gleiche Satzanfänge übereinander, die man vorher nicht gesehen hat. Vielleicht wurde ein Wort beim Setzen verschoben oder übersehen. Möglicherweise hat die Grafikerin irgendwo ein paar Leerzeichen zu viel eingebaut. Eventuell sind durch den Blocksatz im Enddokument die Buchstaben und Wörter zu nahe zusammengerückt. Vielleicht fehlt irgendwo eine Bildunterschrift, eine Überschrift hat das falsche Format, oder das Inhaltsverzeichnis wurde nicht angepasst, weshalb die Seitenzahlen nicht stimmen.

Das – und noch vieles mehr – sind Dinge, die im Eifer des Gefechts passieren können, selbst den erfahrensten Expertinnen. Und wenn es einmal schnell gehen muss, dann steigt auch die Fehlerquote. Aus diesem Grund bin ich ein großer Fan des Endlektorats.“

„Na gut, dann machen wir das! Auf die paar Tage kommt es nun auch nicht mehr an!“

„Richtig! Und vor allem haben wir die paar Tage ohnehin von Anfang an eingeplant. Sollte also niemand zwischendurch auf der Bremse gestanden sein, liegen wir super in der Zeit!“

Wenn das Endlektorat abgeschlossen wurde, kommt das PDF nochmals zu dir, und du gibst dein finales Okay, damit wir in den Druck gehen können.

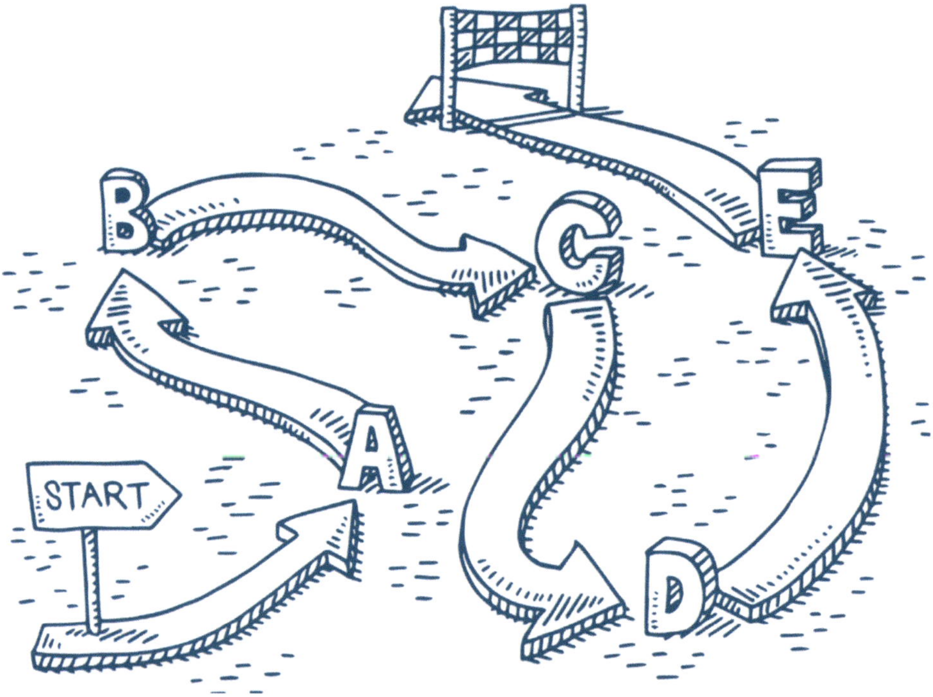
START
B
A
C
D
E

Abschluss – warum vorbei vorbei ist

> „Äääääh Lisa … ich habe jetzt das fertige Dokument vor mir liegen … und ich glaube, da fehlt noch was! Ich habe etwas vergessen! Ich weiß, eigentlich steht es schon da, aber das könnten wir noch einmal in anderen Worten reinschreiben, damit es wirklich jeder versteht!“

„Liebste Sascha, das ist keine gute Idee! Dein Buch ist vollständig. Es ist alles drinnen, was es braucht. Du musst nichts mehr ändern.

Und vor allem erhöht jede Änderung, die wir jetzt noch vornehmen, die Fehleranfälligkeit! Du musst dir vorstellen, es wurde alles lektoriert, gesetzt und eingerichtet. Wir haben anhand der Seitenanzahl die Buchrückenstärke des Covers errechnet, und im Buchinneren hat die Grafikerin gezaubert, damit jeder Buchstabe an der richtigen Stelle sitzt.

Daher muss dir klar sein, dass alles, was wir jetzt noch ändern, Mehrkosten verursacht. Das eine Kapitel, das du ändern möchtest, muss noch einmal lektoriert werden. Wenn du ein ganzes Kapitel neu einfügen willst, sollte alles noch einmal gelesen werden, um Wiederholungen zu vermeiden. Und auch die Grafikerin hat einen ordentlichen Mehraufwand.

Es geht ja nicht nur darum, den neuen Text irgendwie einzufügen. Da muss ein geschmeidiger und perfekter Übergang geschaffen werden. Das kostet. Zeit und Geld.

Wenn es sich also nicht wirklich um einen Notfall handelt, lassen wir bitte alles so, wie es ist. Du kannst dich auf mich verlassen, dass ich mich schon zu Wort gemeldet hätte, wenn du etwas vergessen hättest.

Und auch die Lektorin hat das Buch nicht nur Korrektur gelesen, sondern auch aus dem Gesichtspunkt der Verständlichkeit heraus betrachtet."

Kalte Füße

Kalte Füße vor der Veröffentlichung sind ganz normal. Da kommen tausend Fragen und Unsicherheiten:

- ★ Ist es gut genug?
- ★ Was werden die anderen sagen, wenn sie es lesen?
- ★ Vielleicht finden meine Kolleginnen, dass ich Blödsinn rede/schreibe!
- ★ Ich zeige da schon sehr viel von mir!
- ★ Und noch viel mehr.

Ich kann dich trösten: Diese Phase durchlebt fast jede Autorin. Da kommen die Book-Imposter-Gedanken, das Gefühl, man habe kein Recht, ein Buch zu veröffentlichen und noch mehr. Wenn du an diesem Punkt stehst, dann kommst du zu mir und wir reden drüber.

Wenn ich deine Bedenken nicht zerstreuen kann, habe ich die perfekte Unterstützung für dich: Meine liebe Kollegin und Partnerin in Crime Monika Lexa hat sich als Mental- und Resilienztrainerin auf genau dieses Thema spezialisiert und schon etlichen Jungautorinnen jeden Alters und Geschlechts geholfen.

Und sollte im Buch tatsächlich noch etwas Wesentliches fehlen, kann man natürlich auch nach Fertigstellung der PDF-Druckdatei etwas machen.

Wenn allerdings eine unverschiebbare Deadline vor der Tür steht, solltest du gut abwägen, wie wichtig dir eine bestimmte Information wirklich ist – vor allem, wenn du sie bisher trotz meiner Nachfragen außer Acht gelassen hast.

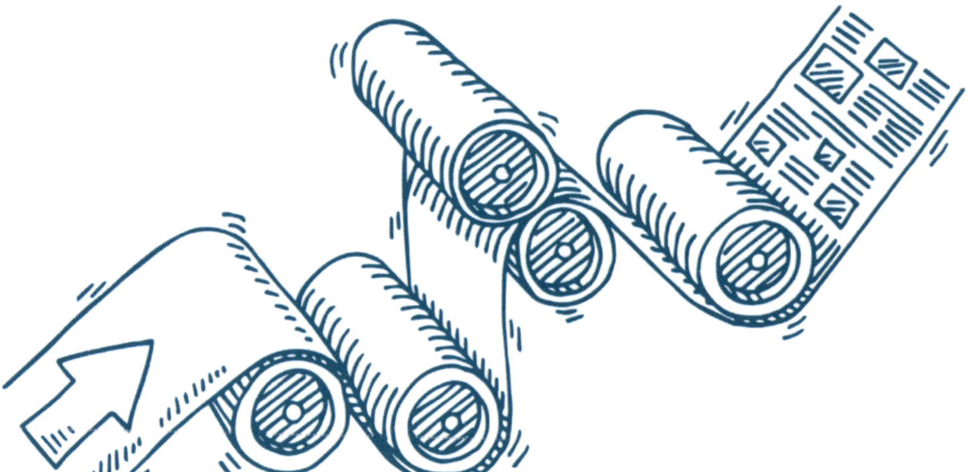

Wege der Veröffentlichung

Wie schon ganz am Anfang geschrieben, gehen wir davon aus, dass du dein Buch über einen On-Demand-Verlag herausbringst. Dennoch möchte ich dir hier kurz die unterschiedlichen Wege der Veröffentlichung vorstellen, die du – je nach Geduld und Ziel – mit deinem Buch einschlagen kannst.

Verlag? Self-Publishing? Oder ganz was anderes?

Diese Frage stellen sich viele Autorinnen erst, wenn es eigentlich schon zu spät für diese Überlegungen ist.

Ja, manche Verlagsagentinnen nehmen dich auch mit einem fertigen Manuskript. Aber bedenke, dass jeder Verlag andere Vorgaben hat, und du dein Werk wahrscheinlich massiv überarbeiten (oder erweitern oder, noch schlimmer, kürzen!) musst.

> *„Also wie jetzt? Self-Publishing? Aber das ist doch ... macht man das? Ist das überhaupt professionell?“*

„Aus meiner Erfahrung – und es sind immerhin über 150 Bücher, die ich in den vergangenen Jahren in die (online) Buchhandlung begleiten durfte – ist an einem GUTEN On-Demand-Verlag nix Verwerfliches.

Im Gegenteil! Ich habe sogar die Bücher einiger Promis auf diesem Weg rausgebracht!“

> *„Oh. Okay. Na dann ... erzähl mal, wie geht das? Und worauf muss ich achten?“*

Worauf du bei der Entscheidung bezüglich Veröffentlichung achten solltest

★ *Für welchen Markt ist dein Buch in erster Linie gedacht?*

Österreich und Deutschland? Auch Schweiz? Übersee?

Achte auf jeden Fall darauf, dass der Verlag deiner Wahl auch dorthin ausliefert, wo dein Markt ist! Das ist gerade bei der Schweiz als Nicht-EU-Land nicht immer gegeben.

★ *Wer soll oder wird es wo kaufen?*

Wirst du dein Buch eher zu deinen Vorträgen und Workshops mitnehmen, oder soll es in erster Linie bestellt werden?

★ *Wie soll dein Buch aussehen?*

Wenn du spezielle Wünsche hinsichtlich Layout und Wording hast, bist du sicher mit einem On-Demand-Verlag besser dran – traditionelle Verlage sind da meist sehr … nun ja, traditionell.

Sollten deine Wünsche allerdings sehr spezifisch sein, was Material oder Cover betrifft, musst du möglicherweise noch auf eine andere Option ausweichen: Druckereien, die auch Vertrieb anbieten.

Wenn du also unbedingt ein Taschenbuch mit Klappen, ein Cover mit Prägung oder Stanzung oder gar Duftlack haben möchtest, dann bist du am besten beraten, nochmal ein bisschen Geld in die Hand zu nehmen und dich an eine Druckerei zu wenden, die darauf spezialisiert ist.

Vorteil: Dort kannst du bei Bedarf auch gleich extra Inlays, Lesezeichen oder andere Materialien drucken lassen, die dem Buch im Direktversand beigelegt werden können.

Nachteil: Es ist nicht gerade billig, dich und dein Buchprojekt auf diese Art zu verwirklichen.

Wenn es in erster Linie schnell und billig sein soll, rate ich dir zu Amazon KDP Publishing, wo man mittlerweile auch Taschenbücher und Hardcover erstellen kann. Du darfst halt keine hohen Ansprüche an die Druckqualität stellen – die kann durchaus schwanken. Und denk daran, dass in der klassischen Amazon-Einstellung (also mit der Produktnummer, die dir von Amazon vorgegeben wird) dein Buch im Buchhandel nicht erhältlich ist. Wenn du im Verzeichnis lieferbarer Bücher (VLB) aufscheinen möchtest, dann musst du deine eigene ISBN[9] kaufen und entsprechend registrieren lassen.

★ *Was planst du mit deinem Buch?*

Wenn das Buch lediglich im Handel erhältlich sein soll, mach dich ruhig auf die Suche nach einem traditionellen Verlag – so du das möchtest und es dir wichtig ist. Wenn du jedoch dein Buch auch selbst verkaufen möchtest – bei deinen Vorträgen, Workshops, etc. –, dann ist es gut, wenn du davon günstige Autorinnenexemplare beziehen kannst. Hier hat Amazon die Nase vorn, allerdings ... siehe oben. Bei manchen On-Demand-Verlagen gibt es Autorinnenexemplare recht günstig, im traditionellen Verlagsbusiness sind sie meist sehr teuer.

★ *Wie bewirbst du dein Buch?*

Viele Menschen denken, dass ein traditioneller Verlag jede Menge Werbung für ihr Buch macht, sie auf Buchmessen o. Ä. schickt und sie promotet. Sorry, aber das ist in den meisten Fällen ein Irrglaube! Ich sage nicht, dass es nie passiert, aber die Norm ist es beileibe nicht. Das heißt: Auf welchem Weg auch immer du dein Buch veröffentlichst, du musst etwas unternehmen, um es bekannt zu machen.

---

[9] Internationale Standardbuchnummer. Die ISBN ist eine spezielle Nummer, die Bücher und andere Veröffentlichungen, wie zum Beispiel Multimedia-Produkte, eindeutig kennzeichnet.

"Ich dachte immer, dass das so läuft: Ich hab ein Buch geschrieben, jetzt kann ich mir einen Ponyhof kaufen!"

„Sorry, nein, so läuft das nicht! Wenn das Buch fertig ist, beginnt die Arbeit erst! Bei manchen funktioniert das super: Ich habe Kundinnen, die fast 40.000 Euro in ihr Buch investiert haben (nein, nicht alles in mich, leider ;-)) und nach einem halben Jahr alles wieder hereingeholt haben.

Das sind keine Einzelfälle. Ich sag immer, ich bin ein bisschen wie die Kaninchenpfote für meine Kundinnen ... ich bringe Glück und Erfolg. Allerdings nur, wenn die Kundinnen mitarbeiten.

Reich werden sie natürlich nicht über den Buchverkauf, sondern ... das findest du ganz hinten unter "Was kann dein Buch für dich tun?".

Andere wiederum nehmen das gute Stück ... und lehnen sich zurück. Komischerweise hört man von diesen Büchern nie wieder was. Was enorm schade ist. Und zwar egal, ob du dir Ghostwriting, Lektorat, Layout und Covergestaltung, etc. geleistet oder ‚nur' wertvolle Lebenszeit investiert hast."

„Und wie soll ich dann mein Buch veröffentlichen?"

„Wie dein Weg aussieht, kannst du selbst entscheiden. Wenn du es nicht kannst, frag andere um Rat, die es wissen.

Aber galoppiere nicht blindlings voran – dort könnte ein Abgrund lauern. Überleg dir, welcher Weg für dich passt, was du brauchst, was du gerne hättest, und was ein Nice-to-have ist. Dann entscheide dich."

Zusammengefasst gibt es folgende Optionen:

★ *Traditioneller Verlag*

Zu traditionellen Verlagen ist nicht viel zu sagen. Außer: Sie sind nicht einfach zu bekommen, zumindest nicht die großen.

Ich selbst hatte das Glück, für meine ersten beiden Bücher relativ problemlos einen Verlag zu finden. Es war ein kleiner Verlag, der von einem sehr ambitionierten Herren betrieben wurde. Ich bekam meine Veröffentlichung, konnte meine Autorinnenexemplare zu einem Preis von circa 10 Euro pro Stück erwerben (also zum Normalpreis mit Buchhandelsrabatt). Und ich wurde tatsächlich einmal zu einer Lesung nach München eingeladen. Doch obwohl ich wirklich fleißig Marketing betrieb, hat mir der Erlös dieses Experiments nicht mehr eingebracht, als man bei einem netten Abendessen in Begleitung ausgibt.

Bei großen Verlagen hast du natürlich die Maschinerie eines Unternehmens hinter dir – da musst du allerdings erst mal reinkommen! Wenn du das wirklich planst, dann rate ich dir dringend zu einer Buchagentin. Und bitte bedenke, dass du im Fall, dass dein Manuskript akzeptiert wird, die Vorgaben erfüllen musst, die der Verlag an dich stellt.

Diese Unternehmen investieren viel in dein Buch. Nicht nur an Kosten für die Fertigstellung, sondern meist auch für den Druck. Daher sind sie nicht sehr risikofreudig, sondern machen das, was aus ihrer Sicht am besten funktioniert.

Wenn du dir Bücher bestimmter Verlage ansiehst, wirst du feststellen, dass die Gestaltung und das Layout immer gleich oder ähnlich sind, immer die gleichen Farben verwendet werden und die Bücher meist die gleiche Größe haben.

Vorteile: Du musst dich nicht um Lektorat, Covergestaltung, Satz, Druck und Vertrieb kümmern.

Nachteile: Eigenexemplare sind meist sehr teuer; du hast wenig Mitspracherecht bei der Gestaltung, der Tonalität und der Länge.

Verlage brauchen ihre Zeit, und du kannst beim Veröffentlichungszeitpunkt deines Buches nicht mitreden.

Achtung: Der Verlag kümmert sich üblicherweise nicht um das Marketing deines Buches, nur um die Veröffentlichung. Manche Verlage fordern sogar aktiv deine Unterstützung beim Marketing ein.

Einer meiner Kunden beschloss mitten im Schreibprozess, dass er doch einen traditionellen Verlag haben wollte.

Also suchte und fand er einen Verlagsagenten, der ihn nach circa einem halben Jahr mit einem Verlag zusammenbrachte.

Das Blöde war, dass der Verlag in etwa die doppelte Zeichenanzahl wollte, als wir bis dato geschrieben hatten – und wir hatten das Buch eigentlich schon als abgeschlossen betrachtet. Noch blöder war, dass der Verlag es am liebsten vorgestern gehabt hätte. Wir setzten uns also gemeinsam auf unsere Hintern und machten es möglich.

Einfach war es nicht, aber irgendwie geil. Warum? Weil wir es hinbekommen haben!

★ *On-Demand-Verlag*

Bei dieser Art der Veröffentlichung sind die Unterschiede bezüglich Qualitätssicherung, Kundenservice und Unterstützung beträchtlich.

Nachdem die Anbieter ihre Bedingungen immer wieder ändern und auch laufend neue Mitbewerber auf den Markt kommen, möchte ich hier keine namentlichen Empfehlungen abgeben.

Nur so viel: Das Unternehmen, das dieses Buch verlegt und veröffentlicht hat, ist aus meiner Sicht sehr empfehlenswert! Ich arbeite mit den Verantwortlichen schon seit vielen Jahren zusammen.

Nachteile: Du musst dich um Lektorat, Covergestaltung, Satz, Druck und Vertrieb selbst kümmern.

Vorteile: Du kannst dein Buch gestalten und die Veröffentlichung timen, wie es für dich ideal passt. Vom fertigen Manuskript bis zum fertigen Buch vergehen wenige Monate oder manchmal sogar nur Wochen.

Autorinnenexemplare sind oft wesentlich günstiger als bei traditionellen Verlagen.

★ *Amazon Kindle Direct Publishing*

Wenn's schnell gehen soll, dann ist KDP definitiv eine gute erste Lösung. Denn Amazon macht nicht nur E-Books, sondern auch Taschenbücher und letzthin sogar Hardcovers.

Du hast einmal Arbeit damit, dein Profil anzulegen und alles auszufüllen, danach brauchst du deine Bücher nur noch hochzuladen. Und wenn du nur ein E-Book möchtest, musst du dazu – außer fürs Cover – nicht einmal einen Designer engagieren.

Allerdings sind diese Bücher üblicherweise nicht im Buchhandel erhältlich. Das liegt daran, dass Amazon keine ISBN verwendet, sondern eine ASIN[10]. Du kannst zwar ein KDP-Buch auch mit einer ISBN anlegen, aber das ist eine andere Geschichte, die hier zu weit führen würde.

Nachteile: Auch hier musst du dich um Lektorat, Covergestaltung, Satz, Druck und Vertrieb selbst kümmern. Die Druckqualität lässt meist zu wünschen übrig.

Vorteile: Einerseits sind die Autorinnenexemplare sehr billig, andererseits sind gerade die Kindle E-Books blitzschnell verfügbar. Wenn es also pressiert, dann ist das eine gute Option.

Auch die Margen für die Bücher sind relativ hoch. Zudem hast du, wenn du dich auskennst, wirklich gute Chancen, aktiv deine Verkäufe nach oben zu bringen.

---

[10] Eine ASIN (Amazon Standard Identification Number) ist eine spezielle Nummer, die Amazon jedem Produkt gibt. Diese Nummer hilft, Produkte auf der Amazon-Website zu finden und zu bestellen.

Ich selbst habe bei Amazon unter einem Pseudonym Bücher veröffentlicht, von denen ich mir auch – zu einem zugegebenermaßen sehr günstigen Preis – Eigenexemplare zuschicken ließ.

Die Bücher sind recht hübsch, haben einen roten Buchrücken. Blöderweise hat jeder Buchrücken einen anderen Farbton oder eine andere Intensität, weil die Dinger halt in Massenfertigung gedruckt wurden.

Tipp: Wenn du mit Amazon arbeiten möchtest, dann such dir eine Expertin, die dich bei diesem Prozess unterstützt.

(Ich bin es leider nicht. Aber – Überraschung! – ich kenne auch da jemanden!)

★ *Reines Self-Publishing*

Was ich damit meine? Das heißt, dass du dich – wie beim Veröffentlichen On Demand – selbst um Lektorat, Covergestaltung, Satz, Druck und Vertrieb kümmerst.

Der nächste Schritt ist jedoch ein anderer: Du checkst dir eine ISBN, lässt dein Buch drucken und kümmerst dich selbstständig um den Vertrieb. Manche Leute gründen auch eigens dafür einen Verlag. Das kann man so machen – muss man aber nicht. Wenn du nicht über ausgezeichnete Kontakte verfügst oder wirklich ganz genau weißt, was du tust, rate ich dir davon ab.

★ „Geschenkbuch“

Eine besondere Form des Self-Publishing ist das Geschenkbuch. Auch hier lässt du dein Buch selbst drucken und verschenkst es „gegen die Portogebühr“.

Dadurch kommst du an Mailadressen potenzieller Kundinnen, die schon die erste Hürde zu deinem „Funnel“ erklommen haben: Sie haben dir Geld geschickt, wenn auch sehr wenig.

Das muss man mögen – für mich fühlt es sich ein bisschen an wie Kaltakquise. Und in weiterer Folge ist es das auch. Denn die meisten Unternehmerinnen, die diese Strategie anwenden, lassen nach Versenden des Buches den Beschenkten anrufen, um weitere Aufträge zu lukrieren.

Tipp am Rande: Was du unter allen Umständen meiden solltest, sind sogenannte Druckkostenzuschuss-Verlage. Um das Geld, das du dort bezahlst, kannst du dein Buch lektorieren, layouten und herausgeben lassen – und noch einen kleinen Urlaub machen.

Ich halte es da mit Woody Allen: „Einem Club, der mich als Mitglied möchte, würde ich niemals beitreten!“ Das Gleiche gilt für Verlage, die dir sagen, dass sie genau dich suchen. Sei auf der Hut und erkundige dich, bevor du in der ersten Euphorie „Ja“ zu etwas sagst, das dir nichts bringt!

Und jetzt? Buchmarketing!

„Okay, dann ist mein Buch fertig, ich habe es im Buchhandel platziert. Und jetzt? Kann ich mich zurücklehnen?“

„Kannst du. Dazu würde ich dir aber nicht raten! Dein Buch verkauft sich nicht von selbst, da gehört schon mehr dazu! Wenn du möchtest, dass dein Buch wirklich gelesen wird, fängt die Arbeit jetzt erst richtig an. Wobei ... wenn du auf mich gehört hast, hast du ja schon längst damit begonnen, andere auf dein Buch aufmerksam zu machen.“

„Ähm ... ja, hab ich jetzt vergessen. Kannst du das bitte nochmal für mich zusammenfassen?“

„Gerne! Beachte aber bitte, dass ich keine Expertin im Buchmarketing bin. Dafür gibt es andere, die das besser machen. Ich arbeite auch immer wieder mal mit jemandem zusammen, der Buchmarketing anbietet.

Allerdings gibt es auch zahlreiche Autorinnen, die ihr Buch ganz einfach allein unter die Menschen bringen! Im Folgenden findest du eine Zusammenfassung von Maßnahmen, die dich dabei unterstützen, dein Buch zu verkaufen. Die Auflistung erhebt keinen Anspruch auf Vollständigkeit. Weil auch hier sehr viel Kreativität und Originalität gefragt sind. Und weil es immer drauf ankommt, welches Buch du geschrieben hast!“

★ Social Media

Ich finde es immer wieder überwältigend, wie viel man mit Social-Media-Postings erreichen kann.

Du kannst auch ein „Making of“ begleitend zum Schreibprozess posten, über deine emotionalen Hochs und Tiefs schreiben, von Hoppalas oder Highlights berichten ... und du kannst auch, sofern du das möchtest, Fotos aus unseren Sessions posten.

Was immer du tust – fang so früh wie möglich damit an und hör nie auf, bis du die Ziele erreicht hast, die du dir für dein Buch gesteckt hast!

Einer meiner Kunden hat ein Exemplar seines Buches an ausgewählte Personen versandt. Diese haben im Gegenzug ein Foto mit sich und dem Buch in den Social Media gepostet und ein paar Worte darüber verloren. Natürlich hatten die ausgewählten Personen selbst viele Follower aus dem richtigen Geschäftsbereich.

Lesungen

Apropos Lesungen: Nicht jedes Buch eignet sich für eine Lesung. Für das gute Stück, das du gerade in Händen hältst, ist eine klassische Lesung zum Beispiel sicherlich weniger geeignet. Ich könnte stattdessen einen kostenlosen oder niedrigpreisigen Einsteiger-Workshop abhalten und das Buch dort anbieten.

Oder ich könnte es so machen wie eine mir gut bekannte Autorin, die ihr Buch ebenfalls mit einem Avatar als Co-Autorin geschrieben und anstatt einer normalen Lesung eine Art Theaterkabarett veranstaltet hat.

> Eine Autorin, die ihre Biografie geschrieben hat, hat ihre Follower während des gesamten Schreibprozesses mit kleinen Ausschnitten aus dem Buch und Einblicken in ihr Leben und ihre Gefühlswelt auf dem Laufenden gehalten. Die Lesung, die sie ebenfalls auf Social Media angekündigt hatte, war völlig überlaufen, und schon in den ersten Wochen hat sich ihr Buch super verkauft.

★ *Medien*

Wenn du die Möglichkeit hast, dein Buch in die Medien zu bringen, dann nutze sie! Auch kleinere Zeitungen können durchaus hilfreich sein, um dein Buch bekannter zu machen. Solltest du bereits über entsprechende Kontakte verfügen, hast du natürlich einen großen Vorteil.

Aber auch als Einsteigerin hast du durchaus die Chance, in Zeitschriften, Zeitungen oder sogar ins Fernsehen zu kommen. Am besten engagierst du dafür eine Medienexpertin. Es gibt allerdings auch immer wieder sehr gute Workshops, in denen du lernst, wie du eigenständig an Medien und Redakteurinnen herantrittst. Auch dafür habe ich Kontakte.

Keine Angst davor – du schaffst das!

Es gibt, wie gesagt, noch viele andere Wege, um dein Buch zu promoten. Aber das würde hier viel zu weit führen.

Abschlussgespräch

„Und dann war's das mit uns? Oder wie?“

„Ja – und nein. Ich habe festgestellt, dass viele meiner Kundinnen wiederkommen. Nach einem Jahr, nach einem Monat. Oder sie wollen gleich das nächste Projekt angehen. Mich freut das natürlich irrsinnig.

Und ich kann dir sagen: Ein Buch ist wie ein Tattoo – hast du eins, willst du noch eins. Und noch eins. Vor allem dann, wenn die Entstehung nicht allzu sehr wehgetan hat. Doch bevor wir weitermachen, möchte ich etwas von dir. Und nein, ich meine nicht die Abschlusszahlung. Also ja, die auch. Aber ich möchte auf etwas anderes hinaus. Ich brauche noch einmal dein Feedback.“

„Wie jetzt? Warum? Wozu?“

„Ganz einfach: Ich möchte mich als Ghostwriterin, aber auch als Mensch weiterentwickeln. Und das geht nur mit Unterstützung von außen. Daher möchte ich von dir wissen, wie es für dich war? War ich gut?" ;-)

Sollten wir tatsächlich ein weiteres Projekt miteinander umsetzen, freue ich mich erstens sehr darüber und sehe es zweitens als Hinweis, dass du mit meiner Arbeit vermutlich zufrieden warst.

Doch um ganz sicherzugehen, würde ich das gerne auch von dir hören bzw. schriftlich haben. Und wenn es für dich okay ist, möchte ich dein positives Feedback auf meine Webseite schreiben, damit es nicht zwischen uns bleibt.

Mit deinem Namen oder anonym – wie es dir lieber ist. Auf der anderen Seite freue ich mich aber auch über ehrliches Feedback deinerseits, in dem du mir mitteilst, was du vielleicht noch gebraucht hättest.

Gab es nur Aha- oder auch Oh-oh-Momente? Wenn ich das weiß, kann ich es ändern. Bei unserem nächsten Projekt und – wenn es nichts sehr Individuelles ist – auch im Umgang mit anderen.

Du kannst mir glauben, dass ich das nicht sehr gerne mache. Dich nach meinen Verfehlungen fragen. Weil auch mir die Karotte lieber ist als der Stock (siehe Box). Aber wie ich zu Beginn des Buches bereits erwähnt habe, ist mir Ehrlichkeit bei einer Zusammenarbeit immens wichtig, also muss ich wohl auch gegebenenfalls mit den Stöcken leben.

Die Theorie über „Stöcke und Karotten" stammt vom Gehirnforscher Rick Hanson und ist eine einfache Art, den sogenannten „negativity bias" zu erklären.

Unser Gehirn ist von Natur aus stärker auf negative Erlebnisse und Informationen fokussiert als auf positive. Das bedeutet, dass wir negative Dinge (die „Stöcke") intensiver wahrnehmen und uns besser an sie erinnern als an positive Dinge (die „Karotten").

Es war für unsere Vorfahren wichtig, auf Gefahren (die „Stöcke") schnell zu reagieren, um zu überleben. Positive Erlebnisse (die „Karotten") waren weniger dringlich für das Überleben.

Wegen dieses Bias' neigen wir dazu, uns mehr auf das Negative zu konzentrieren und es zu verstärken. Das kann dazu führen, dass wir nur den einen kleinen negativen Punkt hören und uns daran aufhängen, selbst wenn wir gleichzeitig Tonnen von positivem Feedback bekommen.

Ich habe ewig gebraucht, bis ich das erste Mal Feedback von einer Kundin eingeholt habe. Weil ich eigentlich gar nicht hören will, wenn ich irgendwo Mist gebaut habe. Doch es ist mir ausreichend wichtig, daher werde ich auch dich befragen.

Ich finde es enorm wichtig, am Ende eines Projekts zu reflektieren.

★ Was war gut?

★ Was hättest du noch gebraucht?

★ Welche Learnings gibt es?

Zudem bekommst auch du von mir Feedback, sofern du das möchtest. In erster Linie – und das lasse ich mir nicht nehmen – sage ich dir, was mir an unserer Zusammenarbeit besonders gefallen hat, was dich als Mensch und als Kundin für mich besonders macht. Solltest du mehr wissen wollen, dann kannst du mich gerne fragen.

Gerade, wenn wir weiterhin zusammenarbeiten wollen – und über 70 Prozent meiner Kundinnen sind Wiederholungstäterinnen –, ist es gut, wenn wir ein Resümee ziehen.

Keine Sorge: Wenn wir es bis hierher geschafft haben, wirst du nichts Unangenehmes von mir hören. Denn sollten sich auf dem Weg bis hierher schon die einen oder anderen Steinchen angesammelt haben, haben wir ziemlich sicher schon darüber geredet.

Nach unserem Abschluss haben wir's für dieses Mal! Natürlich besuche ich gerne deine Buchpräsentation, wenn du magst. Und wenn es notwendig ist, tupfe ich dir auch den Angstschweiß von der Stirn, falls du aufgeregt bist, füttere dich mit Schokolade oder anderer Nervennahrung und stehe bei jedem Schritt hinter dir – außer, wenn du ins Rampenlicht trittst!

Ich bin auch gerne bereit, dein Buch auf meiner Webseite zu präsentieren (sofern du keine Geheimhaltung möchtest) und dich mit Social-Media-Postings zu unterstützen.

Wenn es dir lieber ist, werde ich mich an diesem Punkt unserer Zusammenarbeit aber auch einfach nur in Luft auflösen, dir deine finalen Dokumente zusenden und mich bei dir für die schöne Zeit bedanken.

Damit haben wir den Ablauf eines Buchprojekts aus meiner Sicht durch. Sicherlich findet sich hier oder dort noch ein Fragezeichen – das ist ganz normal.

Wenn du wirklich alles zu 100 Prozent so verstehst, wie ich es meine, würde das bedeuten, dass wir uns ein Gehirn teilen. Und glaub mir: Das willst du nicht!

Wenn du den Großteil verstanden hast, können wir uns beide auf die Schulter klopfen. Und wenn du möchtest, dass auch dein Buch nicht nur verstanden, sondern auch gerne gelesen wird, dann bitte ich dich ins nächste Kapitel weiter.

Platz für deine Notizen

01
Idee
02
Erst-
gespräch
Ziel-
Workshop
07
09
Cover
Layout
Grafik
Freigabe
Grafik
10
End-
Lektorat
11
Upload
Verlag
12
Druck
Veröffent-
lichung

WAS SONST NOCH WICHTIG IST

Was sonst noch wichtig ist

Das solltest du noch wissen

Es gibt viele Themen, die meiner Meinung nach noch angesprochen gehören. Einiges, das ich dir sagen wollte, habe ich wieder rausgeschmissen, weil das Buch sonst meinen eigenen Vorgaben nicht mehr entsprochen hätte: Mein Buch soll geschmeidig zu lesen sein, dich nicht überfordern und dir nicht „TMI" geben, also „too much information".

Es soll dir Spaß machen, dir die Gelegenheit bieten, einiges noch selbst zu entdecken und vor allem: Es geht nicht darum, wie du dein eigenes Buch schreibst, sondern darüber, wie so ein Buchprojekt überhaupt aussieht.

Ich bin jedoch davon überzeugt, dass es noch ein paar Sachen gibt, die du wissen solltest. Zunächst einmal beantworte ich dir die Frage, wie du nun von deinem Buch profitieren kannst, danach gehe ich wie versprochen auf deine mögliche Zielgruppe ein, und schließlich schauen wir uns gemeinsam an, wie viele Wörter ein Buch haben sollte.

Falls ich irgendein wichtiges Thema übersehen habe, scheiß da nix und schreib mir eine Mail. Meine Kontaktdaten findest du gaaaaaanz hinten auf der letzten Seite.

Was kann dein Buch für dich tun?

Ich bin ein großer Fan von emotional intelligentem Verkauf. Wir sind im 21. Jahrhundert angekommen, und es geht nicht an, dass wir immer noch so verkaufen wie in den späten 60er-Jahren, indem wir Angst beim Gegenüber aufbauen – NEIN, auch nicht die FOMO[11]! – oder tricksen.

Wir möchten Kundinnen haben, die nicht nur deshalb zu uns kommen, weil wir sie ausgetrickst oder ihnen FOMO eingeflößt haben, sondern solche, die von selbst den Weg zu uns finden, weil sie uns mögen.

Wir wollen Kundinnen, die glücklich sind mit uns, mit denen uns die Arbeit Spaß macht, mit denen sich die Zusammenarbeit anfühlt wie Nachhausekommen. Die uns weiterempfehlen und schätzen. Menschlich und fachlich.

Und diese Kundinnen bekommst du am besten über ein Buch, in dem du dich so zeigst, wie du wirklich bist. So sympathisch (denke ich mal), offen, als Mensch mit Stärken und kleinen Special Effects. Wenn wir das mit deinem Buch erreichen, dann haben wir es geschafft. Oder zumindest den ersten Schritt gemacht. Denn performen und deine Versprechen halten, das musst du schon selbst!

Wenn du davon abkommst, dass du dein Buch verkaufst, und dich mehr darauf konzentrierst, dass dein Buch dich verkauft, dann hast du unzählige Möglichkeiten.

---

[11] FOMO, kurz für „Fear of Missing Out“, bezeichnet die Angst, etwas Interessantes oder Wichtiges zu verpassen. Es ist das Gefühl, dass andere Menschen tolle Erlebnisse haben oder spannende Dinge tun, während man selbst nicht dabei ist.

Dieses Gefühl kann durch soziale Medien verstärkt werden, weil man dort ständig sieht, was andere gerade machen. FOMO kann dazu führen, dass man sich gestresst oder unzufrieden fühlt, weil man das Gefühl hat, nicht alles mitzuerleben oder nicht überall dabei zu sein.

★ *Dein Buch als Visitenkarte*

Ich kenne Leute, die niemals, NIEMALS ohne ihr Buch das Haus verlassen. Weil man nie weiß, wann man es braucht. Ja, du kannst es wirklich wie eine Visitenkarte einsetzen, Eine, die originell ist und nicht gleich in der Rundablage (= Mistkübel) landet, sobald du dich umdrehst.

Ich gestehe, ich habe eine ganze Lade voller Visitenkarten, teilweise – großteils! – weiß ich nicht einmal mehr, wie die Person, die sie mir gegeben hat, aussieht. Und schon gar nicht, was sie beruflich macht. Weil mir der Kontext fehlt.

Wenn du Menschen, die für dich interessant sind, und die deutlich an dir Interesse zeigen, dein Buch gibst, bleibst du garantiert in Erinnerung.

Schon beim Lesen oder Durchblättern können sich deine zukünftigen Kundinnen ein besseres Bild von dir machen, sie lernen dich kennen, und sie bekommen Lust darauf, dich zu buchen.

Einer meiner Kunden klebte auf sein Buch einen Sticker mit QR-Code und sandte es an die Großunternehmen, von denen er gerne für Vorträge gebucht werden wollte. Hinter dem QR-Code verbarg sich der Link zu einem Ausschnitt aus seinem Programm.

Als er wenige Zeit später nachtelefonierte, konnte er seinen Erfolg kaum fassen: Schon Ende des Jahres hatte er genug Vorträge verkauft, um das ganze nächste Jahr zu füllen!

★ *Dein Buch als Give-away*

Hältst du ab und zu Vorträge oder leitest Workshops? Das ist großartig! Dann musst du die Käuferinnen für dein Buch nicht suchen, sondern hast sie schon vor dir sitzen. Im wahrsten Sinne des Wortes. In deinem Publikum nämlich.

Eine perfekte Gelegenheit, dein Buch als Give-away einzusetzen! Entweder, indem du es als Goodie zu deinem Workshop dazugibst, oder indem du es vor Ort verkaufst.

Ist der Vortrag kostenlos, dann ist es auf jeden Fall sinnvoll, das Buch zu verkaufen. Wenn es sich um einen bezahlten Vortrag oder Workshop handelt, kannst du das Buch einfach in den Gesamtpreis miteinkalkulieren.

Wenn du das Buch verkaufen möchtest, genier dich nicht, es deinem Publikum aktiv anzubieten. Wie sollen die Leute wissen, dass du ein Buch geschrieben hast, wenn du nicht darüber redest?

Vielleicht kannst du ja sogar ein paar Hints zum Buch schon in deinen Vortrag einbauen oder in deinem Workshop auf dein eigenes Buch referenzieren anstatt – wie bei Literaturverweisen üblich – auf die Werke anderer.

Viele Menschen – speziell, wenn ihnen dein Vortrag oder Workshop gefallen hat – freuen sich, dieses gute Gefühl, das du ihnen verschafft hast, mit nach Hause nehmen zu können, vielleicht manches nochmal aufzufrischen oder ihre Partnerinnen, Eltern oder Freundinnen an ihren Erkenntnissen teilhaben zu lassen.

Aus diesem Grund sollte bei jedem deiner Auftritte ein Büchertisch vorhanden sein. Ein gewisser deutscher Speaker hat hier eine besondere Taktik: Zu jeder Veranstaltung bringt er seine gedruckten Werke mit, legt sie auf einen (bemerkenswert großen) Büchertisch und fordert das Publikum auf, sich gegen eine freie Spende zu bedienen. Er gibt netterweise sogar vor, in welcher Höhe die Spende sein sollte.

Eine andere Trainerin gibt ihre Bücher bei jedem Training oder Workshop „ausnahmsweise" um einen Bruchteil des üblichen Preises her.

Sehr großzügig, meinst du? Nicht wirklich – Bücher kosten im Einkauf nicht sehr viel. Und so wird dir pro Stück trotz des niedrigeren Preises wesentlich mehr bleiben als das übliche Autorinnenhonorar. Andererseits kannst du auch mit deinem Verlag einen guten Autorinnenpreis ausverhandeln – und möglicherweise sogar eine professionelle Betreuung für deinen Büchertisch bekommen.

So bleibt dir nach jedem Auftritt noch ein finanzielles Extra aus dem Verkauf deiner Bücher, die Teilnehmerinnen erinnern sich noch besser an dich und deinen Vortrag – und hoffentlich hast du in jedes Buch ein Lesezeichen mit deiner Werbung, einen Flyer oder etwas anderes gelegt! Deiner Fantasie sind kaum Grenzen gesetzt.

Eine andere Möglichkeit wäre, gleich zum Training oder Vortrag eine gewisse Anzahl von Büchern an die Veranstalter mitzuverkaufen. So hast du auch noch eine bessere Verhandlungsbasis für deine Trainings oder Auftritte.

★ *Dein Buch als Türöffner*

Schenk deinen (potenziellen) Kundinnen eines deiner Bücher! Wenn du dein Buch für die Akquise einsetzt und mit einem netten Anschreiben an die Zielperson deiner Wahl verschickst, kannst du ebenfalls davon ausgehen, dass das Buch sicher eher Aufmerksamkeit erregen wird als ein „nacktes“ Brieflein, das möglicherweise nicht einmal die richtige Ansprechperson erreicht.

Oft werden offensichtliche Werbebriefe schon vom Vorzimmerdrachen aussortiert – ein Paket erreicht die Adressatinnen jedoch so gut wie immer!

Bücher, deren Autorin man persönlich kennt, die man eigenhändig überreicht (oder gesendet) bekommen hat, erhalten meist einen besonderen Platz im Regal. Und wenn ein Buch gar signiert und mit einer persönlichen Widmung versehen ist, stehen die Chancen, dass das gute Stück einen Ehrenplatz bekommt, richtig gut.

Dein Buch wird also gesehen – nicht nur von der Person, die es von dir bekommen hat, sondern auch von deren Bekannten, Kundinnen oder Kolleginnen. Oder anders gesagt: Du ziehst deine Runden, ohne den Schreibtisch zu verlassen.

Während überreichte Visitenkarten meist schnell in der nächsten Schublade oder gar im Altpapier landen, wird dein Buch im Umfeld der Empfängerin weiterleben. Es wird von weiteren potenziellen Kundinnen gesehen, und dein Name bleibt in aller Munde.

Ja, es gibt sie vereinzelt immer noch – diese Barbarinnen, die ein Buch wegwerfen –, aber sie sind selten. Und ganz ehrlich: Möchtest du jemanden als Kundin haben, die so etwas Furchtbares tut?

Und sollte wirklich der eine oder die andere dein Buch in einen „Offenen Bücherschrank“ bringen oder weiterschenken, dann hast du ohne großes Zutun deinen potenziellen Kundinnenkreis noch erweitert. Du kannst also nur gewinnen!

★ *Dein Buch als Basis für mehr*

Auf einem Buch kannst du alles aufbauen – vom Workshop über einen Lehrgang bis hin zu einem Einzelcoaching. Und dein Buch kann all das für dich verkaufen.

Eine mir bekannte Psychologin hat ihr Buchmarketing an sich genial aufgebaut: Es gibt zum Buch einen Podcast – zu jedem Kapitel eine Folge. Kostenlos.

Sie hat danach eine sogenannte „Masterclass“ aufgenommen, die ebenfalls als Podcast abzurufen ist. Seltsamerweise ebenfalls kostenlos. Und dann gibt es ein Online-Programm, für das sie Geld verlangt. Du siehst, auch so kann es klappen! Man muss nur Ideen haben!

> Eine meiner Kundinnen hat auf ihrem Buch nicht nur ihren Workshop aufgebaut, sondern auch Videos dazu gedreht, einen Online-Workshop, Arbeitsbücher, ein Hörbuch und vieles mehr erstellt. Sie investierte mehr als 50.000 Euro – und nur ein halbes Jahr später hatte sie alle Ausgaben wieder herinnen. So geht es auch, wenn du kreativ und geschäftstüchtig bist!

Du siehst: Wenn du ein wenig Zeit und Liebe in dein Buch und dein Marketing investierst – durchaus schon, bevor du zu schreiben beginnst! –, bekommst du es hundertfach zurück!

★ *Dein Buch als best Seller für dich*

Zusammengefasst kann man sagen: Sobald du dich von der romantischen Vorstellung trennst, dass du fürderhin als Bestsellerautorin nur mehr die Beine hochlegst und vom Buchverkauf lebst, kannst du dein Buch für dich die Vorarbeit leisten lassen.

So kommst du vom leidigen dich-selbst-Verkaufen weg und kannst dich bzw. deine Dienstleistung stattdessen von deinem Buch anpreisen lassen. Du verwendest also weniger Zeit in unbezahlte Werbe-Arbeit und kannst dafür in dem Bereich performen, in dem du wirklich gut und gerne unterwegs bist.

Und schon hast du deinen best Seller!

Zielgruppe: ALLE

Lassen wir noch einmal kurz unsere liebe Sascha zu Wort kommen:

> *„Also, mein Buch soll JEDER lesen! Deswegen schreib ich es auch für ALLE!"*

Super Ansatz.

Leider nicht.

Doch nun kommt die große Frage, die sich auch bei meinen Kundinnen immer wieder stellt:

> *„Wie definiere ich meine Zielgruppe? Alle geschiedenen Frauen zwischen 35 und 46, die UnternehmerInnen sind? Alle glatzköpfigen Männer unter 50, die einen Labrador und einen Range Rover besitzen?"*

„Liebe Sascha, aus meiner Sicht wird's so nix. Denn nicht alle Menschen in einem bestimmten Alter oder Umfeld haben die gleichen Werte, Einstellungen, Probleme oder Wünsche. Und sie haben nicht einmal annähernd ähnliche Lebenssituationen!"

Ein Blick in die demografische Zielgruppe

Schnappen wir uns doch mal das Vergrößerungsglas und sehen wir uns ein paar der besagten Damen mit den Eckdaten „geschieden, zwischen 35 und 46 und Unternehmerin“ an.

Henriette, 45, hat eine erwachsene Tochter, die sie finanziell unterstützt. Sie lebt in Wien und freut sich darauf, dass Nancy, ihre Tochter, bald in ihr florierendes Immobilienbusiness einsteigt, um ihr unter die Arme zu greifen. Henriettes Exmann war ein notorischer Spieler, der sie mit Schulden in Millionenhöhe zurückgelassen hat, als er vor einigen Jahren fluchtartig das Land verließ.

Dolores, 44, war früher ein Mann. Seit ihrer frühen Jugend hat sie davon geträumt, als Frau zu leben. Mittlerweile hat sie eine Geschlechtsanpassung hinter sich und unterrichtet Yoga in einem kleinen Wellness-Studio auf dem Lande. Auch sie ist geschieden. Seit mehreren Monaten trifft sie sich mit einem jungen Mann, der sehr in sie verliebt ist.

Agnes, 40, lebt mit ihrem kleinen Sohn, ihrer neuen Freundin und drei Hunden in einem Wohnwagen auf dem Grundstück ihres Exmannes, zu dem sie ein recht gutes Verhältnis hat. Sie handelt mit Keramikfiguren und legt jeden Cent zur Seite, um sich eine E-Gitarre kaufen zu können, weil sie den Traum hat, in einer Band zu spielen.

Ruby, 35, ist Coach. Sie lebt seit ihrer Scheidung in einer glücklichen kinderlosen Beziehung. Gemeinsam mit ihrem Partner hat sie eine Firma, und die beiden reisen beruflich quer durch Europa und halten Workshops für große Unternehmen. In ihrer Freizeit geht sie Paragleiten, Mountainbiken und mit Hund weite Strecken spazieren.

Was haben diese Frauen gemeinsam – außer den demografischen Eckdaten?

Vermutlich wenig, und im „echten Leben“ würden sich ihre Wege kaum kreuzen.

„Das bedeutet, die Wahrscheinlichkeit, dass diese Frauen ähnliche Werte, Einstellungen, Probleme oder Wünsche haben, ist eher gering oder kann sich nur durch Zufall ergeben. So wird das wohl nichts. Also schauen wir uns einen anderen Zugang an.

Für wen passt dein Buch wirklich? Wer hat den größten Nutzen davon? Und wer kann die Botschaften in diesem Buch wirklich gut nehmen und umsetzen?"

„Tja, das kommt vermutlich auf den Stil an."

Und jetzt? Wir wissen nicht, welchen Stil wir schreiben wollen, weil wir nicht wissen, für wen. Dumm gelaufen. Lassen wir es doch einfach! Oder ... wir gehen es ganz anders an!

Die emotional Crowd

Mit welchen Menschen fühlst du dich im Alltag am wohlsten? Wenn du so tickst wie die meisten von uns, dann wohl mit solchen, die ähnliche Werte haben wie du, ähnliche Einstellungen, ähnliche Träume wie du hattest oder noch hast.

Wenn du ein Buch schreiben möchtest, um Kundinnen zu gewinnen, wäre es nicht schlau, sich Menschen zu suchen, mit denen du dich wohlfühlst? Die Wahrscheinlichkeit, dass auch sie sich bei dir wohlfühlen, ist relativ hoch – und schon wird das Arbeiten angenehmer und weniger aufwändig.

Menschen suchen Identifikation

Wie ist das bei dir, wenn du einen guten Roman liest? Kann es dir passieren, dass du reinkippst, dass du Stellung beziehst, dich in einer der Figuren wiederfindest? Ja? Gut! Wunderbar!

Dann haben wir beide schon eine gute Basis für unsere Überlegungen. Denn ich kann bei einem Buch – sorry, beim Lesen eines Buches! – ausflippen, weinen, lachen. Ich spüre mein Herz, lebe und fiebere mit. Und ich wünsche mir einen möglichst positiven Ausgang.

„Super! Und was hat das jetzt mit meinem Buch zu tun?“

„Gute Frage! Verdammt viel, um genau zu sein. Denn selbst ein Sachbuch darf berühren und bewegen. Nein, das soll es sogar, denn sonst ist irgendwas schiefgelaufen.“

Es gibt noch immer Menschen, die meinen, dass ein Buch etwas Seriöses sein muss. Dass man möglichst gestelzt daherreden muss, damit man einen guten Eindruck hinterlässt. Auch dann, wenn man eigentlich ganz anders ist. Weil, sonst ...

Ja, was sonst?

Denken wir den Gedanken doch mal zu Ende.

Du schreibst ein Buch, genau in der Art, wie es sich gehört, nämlich seriös und ernsthaft. So, dass es nach menschlichem Ermessen und statistischen Gesichtspunkten so viele Menschen wie möglich erreicht. Das willst du ja, oder? Dieses Buch liegt nun – marketingtechnisch bestens abgesichert – in der Buchhandlung oder im Amazon(as).

Die einen sehen es und scrollen oder gehen gleich weiter, weil ihnen die Aufmachung nicht gefällt. Zu Mainstream, das kennen wir alles schon. Die anderen machen einen echten oder virtuellen Blick ins Buch und stellen fest: Naja, ist sicher ein gutes Sachbuch. Aber davon gibt es viele. Schau'n wir mal, vielleicht kauf ich es irgendwann. Aber KLICK macht es nicht.

Und dann kommen die, denen genau das gefällt, die dein Buch kaufen. Und später auch dich buchen wollen. Deine Leistung in Anspruch nehmen wollen.

Die kommen jetzt zu dir und erwarten sich ... dich – so wie du dich ihnen in deinem Buch präsentiert hast.

Also so, wie du eigentlich nicht bist.

Weil du eben eine Menge Special Effects hast und nicht bist wie alle anderen.

Nun hast du zwei Möglichkeiten: Entweder du machst weiter einen auf seriös und ernsthaft – sehr anstrengend!

Oder du versuchst, deine neuen Kundinnen davon zu überzeugen, dass du in echt eigentlich viel cooler bist.

Geht sicher.

Aber warum nimmst du nicht den einfachen Weg?

Wenn du von Haus aus weißt, wer du bist, wofür du wirklich stehst und welche Werte du vertrittst, kannst du all das in dein Buch einbauen – und die richtigen Kundinnen kommen automatisch zu dir! Die, die einen ähnlichen Humor haben, ähnliche Werte wie du. Die, mit denen Arbeiten Spaß macht.

Nicht alle geschiedenen Frauen zwischen 35 und 46, die Unternehmerinnen sind. Sondern die, die ähnliche Visionen und Ziele haben wie du.

Die, die vielleicht dorthin kommen wollen, wo du stehst. Weil sie auf dem Weg ein paar Schritte hinter dir sind.

Die, mit denen du auch auf einen Kaffee gehen würdest, wenn sie nicht deine Kundinnen wären. Egal, ob sie geschieden sind, Männer oder Frauen, zwischen 35 und 46 oder älter oder jünger, Kinder haben oder nicht, Unternehmerinnen oder Angestellte oder Arbeitssuchende sind.

Die, deren Herzen höherschlagen, wenn sie an deine Geschichten denken, deren Augen leuchten, wenn sie dein Buch lesen.

Deine natürliche Fanbase.

Es gibt wissenschaftliche Studien, die zeigen, dass wir Menschen, die uns ähnlich sind, meist attraktiver finden. Dass wir uns in deren Nähe wohler fühlen. Die meisten von uns kuscheln am liebsten in ihrer menschlichen Komfortzone – gerade dann, wenn sie mit jemandem zusammenarbeiten oder jemanden beauftragen wollen.

Je größer das Projekt, desto wichtiger ist es, dass man sich grün ist.

Doch zurück zum Buch! Wenn wir uns mit der Autorin identifizieren, neigen wir eher dazu, bessere Rezensionen zu geben und das Buch zu empfehlen. Wir sehen leichter über Fehler hinweg und freuen uns, dass wir wieder jemanden gefunden haben, der mit uns in unserer Bubble sitzt. Und wenn wir von derjenigen noch etwas Neues erfahren können – umso besser!

Also: Mach's dir leicht – sei du! Es gibt eines auf der Welt, das du mit Sicherheit besser kannst als alle anderen: DU sein! Also sei du, ziehe Menschen an, die zu dir passen – und rocke es!

Platz für deine Notizen

Wie viele Wörter hat ein Buch?

„In letzter Zeit habe ich mich gefragt: Wie viele Wörter brauche ich eigentlich für ein Buch? Ist eine Buchseite dasselbe wie eine Seite im Manuskript? Wie kann ich das umrechnen?“

„Nun, fangen wir gleich mit der schlechten Nachricht an: Nein, das kannst du nicht. Zumindest nicht so einfach. Warum? Weil es darauf ankommt, wie diese Seite im Buch aussehen soll. Und wie viel auf der originalen Seite im Manuskript steht.“

Nur Text

Wenn du nur Text im Buch haben möchtest, also weder Grafiken noch viel Weißraum, dann ist es relativ einfach. Du suchst dir aus, welches Format dein Buch haben soll.

Wenn du die Schrift, die Schriftgröße oder den Durchschuss (also den Zeilenabstand) veränderst, dann bedenke, dass sich die Seitenanzahl ändert. Eine solche Seite hat zwischen 1.500 und 2.500 Zeichen inklusive Leerzeichen – das ist jedoch nur eine grobe Schätzung und kann im Einzelfall stark abweichen!

Jede Schrift nimmt unterschiedlich viel Platz ein. Das heißt, wenn du zwei unterschiedliche Schriften in, sagen wir, 10 Punkt anlegst, haben sie trotzdem eine unterschiedliche Größe und unterschiedliche Zeilenabstände.

Das liegt daran, dass manche Schriften noch aus dem klassischen Buchsatz kommen und daher ein gewisses „Fleisch“ haben, andere wiederum schon als digitale Schriften entstanden sind.

Denk bitte immer daran, dass nicht jede Schrift für Druckwerke lizensiert ist. Und dass du dir damit wirklich nur einen ersten Überblick verschaffen kannst.

Den Buchsatz solltest du auf jeden Fall einem Profi überlassen. Gerade, wenn du mit deinem Buch deinen Expertenstatus untermauern möchtest, wäre das Erscheinungsbild der letzte Posten, an dem du sparen solltest!

Wie viele Zeichen hat eine grafisch gestaltete Buchseite?

Ha! Jetzt wird es richtig schwierig! Auch wenn man nur mehr Weißraum einfügt, verringert sich die Zeichenanzahl rapide. Hat man dann auch noch Grafiken, etc., braucht man umso weniger Buchstaben, um eine Seite zu befüllen.

Eine Teilnehmerin unseres Lehrgangs meinte einmal, das sei unfairer Wettbewerb und so etwas wie „Zeit schinden, nur auf Seiten umgelegt. Also Seiten schinden“.

Ich sehe das nicht so. Weißraum ist wichtig!

Warum, erfährst du gleich!

Gehirngerechte Aufbereitung

Um einen Text gehirngerecht aufzubereiten, braucht das Auge (und damit das Gehirn) Platz zum Ausruhen; es soll nicht unerbittlich durch den Text geprügelt werden, sondern darf auch einmal innehalten und durchatmen.

Außerdem gibt es für viele Menschen kaum ein größeres Erfolgserlebnis als eine Seite umzublättern. Wenn diese Seiten jedoch in Mini-8-Punktschrift vollgekritzelt sind und dazu weder Abstände noch Weißraum aufweisen, ermüdet nicht nur das Auge, sondern auch der Geist – von der Lesemoral ganz zu schweigen.

Weißt du, wie zäh es ist, für eine einzige Seite eine gefühlte Ewigkeit zu brauchen – wissend, dass du noch 350 vor dir hast?

So spannend kann das Thema gar nicht sein, dass der durchschnittliche Mensch das durchhält!

Und unsere Zielgruppe setzt sich nun mal aus durchschnittlichen Menschen und nicht aus verkappten Genies zusammen.

Abwechslungsreich gestaltete Seiten haben absolut eine Existenzberechtigung in einem Buch – solange sie nicht überhandnehmen. Und natürlich hängt es immer von Zielgruppe und Autorin ab, wie viel Grafisch-Verspieltes Platz finden darf und soll.

Leere Seiten in einem Buch

„Ach, leere Seiten brauche ich also auch?“

„Ja, tatsächlich ergeben sich manche Leerseiten einfach durch den Text. Ein Kapitel fängt in Romanen immer auf einer rechten Seite an. Auch in Sachbüchern ist das mittlerweile üblich. Daher bleibt die linke Seite oft frei, wenn es sich vom Text her nicht anders ausgeht.

Wenn ein größeres Kapitel beginnt, beschließt die Grafikerin möglicherweise, einen Kapiteltrenner einzuziehen. Das kann durchaus hübsch aussehen und hat ebenfalls wieder den Sinn, das Auge und den Geist ein wenig ruhen zu lassen.

Auch ganz vorne im Buch befindet sich ein leeres Blatt – das sogenannte 'Schmutzblatt'[12] – auf dem eventuell Titel und Autorin stehen, wo aber auch Platz für eine Widmung bleiben sollte.“

Vielleicht hast du gar „Platz für Notizen“ in deinem Buch? Weil es ein Buch ist, mit dem die Leserinnen arbeiten sollen? Sieht gut aus – wirft aber deine Seitenkalkulation ganz schön durcheinander!

Wobei – von welchen Seiten reden wir eigentlich? Ich musste feststellen, dass die meisten Menschen mit Begriffen wie „Normseite“ oder „Seite in dieser oder jener Formatierung“ nichts anfangen können. Deshalb schauen wir uns auch das einmal genauer an.

---

[12] Der Name „Schmutzblatt“ kommt aus der Zeit, als Bücher noch keinen Einband hatten. Das Schmutzblatt machte damals seinem Namen alle Ehre – und hielt das Innenleben des Buches sauber.

Was zum Geier ist eine Normseite?

Was eine Normseite ausmacht, das hängt ganz davon ab, wen man fragt. Je nach Definition und Quelle umfasst eine Normseite 1.500 – 1.800 Zeichen, eine Normzeile 55 Zeichen, jeweils inklusive Leerzeichen. Eine Normseite besteht damit aus durchschnittlich 250 Wörtern.

Für uns ist das nicht weiter wichtig – außer wir sind (oder haben) Lektorinnen, die ihre Arbeit nach Normseiten verrechnen. Doch eigentlich arbeiten viele Autorinnen, Übersetzerinnen, Lektorinnen und Verlage heutzutage nicht mehr mit der klassischen Normseite, sondern bestimmen die Länge eines Textes anhand der Zeichenmenge.

So kann ein klassisches Sachbuch zwischen 180.000 und 450.000 Zeichen enthalten, jedoch durchaus auch mehr oder relevant weniger. Das hängt von den Vorstellungen des Verlages ab, aber auch davon, wieviel die Autorinnen sagen bzw. schreiben wollen.

Ich persönlich bevorzuge sowohl beim Schreiben als auch beim Lesen Bücher, die ich zwischen Wien und Salzburg im Zug auslesen kann. Übersetzt für meine deutschen, Schweizer und Südtiroler Leserinnen also in circa dreieinhalb bis vier Stunden. Das müssen deswegen keine dünnen Bücher und schon gar keine Broschüren, sondern können sehr wohl Bücher mit einer schönen offenen grafischen Gestaltung sein.

„Aber in Amerika ist es anders! Da sind die Bücher dick und fett! Wissen die nicht, was sie tun?“

„Ja, ich weiß. Bücher von US-Autorinnen gehen oft ins andere Extrem: Sie sind teilweise richtig dick.

Wenn man genau hinliest, fällt auf, dass sich vieles immer wieder wiederholt. Teilweise sind es Phrasen, die immer wieder gedro..., äh verwendet werden, teilweise wiederholen sich sogar ganze Absätze in etwas anderen Worten.

Zusätzlich gibt es Zusammenfassungen, Zusammenfassungen der Zusammenfassungen und Zitate aus anderen Büchern. Angeblich werden die Autorinnen dort nach Zeichen bezahlt – ob es wirklich stimmt oder ob es sich dabei nur um eine ‚urban legend‘ handelt, weiß ich nicht, und auch Google hat es mir nicht verraten.“

Solche Bücher mögen durch Wiederholung den Lerneffekt anregen – mir gehen sie auf die Nerven! Auch das Herunterbeten von Versuchen und Studien sowie Statistiken erhöht meine Leidenschaft für ein Buch nicht gerade.

Wie es bei dir ist, hängt sicher davon ab, wie du als Leserin tickst.

„Doch zurück zu unserem Thema: Wie ist das jetzt mit der Buchseite? Wann weißt du, dass du genug geschrieben hast, um ein anständiges Buch zu füllen?"

„Genau! Wie viele Wörter brauche ich für ein Buch?“

„Ich habe es schon erwähnt: Was Menschen unter einem ‚richtigen Buch’ verstehen, variiert. Manchen ist alles zu dünn, das dünner ist als die Bibel. Andere wiederum sind mit 100 Seiten zufrieden.

Ich sage: Es muss alles Wichtige drinstehen, das Buch sollte ein bisschen Storytelling sowie das große Warum der

Autorin enthalten, einen roten Faden haben und gut in der Hand liegen.

Ich persönlich bin ein großer Fan des klein-aber-oho-Buches. Also eines schlanken Buches mit allem, was wichtig ist. Abgespeckt und leicht zu lesen. Oder anders gesagt: gehirngerecht!"

Wenn die Gefahr besteht, dass sich das Buch zu dünn anfühlen könnte, kannst du mit dem Papier Boden gutmachen. Hierfür ist es gut, sich vom Verlag oder der Grafikerin beraten zu lassen.

Welches Papier passt zu deinem Buch, welches flufft es haptisch ein bisschen auf, welches Papier passt auch zum Inhalt?

Ich sage einmal, alles ab 200.000 Zeichen inklusive Leerzeichen darf sich getrost Buch nennen, und auch wenn es weniger ist, gibt es keinen Grund zur Panik: Ich durfte schon Bücher begleiten, die über eine Zeichenanzahl von 2.000 nicht hinausgekommen sind und trotzdem wunderbare Werke darstellen. Ich denke da zum Beispiel an ein Minibuch eines Südtiroler Autors oder an Geschenkbücher mit einem sehr hohen Anteil an Illustrationen.

Ein Buch ist ein Buch ist ein Buch

Fazit: Ich kann dir keine klare Antwort auf deine Frage geben, aber ich hoffe, ich konnte ein bisschen Licht in die Dunkelheit der Buchseiten bringen. Und was für dich gut passt, musst du selbst herausfinden. Wichtig ist, dass du dich mit deinem Buch wohlfühlst.

Frequently Asked Questions

Was du schon immer über Ghostwriting wissen wolltest, aber nicht zu fragen wagtest

Du hast dir dieses Buch gekauft, magst aber eigentlich gar nicht so viel lesen? Auch gut! Dann findest du hier eine kurze FAQ-Sequenz. Vielleicht hast du ja Glück und findest hier die Antworten auf deine brennendsten Fragen. Und wenn nicht? Dann musst du doch alles lesen! ;-)

Ich hab mir für diese Fragen nochmal Sascha ins Boot geholt, um die FAQ als Interview zu gestalten. Sascha fragt, und ich werde antworten.

> *„Okay. Fangen wir an! Lisa, wie ist das so mit dem Ghostwriten?"*

„Würdest du die Frage etwas präzisieren?"

> *„Wie funktioniert das – du bist Ghostwriterin, stellst aber immer wieder in den Social Media Bücher vor. Stört deine Kundinnen das nicht, wenn andere wissen, dass sie ihr Buch nicht selbst geschrieben haben?"*

„Es gibt zwei Möglichkeiten.

Entweder schreibe ich dein Buch als Ghost. Das sind die Bücher, die du auf meiner Webseite, in meinen Social-Media-Kampagnen und auch anderswo nicht finden wirst.

Ich gehe so weit zu sagen, dass du den Namen der Bücher oder meiner Kundinnen aus meinem Mund nicht vernehmen wirst. Für solche Fälle gibt's eine Geheimhaltungsvereinbarung.

Oder ich schreibe das Buch offiziell mit dir. Dann gibt es – so du das möchtest – Bilder von einem gemeinsamen ‚Making of', immer wieder kleine Zwischenberichte, vielleicht auch schon den einen oder anderen Sneak Preview.

Und meine volle Unterstützung auch außerhalb des Schreibens.

In diesem Fall bekommst du also zusätzlich eine Marketingunterstützung – bis hin zu meinem Erscheinen bei der Buchpräsentation. Dort halte ich dann dein verschwitztes Händchen oder kredenze Schnaps, falls du vor der Lesung nervös bist.

Du hast also die Wahl: Ghost oder nicht Ghost."

„Ich habe also eine Idee für ein Buch – wie geht's jetzt weiter?"

„Am besten, wir setzen uns einmal zusammen und reden darüber. Dann kann ich dir sagen, ob ich die Richtige für dich bin – wenn ja, werde ich dir ein Angebot machen.

Wenn ich nicht die Richtige bin, kann ich dir sicher eine passende Kollegin empfehlen. Durch die Ghostwriting Academy, die ich Mitte 2019 gemeinsam mit der genialen Monika Lexa ins Leben gerufen habe, kann ich auf einen Pool fähiger, gut ausgebildeter Ghostwriter zugreifen, die sich in ihrer Art zu schreiben und ihren Schwerpunkten wunderbar ergänzen."

„Ach, du machst nicht alles selbst?"

„Nein. Ich habe einen Themenkreis und bestimmte Fachgebiete, in denen ich mich wohlfühle.

Außerdem habe ich diese weltfremde Einstellung, dass ein Buch etwas Positives zum Leben von Menschen beitragen sollte. Sogenannte ‚Enthüllungsbücher' oder politische Themen interessieren mich z. B. absolut nicht. Und akademisches Ghostwriting, das ohnehin illegal ist – in Österreich für die Auftraggeberin wie auch für die Auftragnehmerin –, schon gar nicht."

„Was schreibst du denn dann?“

„Ich bin Spezialistin für alles rund um Verkauf, psychologische und medizinische Themen, Spirituelles, Persönlichkeitsentwicklung und Menschenführung und manchmal auch (Auto-)Biografien.“

„Trittst du mir eigentlich auch mal in den Hintern, damit was weitergeht?“

„Nein. Definitiv nicht. Ich bin ein freundlicher Mensch, aber keine Kindergartentante. Ich erwarte von meinen Kundinnen, dass sie Eigenverantwortung übernehmen. Ich bin immer gerne für Streicheleinheiten und mit einer breiten Schulter da, wenn du ein Tief hast, an dir zweifelst oder es einfach einmal nicht rund läuft.

Was ich jedoch nicht mache, ist dir nachlaufen. Es kostet viel Energie und Kopfarbeit, wenn man für andere mitdenken muss. Wenn du also möchtest, dass dich jemand regelmäßig ans Weitermachen erinnert, dann können wir vielleicht eine Kollegin finden, die besser zu dir passt. ;-)

Ich selbst bin aufgrund der Erfahrungen der letzten Jahre dazu nicht mehr bereit. Entweder du bist ‚all in‘ oder wir lassen es bleiben. Das habe ich auch in meinen AGB festgehalten: Wenn du dich über einen langen Zeitraum totstellst, dann steige ich aus.“

„Und wie lange dauert es, ein Buch fertigzustellen?“

„Ein ‚kleines‘ Buch, bei dem viel Vorarbeit geleistet wurde, kann schon mal in ein paar Monaten fertig sein. Wir reden hier allerdings nur vom Manuskript. Wenn du Covergestaltung und ein cooles Layout haben möchtest, dann leg ruhig noch zwei Monate drauf.

Dickere Bücher – speziell, wenn noch viel Schreibarbeit reinzustecken ist – dauern auch schon mal ein Jahr oder länger.

Wie lange es dauert, hängt auch davon ab, wie viel Zeit du dir nimmst, wie schnell du Feedback gibst und wie wir zusammenarbeiten."

„Ah, okay. Also kann das ganz schön lange dauern. Da kann ich es doch gleich selbst schreiben. Wieso also sollte ich zu dir kommen?"

„Es kann vielfältige Gründe haben, warum du mich oder eine meiner Kolleginnen dein Buch schreiben oder dich von uns beim Schreiben unterstützen lässt.

Vielleicht hast du selbst neben deinem Business so gar keine Zeit dafür. Vielleicht weißt du aber auch, dass Formulierungen nicht ganz deine Stärke sind, oder es ist eventuell auch einfach eine große Herausforderung für dich, dir wirklich regelmäßig die Zeit zum Schreiben zu nehmen.

Möglicherweise weißt du aber auch so viel über dein spezielles Fachgebiet, dass es dir schwerfällt zu entscheiden, was kommuniziert werden muss und was zu viel ist.

Du siehst: Es gibt mannigfaltige Gründe dafür, das Schreiben eines Buches auszulagern.

Und vielleicht hast du auch schon irgendwo ein fertiges Manuskript herumkugeln und brauchst nur jemanden, der es für dich in Szene setzt, also mit der feinen Feder drüber geht, Grafiken, Layout, Cover hinzufügt und es druckfertig gestaltet."

„Warte mal, was meinst du mit ‚druckfertig'?"

„Druckfertig bedeutet für mich: Das Buch ist bereit, gedruckt zu werden. Das beinhaltet die Vorarbeit aus dem Zielworkshop. Begonnen damit, dass du – idealerweise vor Beginn – ein Ziel brauchst und deine Zielgruppe für genau dieses Buch definieren solltest.

Mehr dazu findest du unter dem Punkt ‚Zielgruppe: ALLE!‘ Du brauchst außerdem eine Struktur, den roten Faden, der durch das Buch führt, Storytelling und vieles mehr. Also alles, was ein gutes Manuskript ausmacht.

Wenn wir das alles haben, muss das Buch Korrektur gelesen oder lektoriert werden. Und dann brauchst du noch eine gefällige Form für dein Buch:

- ★ Grafiken
- ★ Infoboxen
- ★ die richtige Schriftart
- ★ rechtliche Auflagen, die erfüllt werden müssen
- ★ ein Inhaltsverzeichnis und Meta-Tags, die deinen Kundinnen dabei helfen, dein Buch zu finden
- ★ ein großartiges Cover
- ★ und vieles mehr

Das alles benötigst du auch, wenn du dein Buch von vorne bis hinten selbst schreibst. Vor allem, wenn du dich für die schnellere und wesentlich lukrativere Option des Eigenverlags entscheidest.“

„Ah, okay. Danke. Ich frage mich aber auch: Wenn du mein Buch schreibst, ist es dann überhaupt mein Buch?“

„Wenn du dir überlegst, ein Buch zu schreiben (oder schreiben zu lassen), dann hast du etwas zu sagen. Du möchtest eine Theorie, eine Methode oder Informationen unter die Leute bringen.

Das heißt auch, du hast dir dazu schon deine Gedanken gemacht – nicht nur in Bezug auf das Buch, sondern auch, weil du in dem Bereich, in den das Buchthema fällt, bereits arbeitest und darin auch Expertin bist.

Du hast deinen Teil der Leistung erbracht.

Die reine Schreibarbeit kannst du also getrost anderen – zum Beispiel mir – überlassen. Immerhin würdest du dein Badezimmer auch nicht selbst fliesen, selbst wenn du genau weißt, wie es nach der Renovierung aussehen soll, oder?"

> „Wenn ich mich nun also dafür entscheide, mein Buch von dir schreiben zu lassen – wie kann ich dir den Inhalt meines Buches übermitteln? Ich habe immerhin (füll hier eine beliebige Zahl ein) Jahre studiert und (hier kommt nochmals eine Zahl) Jahre Praxis und meinen Expertinnenstatus aufgebaut. Wie willst du das anlegen?"

„Hier gibt es zwei gute Nachrichten. Die eine ist die: Um dein Buch zu schreiben, muss ich nicht alles wissen, was du weißt. Sondern nur das, was du deinen Leserinnen vermitteln willst.

Und das hast du sicher in irgendeiner Form parat. Als Vortrag, als Skriptum oder als Workshop, um nur einige Möglichkeiten aufzuzählen.

Und genauso kannst du mir das nötige Wissen auch (großteils) vermitteln. Indem du mich zu deinen Vorträgen oder Workshops einlädst, mir deine Skripten oder deine Doktorarbeit zur Verfügung stellst oder – wenn es keine andere Möglichkeit gibt – mich persönlich oder via Chat briefst."

> „Soweit – so gut. Aber wenn ich dann ewig mit dir herumsitze, dann kann ich es doch wirklich selbst schreiben. Oder?"

„Das kommt darauf an, wie du 'ewig' definierst. Ich habe Kundinnen, die mir – zum Beispiel, wenn ich ihre Biografie schreibe – eine Menge ihrer Zeit widmen müssen (oder dürfen ☺).

Auf der anderen Seite habe ich Bücher aber auch schon nach einem halbstündigen Briefing geschrieben (für das ich übrigens extra nach Deutschland geflogen bin) – und zwar zur vollständigen Zufriedenheit meines Kunden. Wie viel Zeit du mit mir verbringen ‚musst' – sei es elektronisch oder in persona – hängt auch davon ab, wie viele Feedback-Schleifen wir brauchen.

Wenn du also, zum Beispiel als Coach, eine sehr spezifische Ausdrucksweise hast, werden wir sicher mehr Zeit miteinander verbringen, als wenn es dir in erster Linie um die Übermittlung deines Expertinnenwissens und weniger um deinen persönlichen Stil geht."

„Zur Übermittlung habe ich da noch eine Frage. Mir sagen viele Leute, dass sie mich nicht verstehen, wenn ich ihnen mein Wissen vermitteln möchte. Wieso glaubst du, dass du mich dann verstehst – und wie legst du es an, dass auch meine Leserinnen checken, was ich ihnen sagen will?"

„Ich habe viel Erfahrung im Umgang mit den Special Effects von Menschen, und ich habe die Gabe, ihr Wissen in eine Sprache zu übersetzen, die von der Allgemeinheit gut verstanden wird. Ich bin quasi Übersetzerin zwischen Expertin und Leserinnen. Abgesehen davon beschäftige ich mich intensiv mit Wahrnehmungs- und Verarbeitungsmustern, gehirngerechtem Schreiben sowie Kommunikationsmodellen."

„Na gut, Lisa, wenn wir schon dabei sind, von Sensibilität zu sprechen ... wie sieht es denn mit Verschwiegenheit – auch abseits der Geheimhaltung beim Ghostwriting – aus?"

„Natürlich verpflichte ich mich dazu, deine sensiblen Daten so gut zu schützen, wie es mir möglich ist. Dazu bin ich auch per DSGVO verpflichtet.

Zudem kann ich schweigen wie ein Grab, und um nichts in der Welt würden mir die Geheimnisse über die Lippen kommen, die du mir im Zuge unserer Sitzungen anvertraust.

Selbstverständlich gebe ich auch deine Skripten nicht weiter – außer an eine Mitbearbeiterin wie zum Beispiel eine Kollegin, die die Recherche übernimmt, eine Grafikerin oder die Lektorin, die ebenfalls per Vertrag zum Schweigen verpflichtet sind.

Ansonsten wird alles, was du mir schickst oder gibst, gut gesichert verwahrt."

„Eine Sache muss ich noch wissen: Was kostet es, wenn du mein Buch schreibst?"

„Diese Frage kann ich nicht ad hoc beantworten, weil der Preis von vielen Faktoren abhängt. Es ist so, als würdest du mich fragen, was ‚ein Auto' kostet. Welches? Wie viele PS? Welche Marke?

So viel vorab: Ich bin eher ein Mercedes als ein 2CV!

Zudem hängen die Kosten von vielen unterschiedlichen Faktoren ab wie: Hast du schon eine Struktur, etwas Geschriebenes, auf dem man aufbauen kann? Wie konkret sind deine Vorstellungen? Wie viel Vorarbeit und wie viel Recherchearbeit sind zu leisten?

Welche Art von Buch soll es werden? Wie dick soll das Buch werden? Und vieles mehr.

Nach einem Workshop, in dem wir gemeinsam deinen Bedarf abklären, kalkuliere ich die Kosten für dein Buch und sende dir danach einen verbindlichen Kostenvoranschlag zu."

„Das heißt, du machst Pauschalpreise? Wäre ich mit einer Verrechnung nach Stunden nicht besser dran?"

„Ich arbeite ausschließlich mit Pauschalpreisen. So weißt du genau, was du an Kosten zu erwarten hast, und ich muss nicht ständig mit der Stoppuhr herumsitzen. Außerdem passiert die wichtigste Arbeit an einem Buch dann, wenn ich (scheinbar) nicht arbeite. Wenn ich also unterwegs bin, schlafe, meditiere oder mich mit anderen unterhalte.

Dein Buch läuft in dieser Zeit als Hintergrundprogramm bei mir ab, und wenn ich etwas Relevantes wahrnehme, ‚speichere' ich es ab.

Wenn ich dann wieder an meinem Mac sitze, sprudelt es oft aus meinen Fingern. Weil eben die wichtigste Arbeit – der kreative Prozess – schon erledigt ist."

„Nachdem ich mir sicher bin, dass mein Buch ein Erfolg wird, und ich auf jeden Fall eine ganze Menge Bücher verkaufen werde – wie wäre es, wenn ich dich einfach am Verkaufsgewinn beteilige? Dann werden wir gemeinsam reich!"

„Weißt du, ich bin ein bescheidener Mensch. Ich nehme nicht mehr als mir zusteht. Allerdings nehme ich auch nicht weniger.

Und nachdem ich dir aus Erfahrung sagen kann, dass die Wahrscheinlichkeit, allein mit dem Verkauf eines Buches im deutschsprachigen Raum reich zu werden, geringer ist als die Überlebenschance einer Tafel Schokolade in einer Frauen-WG, lasse ich mich darauf nicht ein. Daher rate ich dir auch vehement davon ab, ein Buch zu schreiben, wenn du nur von dessen Verkauf leben willst. Hast du noch weitere Fragen?"

„Nein, jetzt fällt mir beim besten Willen nichts mehr ein. Danke!"

Fertisch!

So – jetzt weißt du alles, was ich auch weiß. Oder zumindest das, was ich dir in diesem Buch unbedingt mitteilen wollte.

Vielleicht weißt du nun auch, ob du ein Buch schreiben möchtest oder nicht.

Wenn ja, dann wünsche ich dir alles Gute und viel Erfolg! Nur Mut, schreib einfach mal drauflos!

Wenn du dich verwurschtelst, kannst du dich ja immer noch bei mir oder einer meiner großartigen Kolleginnen melden.

Und wenn du es allein schaffst, dann feiere ich dich wie blöd. Nur eines: Vergiss nicht, dass jedes Buch – wirklich jedes! – ein Lektorat benötigt! Ich selbst habe das auf die harte Tour gelernt.

Mach's nicht nur gut, mach's besser! Und gerne auch besser als ich!

Solltest du übrigens jetzt noch immer nicht genug von mir haben, findest du ganz hinten ein paar Möglichkeiten, mit mir Kontakt aufzunehmen.

Und solltest du überhaupt Blut geleckt haben und ebenfalls ins Ghostwriting-Business einsteigen wollen, dann findest du dort auch einen Link zu unseren – also Monis und meinen – Lehrgängen. Ich freu mich, wenn wir uns persönlich über den Weg laufen!

Herzliche Grüße, Lisa

Geschafft!

Na, hast du es auch geschafft, alles zu lesen?

Ich hab mein Buch schon fast fertig!

Die Keskin ist mir zwischendurch ganz schön auf den Keks gegangen mit ihren Ideen und ihrer Pitzeligkeit. Aber recht hat sie gehabt. Und jetzt sind wir schon im Layout.

Bald kannst du mein Buch in der Buchhandlung erwerben. Ich hab mir übrigens die Lisa als Co-Autorin ins Boot geholt. Unser Buch heißt „The Write Way" – such einfach nach den Autorinnen: Lisa Keskin und Sascha Beinhart!

Grüezi und gutes Gelingen wünscht dir

Sascha

thank you!

Time to say … danke!

Ich bin am Ende. Zumindest am Ende dieses Buches. Und ich habe ein volles Herz. Denn meine female Buddys waren einfach ein Hammer!

Als erstes möchte ich dem besten Eheweib von allen danken: Monika Lexa.

Moni, du hast mich nicht nur zum Anfangen inspiriert, du hast auch mit scharfem Auge und klugem Kopf alles gelesen und grandioses Feedback gegeben. Danke dafür, dass du wie immer mit viel Humor und Liebe meine Spinnereien verkraftet hast, wenn du mit einem Kommentar wieder einmal recht hattest. Du bist einfach bombig – und so nebenbei die beste Mentaltrainerin, die ich kenne!

Conny Strumberger-Sellner: Conny, du hast das Buch mit deiner fachlichen Kompetenz lektoriert und mich „ich" sein lassen. Danke für deine Unterstützung, dafür, dass du mir immer eine Menge Kommas schenkst, und meinen Füllwörtern den Kampf ansagst!

Babsi Miklosch: Du bist die beste Bürokollegin, die man sich wünschen kann! Danke, dass du so easy-going bist, und dass wir gemeinsam alle Erfolge der jeweils anderen kräftig feiern können.

Spikey Miklosch: Du bist der beste Bürohund ever! Du schnarchendes kleines Muskelpaket mit deiner Vorliebe für alles Essbare (aber am meisten Wassermelone) hast mein Herz in Windeseile erobert!

Ich hab euch alle furchtbar lieb und bin überglücklich, dass ihr in meinem Leben seid!

Hier geht's zu
meiner Webseite

Du hast Lust bekommen,
selbst als Ghostwriter
zu arbeiten?
Hier entlang, bitte!

Kontaktiere deinen Ghostwriter

Konnte ich dich anstupsen, den nächsten Schritt zu machen – entweder zum eigenen Buch oder in deine Zukunft als Ghostwriter?

Dann freu ich mich über deinen Anruf, deine Mail oder jede andere Art der Kontaktaufnahme. (Nur das mit den Rauchzeichen hab ich nicht so drauf.)